America's Energy Future

TECHNOLOGY AND TRANSFORMATION

Committee on America's Energy Future

NATIONAL ACADEMY OF SCIENCES
NATIONAL ACADEMY OF ENGINEERING
NATIONAL RESEARCH COUNCIL
OF THE NATIONAL ACADEMIES

THE NATIONAL ACADEMIES PRESS
Washington, D.C.
www.nap.edu

THE NATIONAL ACADEMIES PRESS   500 Fifth Street, N.W.   Washington, DC 20001

NOTICE: The project that is the subject of this report was approved by the Governing Board of the National Research Council, whose members are drawn from the councils of the National Academy of Sciences, the National Academy of Engineering, and the Institute of Medicine. The members of the committee responsible for the report were chosen for their special competences and with regard for appropriate balance.

Support for this project was provided by the Department of Energy under Grant Number DEFG02-07-ER-15923 and by BP America, Dow Chemical Company Foundation, Fred Kavli and the Kavli Foundation, GE Energy, General Motors Corporation, Intel Corporation, and the W.M. Keck Foundation. Support was also provided by the Presidents' Circle Communications Initiative of the National Academies and by the National Academy of Sciences through the following endowed funds created to perpetually support the work of the National Research Council: Thomas Lincoln Casey Fund, Arthur L. Day Fund, W.K. Kellogg Foundation Fund, George and Cynthia Mitchell Endowment for Sustainability Science, and Frank Press Fund for Dissemination and Outreach. Any opinions, findings, conclusions, or recommendations expressed in this publication are those of the author(s) and do not necessarily reflect the views of the organizations that provided support for the project.

Library of Congress Cataloging-in-Publication Data

America's energy future : technology and transformation / Committee on America's Energy Future, National Academy of Sciences, National Academy of Engineering, and National Research Council of the National Academies.
  p. cm.
 Includes bibliographical references and index.
 ISBN 978-0-309-11602-2 (pbk.) — ISBN 978-0-309-11603-9 (PDF) 1. Power resources—United States. 2. Energy policy—United States. 3. Energy conservation. I. National Academy of Engineering. Committee on America's Energy Future.
 TJ163.25.U6A464 2009
 333.790973—dc22

                                                      2009029730

Copies of this report are available from the National Academies Press, 500 Fifth Street, N.W., Lockbox 285, Washington, DC 20055; (800) 624-6242 or (202) 334-3313 (in the Washington metropolitan area); Internet, http://www.nap.edu.

THE NATIONAL ACADEMIES
Advisers to the Nation on Science, Engineering, and Medicine

The **National Academy of Sciences** is a private, nonprofit, self-perpetuating society of distinguished scholars engaged in scientific and engineering research, dedicated to the furtherance of science and technology and to their use for the general welfare. Upon the authority of the charter granted to it by the Congress in 1863, the Academy has a mandate that requires it to advise the federal government on scientific and technical matters. Dr. Ralph J. Cicerone is president of the National Academy of Sciences.

The **National Academy of Engineering** was established in 1964, under the charter of the National Academy of Sciences, as a parallel organization of outstanding engineers. It is autonomous in its administration and in the selection of its members, sharing with the National Academy of Sciences the responsibility for advising the federal government. The National Academy of Engineering also sponsors engineering programs aimed at meeting national needs, encourages education and research, and recognizes the superior achievements of engineers. Dr. Charles M. Vest is president of the National Academy of Engineering.

The **Institute of Medicine** was established in 1970 by the National Academy of Sciences to secure the services of eminent members of appropriate professions in the examination of policy matters pertaining to the health of the public. The Institute acts under the responsibility given to the National Academy of Sciences by its congressional charter to be an adviser to the federal government and, upon its own initiative, to identify issues of medical care, research, and education. Dr. Harvey V. Fineberg is president of the Institute of Medicine.

The **National Research Council** was organized by the National Academy of Sciences in 1916 to associate the broad community of science and technology with the Academy's purposes of furthering knowledge and advising the federal government. Functioning in accordance with general policies determined by the Academy, the Council has become the principal operating agency of both the National Academy of Sciences and the National Academy of Engineering in providing services to the government, the public, and the scientific and engineering communities. The Council is administered jointly by both Academies and the Institute of Medicine. Dr. Ralph J. Cicerone and Dr. Charles M. Vest are chair and vice chair, respectively, of the National Research Council.

www.national-academies.org

ADRIAN A. FAY, Massachusetts Institute of Technology
SAMUEL FLEMING, Claremont Canyon Consultants
MARK FRANKEL, New Buildings Institute
JIM HARDING, Independent Consultant, Olympia, Washington
JASON HILL, University of Minnesota, St. Paul
NARAIN HINGORANI, Independent Consultant, Los Altos Hills, California
MAURICIO JUSTINIANO, Energetics, Inc.
JON KOOMEY, Lawrence Berkeley National Laboratory
SHELDON KRAMER, Independent Consultant, Grayslake, Illinois
THOMAS KREUTZ, Princeton University
ERIC LARSON, Princeton University
NANCY MARGOLIS, Energetics, Inc.
ALAN MEIER, Lawrence Berkeley National Laboratory
MIKE MESSENGER, Itron, Inc.
STEVE SELKOWITZ, Lawrence Berkeley National Laboratory
CHRISTOPHER WEBER, Carnegie Mellon University
ROBERT WILLIAMS, Princeton University

America's Energy Future Project Director
PETER D. BLAIR, Executive Director, Division on Engineering and Physical Sciences

America's Energy Future Project Manager
JAMES ZUCCHETTO, Director, Board on Energy and Environmental Systems (BEES)

Project Staff
KEVIN D. CROWLEY (*Study Director*), Director, Nuclear and Radiation Studies Board (NRSB)
DANA G. CAINES, Financial Manager, BEES
SARAH C. CASE, Program Officer, NRSB
ALAN T. CRANE, Senior Program Officer, BEES
GREG EYRING, Senior Program Officer, Air Force Studies Board
K. JOHN HOLMES, Senior Program Officer, BEES
LaNITA JONES, Administrative Coordinator, BEES
STEVEN MARCUS, Editorial Consultant
THOMAS R. MENZIES, Senior Program Officer, Transportation Research Board
EVONNE P.Y. TANG, Senior Program Officer, Board on Agriculture and Natural Resources
MADELINE G. WOODRUFF, Senior Program Officer, BEES
E. JONATHAN YANGER, Senior Program Assistant, BEES

Foreword

Energy, which has always played a critical role in our country's national security, economic prosperity, and environmental quality, has over the last two years been pushed to the forefront of national attention as a result of several factors:

- World demand for energy has increased steadily, especially in developing nations. China, for example, saw an extended period (prior to the current worldwide economic recession) of double-digit annual increases in economic growth and energy consumption.
- About 56 percent of the U.S. demand for oil is now met by depending on imports supplied by foreign sources, up from 40 percent in 1990.
- The long-term reliability of traditional sources of energy, especially oil, remains uncertain in the face of political instability and limitations on resources.
- Concerns are mounting about global climate change—a result, in large measure, of the fossil-fuel combustion that currently provides most of the world's energy.
- The volatility of energy prices has been unprecedented, climbing in mid-2008 to record levels and then dropping precipitously—in only a matter of months—in late 2008.
- Today, investments in the energy infrastructure and its needed technologies are modest, many alternative energy sources are receiving insufficient attention, and the nation's energy supply and distribution systems are increasingly vulnerable to natural disasters and acts of terrorism.

All of these factors are affected to a great degree by the policies of government, both here and abroad, but even with the most enlightened policies the overall energy enterprise, like a massive ship, will be slow to change course. Its complex mix of scientific, technical, economic, social, and political elements means that the necessary transformational change in how we generate, supply, distribute, and use energy will be an immense undertaking, requiring decades to complete.

To stimulate and inform a constructive national dialogue about our energy future, the National Academy of Sciences and the National Academy of Engineering initiated a major study in 2007, "America's Energy Future: Technology Opportunities, Risks, and Tradeoffs." The America's Energy Future (AEF) project was initiated in anticipation of major legislative interest in energy policy in the U.S. Congress and, as the effort proceeded, it was endorsed by Senate Energy and Natural Resources Committee Chair Jeff Bingaman and former Ranking Member Pete Domenici.

The AEF project evaluates current contributions and the likely future impacts, including estimated costs, of existing and new energy technologies. It was planned to serve as a foundation for subsequent policy studies, at the Academies and elsewhere, that will focus on energy research and development priorities, strategic energy technology development, and policy analysis.

The AEF project has produced a series of five reports, including this one, designed to inform key decisions as the nation begins a comprehensive examination of energy policy issues this year. Numerous studies conducted by diverse organizations have benefited the project, but many of those studies disagree about the potential of specific technologies, particularly those involving alternative sources of energy such as biomass, renewable resources for generation of electric power, advanced processes for generation from coal, and nuclear power. A key objective of the AEF series of reports is thus to help resolve conflicting analyses and to facilitate the charting of a new direction in the nation's energy enterprise.

The AEF project, outlined in Appendix C, included a study committee and three panels that together have produced an extensive analysis of energy technology options for consideration in an ongoing national dialogue. A milestone in the project was the March 2008 "National Academies Summit on America's Energy Future" at which principals of related recent studies provided input to the AEF study committee and helped to inform the panels' deliberations. A report chronicling the event, *The National Academies Summit on America's Energy Future:*

Summary of a Meeting (Washington, D.C.: The National Academies Press), was published in October 2008.

The AEF project was generously supported by the W.M. Keck Foundation, Fred Kavli and the Kavli Foundation, Intel Corporation, Dow Chemical Company Foundation, General Motors Corporation, GE Energy, BP America, the U.S. Department of Energy, and our own Academies.

Ralph J. Cicerone, President   Charles M. Vest, President
National Academy of Sciences   National Academy of Engineering
Chair, National Research Council   Vice Chair, National Research Council

Preface

The security and sustainability of our nation's energy system have been perennial concerns since World War II. Indeed, all postwar U.S. presidents have focused some attention on energy-supply issues, especially our growing dependence on imported petroleum and the environmental impacts of fossil-fuel combustion—the latter including the direct effects of pollutant emissions on human health and, more recently, the impacts of greenhouse gases, particularly carbon dioxide (CO_2), on global warming.

The United States has made a great deal of progress in reducing traditional gaseous and particulate emissions (e.g., SO_x, NO_x) through regulatory controls and the technology improvements that have followed. But greenhouse gas emissions are only beginning to be addressed in any meaningful way. The United States also needs to lower its dependence on fragile supply chains for some energy sources, particularly petroleum at present and possibly natural gas in the future, and to avoid the impacts of this dependence on our nation's economy and national security.

As a result of these and other factors (described in Chapter 1), such as the nation's increasingly vulnerable transmission and distribution systems, there has been a steadily growing consensus[1] that our nation must fundamentally transform the ways in which it produces, distributes, and consumes useful energy. Given the size and complexity of the U.S. energy system and its reach into all aspects of

---

[1]See, for example: *Lighting the Way: Toward a Sustainable Energy Future*, published by the InterAcademy Council in 2007 (www.interacademycouncil.net/?id=12161); *Ending the Energy Stalemate*, published by the National Commission on Energy Policy in 2007 (www.energy commission.org/ht/d/sp/i/492/pid/492); and *Facing the Hard Truths About Energy*, published by the National Petroleum Council in 2007 (www.npchardtruthsreport.org).

American life, this transformation will be an enormous undertaking; it will require fundamental changes, structural as well as behavioral, among producers and consumers alike. This report lays out the technical opportunities, the uncertainties, and some of the costs and benefits of initiating this transformation in earnest.

Given the massive installed base of long-lived energy production and distribution assets, together with a certain inertia—caused by uncertainties with respect to new technologies and regulations and by the generally slow pace of change in existing industrial practices, public policies, and consumer habits—the challenge that the nation faces not only is great but also will not be met overnight. *As a result, a meaningful and timely transformation to a more sustainable and secure energy system will likely entail a generation or more of sustained efforts by both the public and the private sectors.*

"Business as usual" approaches for obtaining and using energy will be inadequate for achieving the needed transformation. The efforts required will involve not only substantial new investments by the public and private sectors in research, development, demonstration, and deployment—in virtually all aspects of the energy infrastructure—but also new public policies and regulations on energy production, distribution, and use. Our energy system is, after all, much more than a set of technological arrangements; it is also a deep manifestation of society's economic, social, and political arrangements.

The America's Energy Future (AEF) Committee began this study at a moment of rapidly rising prices both in crude oil and in other raw materials that underpin the infrastructure that produces and delivers useful energy. As the study progressed, these prices reached a peak, began to fall steeply in the face of a global recession, and then began to rise again. Because it is virtually impossible to forecast future prices, this report makes no attempt to do so. Nevertheless, it is clear to the committee that market incentives for businesses and individuals to both invest in and deploy new energy technologies will depend most crucially, though not solely, on such prices. The technologies to be deployed must have adequate maturity, market appeal, and capability to meet the desired demands, and their development must be supported by appropriate public policies and regulations governing energy production, distribution, and use.[2]

---

[2]Any substantial change in the demand for key inputs, whether of primary energy stocks or of the resources required to transport and transform them, will strain the existing infrastructure and limit the pace of change.

The committee carefully considered existing and emerging technologies alike, some of which are now fairly well understood in principle though not necessarily deployable at scale or competitive in the marketplace, and it assessed how the deployment of such technologies might enable the nation to achieve meaningful transformation of the energy system over the next few decades. The committee did not, however, consider the opportunities available through conservation efforts or other opportunities through changes in policy or other socioeconomic initiatives. One of the committee's conclusions is that there is no technological "silver bullet" at present that could transform the U.S. energy system through a substantial new source of clean and reasonably priced domestic energy. Instead, the transformation will require a balanced portfolio of existing (though perhaps modified) technologies, multiple new energy technologies, and new energy-efficiency and energy-use patterns. This will in turn require a sustained national will and commitment of resources to develop and deploy these assets where needed.

Throughout this study the committee also paid close attention to the practical problems of developing and deploying new technologies, even assuming that there is the requisite national commitment to do so. An example is the integration of sizable new supplies of electricity from intermittent sources (e.g., wind and solar power) into the nation's electrical transmission and distribution systems. These systems need to be upgraded and continuously improved to enhance their reliability and security, to meet the needs of 21st-century electricity production technologies, and to provide for patterns of use that are more efficient.

Although this report focuses on the U.S. energy system, decision makers will need to take a wider view. It is clear that the country's economic, national security, and environmental goals, especially with respect to energy, cannot be fully achieved without collective international action.[3] Our nation's prosperity depends on global prosperity, our national security is tied to international security, and the achievement of our environmental goals depends on environmental protection actions taken elsewhere. In short, full realization of goals of the United States for transforming its energy sector requires that we find effective mechanisms for working with other nations, many of which face similar challenges. Maintaining an awareness of international developments and cooperating with other countries on research and development, pilot projects, and commercial demonstrations will be key to our own success.

---

[3]Such collective action among nations is not easy to achieve, as it requires broad participation, consequential monitoring, and meaningful compliance by all.

It is beyond the scope of this committee's charge to opine on the priority, relative to other national issues, of initiating and sustaining a national effort to transform our energy sector. However, I personally believe that despite the uncertainties before us, it is a truly urgent matter to begin such a transformation and, moreover, that the technology and knowledge for doing so are at hand. Indeed, the urgency for action to meet the nation's needs in the economic, environmental, and national security arenas as they relate to energy production and use are unique in our history, and delayed action could dramatically increase the challenges we face. But a timely transformation of the energy system is unlikely to happen without finally adopting a strategic energy policy to guide developments over the next decades. *Long-term problems require long-term solutions, and only significant, deliberate, stable, integrated, consistent, and sustained actions will move us to a more secure and sustainable energy system.*

I also believe that we should not allow short-term fluctuations, either in the prices of energy supplies or in geopolitical affairs, to distract us from this critical long-term effort. Creating a more sustainable and secure energy system will require leadership, courage, risk-taking, and ample support, both public and private, but in my view such investments will generate a significant stream of long-term dividends.

Harold T. Shapiro, *Chair*
Committee on America's Energy Future

Acknowledgments

This study could not have been done so well and on such a rapid schedule without the inspired contributions of a large number of individuals and organizations. First and foremost, I thank the committee members and staff for their dedication and hard work. These individuals brought a remarkably diverse array of disciplines, skills, and viewpoints to the study. As a result, our deliberations were intellectually stimulating—sometimes vigorous, but always respectful—as we worked together to develop this consensus report.

The committee initially organized itself into seven subgroups to facilitate information-gathering and, ultimately, the development of Chapters 4–9, which appear in Part 2 of this report:

- Alternative liquid transportation fuels (chaired and staffed, respectively, by Mike Ramage and Evonne Tang)
- Crosscutting and integration issues (Jim Sweeney and Madeline Woodruff)
- Electricity transmission and distribution (Jim Markowsky; Alan Crane and Sarah Case)
- Energy efficiency (Lester Lave; Madeline Woodruff, Greg Eyring, and Tom Menzies)
- Fossil-fuel energy (Lynn Orr and Greg Eyring)
- Nuclear energy (Dick Meserve and Sarah Case)
- Renewable energy (Larry Papay and K. John Holmes, assisted by Mirzayan Science and Technology Policy Graduate Fellows Amy Hee Kim, Dorothy Miller, and Stephanie Wolahan).

I thank these chairs for their able leadership, and I thank the subgroup members, staff, and fellows for their good work. I also express my gratitude to study director Kevin Crowley, who worked tirelessly to keep the entire study moving forward and to help the committee develop and articulate its key findings, which appear in Part 1 of this report.

The subgroups held separate meetings to obtain presentations and to gather the information that now appears in the Part 2 chapters. On behalf of the entire committee, I thank the outside experts who participated in these meetings. They are too numerous to list in this short section but are identified in Appendix B.

I also gratefully acknowledge the consultants who assisted the committee and its three sister panels (see Appendix C) with some of the analyses that were used in this report:

- Anup Bandivadekar, International Council on Clean Transportation
- Peter Biermayer, Sam Borgeson, Rich Brown, Jon Koomey, Alan Meier, and Steve Selkowitz, Lawrence Berkeley National Laboratory
- Anjan Bose, Washington State University
- Steve Dunn, Southwest Energy Efficiency Project
- Adrian A. Fay, Massachusetts Institute of Technology
- Samuel Fleming, Claremont Canyon Consultants
- Mark Frankel, New Buildings Institute
- Jim Harding, Independent Consultant
- Jason Hill, University of Minnesota, St. Paul
- Narain Hingorani, Independent Consultant
- Mauricio Justiniano and Nancy Margolis, Energetics, Inc.
- Sheldon Kramer, Independent Consultant
- Thomas Kreutz, Eric Larson, and Robert Williams, Princeton University
- Mike Messenger, Itron, Inc.
- Christopher Weber, Carnegie Mellon University.

Finally, I thank the many other National Academies staff who helped to make this study a success. Peter Blair and Jim Zucchetto, comanagers of the America's Energy Future Project, provided critical advice and guidance to the committee throughout the project. Mirzayan Science and Technology Policy Graduate Fellow Lawrence Lin and senior program associate Matt Bowen helped with the initial assembly of the massive literature that the committee used, and Matt Bowen also assisted with report review. Anderson Commonweal Intern Stephanie

Oparaugo assisted with research and administrative tasks for the nuclear energy chapter. LaNita Jones and Jonathan Yanger provided critical logistical support of the committee's work. Consultant Steve Marcus edited the report. Stephen Mautner supervised the report's publication by the National Academies Press, Estelle Miller provided design and layout, and Susan Maurizi and Livingston Sheats took responsibility for production editing. All figures in the report were rendered by Danial James Studios of Golden, Colorado.

It has been a great pleasure to work with such a talented and committed group of people. We learned a great deal from our presenters, consultants, and each other during the course of this study. It is my hope that our collective efforts have produced a report that will inform decision making and help engender wise policies and actions among our nation's political and business leaders.

Harold T. Shapiro

Acknowledgment of Reviewers

This report has been reviewed in draft form by individuals chosen for their diverse perspectives and technical expertise, in accordance with procedures approved by the National Research Council's Report Review Committee. The purpose of this independent review is to provide candid and critical comments that will assist the institution in making the published report as sound as possible and to ensure that the report meets institutional standards for objectivity, evidence, and responsiveness to the study charge. The review comments and draft manuscript remain confidential to protect the integrity of the deliberative process. We wish to thank the following individuals for their participation in the review of this report:

Rakesh Agrawal, Purdue University
Philip W. Anderson, Princeton University
R. Stephen Berry, University of Chicago
Thomas Cochran, Natural Resources Defense Council
Michael Corradini, University of Wisconsin, Madison
Paul DeCotis, State of New York, Office of the Governor
David Hawkins, Natural Resources Defense Council
Robert Hirsch, Consultant
Dale Jorgenson, Harvard University
Ernest Moniz, Massachusetts Institute of Technology
Dan Reicher, Google.org
Edward Rubin, Carnegie Mellon University
Christopher Somerville, University of California, Berkeley

James Thorp, Virginia Polytechnic Institute and State University
Carl J. Weinberg, Consultant
John P. Weyant, Stanford University
John Wise, ExxonMobil (retired)
John Wootten, Peabody Energy
Kurt Yeager, Electric Power Research Institute.

Although the reviewers listed above have provided many constructive comments and suggestions, they were not asked to endorse the conclusions or recommendations, nor did they see the final draft of the report before its release. The review of this report was overseen by Elisabeth M. Drake, Massachusetts Institute of Technology, and Robert A. Frosch, Harvard University. Appointed by the National Research Council, they were responsible for making certain that an independent examination of this report was carried out in accordance with institutional procedures and that all review comments were carefully considered. Responsibility for the final content of this report rests entirely with the authoring committee and the institution.

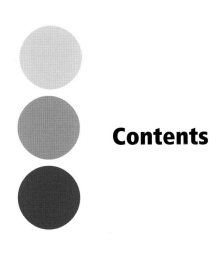

Contents

APPENDIXES

Executive Summary

This report of the Committee on America's Energy Future addresses a potential new portfolio of energy-supply and end-use technologies—their states of development, costs, implementation barriers, and impacts—both at present and projected over the next two to three decades. The report's aim is to inform policy makers about technology options for transforming energy production, distribution, and use to increase sustainability, support long-term economic prosperity, promote energy security, and reduce adverse environmental impacts. Among the wide variety of technologies under development that *might* become available in the future, this report focuses on those with the best prospects of fully maturing during the three time periods considered: 2008–2020, 2020–2035, and 2035–2050.

Eight key findings emerge.

First, with a sustained national commitment, the United States could obtain substantial energy efficiency improvements, new sources of energy, and reductions in greenhouse gas emissions through the accelerated deployment of existing and emerging energy-supply and end-use technologies. These options are described in more detail below and in Chapter 2. Mobilization of the public and private sectors, supported by sustained long-term policies and investments, will be required for the decades-long effort to develop, demonstrate, and deploy these technologies. Moreover, actions taken between now and 2020 to develop and demonstrate several key technologies will largely determine options for many decades to come. Therefore, it is imperative that the technology development and demonstration activities identified in this report be started soon, even though some will be expensive and not all will be successful: some may fail, prove uneconomic, or be overtaken by better technologies.

Second, the deployment of existing energy efficiency technologies is the nearest-term and lowest-cost option for moderating our nation's demand for energy, especially over the next decade. The potential energy savings available from the accelerated deployment of existing energy efficiency technologies in the buildings, transportation, and industrial sectors could more than offset the U.S. Energy Information Administration's (EIA's) projected increases in energy consumption through 2030. In fact, the full deployment of cost-effective energy efficiency technologies in buildings alone could eliminate the need to construct any new electricity-generating plants in the United States except to address regional supply imbalances, replace obsolete power generation assets, or substitute more environmentally benign electricity sources—assuming, of course, that these efficiency savings are not used to support increased use of electricity in other sectors. Accelerated deployment of these technologies in the buildings, transportation, and industrial sectors could reduce energy use by about 15 percent (15–17 quads, that is, quadrillions of British thermal units) in 2020, relative to the EIA's "business as usual" reference case projection, and by about 30 percent (32–35 quads) in 2030 (U.S. energy consumption in 2007 was about 100 quads). Even greater energy savings would be possible with more aggressive policies and incentives. Most of these energy efficiency technologies are cost-effective now and are likely to continue to be competitive with any future energy-supply options; moreover, additional energy efficiency technologies continue to emerge.

Third, the United States has many promising options for obtaining new supplies of electricity and changing its supply mix during the next two to three decades, especially if carbon capture and storage and evolutionary nuclear plants can be deployed at required scales. However, the deployment of these new supply technologies is very likely to result in higher consumer prices for electricity.

- Renewable-energy sources could provide about an additional 500 TWh (500 trillion kilowatt-hours) of electricity per year by 2020 and about an additional 1100 TWh per year by 2035 through new deployments in favorable resource locations (total U.S. electricity consumption at present is about 4000 TWh per year).

- Coal-fired plants with carbon capture and storage (CCS) could provide as much as 1200 TWh of electricity per year by 2035 through repowering and retrofits of existing plants and as much as 1800 TWh per year by 2035 through new plant construction. In combination, the entire existing coal power fleet could be replaced by CCS coal power by 2035.
- Nuclear plants could provide an additional 160 TWh of electricity per year by 2020, and up to 850 TWh by 2035, by modifying current plants to increase their power output and by constructing new plants.
- Natural gas generation of electricity could be expanded to meet a substantial portion of U.S. electricity demand by 2035. However, it is not clear whether adequate supplies of natural gas will be available at competitive prices to support substantially increased levels of electricity generation, and such expansion could expose the United States to greater import dependence and result in increased emissions of carbon dioxide (CO_2).

Fourth, expansion and modernization of the nation's electrical transmission and distribution systems (i.e., the power grid) are urgently needed. Expansion and modernization would enhance reliability and security, accommodate changes in load growth and electricity demand, and enable the deployment of new energy efficiency and supply technologies, especially intermittent wind and solar energy.

Fifth, petroleum will continue to be an indispensable transportation fuel during the time periods considered in this report. Maintaining current rates of domestic petroleum production (about 5.1 million barrels per day in 2007) will be challenging. There are limited options for replacing petroleum or reducing petroleum use before 2020, but there are more substantial longer-term options that could begin to make significant contributions in the 2030–2035 timeframe. Options for obtaining meaningful reductions in petroleum use in the transportation sector include the following:

- *Improving vehicle efficiency.* Technologies to improve vehicle efficiency are available for deployment now, and new technologies continue to emerge.
- *Developing technologies for the conversion of biomass and coal-to-liquid fuels.* By 2035, cellulosic ethanol and coal-and-biomass-

to-liquid fuels with CCS could replace about 15 percent of the fuel currently consumed in the transportation sector (1.7–2.5 million barrels per day of gasoline equivalent) with near-zero life-cycle CO_2 emissions. Coal-to-liquid fuels with CCS could replace about 15–20 percent of current fuel consumption in the transportation sector (2–3 million barrels per day; the lower estimate holds if coal is also used to produce coal-and-biomass-to-liquid fuels) and would have life-cycle CO_2 emissions similar to those of petroleum-based fuels. However, these levels of production would require the annual harvesting of 500 million dry tonnes (550 million dry tons) of biomass and an increase in coal extraction in the United States by 50 percent over current levels, resulting in a range of potential environmental impacts on land, water, air, and human health—including increased CO_2 emissions to the atmosphere from coal-to-liquid fuels unless process CO_2 from liquid-fuel production plants is captured and stored geologically. Commercial demonstrations of the conversion technologies integrated with CCS will have to be pursued aggressively and proven economically viable by 2015 if these technologies are to be commercially deployable before 2020. The development of advanced biomass-conversion technologies will require fundamental advances in bioengineering and biotechnology.

- *Electrifying the light-duty vehicle fleet through expanded deployment of plug-in hybrids, battery electric vehicles, and hydrogen fuel-cell vehicles.* Such a transition would require the development of advanced battery and fuel-cell technologies as well as modernization of the electrical grid to manage the increased demand for electricity.

Sixth, substantial reductions in greenhouse gas emissions from the electricity sector are achievable over the next two to three decades through a portfolio approach involving the widespread deployment of energy efficiency technologies; renewable energy; coal, natural gas, and biomass with carbon capture and storage; and nuclear technologies. Achieving substantial greenhouse gas reductions in the transportation sector over the next two to three decades will also require a portfolio approach involving the widespread deployment of energy efficiency technologies, alternative liquid fuels with low life-cycle CO_2 emissions, and light-duty vehicle electrification technologies.

To enable this portfolio approach in the electricity sector, the viability of two key technologies must be demonstrated during the next decade to allow for their widespread deployment starting around 2020:

- Demonstrate whether CCS technologies for sequestering carbon from the use of coal and natural gas to generate electricity are technically and commercially viable for application to both existing and new power plants. This will require the construction before 2020 of a suite (~15–20) of retrofit and new demonstration plants with CCS featuring a variety of feedstocks, generation technologies, carbon capture strategies, and geologic storage locations.
- Demonstrate whether evolutionary nuclear plants are commercially viable in the United States by constructing a suite of about five plants during the next decade.

A failure to demonstrate the viability of these technologies during the next decade would greatly restrict options to reduce the electricity sector's CO_2 emissions over succeeding decades. The urgency of getting started on these demonstrations to clarify future deployment options cannot be overstated.

Reducing greenhouse gas emissions from the liquid-fuel-based transportation sector in the 2020–2035 timeframe will also require a portfolio approach that includes cellulosic ethanol and coal-and-biomass-to-liquid fuels. Coal-and-biomass-to-liquid fuels can be produced in quantity starting around 2020 but will not have low carbon emissions unless geologic storage of CO_2 is demonstrated to be safe and commercially viable by 2015. Further reductions in greenhouse gas emissions could potentially be achieved in the transportation sector through electrification of the light-duty vehicle fleet, together with the production of electricity and hydrogen in ways that emit little or no CO_2, assuming the availability of suitable batteries or fuel cells. Although substantial reductions in emissions via these pathways are not likely until late in the 2020–2035 period and beyond, the widespread deployment of hydrogen fuel-cell vehicles during that time also holds some hope for more substantial long-term emission reductions in the transportation sector.

Seventh, to enable accelerated deployments of new energy technologies starting around 2020, and to ensure that innovative ideas continue to be explored, the

public and private sectors will need to perform extensive research, development, and demonstration over the next decade. Given the spectrum of uncertainties involved in the creation and deployment of new technologies, together with the differing technological needs and circumstances across the nation, a portfolio that supports a broad range of initiatives from basic research through demonstration will likely be more effective than targeted efforts to identify and select technology winners and losers. High-priority technology demonstration opportunities during the next decade include CCS, evolutionary nuclear power technologies, cellulosic ethanol, and advanced light-duty vehicles. Research and development opportunities during the next decade include advanced batteries and fuel cells, advanced large-scale storage for electrical load management, enhanced geothermal power, and advanced solar photovoltaic technologies.

Eighth, a number of current barriers are likely to delay or even prevent the accelerated deployment of the energy-supply and end-use technologies described in this report. Policy and regulatory actions, as well as other incentives, will be required to overcome these barriers. For technologies to be accepted in the market they must be clearly attractive—in terms of their performance, convenience, and cost—to investors, purchasers, and users. Regulations and standards that target performance characteristics can do a great deal to spur technological development and help improve market attractiveness.

Although the committee has done its best to identify those technologies likely to be available over the next two to three decades, many uncertainties remain on the scientific, technological, and policy frontiers and in energy markets. Consequently, the technology options identified in this report should be considered as important first-step technology assessments rather than as forecasts.

PART 1

1 Context and Challenges

This report assesses the status of energy-supply and end-use technologies[1] in the United States, both at present and over the next two to three decades. It is intended to inform the development of wise energy policies by our nation's decision makers and to provide the technical underpinnings for more detailed explorations of key energy-policy options in the second phase of the National Academies America's Energy Future (AEF) project. The complete study charge is presented in Box 1.1.

This first chapter, which establishes the context for the detailed energy-technology assessments that appear in Part 2 of this report, is divided into five sections. They describe the current U.S. energy system; some challenges that are likely to be encountered in transforming it; the role of technology in this transformation; the AEF Committee's strategy for addressing its study charge; and the report's organization.

---

[1]The AEF Committee uses the term "energy-supply and end-use technologies" in this report to connote the spectrum of technologies involved in the production, distribution, storage, and consumption of energy. These technologies include those that convert primary energy resources (e.g., fossil fuels, nuclear, solar, and wind) into useful forms (e.g., gasoline and electricity); technologies that transmit this energy to consumers (e.g., electrical transmission and distribution systems); technologies that store and utilize this energy (e.g., batteries, motors); and associated technologies, sometimes referred to as "demand-side" technologies, that control energy use (e.g., advanced electricity metering systems, or "smart meters").

BOX 1.1 **Study Charge**

This study will critically evaluate the current and projected state of development of energy-supply, storage, and end-use technologies. The study will not make policy recommendations, but it will analyze where appropriate the role of public policy in determining the demand and cost for energy and the configuration of the nation's energy systems. The committee will develop a "reference scenario" that reflects a projection of current economic, technology cost and performance, and policy parameters into the future. Within that scenario, the committee will evaluate energy technologies with respect to:

- Estimated times to readiness for deployment
- Current and projected costs (e.g., per unit of energy production or savings)
- Current and projected performance (e.g., efficiency, emissions per unit of output)
- Key technical, environmental, economic, policy, and social factors that would enhance or impede development and deployment
- Key environmental (including CO_2 mitigation), economic, energy security, social, and other life-cycle impacts arising from deployment
- Key research and development (R&D) challenges.

The committee may assess the sensitivity of these factors to possible variations in the key economic, technology cost and performance, and policy parameters that define the reference scenario.

The primary focus of the study will be on existing technologies and technologies likely to be available for deployment within the next decade. A secondary focus will be on technologies with longer times to deployment. The study will specifically provide estimates and findings on the following:

- For current technologies and technologies where initial deployment is judged to be within the next decade: estimates of costs, performance, and impacts
- For technologies where deployment is judged likely to be between 10 and 25 years: findings regarding key factors that enhance or impede adoption, implications for costs, and R&D challenges
- For technologies where deployment is judged likely to be greater than 25 years: findings regarding key factors that enhance or impede R&D challenges.

THE CURRENT U.S. ENERGY SYSTEM

The U.S. energy system currently comprises a vast and complex set of interlocking technologies for the production, distribution, and use of fuels and electricity (Boxes 1.2 and 1.3; Figure 1.1[2]). It evolved over the last century in response to a broad set of circumstances: rapidly growing demand for energy, advances in technology, diverse public policies and regulations, and the powerful market forces that have accompanied economic growth and globalization. As a result, the energy system's technologies and production assets are of many different vintages and often rely on aging and increasingly vulnerable infrastructures.

Five critical characteristics of this system stand out:

1.   The United States relies on the burning of carbon-based fossil fuels for more than 85 percent of its energy needs (Figure 1.2).

2.   The burning of fossil fuels has a number of deleterious environmental impacts, among the most serious of which is the emission of greenhouse gases,[3] primarily carbon dioxide (CO_2). At present, the United States emits about 6 billion tonnes (6 gigatonnes) of CO_2 per year into the atmosphere. Emissions have grown by almost 20 percent since 1990 but have recently leveled off somewhat (Figure 1.3). However, CO_2 emissions are projected to increase in the future under the Energy Information Administration's (EIA's) "business as usual" reference case (see Box 2.1 in Chapter 2).

3.   Despite decades of declining energy intensity (i.e., energy consumption per dollar of gross domestic product; see Figure 1.4), the United States still has a higher per capita consumption of energy than either the European Union or Japan (Figure 1.5). And despite improvements in energy efficiency, U.S. energy consumption continues to rise, in part because of

---

[2]Figures 1.1 through 1.12 are grouped under the section titled "America's Energy Past, Present, and Future: An Overview in Charts and Graphs," which starts on page 17.

[3]Greenhouse gases are so named because of their ability to absorb and emit infrared radiation. Water vapor and CO_2 are the most common greenhouse gases in Earth's atmosphere, but methane, nitrous oxides, and chlorofluorocarbons (CFCs) are also greenhouse gases. Recent studies (e.g., IPCC, 2007) indicate a high probability of a link between anthropogenic greenhouse gas emissions and observed effects on global warming, precipitation patterns, ocean acidification, and weather patterns. The National Academies recently initiated "America's Climate Choices," a suite of studies to inform and guide responses to climate change across the nation.

BOX 1.2 ***Primary Energy and Useful Energy***

The energy that powers our civilization is obtained from a number of primary energy sources that exist in nature. These sources fall into two categories: flows of energy and stored energy. Examples of energy flows include sunlight, wind, and waves. Stored energy includes fossil energy (petroleum, natural gas, and coal), bioenergy (contained in biomass), and nuclear energy (stored in atomic nuclei in radioactive elements such as uranium) and the heat stored in Earth's upper crust. Primary energy sources can be converted into *useful* energy that, for example, powers a vehicle, lights a building, or supplies heat for an industrial process, although the conversion process inevitably involves energy losses (which can be quite considerable) and often entails substantial costs.

While the extent of these primary energy sources is usually large, there are a number of technological, economic, environmental, and labor constraints on converting them into useful energy. For example, many remaining domestic supplies of petroleum and natural gas are in difficult-to-access locations. Some are in environmentally sensitive areas. And renewable energy is unevenly distributed across the United States; in some cases, regions with abundant renewable potential are physically distant from demand centers. Such constraints are in fact critical in determining the actual mix of useful energy supplies that are available at particular times. Much of this report deals with the technology options for overcoming some of these constraints.

economic and population growth.[4] U.S. dependence on energy imports continues to rise as well (Figure 1.6). And steady increases in energy use are projected for the future (Figure 1.7) under EIA's business-as-usual reference case.[5]

4. The United States is almost completely dependent on petroleum for transportation—a situation that entails unique energy-security[6] chal-

---

[4]In many cases, energy efficiency gains that could have further moderated per capita energy demand have instead been used to support new demands for energy, for example, through increased size and performance of light-duty vehicles.

[5]These are long-term projections that do not account for short-term demand variations. For example, global consumption of petroleum dropped in 2008 and is projected to drop in 2009 because of the current worldwide economic recession.

[6]The committee uses the term "energy security" to mean protection against disruptions to the energy supply chain that produces, distributes, and uses energy. Such disruptions can result from

---

BOX 1.3 *Resources, Reserves, and Potential*

The terms "resource," "reserve," and "potential" are used throughout this report to describe the primary energy sources that exist in nature and may be tapped to produce useful energy. "Resource" refers to quantities of stored energy—i.e., solid, liquid, or gaseous fuels derivable from petroleum, natural gas, coal, uranium, geothermal, or biomass—that exist in nature and may be feasible to extract or recover, given favorable technology and economics. "Reserve" refers to that portion of the resource that can be economically extracted or recovered with current technology. "Potential" is used instead of resource to describe energy flows—such as from sunlight, wind, or the movement of water—that occur in nature and may be feasible to recover, given favorable technology and economics.

These primary energy sources are very large compared to U.S. energy demand. For example, the energy from sunlight reaching the land surface of the United States is thousands of times greater than the country's current annual consumption of energy; the energy from wind available in the United States is at least an order of magnitude larger; and the energy stored in geothermal, nuclear, and fossil reserves available to the United States is at least thousands of times larger. The challenge is to transform these vast resources into energy forms that are readily usable in a commercially and environmentally acceptable fashion.

---

lenges. The nation relies on coal, nuclear energy, renewable energy (primarily hydropower), and, more recently, natural gas for generating its electricity (Figures 1.8 and 1.9).

5. Many of the energy system's assets are aging: domestic oil and gas reserves are being depleted; currently operating nuclear plants were constructed largely in the 1970s and 1980s, and many coal plants are even older (Figure 1.10); and electrical transmission and distribution systems contain infrastructure and technologies from the 1950s. Renewing or replacing these assets will take decades and require investments totaling several trillion dollars.

---

interruptions in energy imports, for example, or from damage to the energy infrastructure (either through intentional acts or overuse).

Fossil fuels have supported U.S. economic prosperity since the latter part of the 19th century. But their low market prices during most of this period encouraged high levels of energy consumption per capita and generally discouraged the development of alternative sources of energy, with two notable exceptions: hydroelectric and nuclear power, which currently account for about 7 percent and 19 percent, respectively, of U.S. electricity generation. Our nation's dependence on fossil fuels evolved not only because they were available at low market costs[7] but also because their physical and chemical properties are well suited to particular uses: petroleum for transportation; natural gas as an industrial feedstock, for residential and commercial space heating, and, more recently, as a fuel for electric-power generation; and coal for the generation of electricity and as a feedstock for some industrial processes (Figure 1.8). Indeed, most consumer-based, industrial, and governmental activities require, either directly or indirectly, the consumption of fossil fuels.

The current profile of U.S. energy use, summarized in Figures 1.1 and 1.2, shows that nearly 40 percent of the nation's economy is fueled by *petroleum*. More important, nearly all of our nation's transportation needs are being met by petroleum-based fuels, as shown in Figure 1.8, and prospects for alternatives are currently limited. Domestic production of petroleum in the United States peaked[8] in the 1970s and has been in decline for the past three decades. However, improvements in exploration and production technologies have helped to moderate these declines. About 56 percent of the petroleum consumed in the United States in 2008 was imported, in some cases from geopolitically turbulent or fragile regions.[9]

America's enormous appetite for oil, coupled with growth in demand from other countries, puts upward pressures on world prices, increases revenues to oil-exporting nations, and heightens the influence of those nations in world affairs.

---

[7]These market costs often did not account for "externality" costs such as those stemming from the environmental and health impacts of producing, distributing, and consuming energy.

[8]There is a vigorous debate among experts about when we can expect world oil production to peak—and also about the importance of this issue for long-term energy supplies. Some judge that world production has already peaked or will do so in the near future; others argue that world oil production will continue to increase slowly for the foreseeable future or will have a sustained plateau. See, for example, Simmons (2005) and Wood et al. (2004).

[9]Each nation's access to or competition for energy is central to some of the major geopolitical tensions of our time. Using energy more efficiently and developing new domestic energy sources could help reduce U.S. dependence on imports from these unstable regions.

Historically, this influence has often been expressed in directions that are neither supportive of a well-functioning world oil market nor consistent with U.S. interests. Since U.S. consumption of oil is concentrated in key economic activities, such as transportation and home heating, this produces what many consider to be increasingly substantial economic and national security vulnerabilities. The current recession has reduced petroleum demand, and this condition may persist until the economic recovery gets under way. But if history is any guide, the resumption of worldwide economic growth will again raise the demand for petroleum and increase pressures on prices.

Almost a quarter of our nation's economy is fueled by *natural gas*, mostly for residential and commercial space heating and industrial uses, but increasingly for electric-power generation (see Figure 1.8). Indeed, over the past two decades, natural-gas-fired facilities have accounted for a significant fraction of new U.S. baseload power plants (Figure 1.10).[10] About 86 percent of the natural gas used in the United States at present is produced domestically, and much of the remainder comes from Canada. Prices for natural gas in the North American market have been lower than the price for liquefied natural gas[11] on the world market.

During the last 2 years, North American natural gas production from conventional resources has declined. But production from unconventional sources—such as coal beds, tight gas sands (rocks through which flow is very slow), and shale—has increased, largely in response to higher prices and new technology. Recent price declines, however, have reduced the incentive to develop new natural gas production, especially from unconventional sources. If domestic production growth could be continued and production sustained over long periods, North American sources could meet some portion of the potential growth in U.S. demand for natural gas. If, however, growth in domestic natural gas production is limited—by some combination of production declines from existing sources and of less-than-expected growth in new sources—the United States might have to import more natural gas, which would result in increased import dependence and exposure to world prices for liquefied natural gas.

---

[10]Natural gas plants are used to provide both baseload and peaking power, but Figure 1.10 shows only baseload plants.

[11]Liquifying natural gas by cooling it to low temperatures (about −160°C) at close to atmospheric pressures makes it easier to transport and store.

Which of these futures occurs will depend as well on a set of interrelated factors including the following: demand growth, production technology, resource availability, and prices. There is some uncertainty, for example, regarding how easily domestic sources of natural gas production could be expanded and how quickly a global market for natural gas would emerge. If North American natural gas production cannot be expanded to meet demand, then dependence on natural gas imports could mirror that on petroleum.

Figure 1.8 also shows that the burning of fossil fuels—principally coal and secondarily natural gas—accounts for almost 75 percent of the electricity generated in the United States. Coal is abundant in this country and relatively inexpensive compared to other fossil fuels. The United States currently has about 20 years' worth of identified coal reserves in active mines. However, a much larger resource would be available for production if new mines were opened and if the rail infrastructure required to deliver coal—or sufficient long-distance transmission lines for delivery of electricity generated near the mine mouth—could be put in place. Costs of production are low enough that substantial quantities of coal can be produced at current coal prices. However, coal mining has significant environmental impacts, which will limit its suitability in some locations.

The use of fossil fuels to generate energy has a number of deleterious impacts on land resources, water supplies, and the well-being of citizens. Arguably, the most important unregulated environmental impact of fossil-fuel use is the emission of greenhouse gases, primarily CO_2, to the atmosphere. Petroleum use for transportation accounts for about one-third of total annual U.S. emissions of CO_2 (Figures 1.11 and 1.12), and fossil-fuel use for electricity generation accounts for more than another third (Figure 1.11). Coal use causes most of the emissions from electricity production. Natural gas produces about half as much CO_2 as coal per kilowatt-hour of electricity generated, but its emissions can be some 10–20 times higher per kilowatt-hour than those from nuclear or renewables (see Figure 2.15 in Chapter 2).

Although technologies for capturing and storing CO_2 have been demonstrated in nonelectrical applications, they have not yet been shown to be safely deployable at a sufficient scale for coal- and natural-gas-fired power plants. Even if the technology were to be proven for electrical applications, building the necessary infrastructure would require major investments over long periods of time, and substantial new regulations would have to be formulated to address safety, ownership, and liability issues. And, of course, there would be impacts on the price of electricity.

AMERICA'S ENERGY PAST, PRESENT, AND FUTURE
An Overview in Charts and Graphs

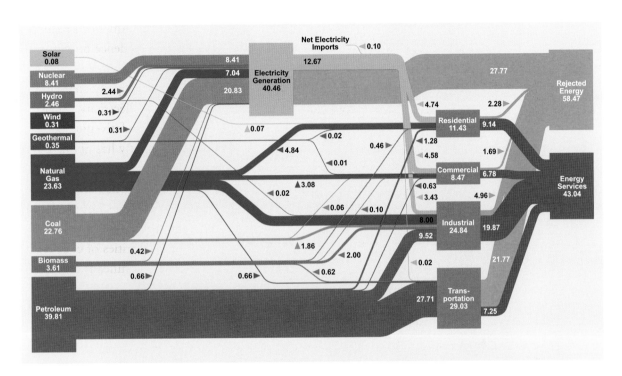

FIGURE 1.1 *Energy consumption in the United States in 2007 in quadrillions of British thermal units (quads). The figure illustrates the delivery of energy from primary fuel sources, which are shown in the boxes on the left side of the figure, to the residential, commercial, industrial, and transportation sectors, which are shown in the boxes at the center-right side of the figure. Energy is delivered to these sectors primarily in three forms: (1) electricity, which is produced principally from coal, natural gas, and nuclear power, and to a much lesser extent from renewable sources (hydro, solar, wind, and biomass); (2) liquid fuels, principally petroleum, with a small contribution from biomass-derived fuels (e.g., corn ethanol); and (3) natural gas for heating and as an industrial feedstock. Small quantities of coal and biomass are also used as industrial feedstocks. The width of the bars indicates the relative contributions of each energy source; the absolute contribution (in quads) is shown by the numerical labels next to each bar. The bar for electricity represents retail electricity sales only and does not include self-generated electricity. The boxes on the right side of the figure show that a total of about 101.5 quads of energy were consumed in the United States in 2007; about 43 quads were used to provide energy services, and more than 58 quads were "rejected" (i.e., not utilized to provide energy services) because of inefficiencies in energy production, distribution, and use.*
Sources: Lawrence Livermore National Laboratory and the Department of Energy, based on data from the Energy Information Administration, 2008a.

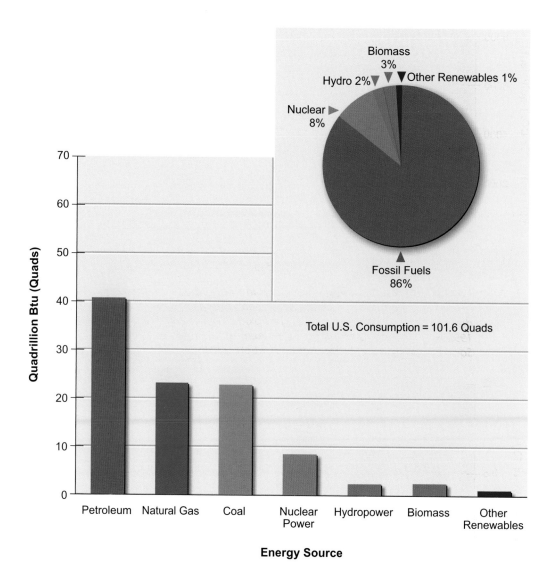

FIGURE 1.2 *Energy consumption in the United States in 2007 by fuel source, in quads (bars) and as percentages (pie chart).*
Source: Energy Information Administration, 2008b.

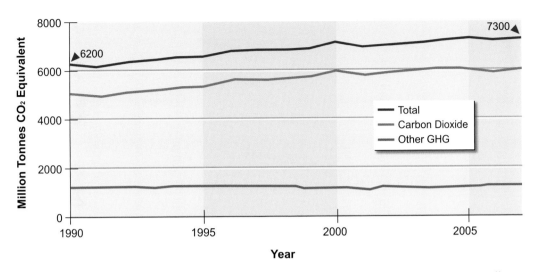

FIGURE 1.3 *Greenhouse gas emissions in the United States, 1990–2007, in millions of tonnes CO$_2$ equivalent. The "other" greenhouse gas (GHG) emissions shown on the diagram include methane and nitrous oxide, converted to CO$_2$-equivalent units. The 1990 and 2007 point estimates have been rounded to two significant figures. Source: Energy Information Administration, 2008b.*

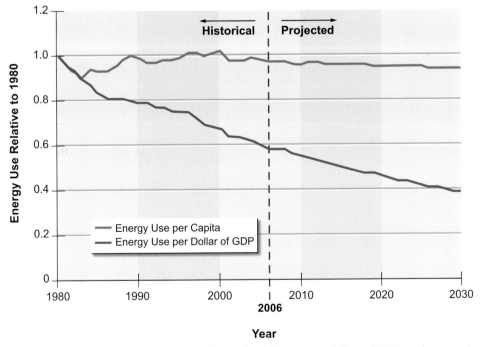

FIGURE 1.4 *Energy use in the United States per dollar of GDP and per capita, with 1980 energy use per dollar of GDP and per capita set to 1.0. Source: Energy Information Administration, 2008b.*

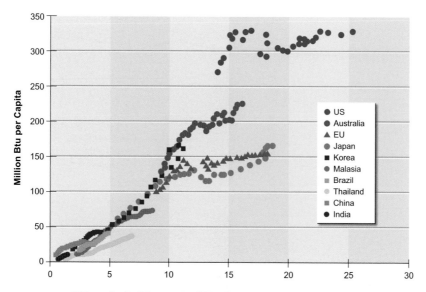

GDP per Capita (Thousands of 2007 Dollars at Purchasing Power Parity)

FIGURE 1.5 *Annual per capita energy use (in million Btu per capita) as a function of gross domestic product (GDP) at purchasing-power parity per capita. A progression over time for several representative countries is shown. GDP is a measure of economic activity. On average, higher per capita energy consumption is associated with increasing per capita GDP; however, in some cases, per capita GDP has increased while energy use has declined.*
Sources: Adapted from Shell International BV, Shell Energy Scenarios to 2050 *(2008), based on data from the International Monetary Fund and British Petroleum.*

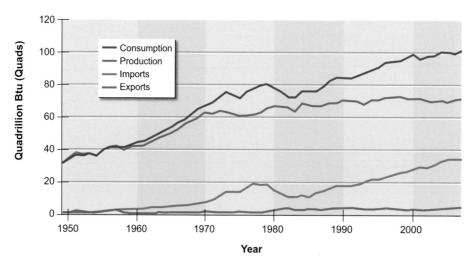

FIGURE 1.6 *Primary U.S. energy consumption, production, imports, and exports, 1949–2007, in quads.*
Source: Energy Information Administration, 2008b.

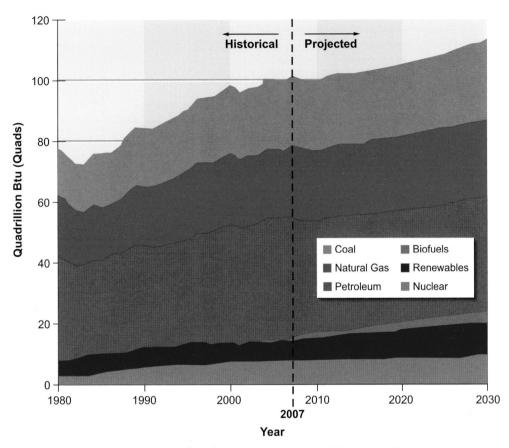

FIGURE 1.7 *Historical (1980–2007) and projected (2008–2030) energy consumption in the United States by primary energy source, in quads. The projected energy use from 2020 to 2030 reflects the U.S. Energy Information Administration's (EIA's) 2008 reference case; this reference case assumes that current policies that affect energy supply and consumption will remain unchanged and that economic growth rates and technology development and deployment trends will continue over the next 20 years. As explained in Box 2.1 in Chapter 2 and in Annex 3.A in Chapter 3, the AEF Committee uses the EIA reference case as the reference scenario for its study.*
Source: Energy Information Administration, 2008b.

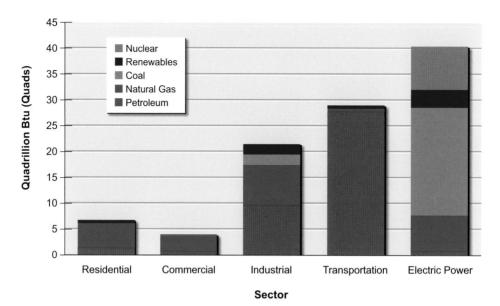

FIGURE 1.8 *Primary energy consumption by production sector and fuel type in the United States in 2007. Energy consumed by the electric power sector is used to produce electricity consumed by the end-use sectors shown in the figure.*
Source: Data from Energy Information Administration, 2008b.

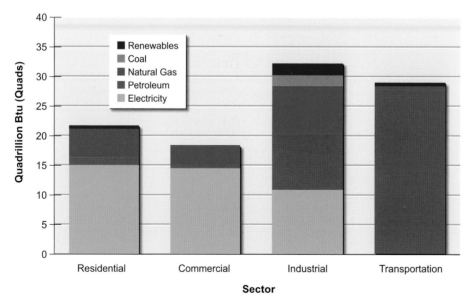

FIGURE 1.9 *Total energy consumption in the United States in 2007, shown by end-use sector and by fuel type. Also shown is each end-use sector's consumption of electricity. Electricity is a secondary energy source and is generated using fossil fuels and nuclear and renewable sources.*
Source: Data from Energy Information Administration, 2008b.

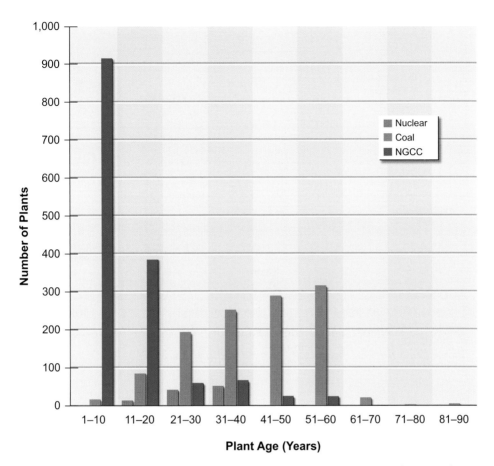

FIGURE 1.10 *Age of U.S. baseload power plants, in years. The age of U.S. nuclear plants, coal plants, and natural gas combined cycle (NGCC) plants is shown in 10-year intervals, as measured from their initial year of operation projected to 2007. Only plants that are used primarily for retail electricity production are shown. Natural gas single-cycle plants are not shown because they are intended for peaking rather than baseload generation. Many of these baseload plants have been upgraded since plant operations commenced. Source: Data from Energy Information Administration (www.eia.gov/cneaf/electricity/ page/capacity/capacity.html).*

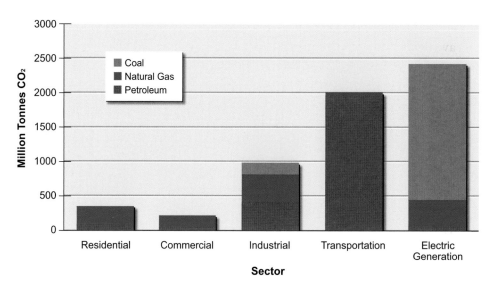

FIGURE 1.11 *Primary CO$_2$ emissions by production sector and fuel type in the United States in 2007 in millions of tonnes per year. Emissions from the electric power sector result from the production of electricity that is consumed by the end-use sectors shown in the figure.*
Source: Data from Energy Information Administration, 2008b.

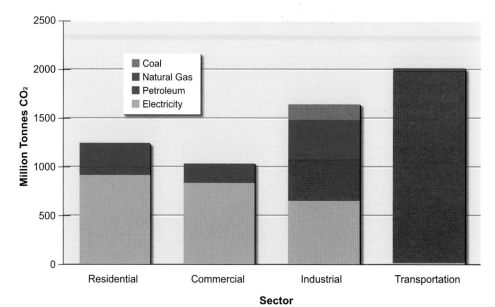

FIGURE 1.12 *Total CO$_2$ emissions in the United States in 2007 by end-use sector and primary energy source, in millions of tonnes per year. Also shown is each end-use sector's consumption of electricity. Electricity is a secondary energy source and is generated using fossil fuels and nuclear and renewable sources.*
Source: Data from Energy Information Administration, 2008b.

Nevertheless, new technologies that use energy more efficiently and that avoid, or capture and safely store, greenhouse gas emissions are essential components of a portfolio of alternatives for transforming energy production and use. *Indeed, failure to develop and implement such technologies will greatly limit the options available for reducing the nation's greenhouse gas emissions to the atmosphere.*

CHALLENGES TO TRANSFORMING ENERGY PRODUCTION AND USE

There is a growing recognition that our nation's current approaches for obtaining useful energy, being largely dependent on fossil fuels, are unsustainable over the long term and that we must therefore transform the manner in which energy is produced, distributed, and consumed. The need to transform the U.S. energy system is motivated by several factors.

- Heightened long-term competition for fossil fuels as a result of worldwide population and economic growth.
- Increasing U.S. reliance on world markets and their vulnerable supply chains for supplies of petroleum (and possibly, in the future, of natural gas).
- Mounting volatility in market prices for fossil fuels. For example, petroleum prices have ranged from about $32 to $147 per barrel over the past 2 years, which has helped to promote volatility in prices for coal and natural gas.
- Growing concerns about the impacts on the environment of burning fossil fuels—especially the impacts of CO_2 emissions to the atmosphere on global warming—and the time spans of such impacts.[12]

The challenge before us is to transform the U.S. energy system in a manner that increases its sustainability, supports long-term economic prosperity, promotes

---

[12]The committee refers in particular to uncertainties in the time-dependent relationships associated with anthropogenic CO_2 emissions and the resulting changes in atmospheric temperatures and sea levels. These uncertainties make it difficult to judge precisely how soon CO_2 emissions must be reduced to prevent major environmental impacts around the world. Many experts judge that there are, at most, just a few decades remaining in which to make these changes.

energy security, and reduces the adverse environmental impacts arising from energy production and use. Such a transformation could, for example, promote sustainability by using energy more efficiently and increasing the use of renewable-energy sources; support long-term economic prosperity by ensuring the availability of adequate supplies of energy; improve energy security by decreasing the nation's reliance on petroleum imports; and reduce adverse environmental impacts by reducing CO_2 emissions to the atmosphere.

Concerns about the sustainability, security, or environmental impacts arising from energy production and use have been reflected in the agendas of all U.S. presidents since Franklin D. Roosevelt. Such concerns were reflected, for example, in:

- Richard Nixon's "Project Independence" (1974)
- Gerald Ford's "Energy Independence Act" (1975)
- Jimmy Carter's "National Energy Plan" (1977)
- Ronald Reagan's "Energy Security" report (1987)
- George H.W. Bush's "National Energy Strategy" (1991)
- Bill Clinton's "Federal Energy R&D for the Challenges of the 21st Century" report (1997)
- George W. Bush's "Reliable, Affordable, and Environmentally Sound Energy for America's Future" report (2001).

Environmental policies and regulations—including, for example, those stemming from the Clean Air Act and the National Environmental Policy Act—have at times focused our nation's attention on energy efficiency and conservation and the use of renewable sources of energy, which has led to dramatic improvements in air quality. In fact, statutes such as the Clean Air Act demonstrate that thoughtful regulation can be a very useful tool for dealing with important externalities—those involved, for example, in the unconstrained emissions of pollutants (e.g., NO_x, SO_x) from the burning of fossil fuels.

These policies and regulations have focused in part on mitigating the environmental impacts of energy production and use, but they have been piecemeal efforts. The fact is that the United States has never implemented a truly comprehensive set of national policies for obtaining and using energy to meet national goals for sustainability, economic prosperity, security, and environmental quality. Instead, as noted previously, the U.S. energy system has developed in response to an array of uncoordinated market forces and shifting public policies.

Yet there has been a growing recognition over the past decade of the need for such comprehensive national policies. Congress made an unsuccessful attempt to pass major energy legislation in 2002, successfully passed such legislation in 2005 (Energy Policy Act of 2005) and in 2007 (Energy Independence and Security Act [EISA] of 2007), and was working on another major energy bill as the present report was being completed. Additionally, the Obama administration recently announced a new national fuel efficiency policy that will accelerate the implementation of EISA fuel economy standards for light-duty vehicles. During this same period there has been a sharp rise in investment activity focused on clean energy—from a few tens of millions of dollars in the late 1990s to hundreds of billions of dollars today.

It is the AEF Committee's judgment that comprehensive and sustained national policies for energy production and use will be needed to achieve a timely transformation to the more sustainable, secure, and environmentally benign energy system envisioned in this report. However, to help shape these policies will require sound and dispassionate technical analyses of the opportunities and challenges before us. Such analyses should address technology capabilities, costs, times to maturity and commercial deployment, and impacts on the environment, economy, and national security. The technical analysis in the present report aims to help support the development of such policies.

THE ROLE OF TECHNOLOGY

Transforming the U.S. energy system as described in the preceding section will require the continued improvement of existing technologies as well as the development and national-scale deployment of new technologies, including:

- Existing and new energy efficiency technologies.
- Existing and new energy-supply technologies—including wind, solar, geothermal, biofuels, and nuclear power.
- Carbon capture and storage (CCS) technologies on a large-enough scale to reduce CO_2 emissions from the burning of fossil fuels.
- Modern electrical transmission and distribution systems to accommodate 21st-century electricity supplies (especially from intermittent resources such as wind and solar), support future growth in electricity demand, and enable national-scale deployments of sophisticated demand-side technologies.

Many different pathways can potentially be pursued to these ends. But identifying pathways that are consistent with the nation's priorities and then taking the actions needed to achieve the desired transformations are among the most difficult challenges of our time.

The national-scale deployment of new technologies will have learning curves and will entail a variety of risks, and such deployments can have unforeseen economic and environmental impacts. Thus, in addition to evaluating the potential contributions of existing and emerging technologies, we also need to understand the nontechnological constraints on their rates of deployment and to decide on the roles of the public and private sectors and current and future generations for shouldering deployment costs and risks. In short, transformation of our nation's energy system will require a sustained national effort involving carefully focused technology research, development, and demonstration; realignments of public policies and regulations; substantial capital investments; and allied resources (materials, infrastructure, and people) in both the public and the private sectors.

Many energy-supply and end-use technologies are ready for significant deployment now, but others will not be available until they have been demonstrated at scale[13] or until important technological barriers have been overcome. Of course, once a technology is ready for deployment, a number of important economic, regulatory and policy, and resource factors will govern the actual pace, scale, and cost of deployment. Especially important in this regard are the prices for fossil fuels and other materials, the availability and costs of specialized resources and capital, and key public policies and regulations that address, for example, renewable-energy portfolio standards, building regulations, corporate average fuel economy (CAFE) standards, and carbon prices.[14] *Because of the uncertainties about how these factors will play out in the decades ahead, the technology-deployment options that are identified in this chapter and in Part 2 of this report should be considered as important first-step technology assessments rather than as forecasts as to which technologies will be implemented and how important each technology will be.*

The committee also recognizes that currently unpredictable developments in

---

[13]The scale of a demonstration should be large enough to give an investor or company the confidence in the technology's economics, performance, and regulatory acceptability to build a commercial plant. The actual scale of demonstration required will vary across technologies.

[14]The term "carbon prices" denotes the costs that would be imposed through statute or regulation for emitting CO_2 and other greenhouse gases to the atmosphere.

technology could have dramatic impacts on future deployment options. There is little doubt that beyond the next few decades, new technologies—which employ, for example, advanced materials and innovative chemical processes not yet in view—could play transformative roles. Along these same lines, better understanding of how geoengineering[15] or the ecology of microbial systems affects climate could yield new insights on managing greenhouse gas emissions from energy production and use. In fact, unexpected breakthroughs might even enable fusion technology to contribute to the U.S. energy supply before 2050. Given the contingent nature of technology development, there will always be uncertainties in future technology pathways.

While the development and widespread deployment of both evolutionary and new technologies will play a central role in transforming the energy system, so too will new public policies and international collective actions that are equitable, efficient, and effective. Such collaborations will be needed not only because of the inherently global nature of the challenges but also because of the differing priorities and capacities of other countries. Market forces alone will not be sufficient to effect this transformation, as market externalities—including social costs not reflected in prices, regulatory constraints, the lack of information for knowledgeable market decision making, and other significant uncertainties—are likely to prevent energy markets from generating fully adequate price signals. Access to and competition for capital will be pertinent as well. And conflicts could arise when individual nations seek their own economic, political, or other national benefits—not necessarily consistent with shared international interests—in addressing issues such as global warming.

Because the energy system is so large, complex, and fully integrated into all aspects of American life, its successful transformation will take the full ingenuity and commitment of the public and private sectors. Moreover, the transformation must engage the routine attention of the public itself. In this sense, the present energy challenge is fundamentally different from historical efforts such as the Manhattan Project and the Apollo Project, which focused on specific technical objectives rather than on a very large and complex societal infrastructure. Those

---

[15]Geoengineering involves the use of technology to change the environment of Earth. For example, the emission of greenhouse gases into the atmosphere from human activity is now judged with very high confidence by climate scientists to cause global warming. Some scientists have proposed geoengineering as a way to reduce global warming—such as by changing the amount of sunlight that reaches the planet's lower atmosphere and surface or by removing greenhouse gases from the atmosphere.

projects were enormous technological and organizational triumphs, to be sure, but they were generally disconnected from the daily lives of the nation's citizens. Nevertheless, particular technological elements—such as CCS, advanced batteries for transportation, advanced geothermal energy for electricity production, and low-cost efficient lighting and solar panels—might very well benefit from focused development and demonstration programs even as the many nontechnological challenges are being addressed.

STRATEGY FOR ADDRESSING THE STUDY CHARGE

The focus of this study, consistent with its charge (Box 1.1), is on energy-supply and end-use technologies—in particular their deployment-readiness, performance, costs, barriers, and impacts. The AEF Committee also assessed the prospects of some other technologies that will be critical both in meeting the anticipated growth of energy demand and in enabling the deeper market penetration of the new energy-supply and end-use technologies themselves. These critical technologies range from CCS, which would support not only the continued use of fossil fuels for electricity generation but also any future production of liquid fuels, to advanced battery, fuel-cell, and hydrogen technologies.

The committee considered technology development and deployment over three time periods—2008–2020, 2020–2035, and 2035–2050—but focused mainly on the first two periods, not only because the more distant future is harder to analyze but also because it depends critically on what occurs (or does not occur) earlier. Notably, the committee found that what can be realized in the two later periods will be contingent on the accomplishments in the critical first period, which is immediately ahead of us. *Indeed, a major message of this report is that the nation can achieve the necessary and timely transformation of its energy system only if it embarks on an accelerated and sustained level of technology development, demonstration, and deployment along several parallel paths between now and 2020.* The cases for such urgent actions are strikingly similar in virtually all of the energy domains addressed in this report, whether they pertain to specific energy-supply technologies, end-use technologies, or electricity transmission and distribution.

In addressing its study charge, the committee avoided reinventing the wheel. Where appropriate, it took advantage of the existing energy literature, which is both extensive and information-rich, to inform its judgments. In some selected cases, the committee performed additional technical analyses to fill gaps in the literature or reconcile conflicting assessments. The approaches that the committee used are described in more detail in Part 2 of this report.

The committee also relied heavily on the reports of the three panels that were created as part of this Phase I study to undertake detailed examinations of energy efficiency technologies, alternative transportation fuels, and renewable-energy technologies. The three panel reports are, specifically:

- *Real Prospects for Energy Efficiency in the United States* (available at http://www.nap.edu/catalog.php?record_id=12621)
- *Liquid Transportation Fuels from Coal and Biomass: Technological Status, Costs, and Environmental Impacts* (available at http://www.nap.edu/catalog.php?record_id=12620)
- *Electricity from Renewable Resources: Status, Prospects, and Impediments* (available at http://www.nap.edu/catalog.php?record_id=12619).

The AEF Committee used these panel reports to inform its judgments about energy supply and cost for the particular technologies involved. Selected members of these panels, including their chairs and vice chairs, also served on the authoring committee for the present report.

The U.S. energy system is so large and complex that the committee was unable, in the time available, to assess the potential for transformation of its every relevant aspect. Note in particular that:

- The focus of the report is on energy-supply and end-use technologies that are most likely, in the judgment of the committee, to have meaningful impacts on the U.S. energy system during the three time periods considered in this study (encompassing the next 40 years or so). However, the committee did not assess the future role of technologies for the exploration, extraction, storage, and transportation of primary energy sources (e.g., fossil fuels), nor did it assess the role of some critial components of a modernized infrastructure—including tankers, roads, pipelines, and associated storage facilities—in delivering these resources from suppliers to consumers.
- The report does not explore in any depth the U.S. energy system at the regional level. Thus, the implications of the dramatic regional heterogeneity in the United States—for example, in energy resource endow-

ments, climates, and prices—on energy-supply and end-use technologies are not considered in any detail.

- The potential energy-supply contributions from the options assessed in this report have been estimated technology by technology. The committee did not, however, conduct an integrated assessment of how these technologies might compete in the marketplace and how that competition and other external factors could affect actual deployment rates and outcomes over time. For example, the successful deployment of energy efficiency technologies could reduce the demand for electricity and the need to deploy additional electricity-generation capacity, except perhaps to correct regional supply imbalances or replace aging assets with more efficient and environmentally benign facilities. Also, the utilization of biomass for liquid fuels production could reduce the supplies of biomass available for electricity generation. *Therefore the potential contributions of the energy-supply and end-use technologies identified in this report should not be viewed as predictions of any specific future mix of primary energy resources and conversion methods.*

- The committee has not made judgments about the relative desirability of the supply options described in this report or about their appropriate pace and scale of deployment. Such decisions are beyond the committee's charge and are the responsibility of policy makers, investors, consumers, and, indeed, all citizens.

- The committee and its panels developed the cost estimates presented in this report by using a range of methodologies (as described in Annex 3.A in Chapter 3 and in Part 2). It derived some of these estimates independently, with the assistance of consultants, whereas other estimates came from assessments documented in the literature. The cost estimates themselves were based on a number of underlying assumptions about commodity prices, construction costs, and fuel, regulatory, and operating costs, as well as on "conditional" assumptions[16] about the success of new-technology deployment. *As a consequence the cost estimates presented in this report should not be used to make* detailed *comparisons across technologies. However, because these estimates are presented as ranges that reflect the principal uncertainties in the underlying*

---

[16]Conditional assumptions posit that new technologies can be successfully deployed within a given time and at a given cost, even though the deployment is the first of its kind.

assumptions, the committee judges that they are sufficiently robust to be useful for rough *comparisons.*

- The report does not provide an evaluation of the full range of options for reducing energy use. Such reductions are generally understood to be obtainable in two ways: (1) deploying technologies to improve the efficiency of energy production and use and (2) conserving energy through behavioral or lifestyle changes (e.g., taking public transportation to work rather than driving).[17] The focus of this report is on the assessment of technologies that address the first factor—improving the efficiency of energy use. It addresses energy conservation only insofar as conservation is affected by the deployment of more energy-efficient technologies. To be sure, conservation is an important option for reducing energy use, but its detailed consideration is well beyond the technological scope of this study. A study on energy conservation would require, for example, an in-depth understanding of how social, economic, and policy factors affect energy consumption.

- The report does not provide forecasts of future prices of primary energy inputs (e.g., for petroleum and coal) or the effects of possible future policies and regulations concerning CO_2 emissions on such prices. Such prices, however, will influence the relative competitiveness of the energy-supply and end-use technologies discussed in this report, and they will affect technology choices and paces of development, especially in the private sector.

REPORT ORGANIZATION

This chapter has briefly discussed the current characteristics of the U.S. energy system, the challenges to improving the system's sustainability and security, the role of technology, and the committee's strategy for addressing its study charge (Box 1.1). The next two chapters complete Part 1 of this report by providing sum-

---

[17]Per capita energy use in a particular country can also be reduced by lowering energy intensity, for example, by importing energy-intensive goods from abroad rather than producing them domestically. However, this approach would not reduce overall energy use and could in some cases even result in increased energy use.

maries of the study's key findings (Chapter 2) and of the technology assessments (Chapter 3) of Part 2.

Part 2 contains six chapters (Chapters 4–9), which document the committee's detailed assessments of energy-supply and end-use technologies. The topics addressed in these chapters are, specifically,

- *Energy efficiency* in transportation, industry, and residential and commercial buildings (Chapter 4)
- Production and use of *alternative transportation fuels*, in particular biofuels as well as fuels derived from converting coal, or mixtures of coal and biomass, into liquids (Chapter 5)
- Production of *renewable energy* such as wind, solar, and geothermal energy, as well as hydropower and biopower (Chapter 6)
- Domestic *fossil-fuel energy*, particularly as coupled with technologies that would capture and safely store CO_2 (Chapter 7)
- Production of electricity from *nuclear energy* (Chapter 8)
- *Electricity transmission and distribution* systems that reliably accommodate intermittent energy supplies such as solar and wind and sophisticated demand-side energy efficiency technologies (Chapter 9).

REFERENCES

EIA (Energy Information Administration). 2008a. Annual Energy Review 2007. DOE/EIA-0384(2007); Washington, D.C.: U.S. Department of Energy, Energy Information Administration.

EIA. 2008b. Annual Energy Outlook 2008. DOE/EIA-0383(2008). Washington, D.C.: U.S. Department of Energy, Energy Information Administration.

IPCC (Intergovernmental Panel on Climate Change). 2007. Climate Change 2007: Synthesis Report. Contribution of Working Groups I, II and III to the Fourth Assessment Report of the Intergovernmental Panel on Climate Change. Geneva: IPCC.

Simmons, M.R. 2005. Twilight in the Desert: The Coming Saudi Oil Shock and the World Economy. Hoboken, N.J.: John Wiley & Sons, Inc.

Wood, J.H., G.R. Long, and D.F. Morehouse. 2004. Long-Term World Oil Supply Scenarios: The Future Is Neither as Bleak nor as Rosy as Some Assert. Energy Information Administration. Available at http://www.eia.doe.gov/pub/oil_gas/petroleum/feature_articles/2004/worldoilsupply/oilsupply04.html.

2

Key Findings

This chapter presents eight key findings from the AEF Committee's detailed analysis of existing and new energy-supply and end-use technologies presented in Part 2 of this report. These findings identify options for the *accelerated* deployment of these technologies during the next two to three decades, and they also identify needs for supporting research, development, and demonstration. Pursuing such options would, in the committee's judgment, hasten the transformation of the U.S. energy system, as described in Chapter 1.

By "accelerated," the committee means deployment of technologies at a rate that would exceed the "reference scenario" deployment pace (Box 2.1) but at a less dramatic rate than an all-out or "crash" effort, which could require disruptive economic and lifestyle changes that would be challenging to initiate and sustain. By contrast, accelerated technology deployments could likely be achieved without substantial disruption, although some changes in the behavior of businesses and consumers would be needed. Moreover, many of these changes could involve new costs and higher prices for end users.

The accelerated-deployment options identified in this chapter are based on the committee's judgments regarding two important factors: (1) the readiness of evolutionary and new technologies for commercial-scale deployment and (2) the pace at which such technologies could be deployed without the disruptions associated with a crash effort. In estimating these factors, the committee considered the maturity of a given technology together with the availability of the necessary raw materials, human resources, and manufacturing and installation capacity needed to support its production, deployment, and maintenance. In some cases, estimates of the evolution of manufacturing and installation capacity were based on the documented rates of deployments of specific technologies from the past.

BOX 2.1 *Reference Scenarios*

The statement of task for this study (Box 1.1) called for the development of a reference scenario "that reflects a projection of current economic, technology cost and performance, and policy parameters into the future." The AEF Committee decided to meet this requirement by adopting the Energy Information Administration's (EIA's) reference case for U.S. energy supply and consumption, which is the most commonly cited scenario for the U.S. energy system. It provides estimates of past, current, and future energy supply and consumption parameters by assuming that current energy policies remain unchanged and then extrapolating economic growth rates and technology-development trends into the future. In other words, the EIA reference case represents a business-as-usual and policy-neutral projection.

The EIA updates this reference case annually and presents it in the agency's Annual Energy Outlook reports. In this study, the committee uses the 2008 update (EIA, 2008), which reflects U.S. energy supply and consumption through 2007 and future projections through 2030, as its primary reference scenario. However, in limited cases the 2009 update (EIA, 2009a) was used, and explicitly noted in this report, when it was considered to be more indicative of current conditions.

The EIA's Annual Energy Outlook reports can be accessed at www.eia.doe.gov/oiaf/aeo/. Selected energy supply and consumption estimates from the 2008 update are shown in the three tables that follow.

TABLE 2.1.1 Reference Scenario Estimates of Electricity Consumption and Supply

	2007	2020	2030
Electricity Consumption (terawatt-hours)			
Residential	1400	1500	1700
Commercial	1300	1700	1900
Industry	1000	1100	1000
Transportation	6	8	9
Electricity Supply (terawatt-hours)			
Coal	2000	2300	2800
Petroleum	48	52	56
Natural gas	680	610	500
Nuclear power	800	870	920
Renewables			
Conventional hydropower	260	300	300
Onshore wind	38	100	120
Offshore wind	0	0	0
Solar photovoltaic	0.08	0.52	1.0
Concentrating solar power	0.92	2.0	2.2
Geothermal	16	24	31
Biopower	12	78	83

Note: Estimates have been rounded.
Source: EIA, 2008.

TABLE 2.1.2 Reference Scenario Estimates of Natural Gas Consumption and Supply

	2007	2020	2030
Natural Gas Consumption (trillion cubic feet)			
Residential	4.7	5.2	5.2
Commercial	3.0	3.4	3.7
Industrial	6.6	6.9	6.9
Electric power	6.8	5.9	5.0
Transportation	0.02	0.07	0.09
Natural Gas Supply (trillion cubic feet)			
Domestic production	19	20	19
Net imports	3.8	3.6	3.2

Note: Estimates have been rounded.
Source: EIA, 2008.

TABLE 2.1.3 Reference Scenario Estimates of Liquid Fuels Consumption and Supply

	2007	2020	2030
Liquid Fuels Consumption (million barrels per day)			
Residential and commercial	1.1	1.1	1.1
Industrial	5.1	4.8	4.7
Transportation	14	16	17
Electric power	0.25	0.26	0.28
Liquid Fuels Supply (million barrels per day)			
Petroleum			
Domestic production	5.1	6.2	5.6
Net imports	10	9.8	11
Natural gas plant liquids	1.8	1.7	1.6
Net product imports	2.1	1.4	1.3
Ethanol	0.44	1.4	2
Biodiesel	0.03	0.07	0.08
Biomass-to-liquids	0	0.14	0.29
Coal-to-liquids	0	0.15	0.24
Biomass-and-coal-to-liquids	Not considered		

Note: Estimates have been rounded.
Source: EIA, 2008.

FINDING 1: TECHNOLOGY DEPLOYMENT OPTIONS

With a sustained national commitment, the United States could obtain substantial energy efficiency improvements, new sources of energy, and reductions in greenhouse gas emissions through the accelerated deployment of existing and emerging energy-supply and end-use technologies, as described in some detail in Findings 2–5 in this chapter. Many energy efficiency and energy-supply technologies are ready for deployment now. But some emerging technologies will first require demonstration, either to prepare them for widespread commercial deployment starting about 2020 or to assess their readiness for deployment.

The U.S. energy system encompasses a large and complex installed base of energy-supply and end-use technologies. Transforming this system to increase sustainability, promote economic prosperity, improve security, and reduce environmental impacts as envisioned in Chapter 1 will require sustained national efforts to change the ways in which energy is produced, distributed, and used. The good news from the AEF Committee's assessment is that there are many practical options for obtaining energy savings, new supplies of energy, and reductions in greenhouse gas emissions through widespread and sustained deployments of existing and emerging energy-supply and end-use technologies. The most important of these options are described in Findings 2–5.

The United States cannot continue to muddle along on its current course if it hopes to transform its energy system. Indeed, both the public and the private sectors will have to be mobilized to achieve the necessary deployments in the decades ahead. Moreover, there is no "silver bullet" technology that can be deployed to overcome U.S. energy challenges. Contributions will be needed from the full array of currently available and emerging technologies:

- Numerous energy-supply and end-use technologies—energy efficiency, certain renewable-energy sources, and transmission and distribution (T&D) technologies—which can be deployed now and at relatively rapid rates with the appropriate mix of incentives.[1]

---

[1]Such incentives might include carbon taxes, cap and trade systems for CO_2 emissions, and tax credits for investments in energy efficiency or renewable-energy sources. In addition, regulations that require increased energy efficiency in the buildings, transportation, and industrial

- Evolutionary nuclear energy technologies, already being deployed in some other countries, which are ready for deployment in the United States. However, their commercial viability in the United States will first need to be demonstrated.
- Some emerging technologies, such as carbon capture and storage (CCS), for which sustained programs of development and commercial-scale demonstration will be needed during the next decade to ready the most promising among them for widespread deployment starting around 2020.

Expanding the deployment of coal with CCS, renewable energy, and evolutionary nuclear energy technologies may require continuing strong financial and regulatory pushes and new policy initiatives.[2] But many of the technologies identified in this report will require decades-long lead times for development, demonstration, and deployment. *Therefore it is imperative that these activities be started immediately even though some will be expensive and not all will be successful: some may fail, prove uneconomic, or be overtaken by better technologies.* Some failures are an inevitable part of learning and development processes. Long-term success requires that we stay the course and not be distracted by the inevitable short-term disappointments. To help ensure that the potential benefits outweigh the risks, investments in new technology demonstrations must be carefully chosen so as to produce results that usefully inform the deployment decision-making process.

Although it is beyond the committee's charge to recommend policy actions, it notes that the effective transformation of the energy system will require long-term investment in new energy technologies, policies that encourage such investment, and acceptance of the inevitable disappointments that will punctuate our long-term success.

sectors could play a key role both in moderating the demand for energy and stimulating related R&D.

[2]In addition to the incentives listed in Footnote 1, other possible actions include expanding renewable-energy portfolio standards to promote the deployment of renewable energy and providing federal loan guarantees to promote construction of a handful of evolutionary nuclear plants. Some of these actions are already under way.

FINDING 2: ENERGY SAVINGS FROM IMPROVED EFFICIENCY

The deployment of existing energy efficiency technologies is the nearest-term and lowest-cost option for moderating our nation's demand for energy, especially over the next decade. The committee judges that the potential energy savings available from the accelerated deployment of existing energy-efficiency technologies in the buildings, transportation, and industrial sectors could more than offset the Energy Information Administration's projected increases in U.S. energy consumption through 2030.

The deployment of energy efficiency technologies[3]—especially of mature technologies in the buildings, transportation, and industrial sectors—is the nearest-term and lowest-cost option for extending domestic supplies of energy. Many energy efficiency savings can be obtained almost immediately by deploying currently available technologies. In contrast, providing new energy supplies typically takes many years. Moreover, energy efficiency has broader societal benefits beyond saving energy. Society is giving more attention to the environment and other externalities as exemplified, for example, by concerns about the impacts of carbon dioxide (CO_2) emissions on global climate change. Laws and regulations, from the Endangered Species Act to the Clean Air Interstate Rule, inevitably slow the development of new energy supplies. In contrast, efficiency involves few emissions, endangers no species, and does not destroy scenic vistas.

To achieve such benefits, however, the efficiency savings must translate into actual reductions in energy consumption. This has been a particular issue in the transportation sector, where efficiency improvements that could have been used to raise vehicle fuel economy were instead offset by higher vehicle power and increased size.

Efficiency savings are realized at the site of energy use—that is, at the residence, store, office, factory, or transportation vehicle. The efficiency supply curves shown later in this chapter demonstrate that many energy efficiency investments cost less than delivered electricity, natural gas, and liquid fuels; in some cases, those costs are substantially less. In the electricity sector, many efficiency investments even cost less than transmission and distribution costs, which are typically

---

[3]As noted in Chapter 1, the committee draws a sharp distinction between energy efficiency and energy conservation. Conservation can be an important strategy for reducing energy use, but it generally does not involve technology deployment and is therefore not addressed in this report.

4–6¢/kWh for a residential customer and about half that for large commercial and industrial customers. Chapter 4 also shows that many energy efficiency projects with a rate of return of 10 percent or more could be undertaken by industry. Although most companies do not consider this rate of return attractive, it is nevertheless an attractive investment for society.

The greatest capability for energy efficiency savings is in the buildings sector, which accounted for about 70 percent of electricity consumption in the United States in 2007 (2700 TWh out of approximately 4000 TWh in total). Improvements in the energy efficiency of residential and commercial buildings—through the accelerated deployment of efficient technologies for space heating and cooling, water heating, lighting,[4] computing, and other uses—could save about 840 TWh per year by 2020 (Figure 2.1), which exceeds the EIA's projected increase in electricity demand of about 500 TWh for residential and commercial buildings by the year 2020 (EIA, 2008) (see Table 2.1.1 in Box 2.1). Further continuous improvements in building efficiency could save about 1300 TWh of electricity per year by 2030 (Figure 2.1), which also exceeds the EIA-projected reference scenario increase in electricity demand of about 900 TWh per year. In addition, improvements in building efficiency could save 2.4 quads of natural gas annually by 2020 and 3 quads of natural gas annually by 2030 (Figure 2.2).

There are many examples of cost-effective efficiency investments that could be made in the buildings sector to save energy. For example, an approximate 80 percent increase in energy efficiency—translating to nearly a 12 percent decrease in overall electricity use in buildings—could be realized immediately by replacing incandescent lamps with compact fluorescent lamps or light-emitting diodes. Energy savings between 10 and 80 percent could be realized by replacing older models of such appliances as air conditioners, refrigerators, freezers, furnaces, and hot water heaters with the most efficient models. Such replacements would not occur as quickly as replacing lamps because it is usually cost-effective to replace appliances only when they near the end of their service lives. The same is true for motor vehicles. Buildings last decades, so the energy savings benefits of new buildings will take decades to realize. However, there are cost-effective retrofits that could be installed immediately.

---

[4]On June 26, 2009, the Obama administration issued a final rule to increase the energy efficiency of general service fluorescent lamps and incandescent reflector lamps. The changes will take effect in 2012.

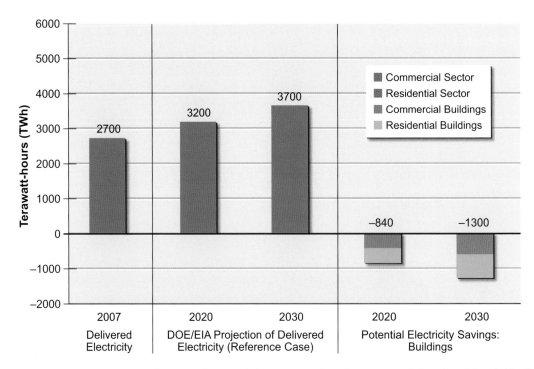

FIGURE 2.1 *Estimates of potential energy savings in commercial and residential build-ings in 2020 and 2030 (relative to 2007) compared to projected delivered electricity. The commercial and residential sectors are shown separately. Current (2007) U.S. delivered electricity in the commercial and residential sectors, which is used primarily in buildings, is shown on the left, along with projections for 2020 and 2030. To estimate savings, an accelerated deployment of technologies as described in Part 2 of this report is assumed. Combining the projected growth with the potential savings results in lower electricity consumption in buildings in 2020 and 2030 than exists today. The industrial and transpor-tation sectors are not shown. Delivered energy is defined as the energy content of the electricity and primary fuels brought to the point of use. All values have been rounded to two significant figures.*
Sources: Data from Energy Information Administration (2008) and Chapter 4 in Part 2 of this report.

In fact, the *full* deployment of cost-effective[5] energy efficiency technologies in buildings alone could eliminate the need to build any new electricity-generating plants in the United States—except to address regional supply imbalances, replace obsolete power-generation assets, or substitute more environmentally benign elec-tricity sources—assuming, of course, that these efficiency savings would not be used to support greater electricity use in other sectors.

---

[5]See the section titled "Energy Efficiency" in Chapter 3 for a definition of "cost-effective."

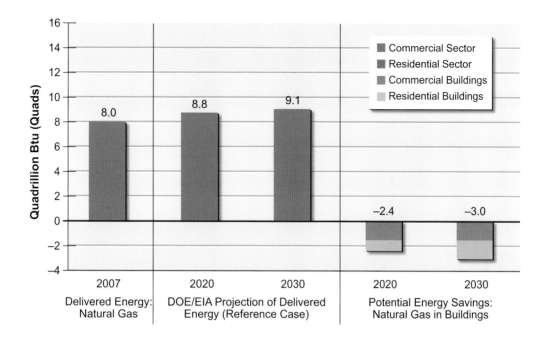

FIGURE 2.2 *Estimates of potential natural gas savings in commercial and residential buildings in 2020 and 2030 (relative to 2007) compared to delivered energy from natural gas. The commercial and residential sectors are shown separately. Current (2007) U.S. delivered energy from natural gas in the commercial and residential sectors, which is used primarily in buildings, is shown on the left, along with projections for 2020 and 2030. To estimate savings, an accelerated deployment of technologies as described in Part 2 of this report is assumed. Combining the projected growth with the potential savings results in lower natural gas consumption in buildings in 2020 and 2030 than exists today. The industrial and transportation sectors are not shown. Delivered energy is defined as the energy content of the electricity and primary fuels brought to the point of use. All values have been rounded to two significant figures.*
Sources: Data from Energy Information Administration (2008) and Chapter 4 in Part 2 of this report.

Opportunities for achieving substantial energy savings exist in the industrial and transportation sectors as well. For example, deployment of energy efficiency technologies in industry could reduce energy use in manufacturing by 4.9–7.7 quads per year (14–22 percent) in 2020[6] relative to the EIA reference case projection (Figure 2.3). Most of these savings would occur in the pulp and paper, iron

---

[6]These identified savings would provide industry with an internal rate of return on its efficiency investments of at least 10 percent or exceed the company's cost of capital by a risk premium. See Chapter 4 for additional discussion.

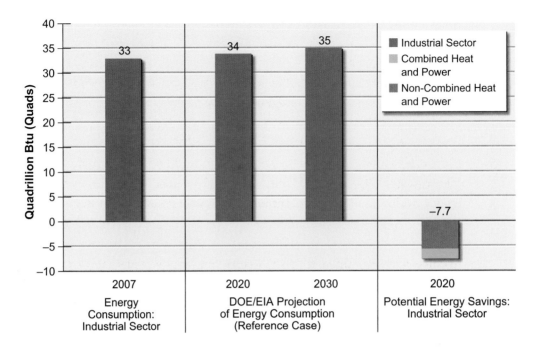

FIGURE 2.3 *Estimates of potential energy savings in the industrial sector in 2020 (relative to 2007) compared to total delivered energy in the industrial sector. Current (2007) U.S. delivered energy in the industrial sector is shown on the left, along with projections for 2020 and 2030. To estimate savings, an accelerated deployment of technologies as described in Part 2 of this report is assumed. Combining the projected growth with the potential savings results in lower energy consumption in the industrial sector in 2020 (7.7 quads) than exists today. A more conservative scenario described in Chapter 4 could result in energy savings of 4.9 quads. The committee did not estimate savings for 2030. Delivered energy is defined as the energy content of the electricity and primary fuels brought to the point of use. All values have been rounded to two significant digits. Sources: Data from Energy Information Administration (2008) and Part 2 of this report.*

and steel, and cement industries. The increased use of combined heat and power in industry is estimated to contribute a large fraction of these potential savings—up to 2 quads per year in 2020.

In the transportation sector, energy savings can be achieved by increasing the efficiencies with which liquid fuels (especially petroleum) are used and by shifting the energy source for part of the light-duty vehicle (LDV) fleet from petroleum to electric power. Of course, the environmental impacts of such a fuel shift are dependent on how electricity (or hydrogen, if fuel-cell vehicles are produced) is generated. Moreover, electrification of LDVs will increase the overall demand for electricity. Shifting this electricity demand to off-peak times (e.g., at night), through the use of demand-side technologies such as smart metering, may reduce the need for new power-plant construction and improve the utilization of current baseload power plants.

Improvements in the efficiency of today's spark-ignition and diesel engine LDVs, combined with increased use of hybrid and other advanced vehicle technologies, could reduce these vehicles' fuel consumption beyond 2020 to below that projected by the EIA (EIA, 2008). The EIA projection, which incorporates the increased fuel-economy standards mandated by the Energy Independence and Security Act (EISA) of 2007, equates to a 30 percent reduction in average fuel consumption (and a 40 percent increase in average fuel efficiency) in new LDVs in 2020 over today's consumption.[7] Exceeding this EIA projection is possible, but only if vehicle manufacturers focus on increasing vehicle fuel economy as opposed to their historic emphasis on increasing vehicle power and size. Figure 2.4 shows projections (described in Chapter 4) that illustrate how improvements in LDV fuel efficiency beyond that projected by the "no-change" reference scenario could further reduce total fuel consumption. These efficiency improvements, which include plug-in hybrid vehicles but not (fully) battery-electric vehicles or hydrogen fuel-cell vehicles, could reduce gasoline consumption by about 1.4 million barrels per day in 2020 and 5.6 million barrels per day in 2035. Of course, these fuel-efficient vehicles will have to be acceptable to consumers. Improvements are also possible in fuel consumption for freight shipping, but projected growth in airline travel is likely to offset improvements in aviation technologies.

Many energy efficiency technologies save money and energy. The cost of conserved energy (CCE) is a useful way to compare the cost of an energy efficiency technology to the cost of electricity and natural gas.[8] The range of

---

[7]The EIA (2008) reference case incorporates the EISA corporate average fuel economy (CAFE) standard of 35 miles per gallon (mpg) by 2020. The EIA reference case projects that the fuel economy of new vehicles will reach 36.6 mpg in 2030. As is noted in Chapter 1, the Obama administration recently announced a new national fuel efficiency policy that requires an average fuel economy standard of 35.5 mpg for new light-duty vehicles in 2016.

[8]CCE is defined as the levelized annual cost of an energy efficiency measure—that is, the cost of a new technology, or the incremental cost for a more efficient technology compared with a less efficient one—divided by the annual energy savings in kilowatt-hours or British thermal units over the lifetime of the measure. (The levelized annual costs do not include the costs for public policies and programs aimed at stimulating adoption of energy efficiency measures.) The CCE is expressed here in cents per kilowatt-hour (¢/kWh) for electricity efficiency measures and dollars per million British thermal units ($/million Btu) for natural gas efficiency measures. The CCEs presented in this report were computed using a real discount rate of 7 percent.

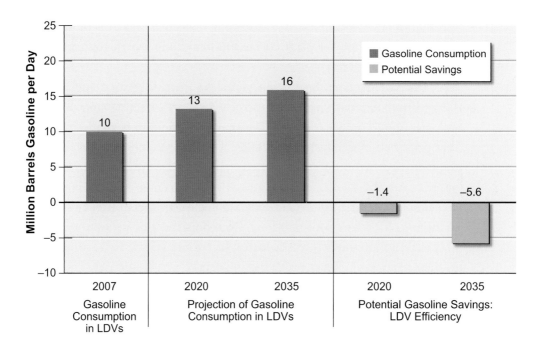

FIGURE 2.4 *Estimates of potential for gasoline consumption reduction in the U.S. light-duty vehicle (LDV) fleet in 2020 and 2035 (relative to 2007). Current (2007) U.S. gasoline consumption in LDVs is shown on the left. This consumption estimate, which was developed by the committee, includes gasoline-equivalent diesel fuel consumption in LDVs as well as fuel consumption in LDVs between 8,500 and 10,000 lb weight (the new Environmental Protection Agency upper limit on light trucks). Projected gasoline consumption in LDVs in 2020 and 2035 is shown by the middle set of bars. The projected consumption shown is an illustrative, no-change baseline scenario, where any efficiency improvements in powertrain and vehicle are offset by increases in vehicle performance, size, and weight. This baseline is described in more detail in Chapter 4 in Part 2 of this report. To estimate savings, an accelerated deployment of technologies as described in Part 2 of this report is assumed. Specifically, fuel efficiency improvements result from an optimistic illustrative scenario in which the corporate average fuel economy (CAFE) standards of the Energy Independence and Security Act of 2007 are met in 2020. This scenario assumes that fuel economy for new LDVs continues to improve until it reaches, in 2035, double today's value. Combining the projected growth in vehicle fleet size with the potential efficiency savings results in only slightly higher gasoline consumption in vehicles in 2020 and 2035 than exists today. A more conservative illustrative scenario, which results in savings of 1.0 and 4.3 million barrels of gasoline per day in 2020 and 2035, respectively, is also shown in Part 2 of this report. Beyond 2020, a 1 percent compounded annual growth in new vehicle sales and annual mileage per vehicle, combined, is assumed. Gasoline consumption can be further reduced if vehicle use (vehicle miles traveled) is reduced. All values have been rounded to two significant figures. Source: Data from Chapter 4 in Part 2 of this report.*

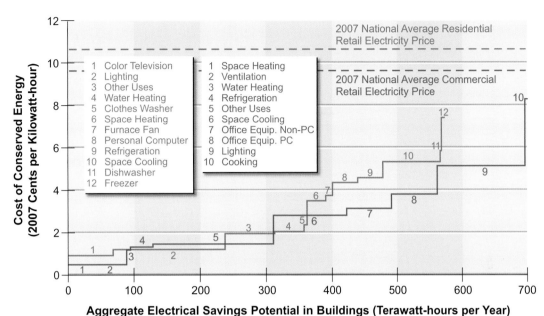

FIGURE 2.5 *Estimates of the cost of conserved energy (CCE) and energy savings potential for electricity efficiency technologies in buildings in 2030. The CCEs for potential energy efficiency measures (numbered) are shown versus the ranges of potential energy savings for these measures. The total savings potential is 567 TWh per year in the residential sector and 705 TWh per year in the commercial sector. Commercial buildings (red solid line) and residential buildings (blue solid line) are shown separately. For comparison, the national average 2007 retail price of electricity in the United States is shown for the commercial sector (red dashed line) and the residential sector (blue dashed line). For many of the technologies considered, on average the investments have positive payback without additional incentives. CCEs include the costs for add-ons such as insulation. For replacement measures, the CCE accounts for the incremental cost—for example, between purchasing a new but standard boiler and purchasing a new high-efficiency one. CCEs do not reflect the cost of programs to drive efficiency. All costs are shown in 2007 dollars. Sources: Data from Brown et al. (2008) and Chapter 4 in Part 2 of this report.*

CCE for electricity savings from commercial and residential buildings is shown in Figure 2.5. The range of CCE for electricity savings from commercial buildings is 0.5–8.4¢/kWh, with a weighted average of 2.7¢/kWh. However, nearly all of the efficiency savings are achievable at a CCE of 5¢/kWh or less. The range of CCE for electricity savings from residential buildings is 0.9–7.4¢/kWh, with a weighted average of 2.7¢/kWh. More than 80 percent of the potential savings are achievable at a CCE of 5¢/kWh or less. For comparison purposes, the average retail price of electricity in the residential and commercial sectors in 2007 was about

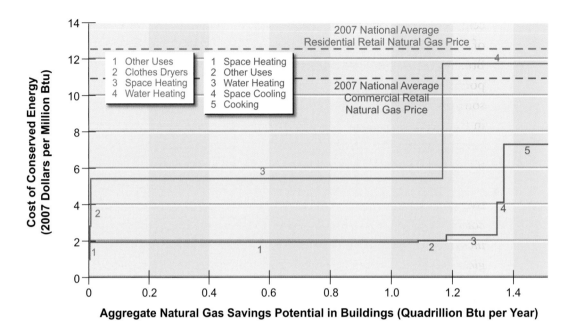

FIGURE 2.6 *Estimates of the cost of conserved energy (CCE) and energy savings potential for natural gas efficiency technologies in buildings in 2030. The CCEs for potential energy efficiency measures (numbered) are shown versus the ranges of potential energy savings for these measures. The total savings potential is 1.5 quads per year in the residential sector and 1.5 quads per year in the commercial sector. Commercial buildings (red solid line) and residential buildings (blue solid line) are shown separately. For comparison, the national average 2007 retail price of natural gas in the United States is shown for the commercial sector (red dashed line) and the residential sector (blue dashed line). For many of the technologies considered, on average the investments have positive payback without additional incentives. CCEs include the costs for add-ons such as insulation. For replacement measures, the CCE accounts for the incremental cost—for example, between purchasing a new but standard boiler and purchasing a new high-efficiency one. CCEs do not reflect the cost of programs to drive efficiency. All costs are shown in 2007 dollars.*
Sources: Data from Brown et al. (2008) and Chapter 4 in Part 2 of this report.

10¢/kWh.[9] In other words, it is substantially cheaper for a customer to save electricity rather than purchase electricity, even if these savings require up-front costs.

The range of CCE for natural gas savings from commercial and residential buildings is shown in Figure 2.6. The range of CCE from commercial buildings is $1.9–7.4/million Btu, with a weighted average of $2.5/million Btu. Nearly 80 per-

---

[9]The figures were 10.65¢/kWh for residential and 9.65¢/kWh for commercial; see http://www.eia.doe.gov/cneaf/electricity/epa/epat7p4.html.

cent of the potential savings are achievable at a CCE of $2/million Btu. The range of CCE for natural gas savings from residential buildings is $1.1–11.8/million Btu, with a weighted average of $6.9/million Btu or less. Nearly 80 percent of the potential savings are achievable at a CCE of $5/million Btu or less. For comparison purposes, the retail price of natural gas in 2007 was about $12.7/million Btu in the residential sector and $11/million Btu in the commercial sector. Again, it is substantially cheaper for a customer to save natural gas rather than purchase natural gas, even if these savings require up-front costs.

The energy efficiency savings identified in this report are highly cost-effective with short payback periods. *Substantially greater energy efficiency savings could likely be obtained with a more aggressive mix of policies, regulations, and incentives to encourage an even wider deployment of energy efficiency technologies.* However, it should be noted that businesses and consumers have historically been resistant to making even modest up-front investments in such technologies (Box 2.2). New approaches may be required to break these patterns.

FINDING 3: OPTIONS FOR INCREASING ELECTRICITY SUPPLIES AND CHANGING THE SUPPLY MIX

The United States has many promising options for obtaining new supplies of electricity and changing its supply mix during the next two to three decades, especially if carbon capture and storage and evolutionary nuclear energy technologies can be deployed at required scales. However, the deployment of these new supply technologies is very likely to result in higher consumer prices for electricity.

The U.S. supply of electricity in 2007, about 4000 TWh,[10] was obtained from the following sources (EIA, 2009b):[11]

- 2000 TWh from coal-fired power plants
- 810 TWh from nuclear power plants

---

[10]This estimate is for electricity supplied to the grid. The electricity delivered to the consumer is slightly lower because of losses in the transmission and distribution system. In 2007, these losses were estimated to be about 9 percent based on sales of electricity.
[11]These numbers have been rounded from the EIA estimates.

BOX 2.2   *Energy Efficiency and the Behavioral Gap*

A key finding of the present report is that there are substantial opportunities to reduce energy use through the widespread deployment of energy efficiency technologies in the buildings, transportation, and industrial sectors. The costs of deploying many of these technologies are much less than the costs to purchase energy; in fact, in these cases deployment saves money as well as energy. In spite of such advantages, many consumers are reluctant to make the necessary investments to deploy these technologies. Why the apparent dichotomy persists is the subject of ongoing research, which has already identified several reasons.

One reason for the behavioral gap between economically optimal technology choices and actual choices is the low salience of energy efficiency for consumers. That is, consumers in this case do not reflect the neoclassical economic model of the optimizing consumer. Although real-world consumers may recognize that purchasing an energy-efficient technology would be economically beneficial, the net benefits are usually so small relative to family budgets that individuals do not take the time to gather and analyze the requisite information.

Another reason for the gap has to do with the difficulty of changing consumers' purchasing and use habits. Preferences learned from parents, neighbors, and friends may change only very slowly, if at all. Also, most consumers do not calculate life-cycle costs when making purchases; instead, they focus primarily on first-purchase costs. Producers who understand this bias may be reluctant to design and market energy-efficient products unless forced to do so by governmental regulation.

Part of the behavioral gap is also based on economic-incentive issues—e.g., landlords of residential rental units are not motivated to pay for technologies that are more efficient when their tenants pay the utility bills. There are also historical path dependencies. For example, many existing building codes were developed when energy costs were not seen as important; these codes were optimized for safety, not for minimum life-cycle costs. Consumers also pay attention to product characteristics that tend to be ignored by analysts. They resisted buying early-generation compact fluorescent lamps, for instance, because they did not like the color of the light produced.

Continuing research is needed to more fully understand these and other reasons for the behavioral gap and to devise appropriate strategies for closing it.

- 690 TWh from natural-gas-fired power plants
- 320 TWh from renewable-energy sources, mostly hydropower (250 TWh), wind (34 TWh), geothermal (15 TWh), and biopower (8.7 TWh)
- 180 TWh from combined-heat-and-power plants, fed primarily by natural gas and coal
- 57 TWh from oil-fired power plants.

Through the deployment of new technologies and the repowering of current assets, the United States has many promising options both for increasing its electricity supply and for changing its electricity-supply mix. These estimates of new electricity supplies using different energy sources and technologies were derived independently and should not be added to obtain a total new supply estimate. As noted in Chapter 1, the AEF Committee has not conducted an integrated assessment of how these energy-supply technologies would compete in the marketplace or of how that competition and other external factors would affect deployment success.

Renewable-energy sources (Figure 2.7) could provide about an additional 500 TWh of electricity per year by 2020 and about an additional 1100 TWh per year by 2035 through new deployments in favorable locations. These levels exceed the amounts of new electricity supplies that are likely to be available from new nuclear-power generation or new coal-power generation with CCS in 2020 or from new nuclear power generation in 2035. However, expansion of transmission capabilities would be required to transport new electricity supplies from renewable resources to demand centers and regional energy markets. Backup supplies of electricity, or the capability to store energy during times when electricity production exceeds demand, would be needed when renewable sources were unavailable. Given current cost structures for renewable energy (discussed later in this chapter), policies such as renewable portfolio standards and tax credits would likely need to be continued, and possibly expanded, to obtain these new supplies.

Coal-fired plants with CCS (Figure 2.8) could provide as much as 1200 TWh from repowering and retrofit of existing plants and as much as 1800 TWh from new plants. In combination, the entire existing coal power fleet (which currently delivers about 2000 TWh of electricity per year) could be replaced by CCS coal power by 2035. However, successful commercial-scale demonstrations of CCS technologies would be required during the coming decade to realize this potential. (A brief discussion of CCS demonstration needs and constraints is provided under Finding 6; additional information is

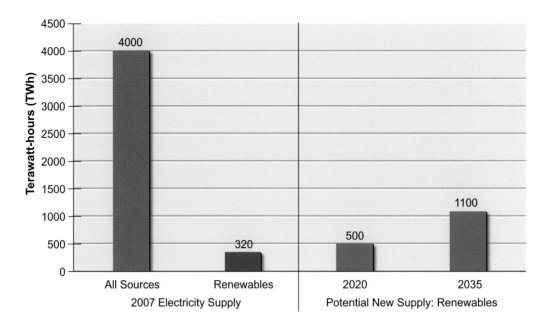

FIGURE 2.7 *Estimates of potential new electricity supply from renewable sources in 2020 and 2035 (relative to 2007) compared to current supply from all sources. The total electricity supplied to the U.S. grid in 2007 is shown on the left (in green). The supply generated by renewable sources (including conventional hydropower) is shown in red. Potential new supply shown is in addition to the currently operating supply. To estimate future supply, an accelerated deployment of technologies as described in Part 2 of this report is assumed. Potential new electricity supply does not account for future electricity demand or competition among supply sources. All values have been rounded to two significant figures.*
Sources: Data from Energy Information Administration (2008) and Part 2 of this report.

available in Chapter 7 in Part 2 of this report.) In addition, it will be necessary to assess the full implications, including the environmental externalities, of any very large expansion in coal production and use. Given the projected costs of CCS, the widespread deployment of CCS technologies will likely require new governmental policies that provide a regulatory or CO_2 price push. These deployments would reduce the environmental impacts of electricity generation and thereby provide indirect economic benefits to consumers, though such benefits are difficult to quantify.

Nuclear plants (Figure 2.9) could provide an additional 160 TWh of electricity per year by 2020 and about 850 TWh by 2035 through the modification of current plants to increase power output (referred to as "uprating")

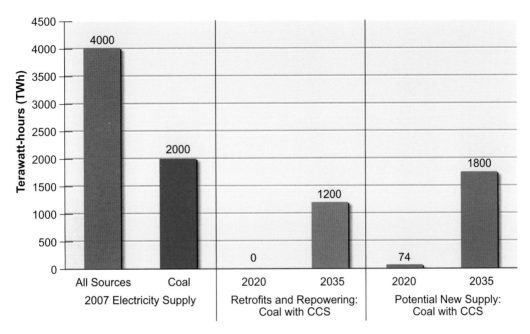

FIGURE 2.8 *Estimates of potential electricity supply, in 2020 and 2035 (relative to 2007) compared to supply from all sources, from new coal-fired plants with carbon capture and storage (CCS) and from plants retrofitted or repowered to add CCS. The total electricity supplied to the U.S. grid in 2007 is shown on the left (in green). The supply generated by coal is shown in red. To estimate future supply, an accelerated deployment of technologies as described in Part 2 of this report is assumed. The potential supply from new coal plants built with CCS is shown in blue; the potential supply from retrofitting and repowering currently operating plants to add CCS is shown in orange. Potential new supply with CCS and potential retrofits with CCS compete for the same CO_2 storage sites and other enabling elements. The simultaneous realization of both estimates of potential 2035 deployment is not anticipated because of this competition. Over the next decade CCS technologies will need to be successfully demonstrated to achieve the potential supply shown from coal plants with CCS in 2035. A strong policy push will also be required to realize the 2020 supply estimate. The AEF Committee assumed an average capacity factor of 85 percent for coal plants with CCS. Potential new electricity supply does not account for future electricity demand, fuel availability or prices, or competition among supply sources. All values have been rounded to two significant figures.*
Sources: Data from Energy Information Administration (2008) and Chapter 7 in Part 2 of this report.

and through new-plant construction. These amounts would be in addition to the 800 TWh produced by currently operating plants and do not account for possible plant retirements, which are shown by the negative 2035 supply estimate in Figure 2.9. The original (40-year) operating licenses of current plants are now beginning to expire. For the majority of these plants, license extensions

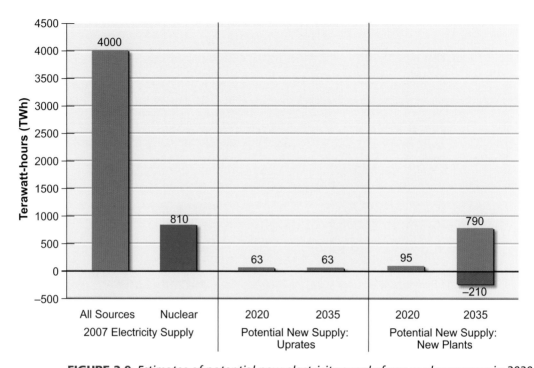

FIGURE 2.9 *Estimates of potential new electricity supply from nuclear power in 2020 and 2035 (relative to 2007) compared to supply from all sources. The total electricity supplied to the U.S. grid in 2007 is shown on the left (in green). The supply generated by nuclear power is shown in red. Over the next decade, the first few nuclear plants will need to be constructed and operated successfully to achieve the potential supply shown from nuclear power in 2035. To estimate supply, an accelerated deployment of technologies as described in Part 2 of this report is assumed. Current plants are assumed to be retired at the end of 60 years of operation, resulting in a reduced electricity supply from nuclear power in 2035 (shown by the negative valued red bar). However, operating license extensions to 80 years are currently under consideration, and it is possible that many of these plants may not be retired by 2035. The AEF Committee assumed an average capacity factor of 90 percent for nuclear plants. Potential new electricity supply does not account for future electricity demand, fuel availability or prices, or competition among supply sources. All values have been rounded to two significant figures. Sources: Data from Energy Information Administration (2008) and Part 2 of this report.*

for an additional 20 years (to allow for a total of 60 years of operation) have been approved or are being processed. This will allow more electricity to be generated over the operating life of each of these plants. The negative 2035 supply estimate shown in Figure 2.9 illustrates potential supply losses resulting from the retirement of plants when these 60-year licenses expire; however, it is possible that some plants will receive license extensions for up to an additional 20 years (to allow for a total of 80 years of operation), decreasing these potential supply losses. A discussion of nuclear-plant demonstration needs and constraints is included under Finding 6; additional information is provided in Chapter 8 in Part 2 of this report. Existing federal loan guarantees (described in Chapter 8) will probably be essential for constructing at least some of the first few new nuclear plants in the United States.

Natural gas generation of electricity could be expanded to meet a substantial portion of U.S. electricity demand—if there were no concerns about the behavior of world natural gas markets and prices and about further increasing CO_2 emissions and U.S. import dependence. In fact, lower capital cost and shorter construction times favor natural gas over coal or nuclear power plants for new electric-power generation (see Figure 1.10). But it is not clear whether natural gas supplies at competitive prices would be adequate to support substantially increased levels of electricity generation. The role of natural gas will likely depend on the demand for electricity, the magnitude of growth in domestic natural gas production, the demand for natural gas for other uses (e.g., as an industrial feedstock or for space heating), and controls on CO_2 emissions. If growth in new domestic natural gas production were sufficient to offset declines in production from existing fields and could be sustained for extended periods, domestic resources could be used to support expanded electricity production. If domestic supplies could not be increased, liquefied natural gas imports would be needed, thereby exposing the U.S. market to increased import dependence and to international prices. Increased import dependence has important energy-security implications, as discussed in Chapter 1.

Although the potential picture with these new supplies is promising, they will likely result in higher electricity prices.[12] Estimates of the levelized cost of electricity (LCOE; Box 2.3) for new baseload and intermittent electricity generation in 2020 are shown in Figure 2.10. Descriptions of the methods and assumptions that were used to estimate these LCOEs are provided in Annex 3.A in Chapter 3 and in the Part 2 chapters. It is important to recognize that estimating future costs is notoriously difficult. The estimates are strongly dependent on the judgments of the experts who make them and are based on a necessarily lim-

---

[12]The deployment of new generating capacity, whether from an existing or a new technology, generally results in an increase in the cost of electricity. This is because the embedded costs (i.e., the "book values") of existing generating assets are typically at least an order of magnitude less than those of the new generating assets (whether for replacement or supplementation).

BOX 2.3 *Levelized Cost of Electricity*

The levelized cost of electricity (LCOE) is defined as the average cost of generating a unit of electricity over the generating facility's service life. The LCOE is computed by dividing the present value of the estimated full life-cycle costs of the generating facility by its estimated lifetime electricity production. The result is usually expressed in terms of cents per kilowatt-hour.

The full life-cycle costs of the generating facility include:

- Capital costs for construction
- Financing costs
- Operations and maintenance costs
- Fuel costs
- Decommissioning costs.

Facility lifetime is typically taken to be between 20 and 40 years, depending on the generating technology.

The LCOE is less than the cost of electricity to the consumer (i.e., less than the retail price) because it does not include the costs of transmission and distribution or the electricity generator's profit. These additional costs can typically add several cents per kilowatt-hour to the wholesale cost of electricity.

---

ited understanding about how future events might unfold. Consequently, such estimates usually have large uncertainties. Given these uncertainties and the particular methodologies used to estimate LCOEs in this report, differences in LCOEs of 2¢/kWh or less are probably not significant.[13]

Figure 2.10 shows both that there is a range of LCOE values for each technology and that the ranges for many different technologies are overlapping. For comparison purposes, consider that the EIA-estimated average wholesale price of electricity[14] in 2007 was about 6¢/kWh and is forecast to

---

[13]It was difficult to obtain consensus within the committee about how to estimate LCOEs for different technologies on exactly comparable bases given the large number of assumptions that had to be made about costs, performance, and expected lifetimes for each technology. Consequently, the estimates shown in Figure 2.10 should be considered approximations.

[14]The wholesale price of electricity represents the price of electricity supplied at the busbar. It does not include the prices for transmission and distribution. As noted previously, the average retail price for electricity in 2007 was about 10¢/kWh.

remain at that level through 2030 under the agency's reference case projection (EIA, 2008). The LCOEs for most new electricity sources in 2020 shown in Figure 2.10 are higher than the EIA-projected wholesale cost. The clear exceptions are coal without CCS, some biopower for baseload generation, and onshore wind for intermittent generation. The cost for electricity from natural gas strongly depends on gas prices as shown in Figure 2.10. However, biopower can provide limited new supplies of electricity, and wind power can have large electrical-transmission and distribution costs because power generation sources are spatially distributed. Additionally, generation of electricity using natural gas and coal without CCS might not be environmentally acceptable, and the price for electricity from natural gas could increase substantially, of course, if there were large price increases for this fuel.

The LCOEs shown in Figure 2.10 represent what the AEF Committee judges to be reasonable cost ranges based on available information. Actual LCOEs could be different from those shown in the figure, however, for reasons such as unanticipated future changes in fuel prices, higher- or lower-than-expected costs for deploying and operating new technologies, costs arising from deployments at particular locations, and other regional cost differences. Obviously, the LCOEs for some technologies would be affected more than others by these factors; natural gas combined cycle (NGCC) plants, for example, utilize natural gas as a fuel, and recent prices for this fuel have been volatile. On the other hand, fuel costs for nuclear plants are only a small part of electricity generation costs. Wind, solar, hydro, and geothermal power have no fuel charges and their deployment costs are well established, especially for onshore wind and solar. Still, the potential outputs of solar arrays and wind turbines can vary greatly because of local conditions, so these technologies will have site-dependent cost impacts.

The overlapping LCOE estimates shown in Figure 2.10 make it difficult to pick winners and losers, suggesting the need to proceed on parallel tracks for demonstrating and deploying technologies. The results for electricity from natural gas strengthen this conclusion: given the low and high prices of natural gas in recent years, the LCOE for NGCC can be one of the lowest-cost—or one of the highest-cost—sources of electricity, as shown in Figure 2.10. Given the variability of fuel prices over the decades-long lives of these plants, it is impossible to be confident that a particular technology will have the lowest cost or even a reasonably low cost. Although the committee, along with most observers, concluded that over the 30-year life of an NGCC plant the price of

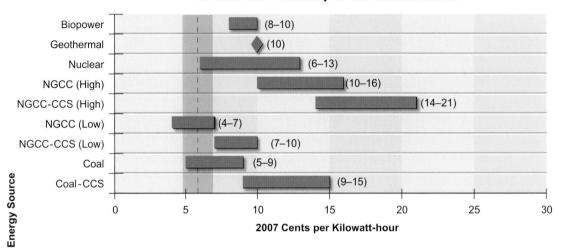

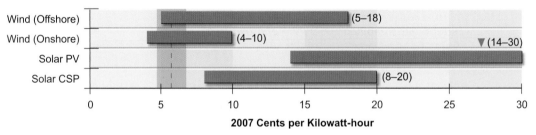

natural gas would be likely to rise, the year-to-year variations could also be large because of changes in the balance between demand and supply.

Figure 2.10 indicates that the LCOE range for nuclear plants is comparable with those for coal with CCS and certain renewable-energy sources, such as offshore wind and concentrating solar power. The bottom of the LCOE range for nuclear is for plants built with federal loan guarantees. At present, such guarantees are available only for the first few plants. The bottom of the LCOE range for wind, corresponding to class 7 wind sites, extends below the range for nuclear. However, nuclear and fossil-fuel electricity generation provide baseload power, whereas most renewable sources provide intermittent power, which reduces their value in the electricity system. The costs of integrating intermittent renewables such as wind and solar into the grid are generally low if they provide less than about 20 percent of total electricity generation (see Chapter 6), except when expensive transmission capacity must be added to bring power to demand centers.

FIGURE 2.10 *Estimates of the LCOE at the busbar for new baseload and intermittent generating sources in 2020. The horizontal bars represent the AEF Committee's judgments regarding plausible ranges of costs, given the uncertainties in fixed costs for each technology. Baseload electric power includes generating options with capacity factors above 75 percent; intermittent electric power includes generating options with capacity factors between about 25 and 40 percent. The vertical shaded bar shows the approximate range of average U.S. wholesale electricity prices across NERC regions in 2007; the dashed vertical line shows the average value in 2007, which was 5.7¢/kWh. Coal prices are assumed to be $1.71/GJ. Natural gas prices are shown for two cases: $6/GJ (low price case) and $16/GJ (high price case). The lower LCOE for nuclear power (6–8¢/kWh) includes federal loan guarantees. When installed at the point of energy use, such as on a residential rooftop, PV competes with the retail cost of electricity rather than with wholesale electricity prices. The cost estimates for different generating technologies were derived independently, with transmission and distribution costs not included explicitly in the estimates. These transmission and distribution costs are likely to be significant, however, for example, when installations are located far from load centers. Intermittent technology costs do not account for plants that must be kept available to assure adequate power supplies when the intermittent source is unavailable. All costs are in 2007 dollars. Estimated costs should be considered approximations.*
Note: CCS = carbon capture and storage; CSP = concentrating solar power; LCOE = levelized cost of electricity; NERC = North American Electric Reliability Corporation; NGCC = natural gas combined cycle; PV = photovoltaics.
Sources: Data from Energy Information Administration (www.eia.doe.gov/cneaf/electricity/wholesale/wholesalet2.xls) and Part 2 of this report.

Figure 2.10 also shows that solar photovoltaic (PV) technologies are a higher-cost option for generating electricity than most other renewables. However, when installed at the point of energy use, such as on a residential rooftop, PV competes with the retail cost of electricity and are therefore more cost competitive for a purchasing customer. Additional R&D work on this technology, particularly to find new materials and manufacturing methods to lower these costs, will be necessary if it is to be more cost competitive and, as a result, more widely deployed.

Although the LCOE is generally informative for assessing technology costs, many other factors will also influence technologies' competitiveness in the marketplace. Some of these factors have already been mentioned: fuel prices over the life of the generating asset, environmental regulations, costs of competing technologies, and, for technologies that are not yet commercial, uncertainties in construction and operation costs.

The deployment of new electricity-supply technologies will have a range of impacts beyond higher costs. They could include, for example, increased water consumption, especially for large baseload generating plants (see Chapters 7 and 8); health effects from pollutant emissions; and the siting and construction of facilities that are sometimes viewed as undesirable. Such facilities include electricity transmission lines, CO_2 pipelines, coal and uranium mines as well as coal and nuclear power plants, and waste-disposal facilities for mine tailings, fly ash, and used nuclear fuel. Even renewable-energy facilities such as wind plants could be difficult to site because of potentially degraded vistas and other environmental impacts. These kinds of deployment challenges should not be underestimated.

FINDING 4: MODERNIZING THE NATION'S POWER GRID

Expansion and modernization of the nation's electrical transmission and distribution systems (i.e., the power grid) are urgently needed to enhance reliability and security, accommodate changes in load growth and electricity demand, and enable the deployment of new energy efficiency and supply technologies, especially to accommodate future increases in intermittent wind and solar energy.

The nation's electrical transmission and distribution systems require expansion and modernization for several reasons:

- Increasing congestion threatens reliability and prevents the efficient transmission of electricity to areas where it is needed.
- Transmission systems are subject to cascading failures—resulting, for example, from human error, natural disasters, and terrorist attacks—that can lead to widespread and lengthy outages.
- Current systems have limited ability to accommodate new sources of electricity supply, especially intermittent sources, and sophisticated demand-side technologies such as advanced electricity metering technologies, sometimes referred to as "smart meters."

Modernization of these systems would have a number of economic, national security, and social benefits, among them:

- Reduced need for new transmission lines because systems could be operated more efficiently.
- Improved reliability and more rapid recovery from system disturbances.
- Ability to accommodate an expanded generation base, especially from intermittent wind and solar energy and from generation sources that are located at a distance from load-demand centers, which would help meet projected growth in future demand and deliver power to areas where it is needed.
- Ability to provide real-time electricity price information that could motivate consumers to use electricity more efficiently, thereby moderating future growth in electricity demand.

Some near-term expansion and modernization options include the deployment of modern power electronics and sensors, advanced control technologies, higher-capacity conductors, dispatchable energy storage, and other "smart" technologies.[15] Over the long term, new power storage and load-management strategies must be developed to accommodate the intermittent nature of solar and wind power.

The technologies needed to modernize and, where necessary, expand the transmission and distribution system are largely available now. Installing these technologies concurrently—that is, expanding and modernizing these systems simultaneously—would offer substantial cost savings. The committee estimates (see Chapter 9) that it would cost (in 2007 dollars) $175 billion for expansion and $50 billion for modernization of the transmission system when they are done concurrently, compared to $175 billion for expansion and $105 billion for modernization when done separately—a cost savings of $55 billion with simultaneous expansion and modernization. The committee also estimates that it would cost $470 billion for expansion and $170 billion for modernization of the distribution system when they are done concurrently, compared to $470 billion for expansion and $365 billion for modernization when done separately—a cost savings of $195 billion.

---

[15]That is, technologies that allow the transmission and distribution systems to rapidly and automatically adjust to changing conditions without the need for human intervention.

FINDING 5: CONTINUED DEPENDENCE ON PETROLEUM

Petroleum will continue to be an indispensable transportation fuel during the time periods considered in this report, but maintaining current rates of domestic petroleum production will be challenging. There are limited options for replacing petroleum or reducing petroleum use before 2020, but there are more substantial longer-term options that could begin to make contributions in the 2030–2035 timeframe. The options include increasing vehicle efficiency, replacing imported petroleum with other liquid fuels produced from biomass and coal that have CO_2 emissions similar to or less than that of petroleum-based fuels, and electrifying the light-duty vehicle fleet.

The United States consumed about 21 million barrels of liquid fuels[16] per day in 2007. Domestic consumption of liquid fuels is projected to increase to about 22 million barrels per day in 2020 and about 23 million in 2030 (EIA, 2008). In 2007, about 14 million barrels of liquid fuels per day were used in the transportation sector, of which about 9 million barrels were consumed by LDVs.

The best near-term option for reducing dependence on imported petroleum is through greater vehicle efficiency. The EISA requires a 40 percent increase in fuel economy for new LDVs by 2020. This could eventually result in a savings of about 1.4 billion barrels of gasoline per year (60 billion gallons of gasoline per year or about 164 million gallons of gasoline per day) when these fuel economy standards are fully realized in the on-the-road fleet. As noted previously, the Obama administration recently announced a new policy that requires an average fuel economy standard of 35.5 miles per gallon for new LDVs in 2016. As explained in Chapter 4, further efficiency gains are projected after 2020.

Reducing dependence on imported petroleum by substituting domestically produced liquid fuels would seem to be a good strategy, but the near-term options are limited. Just maintaining current rates of domestic petroleum production (about 5.1 million barrels per day in 2007) over the next two to three decades will be challenging. Petroleum production in current fields is declining, and it will be difficult to increase domestic production even with favorable developments in technology, prices, and access to new resources. Nevertheless, continued devel-

---

[16]Including 15.2 million barrels of crude oil, 2.1 million barrels of import products such as gasoline and jet fuel, and 3.5 million barrels of other liquid fuels such as natural gas liquids, ethanol, and biodiesel.

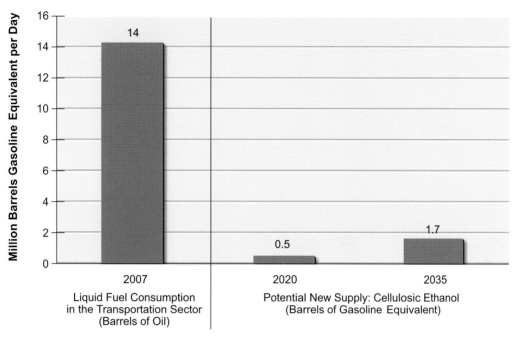

FIGURE 2.11 *Estimates of the potential cellulosic ethanol supply in 2020 and 2035 (relative to 2007) compared to total liquid fuel consumption. The current (2007) U.S. liquid fuel consumption, in barrels of oil, for transportation is shown on the left (in green). To estimate supply, an accelerated deployment of technologies (as described in Part 2 of this report) and the availability of 500 million dry tonnes per year of cellulosic biomass for fuel production are assumed after 2020. Potential liquid fuel supplies are estimated individually for each technology, and estimates do not account for future fuel demand, competition for biomass, or competition among supply sources. Potential supplies are expressed in barrels of gasoline equivalent. One barrel of oil produces about 0.85 barrels of gasoline equivalent of gasoline and diesel. All values have been rounded to two significant figures.*
Sources: Data from Energy Information Administration (2008) and Chapter 5 in Part 2 of this report.

opment of domestic resources will be essential to help prevent increases in U.S. import dependence.

Substituting other domestically produced liquid fuels could further reduce petroleum imports. Ethanol is already being made from corn grain in commercial quantities in the United States, but corn ethanol is likely to serve only as a transition fuel to more sustainable biofuels production, given the social and environmental concerns about using corn for fuel. The most promising substitutes before 2020 are cellulosic ethanol (Figure 2.11) and fuels produced from coal (coal-to-liquid fuels; Figure 2.12) and mixtures of coal and biomass (biomass-and-coal-to-

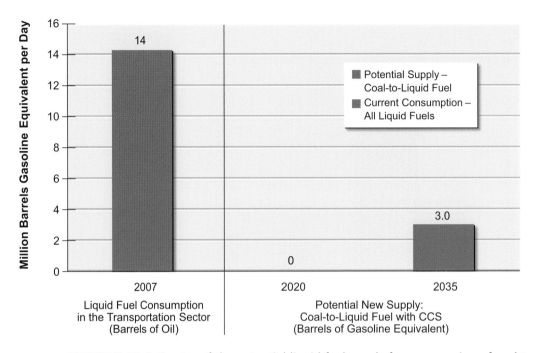

FIGURE 2.12 *Estimates of the potential liquid fuel supply from conversion of coal to liquid fuels in 2020 and 2035 (relative to 2007) compared to total liquid fuel consumption. The current (2007) U.S. liquid fuel consumption, in barrels of oil, for transportation is shown on the left (in green). To estimate supply, an accelerated deployment of technologies as described in Part 2 of this report is assumed for coal-to-liquid fuel (CTL) with carbon capture and storage (CCS). It is assumed that CTL without CCS would not be deployed. There is uncertainty associated with the technical potential for CCS. CCS technologies will need to be successfully demonstrated over the next decade if they are to be used for liquid fuel production in 2035. The volume of liquid fuel estimated to be available in 2020 and 2035 depends primarily on the rate of plant deployment. Potential liquid fuel supplies are estimated individually for each technology, and estimates do not account for future fuel demand or competition among supply sources. Potential supplies are expressed in barrels of gasoline equivalent. One barrel of oil produces about 0.85 barrels of gasoline equivalent of gasoline and diesel. All values have been rounded to two significant figures.*
Sources: Data from Energy Information Administration (2008) and Chapter 5 in Part 2 of this report.

liquid fuels; Figure 2.13). Cellulosic ethanol is in the early stages of demonstration, but coal-to-liquid fuels are being commercially produced today (but without geologic storage of CO_2) outside the United States. Coal-to-liquid fuels technologies could be deployed domestically, but these technologies would have to be integrated with CCS to produce fuels with CO_2 emissions similar to or less than those from petroleum-based fuels.

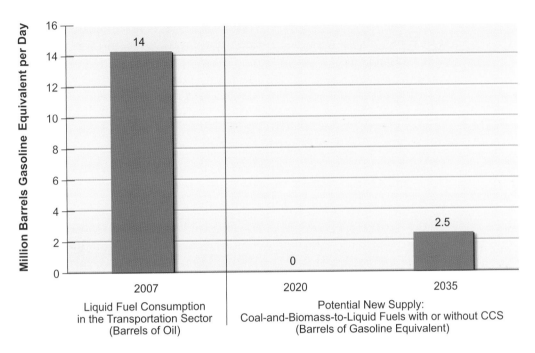

FIGURE 2.13 *Estimates of the potential liquid fuel supply from conversion of coal and biomass to liquid fuels in 2020 and 2035 (relative to 2007) compared to total liquid fuel consumption. The current (2007) U.S. liquid fuel consumption, in barrels of oil, for transportation is shown on the left (in green). To estimate supply, an accelerated deployment of technologies as described in Part 2 of this report is assumed. A mix of 60 percent coal and 40 percent biomass (on an energy basis) is assumed as well. The volume of liquid fuels estimated to be available in 2020 and 2035 depends primarily on the rate of plant deployment and also assumes availability of 500 million dry tonnes per year of cellulosic biomass for fuel production after 2020. The supply of cellulosic ethanol estimated in Figure 2.11 cannot be achieved simultaneously with this coal-and-biomass-to-liquid fuel (CBTL) supply, as the same biomass is used in each case. There is uncertainty associated with the technical potential for carbon capture and storage (CCS). CCS technologies will need to be successfully demonstrated over the next decade if they are to be used for liquid fuel production in 2035. Potential liquid fuel supplies are estimated individually for each technology, and estimates do not account for future fuel demand, competition for biomass, or competition among supply sources. Potential supplies are expressed in barrels of gasoline equivalent. One barrel of oil produces about 0.85 barrels of gasoline equivalent of gasoline and diesel. All values have been rounded to two significant figures. Sources: Data from Energy Information Administration (2008) and Chapter 5 in Part 2 of this report.*

Beyond 2020, more advanced biofuels—with higher energy content and greater compatibility with the existing transportation-fuel infrastructure—might become available. However, additional research, development, and demonstration will be required to ready these technologies for widespread commercial deployment.

By 2035, cellulosic ethanol and coal-and-biomass-to-liquid fuels *with CCS* could replace 1.7–2.5 million barrels per day of gasoline equivalent—about 12–18 percent of the current liquid fuel consumption in the transportation sector—with near-zero life-cycle CO_2 emissions. Coal-to-liquid fuels *with CCS* could replace 2–3 million barrels per day of gasoline equivalent (the 2 million barrels per day estimate assumes that some coal is diverted to produce coal-and-biomass-to-liquid fuels)—about 14–21 percent of current liquid fuels consumption in the transportation sector—and would have life-cycle CO_2 emissions similar to those of petroleum-based fuels (Figures 2.11–2.13). However, commercial demonstration of these technologies would have to be started immediately and pursued aggressively to achieve that level of production by 2035. In addition, the annual harvesting of up to 500 million dry tonnes (550 million dry tons) of biomass and an increase in U.S. coal extraction by 50 percent over current levels would be required to provide the necessary feedstock supply for this level of liquid fuel production.

These expanded levels of liquid fuel production could have a range of environmental impacts on land, water, air, and human health. Moreover, the production of liquid fuel from coal would increase CO_2 emissions to the atmosphere unless conversion plants were equipped with CCS. Although CO_2 from the off-gas streams of conversion plants could be readily captured using commercially available technologies, engineered geologic storage of captured CO_2 has not yet been demonstrated at the needed scales. Additional discussion of CCS technologies is provided under Finding 6.

Coal-to-liquid fuel production, with or without CCS, is the least expensive option for producing alternative liquid fuels (less than or equal to $70 per barrel; see Figure 2.14), although such production raises important health and environmental issues, as noted above. Deploying cellulosic ethanol would be economically competitive only with petroleum prices above about $115 per barrel.

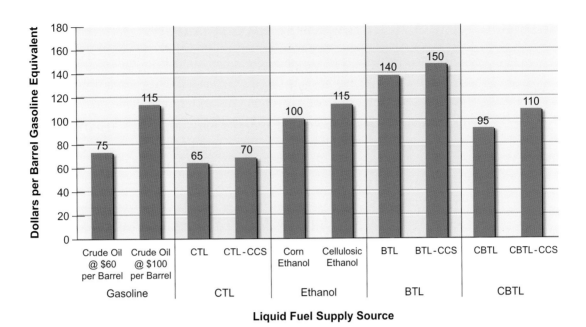

FIGURE 2.14 *Estimated gasoline-equivalent costs of alternative liquid fuels. For comparison, the costs of gasoline at crude oil prices of $60 per barrel and $100 per barrel are shown on the left. Estimated costs assume that a zero price is assigned to CO_2 emissions. Liquid fuels would be produced using biochemical conversion to produce ethanol from Miscanthus or using thermochemical conversion via Fischer-Tropsch or methanol-to-gasoline. All costs are in 2007 dollars and are rounded to the nearest $5.*
Note: BTL = biomass-to-liquid fuel; CBTL = coal-and-biomass-to-liquid fuel; CCS = carbon capture and storage; CTL = coal-to-liquid fuel.
Source: Data from Chapter 5 in Part 2 of this report.

Additional reductions in petroleum imports would be possible by increasing the electrification of the vehicle fleet. The widespread deployment of electric and/or hydrogen fuel cell vehicles between 2035 and 2050 could lead to further and possibly substantial long-term reductions in liquid fuel consumption in the transportation sector. The National Research Council (2008), for example, estimated the potential reduction in petroleum use in 2050 from the deployment of hydrogen fuel-cell LDVs under a best-case scenario to be about 70 percent below the projected petroleum consumption of a fleet of comparable gasoline-fueled vehicles. The LDV fleet turns over every one to two decades, so the introduction of higher-efficiency vehicles would have relatively low impacts on petroleum use and CO_2 emissions from the transportation sector until sometime after the 2020–2030 period.

FINDING 6: REDUCING GREENHOUSE GAS EMISSIONS

Substantial reductions in greenhouse gas emissions from the electricity sector are achievable over the next two to three decades. They can best be realized through a portfolio approach involving the widespread deployment of multiple technologies: energy efficiency; renewable energy; coal, natural gas, and biomass with carbon capture and storage; and nuclear. However, to enable this portfolio approach, the viability of the following two technologies must be demonstrated during the next decade to make them ready for widespread commercial deployment starting around 2020: (1) the technical and commercial viability of CCS for sequestering CO_2 from electricity production and (2) the commercial viability of evolutionary nuclear plants in the United States. Achieving substantial greenhouse gas reductions in the transportation sector over the next two to three decades will also require a portfolio approach involving the widespread deployment of energy efficiency technologies, alternative liquid fuels with low life-cycle CO_2 emissions, and light-duty-vehicle electrification technologies.

As noted in Chapter 1, the United States emits some 6 billion tonnes (6 gigatonnes) of CO_2 into the atmosphere each year (see Figure 1.3); about 5.6 gigatonnes are attributable to the energy system. The potential for reducing greenhouse gas emissions from this system before 2020 is limited, but the potential for reducing emissions after 2020 is significant, especially in the electricity sector, if certain technologies can be successfully deployed at commercial scales.

Electricity is produced in stationary facilities, which in principle makes it easier to effectively monitor and control their greenhouse gas emissions. The options for reducing the electricity sector's emissions are apparent from an inspection of Figure 2.15, which provides estimates of life-cycle CO_2-equivalent[17] (denoted CO_2-eq) emissions per kilowatt-hour of electricity produced. Coal and natural gas plants are by far the largest emitters of greenhouse gases from electricity generation. In fact, their CO_2-eq emissions are far higher than those of any of the other technologies represented. As shown in Figure 1.8, coal and natural gas plants collectively supplied about 70 percent of electricity demand in 2007.

Achieving substantial reductions in CO_2 emissions from the U.S. electricity

---

[17]CO_2 equivalent expresses the global warming potential of a greenhouse gas in terms of CO_2 quantities.

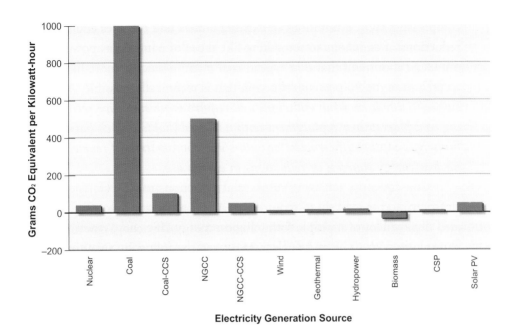

FIGURE 2.15 *Estimated greenhouse gas emissions from electricity generation. Estimates are in units of grams of CO$_2$-equivalent (CO$_2$-eq) emissions per kilowatt-hour of electricity produced. Estimates for all technologies (with the exception of coal, coal-CCS, NGCC, and NGCC-CCS) are life-cycle estimates, which include CO$_2$-eq emissions due to plant construction, operation, and decommissioning, levelized across the expected output of electricity over the plant's lifetime. For coal, coal-CCS, NGCC, and NGCC-CCS, only emissions from the burning of the fossil fuels are accounted for. A 90 percent capture fraction is assumed for CCS technologies. Negative CO$_2$-eq emissions mean that on a net life-cycle basis, CO$_2$ is removed from the atmosphere. For example, the negative CO$_2$ emissions for biopower result from an estimate that the sequestration of biomass carbon in power-plant char and the buildup of carbon in soil and roots will exceed the emissions of carbon from biofuel production. The life-cycle CO$_2$ emission from biofuels includes a CO$_2$ credit from photosynthetic uptake by plants, but indirect greenhouse gas emissions, if any, as a result of land-use changes are not included.*
Note: CCS = carbon capture and storage; CSP = concentrating solar power; NGCC = natural gas combined cycle; PV = photovoltaics.
Sources: Data from Part 2 of this report and from NAS-NAE-NRC (2009a).

sector will be possible only if existing coal plants and natural gas plants are retrofitted or repowered with CCS technologies or are retired.[18] However, retrofitting these plants will require diversion of some of their energy input to capturing and

---

[18]Comparable actions at existing fossil-fuel plants in other countries will also be required to achieve substantial reductions in worldwide CO$_2$ emissions.

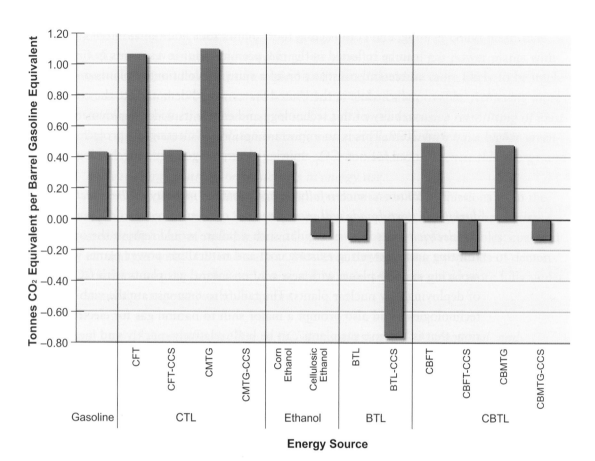

FIGURE 2.16 *Estimated net life-cycle CO$_2$-equivalent (CO$_2$-eq) emissions for production, transportation, and use of alternative liquid transportation fuels. Emissions are shown in units of tonnes of CO$_2$ equivalent per barrel of gasoline equivalent produced from biomass, coal, or a combination of coal and biomass. For comparison, the CO$_2$-eq emissions for gasoline are shown on the left. Negative CO$_2$-eq emissions mean that on a net life-cycle basis, CO$_2$ is removed from the atmosphere; for example, the negative CO$_2$ emissions for BTL and cellulosic ethanol result from an estimate that the sequestration of biomass carbon in power-plant char or the buildup of carbon in soil and roots will exceed the emissions of carbon in biofuel production. Growing perennial crops for cellulosic fuels provides CO$_2$ benefits because these crops store carbon in the root biomass and the associated rhizosphere, thereby increasing soil carbon sequestration. The precise value of CO$_2$-eq emissions from CBTL depends on the ratio of biomass to coal used. Indirect land-use effects on CO$_2$ emissions are not included.*
Note: BTL = biomass-to-liquid fuel; CBFT = coal-and-biomass-to-liquid fuel, Fischer Tropsch; CBMTG = coal-and-biomass-to-liquid fuel, methanol-to-gasoline; CBTL = coal-and-biomass-to-liquid fuel; CCS = carbon capture and storage; CFT = coal-to-liquid fuel, Fischer-Tropsch; CMTG = coal-to-liquid fuel, methanol-to-gasoline; CTL = coal-to-liquid fuel.
Sources: Data from Chapter 5 in Part 2 of this report and from NAS-NAE-NRC (2009b).

produced with petroleum. As noted under Finding 5, however, alternative liquid fuels can only substitute for a portion of petroleum use. Moreover, geologic storage of CO_2 from coal-to-liquid fuel and coal-and-biomass-to-liquid fuel production would have to be demonstrated to be safe and commercially viable by 2015 for these fuels to be produced in quantity starting around 2020.

Further reductions in greenhouse gas emissions from the transportation sector will have to be achieved through greater vehicle efficiency and, if greenhouse gas emissions from the electricity sector can be reduced, through electrification of the LDV fleet (as discussed under Finding 5). However, substantial reductions in emissions via these pathways are not likely to occur until late in the 2020–2035 period or beyond. As is the case for liquid fuel supply, the widespread deployment of electric or hydrogen fuel-cell vehicles between 2035 and 2050 holds some hope for more substantial long-term reductions in greenhouse gas emissions in the transportation sector, again depending on how the electricity and hydrogen are generated. As noted previously, the National Research Council (2008) estimated the potential reduction in petroleum use in 2050 from the deployment of hydrogen fuel-cell LDVs under the best-case scenario to be about 70 percent below the projected petroleum consumption of a fleet of comparable gasoline-fueled vehicles.

FINDING 7: TECHNOLOGY RESEARCH, DEVELOPMENT, AND DEMONSTRATION

To enable accelerated deployments of new energy technologies starting around 2020, and to ensure that innovative ideas continue to be explored, the public and private sectors will need to perform extensive research, development, and demonstration over the next decade. Given the spectrum of uncertainties involved in the creation and deployment of new technologies, together with the differing technological needs and circumstances across the nation, a portfolio that supports a broad range of initiatives from basic research through demonstration will likely be more effective than targeted efforts to identify and select technology winners and losers.

As discussed in some detail in Part 2 of this report, the next decade offers opportunities to gain knowledge and early operating experience that in turn could enable widespread deployments of new energy-supply technologies beginning around 2020. These technology-development opportunities include:

- The full range of energy efficiency technologies in the buildings, transportation, and industrial sectors.
- Coal and natural gas with CCS (see Finding 6 and Chapter 7 for details).
- Evolutionary nuclear power (see Finding 6 and Chapter 8).
- Integrated gasification combined cycle, ultrasupercritical pulverized coal, and oxyfuel plants to improve the efficiency and performance of coal-generated electricity, pursued in coordination with research, development, and demonstrations on advanced materials and CCS technologies (see Chapter 7).
- Thermochemical conversion of coal and coal-and-biomass mixtures to liquid fuels, integrated with CCS, at commercial scale. If decisions to proceed with such demonstrations are made soon, and if CCS is shown to be safe and viable by about 2015, these technologies could be commercially deployable within a decade under favorable economic conditions (see Chapter 5).
- Research and development on cellulosic-conversion methods, followed by demonstration of cellulosic ethanol production at commercial scale, to achieve proof of principle and prepare this technology for widespread deployment (see Chapter 5).
- Advanced LDVs, including plug-in hybrids and battery-electric and fuel-cell vehicles. Demonstrations of on-the-road vehicles are critical to getting real-world data on performance and service lives (see Chapter 4).

R&D will help to ensure the success of future new-technology deployments and especially to ensure that the technology pipeline remains full in the decades ahead. Significant investments in R&D over the next decade, by the public and the private sector alike, will be required for bringing some of the technologies described in this report to the point that they are cost-effective and ready for widespread deployment. The needed areas of R&D include:

- Advanced biosciences—genomics, molecular biology, and genetics—to develop biotechnologies for converting biomass to lipid, higher-alcohol, and hydrocarbon fuels that can be integrated directly into existing transportation infrastructures.
- Advanced technologies for producing alternative liquid fuels from renewable resources—such as fuel production from CO_2 feedstocks

(e.g., algae biofuels). Such fuels are needed to expand options for reducing petroleum use.

- Advanced technologies for the production of biomass that provides sustainable yields, minimizes competition with food and feed crops, and offers substantial greenhouse-gas-reduction benefits.
- Advanced PV materials and manufacturing methods to improve efficiencies and to lower costs. The deployed efficiency of current PV materials is greater than 10 percent, which is much higher than the field efficiency of plants for biomass. Although biomass is a compact form of chemical energy storage, its production requires a great deal of land and energy and it has to be harvested and processed to make electricity or liquid fuels, whereas the electricity from PV cells can be used directly.
- Advanced batteries and fuel cells for LDVs.
- Advanced large-scale storage for wind energy and electrical-load management.
- Enhanced geothermal power.
- Advanced technologies for extracting petroleum from shale and for harvesting natural gas from hydrates.
- Alternative fuel cycles that would allow for greater utilization of the energy content of nuclear fuel and the minimization of very-long-lived radioactive waste from nuclear power generation.
- Further exploration of geoengineering options.

R&D in other scientific fields that are not addressed in this report will likely provide important support for the development and deployment of new energy-supply and end-use technologies. For example, researchers' efforts to better understand the interactions between patterns of energy use and climate systems—including, for example, the ecology of microbial systems—could support the development of more effective means to capture, store, and recycle CO_2 from energy production. Additionally, social science research on how households and businesses make decisions could lead to more effective measures to encourage energy efficiency.

Finally, attractive technology options will likely emerge from innovation pathways that are essentially unforeseen today—some examples are cited in Part 2 of this report—underscoring the need for a continuing focus on and investments in basic research. Some breakthrough technologies are probably not even on the present horizon; in fact, they may not become apparent until the final time period considered in this report (2035–2050) or later. However, it is very likely that some of the potential breakthrough technologies that are indeed visible on today's horizon—for example, superconducting materials, second- and third-generation PV technologies, and advanced batteries—may begin to develop and have an important influence on technology trends during the first two time periods (2008–2020 and 2020–2035) considered in this study. Achieving such breakthroughs will require sustained federal support for basic scientific research, both in the physical and in the biological sciences, and private-sector "venture-backed" support for early-stage energy R&D.

The Department of Energy (DOE) has been the primary catalyst for basic energy research in the United States, primarily through its Office of Science. There are substantial opportunities in the years ahead for this office to increase the support of such activities and to ensure their coordination by partnering with the DOE's energy offices and with other basic-research agencies such as the National Science Foundation.

FINDING 8: BARRIERS TO ACCELERATED TECHNOLOGY DEPLOYMENT

A number of current barriers are likely to delay or even prevent the accelerated deployment of the energy-supply and end-use technologies described in this report. Policy and regulatory actions, as well as other incentives, will be required to overcome these barriers.

The assessments provided in the forgoing sections reflect the AEF Committee's judgments about the potential contribution of new energy technologies *if* the accelerated-deployment options identified in this report are actively pursued. However, a number of potential barriers could influence these options and, in turn, affect the actual scale and pace of the implementation of the technologies. Some of the barriers are purely market driven: technologies must be clearly attractive to potential investors, purchasers, and users. They must also provide improvements, relative to existing technologies, in terms of performance, convenience, and cost attributes; of course, they must also meet relevant performance standards and regulations.

In the course of this study, the AEF Committee identified several policy and regulatory barriers to the deployment of the energy-supply and end-use technologies that were examined. Some of these barriers have already been identified in this chapter, and additional ones are described in Part 2. But because the following barriers crosscut many of the technologies examined in this report, the committee considers them to be impediments to future deployment success:

- *Lack of private-sector investments for technology deployment*, ranging from relatively low-cost energy efficiency devices to capital-intensive facilities, because of uncertainties about a technology's return on investment, its viability and cost-effectiveness, the future costs of fuels, and other raw-material and construction costs. The mobilization of trillions of dollars of new capital between now and 2050 will be needed to transform our nation's energy system, but such capital may be difficult to obtain from the private sector if the noted uncertainties are not attenuated. The current economic downturn further complicates matters: the limited availability of resources, especially capital, and the reduction in energy demand may be additional barriers to new-technology deployment.

- *The low turnover rate of the energy system's capital-intensive infrastructure,* which makes rapid change difficult. Failure to take advantage of windows of opportunity to deploy new technologies as infrastructure turns over could lock in older technologies for decades, and this difficulty is compounded by the long lead times for deploying new technologies, especially capital-intensive technologies. Thus, there is a premium on modifying or retrofitting existing infrastructure and on pushing new technologies to be ready for deployment when assets reach the end of their service lives. There are some technology "lock-ins," however, that might not allow for future modification or improvements. Examples include new coal plants that cannot be easily retrofitted with CCS[19] and new buildings that are not designed to use energy efficiently over their lifetimes.

- *Resource and supply barriers to technology deployment.* They range, for example, from the limited availability of industrial capacity and skilled personnel for deploying the technologies to the availability of the biomass needed to expand the domestic production of liquid fuels.

---

[19]This problem is not restricted to the United States alone. It will be an especially critical issue in countries, such as China, that are building new coal plants at very high rates.

Some of these barriers can be overcome with the right market and regulatory signals.

- *Uncertainties arising from the nature and timing of public policies and regulations related to carbon controls.* There is no authoritative guidance on best available technologies for CCS that could be used to guide deployment. Such guidance might be similar to New Source Performance Standards developed under the Clean Air Act for criteria pollutants. The initial rates of deployment of reduced-carbon technologies (energy efficiency, renewable-energy sources, nuclear energy, and coal with CCS) can be accelerated by such guidelines, by a better alignment of incentives, and by some selected direct public investments.

- *Coupling the commercial deployment of energy-supply technologies with key supporting technologies.* Examples include CCS both for electric-power generation and the production of transportation fuels; adequate dispatchable energy supplies or storage[20] for advanced and expanded transmission and distribution systems; and advanced batteries for plug-in hybrid and battery-electric vehicles. Successful demonstration of the key supporting technologies will clearly be required, but so too will a better alignment of incentives and the resolution of a number of economic, legal, and policy questions.

- *The regional ownership and regulation of the transmission and distribution systems* in the United States make it difficult to implement nationwide modernizations. Although there are exceptions in some regions, the current regulatory system is not designed to adopt available and future innovations in the national transmission system because of fractured jurisdictions at the local, regional, and national levels, as well as an institutional culture that emphasizes quantity of service over reliability, quality, efficiency, and security. Additionally, the methods for assessing returns on private investment in the transmission system are unclear because, owing to the dispersed nature of electricity transmission, reliability and societal benefits extend beyond a single region.

- *The lack of energy efficiency standards for many products* means that in many cases individual consumers must take the initiative to acquire

---

[20]Dispatchable energy storage is a set of technologies for storing or producing electricity that can be deployed quickly (dispatched) into the grid when other power sources become unavailable. These technologies are described in Chapter 9.

information about the costs and benefits of available energy efficiency technologies. Most consumers are unwilling or ill equipped to do so (see Box 2.2).

Overcoming these barriers will require a judicious mix of policies, regulations, and market incentives. A full analysis of the barriers, as well as of the means to overcome them, is beyond the scope of this AEF Phase I study. The National Academies will address many of these issues, however, in the project's Phase II.

REFERENCES

Brown, R., S. Borgeson, J. Koomey, and P. Bremayer. 2008. U.S. Building-Sector Energy Efficiency Potential. Berkeley, Calif.: Lawrence Berkeley National Laboratory.

EIA (Energy Information Administration). 2008. Annual Energy Outlook 2008. DOE/EIA-0383(2008). Washington, D.C.: U.S. Department of Energy, Energy Information Administration.

EIA. 2009a. Annual Energy Outlook 2009. DOE/EIA-0383(2009). Washington, D.C.: U.S. Department of Energy, Energy Information Administration.

EIA. 2009b. Annual Energy Review 2008. DOE/EIA-0384(2008). Washington, D.C.: U.S. Department of Energy, Energy Information Administration.

NAS-NAE-NRC (National Academy of Sciences-National Academy of Engineering-National Research Council). 2009a. Electricity from Renewable Resources: Status, Prospects, and Impediments. Washington, D.C.: The National Academies Press.

NAS-NAE-NRC. 2009b. Liquid Transportation Fuels from Coal and Biomass: Technological Status, Costs, and Environmental Impacts. Washington, D.C.: The National Academies Press.

National Research Council. 2008. Transitions to Alternative Transportation Technologies—A Focus on Hydrogen. Washington, D.C.: The National Academies Press.

3

Key Results from Technology Assessments

This chapter summarizes the detailed assessments presented in Part 2 of this report, organized by subject and chapter as follows:

- Energy efficiency (Chapter 4)
- Alternative transportation fuels (Chapter 5)
- Renewable energy (Chapter 6)
- Fossil-fuel energy (Chapter 7)
- Nuclear energy (Chapter 8)
- Electricity transmission and distribution (Chapter 9).

The chapter annex, Annex 3.A, describes the key methods and assumptions that were used to develop the energy supply, savings, and cost estimates in this report. Additional detailed supporting information can be found in Part 2 of this report and in the following National Academies reports derived from this America's Energy Future (AEF) Phase I study:

- *Real Prospects for Energy Efficiency in the United States* (NAS-NAE-NRC, 2009c; available at http://www.nap.edu/catalog. php?record_id=12621)
- *Liquid Transportation Fuels from Coal and Biomass: Technological Status, Costs, and Environmental Impacts* (NAS-NAE-NRC, 2009b; available at http://www.nap.edu/catalog.php?record_id=12620)
- *Electricity from Renewable Resources: Status, Prospects, and Impediments* (NAS-NAE-NRC, 2009a; available at http://www.nap.edu/ catalog.php?record_id=12619).

ENERGY EFFICIENCY

The potential for increasing energy efficiency—that is, for reducing energy use while delivering the same energy services—in the United States is enormous. Technology exists today, or is expected to be developed over the normal course of business between now and 2030, that could save about 30 percent of the energy used annually in the buildings, transportation, and industrial sectors. These savings could easily repay, with substantial dividends, the investments involved. In particular, if energy prices were high enough to motivate investment in energy efficiency or if public policies had the same effect, energy use could be lower by 15–17 quads (about 15 percent) in 2020 and by 32–35 quads (about 30 percent) in 2030 than the reference case projection of the U.S. Department of Energy's Energy Information Administration (EIA). The opportunities for achieving these savings reside in hundreds of technologies, many of them already commercially available and others just about to enter the market.

This section summarizes the capability of energy efficiency technologies to reduce energy use or moderate its growth. Technologies that pay for themselves (in reduced energy costs) after criteria have been applied to reflect experience with consumer and corporate decision making are considered cost-effective. For the buildings sector, supply curves were developed that reflect implementation of efficiency technologies in a logical order, starting with lowest-cost technological options. Using discounted cash flow[1] and accounting for the lifetimes of technologies and infrastructures involved, the reported efficiency investments in buildings generally pay for themselves in 2–3 years. For the industrial and transportation sectors, the AEF Committee relied on results from the report by the America's Energy Future Panel on Energy Efficiency Technologies (NAS-NAE-NRC, 2009c).[2] For industry, the panel reported industry-wide potential for energy savings reflecting improvements that would offer an internal rate-of-return on the efficiency investment of at least 10 percent. For transportation (which addresses fewer technologies and thus includes more in-depth assessments of each), the panel focused on how the performance and costs of vehicle technologies might evolve relative to one another (and the capability of these technologies to reduce fleet fuel consumption).

---

[1]The discounted cash flow approach describes a method of valuing a project, company, or asset such that all future cash flows are estimated and discounted to give their present values.

[2]Further details on these estimates can also be found in Chapter 4 in Part 2 of this report.

The panel examined the available energy efficiency literature and performed additional analyses. For each sector, comparisons were made to a "baseline" or "business as usual" case to estimate the potential for energy savings. These are described in Annex 3.A.

Buildings Sector

About 40 percent of the primary energy used in the United States, and fully 73 percent of the electricity, is used in residential and commercial buildings. Diverse studies for assessing this sector's energy-savings potential, although they take many different approaches, are remarkably consistent and have been confirmed by the supply curves developed for this report. The consensus is that savings of 25–30 percent relative to current EIA (2008) reference case projections could be achieved over the next 20–25 years. These savings, which would come principally from technologies that are more efficient for space heating and cooling, water heating, and lighting, could hold energy use in buildings about constant even as population and other drivers of energy use grow. Moreover, the savings could be achieved at a cost per energy unit that would be lower than current average retail prices for electricity and gas.[3] For the entire buildings sector, the supply curves in Chapter 2 of this report (Figures 2.5 and 2.6) as well as in the panel report (NAS-NAE-NRC, 2009c) show that a *cumulative* investment of $440 billion[4] in existing technology between 2010 and 2030 could produce an *annual* savings of $170 billion in reduced energy costs.

Advanced technologies just emerging or under development promise even greater gains in energy efficiency. They include solid-state lighting (light-emitting diodes); advanced cooling systems that combine measures to reduce cooling requirements with emerging technologies for low-energy cooling, such as evaporative cooling, solar-thermal cooling, and thermally activated desiccants; control sys-

---

[3]The average residential electricity price in the United States in 2007 was 10.65¢/kWh (in the commercial sector, the average price was 9.65¢/kWh). The average residential price for natural gas in the United States in 2007 was $12.70/million Btu (in the commercial sector, the average price was $11/million Btu).

[4]The investments include both the full add-on costs of new equipment and measures (such as attic insulation) and the incremental costs of purchasing an efficient technology (e.g., a high-efficiency boiler) compared with purchasing conventional-counterpart technology (e.g., a standard boiler). These investments would be made instance-by-instance by the individuals and public or private entities involved. The costs of policies and programs that would support, motivate, or require these improvements are not included.

tems for reducing energy use in home electronics; "superwindows" with very low U-values;[5] dynamic window technologies that adjust cooling and electric lighting when daylight is available; and very-low-energy houses and commercial buildings that combine fully integrated design with on-site renewable-energy generation.

Transportation Sector

The transportation sector, which is almost solely dependent on petroleum, produces about one-third of the U.S. greenhouse gas emissions[6] arising from energy use. The sector is dominated by use of the nation's highways, for both freight and passengers.

Current technologies offer many potential improvements in fuel economy, and they become increasingly competitive and attractive as fuel prices rise. Reductions in fleet fuel consumption over the next 10–20 years will likely come primarily from improving today's spark-ignition (SI), diesel, and hybrid vehicles that are fueled with petroleum, biofuels, and other nonpetroleum hydrocarbon fuels.

Over the subsequent decade, plug-in hybrid vehicles (PHEVs) that use electricity plus any of the fuels just mentioned may be deployed in sufficient volume to have a significant effect on petroleum consumption. Longer term, after 2030, major sales of hydrogen fuel-cell vehicles (HFCVs) and battery-electric vehicles (BEVs) are possible.

- *Light-duty vehicles*. Power-train improvements for LDVs offer the greatest potential for increased energy efficiency over the next two decades. Technologies that improve the efficiency of SI engines could reduce average new-vehicle fuel consumption by 10–15 percent by 2020 and a further 15–20 percent by 2030. Turbocharged diesel engines, which are some 10–15 percent more efficient than equal-performance SI engines, could steadily replace nonturbocharged engines in the SI fleet. Improvements in transmission efficiency and reductions in rolling resistance, aerodynamic drag, and vehicle size, power, and weight can all increase vehicle fuel efficiency.

---

[5]U-values represent how well a material allows heat to pass through it. The lower the U-value, the greater a product's ability to insulate.

[6]In this report, the cited quantities of greenhouse gases emitted are expressed in terms of CO_2-equivalent (CO_2-eq) emissions.

Currently, corporate average fuel economy (CAFE) standards for new LDVs are targeted to reach 35 miles per gallon by 2020, which would equate to a 40 percent improvement in average new-vehicle fuel efficiency (and a 30 percent reduction in average fuel consumption).[7] Achieving this goal, and further improving fuel efficiency after 2020, will require that the historic emphasis on ever-increasing vehicle power and size be reversed in favor of fuel economy.

Gasoline hybrid-electric vehicles (HEVs) currently offer vehicle fuel-consumption savings of as much as 30 percent over SI engines. Thus it is likely that meeting the new CAFE standards by 2020 will require a large fraction of new vehicles to be HEVs or smaller, less powerful vehicles. PHEVs and BEVs could begin to make a large impact beyond 2020; however, the success of these technologies is crucially dependent on the development of batteries with much higher performance capabilities than today's batteries, and with lower costs. Research and development on battery technology continues to be a high priority.

If they could be equipped with batteries that powered the vehicle for 40–60 miles, gasoline PHEVs could reduce gasoline/diesel consumption by 75 percent. While HEVs mainly improve performance or fuel economy, PHEVs actually get most of their energy from the electric grid.

Improvements in battery and fuel-cell technologies are expected to pave the way for possible large-scale deployments of BEVs and HFCVs in the 2020–2035 period. Because BEVs and HFCVs could reduce and ultimately eliminate the need for petroleum in transportation, they could also reduce and possibly even eliminate LDV tailpipe greenhouse gas emissions.

- *Freight transportation.* Future technologies for heavy-duty trucks include continuously variable transmissions and hybrid-electric systems to modulate auxiliaries (such as air-conditioning and power steering) and reduce idling. Significant reductions in aerodynamic drag are also possible. Reductions in fuel consumption of 10–20 percent in heavy- and medium-duty vehicles appear feasible over the next decade or so.

[7]As noted in Chapters 1 and 2, the Obama administration recently announced new policies that will accelerate the implementation of these fuel economy standards.

Rail is about 10 times more energy-efficient than trucking, so shifting freight from trucks to rail can offer considerable energy savings.

- *Air transportation.* The latest generation of airliners offers a 15–20 percent improvement in fuel efficiency.[8] The newer airplanes, however, are likely to do little more than offset the additional fuel consumption caused by projected growth in air travel over the next several decades.

- *Long-term system-level improvements.* Examples of system-level innovations that could substantially improve efficiency include the utilization of intelligent transportation systems to manage traffic flow; better land-use management; and greater application of information technology in place of commuting and long-distance business travel.

Industrial Sector

Estimates from independent studies using different approaches agree that the potential for cost-effective reduction in energy use by industry range from 14 to 22 percent—about 4.9 to 7.7 quads—by 2020, compared with current EIA reference case projections. Most of the gains will occur in energy-intensive industries, notably chemicals and petroleum, pulp and paper, iron and steel, and cement.[9] Growth in the energy-efficient option of combined heat and power production is also likely to be significant. Beyond 2020, new technologies such as novel heat and power sources, new products and processes, and advances in recycling could bring about even greater gains in energy efficiency. Important progress might also come from adapting new technology (such as fuel cells for combined heat and power generation) and adopting alternative methods of operation (e.g., "on-demand" manufacturing).

- *Chemicals and petroleum.* Technologies for improving energy efficiency include high-temperature reactors, corrosion-resistant metal- and ceramic-lined reactors, and sophisticated process controls. Cost-effective improvements in efficiency of 10–20 percent in petroleum refining by 2020 are possible.

---

[8]Increases in passenger airliner efficiency will also benefit air freight transport.

[9]Further details on the potential improvements in these industries can be found in Chapter 4 in Part 2 of this report and in the report of the America's Energy Future Panel on Energy Efficiency Technologies (NAS-NAE-NRC, 2009c).

- *Pulp and paper.* The industry could use more waste heat for drying, advanced water-removal and filtration technologies, high-efficiency pulping processes, and modernized lime kilns. Estimates of cost-effective gains in energy efficiency by 2020 range from 16 to 26 percent.
- *Iron and steel.* Promising advances in technology that could be available by 2020 involve electric-arc furnace (EAF) melting, blast-furnace slag-heat recovery, integration of refining functions, and heat capture from EAF waste gas. The American Iron and Steel Institute recently announced a goal of using 40 percent less energy for iron and steel production by 2025 compared with 2003.
- *Cement.* Major energy savings would require significant upgrades to an advanced dry-kiln process. Efficiency could also be enhanced with advanced control systems, combustion improvements, indirect firing, and optimization of certain components. A combination of these changes could yield a reduction in energy use of about 10 percent. In addition, changing the chemistry of cement to decrease the need for calcination could result in reduced energy use of another 10–20 percent. Advanced technologies for yielding further improvements are under development. Overall savings of 20 percent are possible by 2020.

A set of crosscutting technologies exists that could improve energy efficiency in a wide range of industrial applications. This includes the expansion of combined heat and power systems; separation processes based on membranes and other porous materials; advanced materials that resist corrosion, degradation, and deformation at high temperatures; controls and automation; steam- and process-heating technologies that improve quality and reduce waste; high-efficiency fabrication processes that improve yields and reduce waste; remanufacturing of products for resale; and sensor systems that reduce waste by improving control.

Barriers to Deployment and Drivers of Efficiency

Numerous barriers impede deployment of energy efficiency technologies in each of the sectors previously discussed. In the buildings sector, regulatory policies do not usually reward utility investments in energy efficiency; building owners in rental markets and builders are not responsible for paying energy costs and thus lack incentives to make investments that reduce energy use; information about the energy costs of specific appliances and equipment is often not readily available; and access to capital for such investments is limited. Drivers for greater efficiency—that is, for overcoming these barriers—could include rising energy costs, growing environmental awareness, improved and publicized building codes and appliance efficiency standards, and state and local utility programs.

In the transportation sector, barriers that limit energy efficiency include the lack of clear signals about future oil prices (expectations for future prices strongly affect technology and investment decisions) and the lack of sufficient production capability to manufacture energy-efficient vehicles across vehicle platforms.

The barriers to deployment in the industrial sector include the technical risks of adopting a new industrial technology; high investment costs for industrial energy efficiency improvements; intra-firm competition for capital, which may favor improvements in products and processes over energy efficiency; the lack of specialized knowledge about energy efficiency technologies; and unfavorable provisions of the tax code.

These barriers are formidable, and sustained public and private support will be needed to overcome them. Particular attention must be paid to infrastructure, industrial equipment, and other long-lived assets in order to ensure that energy efficiency technologies and systems are put into place when these assets are constructed or renewed.

Meanwhile, there are several drivers for greater efficiency. They include expected increases in energy prices and concern about availability of fuels and electricity; more stringent air-quality standards, which raise the prices of pollution allowances; demand charges and demand-response incentives; collateral benefits such as higher product quality and productivity; and corporate sustainability initiatives.

In general, substantial energy savings in all sectors will be realized only if efficient technologies and practices achieve wide use. Experience demonstrates that these barriers can be overcome with the aid of well-designed policies. Many policy initiatives have been effective, including efficiency standards (vehicle and appliance) combined with U.S. Department of Energy R&D on efficient equipment; promotion of combined heat and power, largely through the Public Utilities Regulatory and Policy Act of 1978; the ENERGY STAR® product-labeling program; building-energy codes; and utility- and state-sponsored end-use efficiency programs. These initiatives have already resulted in a nearly 13-quad-per-year reduction in primary energy use.

ALTERNATIVE TRANSPORTATION FUELS

The U.S. transportation sector consumed about 14 million barrels of oil per day in 2007, 9 million of which was used in light-duty vehicles. Total U.S. liquid fuels consumption in 2007 was about 21 million barrels per day, about 12 million of which was imported. The nation could reduce its dependence on imported oil by producing alternative liquid transportation fuels from domestically available resources to replace gasoline and diesel, and thereby increase energy security and reduce greenhouse gas emissions.

Two abundant domestic resources with such potential are biomass and coal. The United States has at least 20 years' worth of coal reserves in active mines and probably sufficient resources to meet the nation's needs for well over 100 years at current rates of consumption. Biomass can be produced continuously over the long term if sustainably managed, but the amount that can be produced at any given time is limited by the natural resources required to support biomass production. However, a robust set of conversion technologies needs to be developed or demonstrated and brought to commercial readiness to enable those resources to be converted to suitable liquid transportation fuels.

Biomass Supply

Biomass for fuels must be produced sustainably to avoid excessive burdens on the ecosystems that support its growth. Because corn grain is often used for food, feed, and fiber production, and also because corn grain requires large amounts of fertilizer, the committee considers corn grain ethanol to be a transition fuel to cellulosic biofuels or other biomass-based liquid hydrocarbon fuels (for example, biobutanol and algal biodiesel). About 365 million dry tonnes (400 million dry tons) per year of cellulosic biomass—dedicated energy crops, agricultural and forestry residues, and municipal solid wastes—could potentially be produced on a sustainable basis using today's technology and agricultural practices, and with minimal impact on U.S. food, feed, and fiber production or the environment. By 2020, that amount could reach 500 million dry tonnes (550 million dry tons) annually. A key assumption behind these estimates is that dedicated fuel crops would be grown on idle agricultural land in the Conservation Reserve Program. The size of the facilities for converting biomass to fuel will likely be limited by the supply of biomass available from the surrounding regions.

Producers will likely need incentives to grow biofeedstocks that not only do not compete with other crop production but also avoid land-use practices that cause significant net greenhouse gas emissions. Appropriate incentives can encourage lignocellulosic biomass production in particular. To ensure a sustainable biomass supply overall, a systematic assessment of the resource base—which addresses environmental, public, and private concerns simultaneously—is needed.

Conversion Technologies

Two conversion processes can be used to produce liquid fuels from biomass: biochemical conversion and thermochemical conversion.

Biochemical Conversion

Biochemical conversion of starch from grains to ethanol has already been deployed commercially. Grain-based ethanol was important for stimulating public awareness and initiating the industrial infrastructure, but cellulosic ethanol and other advanced cellulosic biofuels have much greater potential to reduce U.S. oil use and CO_2 emissions and have minimal impact on the food supply.

Processes for biochemical conversion of cellulosic biomass into ethanol are in the early stages of commercial development. But over the next decade, improvements in cellulosic ethanol technology are expected to come from evolutionary developments gained from commercial experience and economies of scale. Incremental improvements of biochemical conversion technologies can be expected to reduce nonfeedstock costs by about 25 percent by 2020 and about 40 percent by 2035. In terms of transport and distribution, however, an expanded infrastructure will be required because ethanol cannot be transported in pipelines used for petroleum transport.

Studies have to be conducted to identify the infrastructure that will be needed to accommodate increasing volumes of ethanol and to identify and address the challenges of distributing and integrating these volumes into the fuel system. Also, research on biochemical conversion technologies that convert biomass to fuels more compatible with the current distribution infrastructure could be developed over the next 10–15 years.

If all the necessary conversion and distribution infrastructure were in place, 500 million dry tonnes of biomass could be used to produce up to 30 billion gallons of gasoline-equivalent fuels per year (or 2 million barrels per day [bbl/d]). However, potential fuel supply does not translate to amount of actual supply. When the production of corn grain ethanol was commercialized, U.S. production capacity grew by 25 percent each year over a 6-year period. Assuming that the rate of building cellulosic ethanol plants would exceed that of building corn grain ethanol plants by 100 percent, up to 0.5 million bbl/d of gasoline-equivalent cellulosic ethanol (1 barrel of oil produces about 0.85 barrel of gasoline equivalent) could be added to the fuels portfolio by 2020. By 2035, up to 1.7 million bbl/d of gasoline equivalent could be produced in this manner, resulting in about a 20 percent reduction in oil used for LDVs at current consumption levels.

Thermochemical Conversion

Without geologic CO_2 storage, technologies for the indirect liquefaction of coal to transportation fuels could be commercially deployable today, but life-cycle greenhouse gas emissions would be more than twice the CO_2 emissions of petroleum-based fuels. Requiring geologic CO_2 storage with these processes would have a relatively small impact on engineering costs and efficiency. However, the viability of geologic CO_2 storage has yet to be adequately demonstrated on a large scale in the United States, and unanticipated costs could occur. Although enhanced oil recovery could present an opportunity for early demonstrations of carbon capture and storage (CCS), that storage would be small compared with the large amounts of CO_2 that would be captured if coal-to-liquid fuels production became widely deployed, potentially in the gigatonne-per-year range.

Liquid fuels produced from thermochemical plants that use only biomass feedstock are more costly than fuels produced from coal, but biomass-derived fuels can have life-cycle CO_2 emissions that are near zero without geologic CO_2 storage or highly *negative* emissions with geologic CO_2 storage. To make such fuels competitive, the economic incentive for reducing CO_2 emissions has to be sufficiently high.

When biomass and coal are co-fed in thermochemical conversion to produce liquid fuels, the process allows a larger scale of operation and lower capital costs per unit of capacity than would be possible with biomass alone. If 500 million dry tonnes of biomass were combined with coal (60 percent coal and 40 percent biomass on an energy basis), production of 60 billion gallons of gasoline-equivalent fuels per year (4 million bbl/d) would be technically feasible. That amount of fuel represents about 45 percent of the current volume (140 billion gal/yr or 9 million bbl/d) of liquid fuel used annually in the United States for LDVs. Moreover, when biomass and coal are co-fed, the overall life-cycle CO_2 emissions are reduced because the CO_2 emissions from coal are countered by the CO_2 uptake by biomass during its growth. Combined coal-and-biomass-to-liquid fuels without geologic

CO_2 storage have life-cycle CO_2 emissions similar to those of gasoline; with geologic CO_2 storage, these fuels have near-zero life-cycle CO_2 emissions.

A program to aggressively support first-mover commercial coal-to-liquid and coal-and-biomass-to-liquid fuel production plants with integrated geologic CO_2 storage would have to be undertaken immediately if the United States were to produce fuels with greenhouse gas emissions similar to or less than petroleum-based fuels to address energy security in the near term.

Whether thermochemical conversion involves coal alone or coal and biomass combined, the viability of CO_2 geologic storage is critical to its commercial implementation. This means that large-scale demonstrations of and the establishment of regulatory procedures for CO_2 geologic storage would have to be aggressively pursued in the next few years if thermochemical conversion plants integrated with CCS are to be ready for commercial deployment in 2020 or sooner. If such demonstrations are initiated immediately, and geologic CO_2 storage is proven viable and safe by 2015, the first commercial plants could be operational in 2020.

Because plants for the conversion of combined coal and biomass into liquids are much smaller than those that convert coal alone, and because they will probably have to be sited in regions that are close to coal and biomass supplies, build-out rates will be lower than for the cellulosic plants discussed above. The committee estimates that at a 20 percent growth rate until 2035, 2.5 million barrels per day of gasoline equivalent could be produced in combined coal and biomass plants. This would consume about 270 million dry tonnes (300 million dry tons) of biomass per year—thus tapping less than the total projected biomass availability—and about 225 million tonnes of coal.

Given the vast coal resource in the United States, the actual supply of such fuel will be limited by its market penetration rather than feedstock availability. At a build rate of two to three plants per year, in 20 years up to 3 million bbl/d of gasoline equivalent could be produced from about 525 million tonnes of coal each year. However, all costs and social and environmental impacts of the associated level of coal production—an increase of about 50 percent—would have to be considered. At a build out of three plants starting up per year, five to six plants would be under construction at any one time.

Costs, Barriers, and Deployment

The committee estimated the costs of cellulosic ethanol, coal-to-liquid fuels with or without geologic CO_2 storage, and coal-and-biomass-to-liquid fuels with or

TABLE 3.1 Estimated Costs of Different Fuel Products With and Without a CO_2-Equivalent (CO_2-eq) Price of $50 per Tonne

Fuel Product	Cost Without CO_2-eq Price ($/bbl gasoline equivalent)	Cost With a CO_2-eq Price of $50 per Tonne ($/bbl gasoline equivalent)
Gasoline at crude price of $60/bbl	75	95
Gasoline at crude price of $100/bbl	115	135
Cellulosic ethanol	115	110
Biomass-to-liquids without CCS	140	130
Biomass-to-liquids with CCS	150	115
Coal-to-liquids without CCS	65	120
Coal-to-liquids with CCS	70	90
Coal-and-biomass-to-liquids without CCS	95	120
Coal-and-biomass-to-liquids with CCS	110	100

Note: The numbers in this table are rounded to the nearest $5. Estimated costs of fuel products for coal-to-liquids conversion represent the mean costs of products from Fischer-Tropsch and methanol-to-gasoline conversion processes.

without geologic CO_2 storage using a consistent set of assumptions (shown in Table 3.A.1 at the end of this chapter). Although those estimates do not represent predictions of future prices, they allow comparisons of fuel costs relative to each other. As shown in Table 3.1, coal-to-liquid fuels with CCS can be produced at a cost of $70/bbl of gasoline equivalent and thus are competitive with $75/bbl gasoline. In contrast, the costs of fuels produced from biomass without geologic CO_2 storage are $115/bbl of gasoline equivalent for cellulosic ethanol produced by biochemical conversion and $140/bbl for biomass-to-liquid fuels produced by thermochemical conversion. The costs of cellulosic ethanol, and coal-and-biomass-to-liquid fuels with CCS, become more attractive if a CO_2 price of $50 per tonne is included.

Attaining supplies of 1.7 million bbl/d of biofuels, 2.5 bbl/d of coal-and-biomass-to-liquid fuels, or 3 million bbl/d of coal-to-liquid fuels will require the permitting and construction of tens to hundreds of conversion plants, together with the associated fuel transportation and delivery infrastructure. Given the magnitude of U.S. liquid-fuel consumption (14 million barrels of crude oil per day in the transportation sector in 2007) and the scale of current petroleum imports (about 56 percent of the petroleum used in the United States in 2008 was imported), a business-as-usual approach for deploying these technologies would be insufficient to address the need to develop alternative liquid transportation fuels, particularly because the development and demonstration of technology, the construction of plants, and the implementation of infrastructure require 10–20 years per cycle. In addition, investments in alternative fuels have to be protected against crude oil price fluctuations.

Because geologic CO_2 storage is key to producing liquid fuels from coal with life-cycle greenhouse gas emissions comparable to those of gasoline, commercial demonstrations of coal-to-liquid and coal-and-biomass-to-liquid fuel technologies integrated with geologic CO_2 storage would have to proceed immediately if the goal is to deploy commercial plants by 2020. Moreover, detailed scenarios for market-penetration rates of biofuels and coal-to-liquid fuels would have to be developed to clarify the hurdles preventing full feedstock utilization and to establish the enduring policies required to overcome them. Further, current government and industry programs would have to be evaluated to determine whether emerging biomass- and coal-conversion technologies could further reduce U.S. oil consumption and CO_2 emissions over the next decade.

Other Transportation Fuels

Technologies for producing transportation fuels from natural gas—such as gas-to-liquid diesel, dimethyl ether, and methanol—have been deployed or will be ready for deployment by 2020. But only if large supplies of natural gas were available at acceptable costs—for example, from natural gas hydrates—would the United States be likely to use natural gas as the feedstock for transportation fuel production.

Hydrogen has considerable potential, as discussed in *Transitions to Alternative Transportation Technologies—A Focus on Hydrogen* (NRC, 2008) and *The Hydrogen Economy: Opportunities, Costs, Barriers, and R&D Needs* (NRC, 2004). Hydrogen fuel-cell vehicles could yield large and sustained reductions in U.S. oil consumption and greenhouse gas emissions, but it will take several decades to realize these potential long-term benefits.

RENEWABLE ENERGY

The level of electricity generation from renewable resources has risen significantly over the past 20 years. Nonhydroelectric renewable sources, however, still provide a very small proportion of the U.S. total (about 2.5 percent of all electricity generated). In the 2008 reference-case estimates of the Energy Information Administration (EIA, 2008), the contribution of nonhydroelectric renewables was projected to be about 7 percent of total electricity generation by 2030. But the AEF Committee found that with a sustained effort and accelerated deployment, nonhydroelectric renewables could collectively provide 10 percent of the nation's electricity generation by 2020 and 20 percent or more by 2035. With current hydropower included, more than 25 percent of electricity generation could come from renewables by 2035.

Generation Capacity and Resource Base

Renewables currently represent a small fraction of total electricity generation. According to the EIA, conventional hydroelectric power is the largest source of renewable electricity in the United States, generating about 6 percent (almost 250,000 GWh out of a total 4 million GWh) of electricity produced by the electric power sector in 2007.[10]

The largest growth rates in renewable resources for electricity generation are currently in wind power and solar power. Though wind power in 2007 represented less than 1 percent of total electricity generation, wind electricity grew at a 15.5 percent compounded annual growth rate over the 1990–2007 time period and at a 25.6 percent rate between 1997 and 2007.

In 2007, wind power supplied over 34,000 GWh, almost 8,000 GWh more than in the year before. An additional 8,400 MW of capacity was added in 2008, representing an additional yearly generation of 25,000 GWh (assuming a 35 percent capacity factor). Total wind power capacity at the end of 2008 was approximately 25,000 MW. However, the overall economic downturn at that time caused financing for new wind power projects and orders for turbine components to slow, and layoffs in the wind turbine-manufacturing sector began. Thus, 2009 recently looked to be considerably smaller in terms of new capacity than 2008. However, recent data reveal that 2.8 GW of new wind power generation capacity was installed in the first quarter of 2009. Over the long term, the impacts of state renewable portfolio standards and the federal production tax credit will continue to spur installation of wind power capacity.

---

[10]The electric power sector includes electricity utilities, independent power producers, and large commercial and industrial generators of electricity.

Central-utility electricity generation from concentrating solar power (CSP) and photovoltaics (PV) combined to supply 600 GWh in 2007, 0.01 percent of the total electricity generation for the United States. This level has been approximately constant since 1990. However, it does not account for the increase in residential and other small-PV installations, the sector that has displayed the largest growth rate for solar electricity. Solar PV in the United States has grown at a compounded annual growth rate of more than 40 percent from 2000 to 2005, with an installed generation capacity of almost 480 MW that, assuming a 15 percent capacity factor, produces approximately 630 GWh.

The United States has sufficient renewable-energy resources to significantly expand the amount of electricity generated from them. Solar in particular, followed by wind, offers the greatest potential among the domestic renewable resources. Solar energy derived from sunlight reaching Earth's surface could produce many times the current and projected future electricity consumption. And the total estimated electrical energy derivable from the continental U.S. wind resource in Class 3 and higher wind-speed areas is 11 million GWh per year—far greater than the estimated 2007 electricity generation of about 4 million GWh. But these numbers, which represent the total resource base, exceed what can be developed at an acceptable cost. Moreover, the resource bases for wind and solar energy are not evenly distributed, spatially as well as temporally, and they are more diffuse compared to fossil and nuclear energy sources. Finally, though the size of the resource base is impressive, there are many technological, economic, and deployment-related constraints on using sources of renewable energy on a large scale.

Technologies

Several renewable-energy technologies are available for deployment or are under active development.

- *Wind.* Turbine technology has advanced substantially in recent years. Future development will be evolutionary and will focus on improved efficiency and lower production costs. Major objectives are to increase the capacity factors and improve integration into the electric grid.
- *Solar photovoltaics.* The two major types of PV are silicon flat plates and thin films on various substrates. The former are more mature, with primary development objectives being higher efficiency and lower production costs. Thin films have the potential for substantial cost advan-

tages and can use a wider array of materials, but they are less well developed.

- *Concentrating solar power.* The three main options are parabolic troughs, power towers, and dish-Stirling engine systems. The first two are now the lowest-cost utility-scale solar electricity technology for regions of high solar flux. Design improvements and advances in high-temperature and optical materials are the major paths to cost reduction.

- *Geothermal.* Conventional geothermal, which relies on hydrothermal sources within 3 km of the surface to drive a heat engine, is a fairly mature technology, but it has a rather limited resource base. A study of the western United States found that 13 GW of electrical power capacity exists in identified geothermal resources in this region. Greatly expanding that base will require enhanced geothermal systems to mine heat down to a depth of 10 km. Such systems, however, face many technical challenges and are not now in operation.

- *Hydropower.* Conventional hydropower is the least expensive source of electricity. The technology is well developed, and objectives are to increase efficiency and reduce impacts on associated water bodies, as efforts to expand are likely to be limited by environmental concerns. Hydrokinetic technologies produce electricity using currents, tides, and ocean waves; many designs and demonstration plants exist, but there are no commercial deployments.

- *Biopower.* There are three main sources: wood/plant waste, municipal solid waste/landfill gas, and other (e.g., agriculture waste, used tires). A variety of technologies may be used to produce electricity, including current technologies based on the steam-Rankine cycle and future applications involving gasification combined-cycle plants. The use of biomass for biopower competes with its use for alternative liquid fuels.

Deployment Potential

Between now and 2020, there are no technological constraints to accelerated deployment of the major renewable resources with existing technologies. However, there are other kinds of barriers. The main ones currently include the cost-competitiveness of existing technologies relative to most other sources of electricity (with no prices assigned to carbon emissions or other externalities); markets not sufficiently shaped so as to allow the existing technologies to reach full scale and thus realize economies of scale; the lack of sufficient transmission capacity to move distant resources to demand centers; and the absence of sustained policies. Also, continued research to reduce costs and increase efficiencies is needed.

A reasonable target is that 20 percent of all electricity be supplied by renewable resources—including hydropower—by 2020. This would mean that approximately 10 percent of electricity generation would be from nonhydropower renewables. Continued accelerated deployment and sustained policies could permit nonhydropower renewables to reach 20 percent of total U.S. electricity generation by 2035.

The most in-depth scenario for increased renewables penetration into the electricity sector is the Department of Energy's (DOE's) 20 percent wind-penetration scenario (DOE, 2008; see Chapter 6 in Part 2 of this report for details), which includes an assessment of wind resources and available technologies; manufacturing, materials, and labor requirements; environmental impacts and siting issues; transmission and system integration; and market requirements. The scenario requires that installations reach an annual rate of about 16,000 MW by 2018, almost double the current annual deployment in the United States but less than the current global deployment of 27,000 MW. The committee considered this projected installation rate together with the reliability of wind facilities, and it concluded that this level of wind power deployment would be achievable with accelerated deployment as defined in Chapter 2.

Another accelerated deployment scenario for reaching 20 percent non-hydropower renewables is reliance on multiple renewable sources. Obtaining 20 percent of electricity generation solely from wind power would be a challenge because the 20 percent refers to an annual average, whereas wind power is inter-mittent. Balancing wind with multiple renewable resources—including solar, which does not normally peak when wind does, and baseload power from geothermal and biomass—could mitigate the temporal variability in generation. Reaching the goal of 20 percent nonhydropower renewables by 2035 could be achieved by adding 9.5 GW per year of wind power and a total of 70 GW of solar PV and 13 GW each of geothermal and biomass. Using multiple renewable resources to reach this level would take advantage of the geographical variability in the resource base.

Relying on multiple renewable resources would not eliminate the need to expand transmission capacity or make other improvements in the electricity infra-structure to enable the integration of renewables, nor would it reduce the magni-tude of costs. However, such an approach to reaching 20 percent nonhydropower renewables could offer other attributes, such as providing baseload generation and combining different intermittent renewables to reduce the temporal variability in generation. The installation rate for wind under this option is approximately the current rate of deployment, and the installation rates of the other renewables technologies are consistent with the accelerated-deployment definition.

Greatly expanding electricity generation from renewable sources will require changes in the present electric system because of the intermittency, spatial distribution, and scalability of renewable resources. Integrating an additional 20 percent of renewable electricity, whether it comes from wind, solar, or some combination of renewable sources, requires expansion of the transmission system (to enable the power to reach demand centers and regional electricity markets) as well as large increases over current levels in manufacturing, employment, and investment. Further, although electricity storage is not needed, integrating intermittent renewables up to the 20 percent level would also require improvements in the electricity distribution system and fast-responding backup electricity generation.

Integrating renewables at a much greater level so that they account, say, for more than 50 percent of U.S. electricity generation would require scientific advances and major changes in electricity production and use. It would also necessitate the deployment of electricity storage technologies to offset renewables' intermittency. More details on deployment are available in Chapter 6 in Part 2 of this report, and an extensive discussion is presented in the panel report *Electricity from Renewable Resources: Status, Prospects, and Impediments* (NAS-NAE-NRC, 2009a).

Cost

Given the experience with renewables over the past 20–30 years, it is clear that their economics have generally not been attractive compared to most competing sources of electricity. The most favorable technology out to 2020 is onshore wind; with a federal production tax credit for renewables, or with high natural gas prices, wind is competitive with electricity generation from natural gas. Solar PV presents a different economic picture. It is much more expensive than current sources of electricity generated by centralized generating facilities, but PV installed for residential and commercial consumers provides electricity directly to the consumer.

Thus, the economics for a so-called distributed renewable generation source (termed a "distributed" source because the electricity generation occurs on the distribution side of the electricity system) depends on costs being competitive with retail electricity prices. Many residential and commercial systems are unlikely to have high capacity factors, given that such systems would be installed on roofs that are not currently designed to maximize sun exposure. Additionally, the full electricity distribution system and centralized power sources are still required for periods when electricity generation from distributed sources is not available. However, if electricity prices continue to increase and more utilities adopt time-of-day pricing (which charges the highest rate during the middle of the day), solar PV could become more widely competitive.

Nearly all of the costs associated with renewable energy are in the manufacture and installation of the equipment; fuel costs during operation—except for biomass—are zero. Economies of scale occur primarily during equipment manufacturing for nonhydropower renewable technologies and much less so with respect to plant size. The plants, however, can be built quickly and incrementally compared to conventional coal and nuclear electricity plants, allowing utilities and developers to begin recouping costs much more quickly. Thus, technological innovations will play a major role in how costs for renewables evolve in the future.

One estimate of the costs of obtaining 20 percent of electricity from renewables is provided by the DOE 20 percent wind energy study (DOE, 2008) referred to earlier and discussed in greater detail in Chapter 6 of this report and in NAS-NAE-NRC (2009a). Though this is a single study on the costs, it was developed with contributions from a wide array of stakeholders in the electric utility industry, wind power developers, engineering consultants, and environmental organizations. The study, which was externally peer reviewed (as mandated by the U.S. Office of Management and Budget), considered the direct costs both of installing the generating capacity and of integrating this power into the electricity system. Overall, it projects that increases in wind power generation costs (capital, operation, and maintenance expenses) in net present value would be approximately $300 billion—covering the installation of approximately 300 GW of new wind power capacity, of which about 250 GW would be installed onshore and 50 GW installed offshore. The total number of wind turbines required is estimated to be about 100,000. Estimates of the transmission costs range widely, from the $23 billion estimated within the DOE (2008) study to American Electric Power's $60 billion estimate (AEP, 2007) to the recent estimate of $80 billion by the Eastern Wind Integration and Transmission Study/Joint Coordinated System Planning Study (JCSP, 2009) for integrating 20 percent in the eastern part of the United States.

Barriers to Deployment

The major barrier to greater deployment of renewable electricity sources has been their high costs. And recent capacity limitations—in personnel, materials, and manufacturing—have raised the costs of PV and wind power projects even higher. Moreover, the variability of renewable energy makes integration into the electric power system more difficult as deployment grows. Integrating renewables at levels approaching 20 percent of all electricity generation requires not only greater transmission capacity but also the increased installation of fast-responding generation to provide electricity when renewables are not available. Expansion of the transmission system, improving its flexibility through advanced control technologies, and co-siting with other renewable or conventional generation can help this integration. Expansion of the transmission system also gives providers of renewable electricity access to regional wholesale electricity markets, thus improving its marketability. However, at a high level of renewable technology deployment, land-use and other local impacts would become quite important. In the past, such impacts have provoked local opposition to the siting of renewable electricity-generating facilities and associated transmissions lines, and opposition is likely to occur in the future. This represents an additional potential barrier.

In order to facilitate investment in the face of high costs and, as a result, allow renewable electricity generation to meet its potential, consistent and long-term policies are essential. As is shown in Chapter 6, the on-and-off nature of the federal production tax credit has direct impacts, positive and negative, respectively, on the installation of new renewable-energy generation facilities. The 20-percent-by-2030 target can be reached, but substantial increases in manufacturing capacity, employment, investment, and installation will be needed.

Impacts

Renewable-energy sources have significantly smaller lifetime emissions of CO_2 and criteria pollutants per kilowatt-hour than does fossil energy, although renewables' emissions are about the same as those of nuclear power (see Figure 2.15). Renewable electricity technologies (except biopower, some geothermal, and high-temperature solar technologies) also use significantly less water than do nuclear and gas- or coal-fired electricity technologies. On the other hand, land-use requirements are substantially higher for renewables but could be mitigated to some degree by multiuse features that allow some of the land to be devoted, say, to agricultural activities. However, land-use and related issues associated with renewables deployment, such as noise and aesthetics, often fall to local jurisdictions for approvals, and the resulting procedures can be controversial.

FOSSIL-FUEL ENERGY

Fossil fuels—petroleum, natural gas, and coal—have been the dominant energy source in the United States for decades past and will continue to be a major source for decades to come. At present, they collectively supply about 85 percent of the nation's primary energy (see Figure 1.2).

Resource Base for Petroleum and Natural Gas

Worldwide, the amount of petroleum and natural gas that could ultimately be produced is very large, but most of this resource is located outside the United States. In 2008, the United States imported about 56 percent of the petroleum it consumed, a drop from the peak of 60 percent in 2006. This drop can be attributed mainly to the growth in production of a half million barrels per day from the deepwater Gulf of Mexico, illustrating that domestic production depends on the ability to develop discovered resources to make up for the decline from existing fields.

Maintaining domestic petroleum production at current levels over the long run will be very challenging, however. Production of petroleum from U.S. unconventional resources (primarily oil shales), which is not likely to occur in significant volumes before 2020, will be more expensive than that from conventional oil sources and may have more negative environmental impacts. In any case, because U.S. crude oil reserves and production are 2 percent and 8 percent, respectively, of world levels, the actions of other countries could have greater effects than those of the United States on world oil production. By contrast, because U.S. petroleum consumption is 24 percent of world consumption, changes in U.S. demand are a significant factor in determining world demand. Growing demand in other countries could, however, offset any downward price pressures resulting from reduced U.S. demand.

Natural gas is the cleanest of the fossil fuels and has the lowest greenhouse gas emissions per unit of energy (emitting about half of the CO_2 of coal when burned for electricity generation). While the U.S. natural gas resource base is only about 9 percent of the known world total, some 86 percent of the natural gas consumed in the United States is produced domestically, with much of the remainder coming from Canada. In recent years, natural gas production from conventional resources has continued to decline, but production from unconventional resources such as coal beds, tight gas sands (rocks through which flow is very slow), and particularly from natural gas shales has increased. Higher natural gas prices in 2007 and 2008 led to expanded drilling in tight gas sands and gas shales, which increased total U.S. gas production by about 9 percent in 2008 after a decade of its being roughly constant.

If the increase in domestic natural gas production continues and is sustained over long periods, some portion of potential growth in domestic demand for natural gas could be accommodated. If, on the other hand, growth in U.S. natural gas production is limited by a combination of production declines from existing resources and modest growth from new resources, the United States may have to import liquefied natural gas (LNG) at prices subject to international market fluctuations. Which of these futures occurs will depend on some combination of linked factors that include the magnitude of demand growth, production technology, resource availability, and price.

About 12 percent of U.S. petroleum resources and 20 percent of U.S. natural gas resources are believed to lie in areas that, for a variety of policy reasons, are currently off-limits. These estimates are highly uncertain, however, and the technologies for exploration and production (which might permit more of these activities elsewhere) have advanced considerably since the estimates were made. Further, estimates of production from the restricted areas are moderate—for petroleum, they are on the order of several hundred thousand barrels per day by the mid-2020s (compared to current domestic production of 5.1 million barrels per day). The contribution to gas production from these areas could be about 1.5 trillion cubic feet per year in the 2020–2030 period, compared to current domestic production of 19.3 trillion cubic feet per year.

The issue for policy makers is to balance the energy security and economic benefits of developing these currently off-limits resources against the potentially negative environmental impacts. Most observers believe that the effect of incremental U.S. oil production from restricted areas on world oil price would be small, but because natural gas markets are more regional, they might respond differently; increased natural gas production from restricted areas could potentially offset the need for LNG imports.

Resource Base for Coal

U.S. recoverable reserves of coal are well over 200 times the current annual production of 1 billion tonnes, and additional identified resources are much larger. Thus the coal resource base is unlikely to constrain coal use for many decades to come. Rather, environmental, economic, geographic, geologic, and legal issues will likely be the primary constraints. Of particular concern regarding the greenhouse gas problem is that burning coal to generate electricity produces about 1 tonne of CO_2 equivalent per megawatt-hour, about twice the amount produced by natural gas. If CCS technologies were successfully developed, it is possible that future coal consumption could remain at current levels or increase (as a result, for example, of demand from a new coal-to-liquid-fuels industry), even if policies were put in place to constrain greenhouse gas emissions. On the other hand, if practical CCS technologies fail to materialize, coal use would be severely curtailed in a carbon-constrained world.

Fossil Energy Use for Electric Power Generation

In 2006, about 52 percent of U.S. electricity was generated from coal and 16 percent from natural gas. Many of these plants could operate for 60 years or more, and there is great reluctance on the part of plant operators to shorten their period of operation, given that new plants would require large amounts of capital and new permitting. Yet significant mitigation of U.S. greenhouse gas emissions will require dramatic reductions in the emissions from these plants. Alternatives include (1) retiring the plant; (2) raising the generating efficiency, thereby reducing greenhouse gas emissions per unit of electricity produced; (3) retrofitting with CO_2 "post-combustion" capture capability; or (4) repowering/rebuilding at the site, resulting in an entirely new or mostly new unit.

The two principal technologies for future coal-burning power plants are pulverized coal (PC) and integrated gasification combined cycle (IGCC), though the possibility of coal combustion with pure oxygen (oxyfuel) instead of air would simplify subsequent CO_2 capture. This option is also being investigated and may be competitive in the future. PC units now produce nearly all of the coal-based electric power in the United States. PC plants with 40–44 percent efficiency[11]

---

[11]Potential PC efficiencies as high as 48 percent have been estimated in the literature. This would require steam pressures and temperatures of 5000 psi and 1400°F main steam, 1400°F reheat, whereas the most robust current ultrasupercritical plants operate at pressures of around

(ultrasupercritical plants) could be achieved in the 2020–2035 period, as compared with a typical efficiency of 34–38 percent for older subcritical and supercritical steam plants. Replacing a 37 percent efficient plant with a 42 percent efficient plant, for example, would reduce CO_2-eq emissions and fuel consumption per kilowatt-hour of output by about 12 percent. To reduce emissions more dramatically in PC plants, CCS would be required.

Retrofitting for 90 percent CO_2 capture at existing PC plants with technology available today would require capital expenditures approaching those of the original plant itself; and 20–40 percent of the plant's energy would be diverted for separation, compression, and transmission of the CO_2, thereby significantly reducing thermal efficiency and increasing the levelized cost of electricity. In addition, retrofits face the added problems of site constraints and steam-management limitations, rendering the feasibility of installing CO_2 capture retrofits in existing plants highly plant dependent. Also, the optimum percentage of CO_2 capture in a retrofitted coal plant could be lower than that of a new coal plant. In any case, further engineering analyses to establish the shape of these cost-versus-percent-capture curves would aid policy analysis considerably.

Electricity demand and CO_2 price will have a strong effect on the rate of introduction of new coal plants. If the CO_2 price is zero and electricity demand stays relatively flat (as a result of increasing end-use efficiency, for instance), hardly any of the existing PC plants will be retired or modified and very few new plants will be built.

New natural gas combined cycle (NGCC) plants compete with new coal plants. Favoring natural gas plants are their lower capital costs and shorter construction times, but of primary importance is the price of natural gas. For example, in the committee's calculations, at a price of $6 per million British thermal units (Btu), NGCC plants have the lowest levelized cost of electricity (LCOE) of any baseload generating option, while at $16/million Btu they have the highest LCOE (see Figure 2.10 in Chapter 2). (Over the course of this study, U.S. natural gas prices have risen above $13/million Btu and fallen to below $4/million Btu.) Future rules governing greenhouse gas emissions and the pace at which CCS technologies can be commercialized will also affect the coal-gas competition.

If domestic natural gas (e.g., from shale gas deposits) proves plentiful, and

---

4640 psi and temperatures of 1112–1130°F. Thus, achieving this potential efficiency would require major R&D breakthroughs. In addition, operating plants often do not realize their full design efficiencies.

confidence grows that prices will remain in the range of $7–9/million Btu or lower for decades, as some commentators think may happen,[12] then NGCC plants with CCS could compete economically with PC and IGCC plants with CCS. In such a world, the cheapest way to gain large CO_2 reductions would be to use NGCC with CCS to replace existing and future coal units over time.

Although a large shift in this direction would increase natural gas demand significantly and put upward pressure on prices, the committee still considers it wise to plan for a broad range of future natural gas prices and domestic availabilities. Consequently, the committee envisions some CCS projects involving NGCC technology being part of the recommended 10 GW of CCS demonstrations (see Chapter 7 in Part 2). The committee did not make a judgment about the mix of NGCC, PC, and IGCC plants with CCS that would be appropriate.

The committee compared the costs of new PC and IGCC plants, with and without CCS, built with components available today and with various prices for CO_2 emissions. (It also considered as feedstocks not only coal but also natural gas, biomass, and biomass and coal in combination.) If no price is put on CO_2 emissions, PC without CCS is the cheapest option. However, the extra cost to add CCS to IGCC is less than the extra cost to add CCS to PC, because in IGCC, CO_2 is captured at high pressure[13] after gasification but before power generation (pre-combustion capture). For bituminous coal—at a price of $50 per tonne of CO_2 emitted—IGCC with CCS is the cheapest of the four options, although all have a higher cost than current plants. These cost estimates, and similar estimates for the capture of CO_2 from natural gas plants and low-rank coal plants, have significant uncertainties particularly in fuel costs, capital costs for first-of-a-kind plants, and the costs of CO_2 capture and storage technologies.

Based on historical experience, and assuming that all goes well in the development and operation of CCS demonstrations from pilot plants to commercial scale, 10 GW of demonstration fossil-fuel CCS plants could be operating by 2020 with a strong policy driver (e.g., a CO_2 emissions price of about $100 per tonne or comparably strong regulation), but not a crash program. With similar assumptions, 5 GW per year could be added between 2020 and 2025, and a further 10–20 GW per year from 2025 to 2035, resulting in a total of 135 to 235 GW

[12]CERA, "Rising to the Challenge: A Study of North American Gas Supply to 2018," www.cera.com/aspx/cda/public1/news/pressReleases/pressReleaseDetails.aspx?CID=10179.

[13]However, additional compression is still needed before the CO_2 can be injected underground.

of fossil-fuel power with CCS in 2035. Whether any coal plants and natural gas plants without CCS would still be operating in 2035 would depend on the nature of greenhouse gas policies at that time.

Carbon Capture and Storage

CCS technologies have been demonstrated at commercial scale, but no large power plant today captures and stores its CO_2. The few large storage projects now under way are all coupled to CO_2 capture at nonpower facilities; for example, in one offshore operation in Norway, 50 million standard cubic feet per day of CO_2 (1 million tonnes per year) are separated from natural gas before the fuel is inserted into the European grid; the CO_2 is injected under the North Sea.

CO_2 storage could be implemented in oil and gas reservoirs, deep formations with salt water, and deep coal beds. Specific sites would have to be selected, engineered, and operated with careful attention to safety. In particular, the deep subsurface rock formations that hold the CO_2 must allow injection of large total quantities at sufficient rates and have geologic layers that prevent, over centuries to millennia, the upward migration of injected CO_2. Current surveys suggest that the available storage within 50 miles of most of the major U.S. sources of CO_2 would be more than sufficient to handle all emissions for many decades and that up to 20 percent of current emissions could be stored at estimated costs of $50 per tonne of CO_2 or less. However, given the large volumes of CO_2 involved, the storage challenges should not be underestimated. At typical densities in the subsurface, a single 1 GW coal-fired plant would need to inject about 300 million standard cubic feet of CO_2 per day, or a volume flow equivalent to about 160,000 barrels per day—comparable to the petroleum production from a large oil field.

Too little is known at present to determine which power-generation technologies and which storage options could best produce electricity after 2020 if carbon emissions were constrained. Reliable cost and performance data are needed, both for capture and storage, and they can be obtained only by construction and operation of full-scale demonstration facilities. Such demonstrations could assure vendors, investors, and other private-industry interests that power plants that incorporate advanced technologies, and the associated storage facilities, could be built and operated in accordance with commercial criteria. Because of the variety of coal types and the myriad of technology-conversion options for coal, natural gas, and biomass fuels, a diverse *portfolio* of demonstrations of CO_2 capture technology will actually be required. Similarly, to sort out storage options and gain experience with their costs, risks, environmental impacts, legal liabilities, and regulatory and management issues, it will be necessary to operate a number of large-scale storage projects in a variety of subsurface settings.

The investments in this portfolio of CCS demonstrations will certainly be large, but there is no benefit in waiting to make them. The committee judges that the period between now and 2020 could be sufficient for acquiring the needed information on CCS viability, provided that the deployment of CCS demonstration projects proceeds as rapidly as possible. If these investments are made now, 10 GW of CCS projects could be in place by 2020. If not, the ability to implement CCS will be delayed.

Fossil Energy Use for Transportation

About 95 percent of the energy for transportation comes from crude oil, of which about 56 percent is imported. The transportation sector also generates about one-third of U.S. greenhouse gas emissions, which are difficult to eliminate from moving vehicles. Coal-to-liquid and natural-gas-to-liquid technologies with CCS can produce liquid transportation fuels with no more greenhouse gas emissions than those of crude oil. Other technologies to replace petroleum in the transportation sector are described in the "Energy Efficiency" and "Alternative Transportation Fuels" sections of this chapter.

Impacts and Barriers to Deployment

The widespread use of fossil fuels in the United States creates a substantial array of environmental impacts, most of which (with the notable exception of greenhouse gas impacts) have been addressed in principle by a broad array of laws and regulations over the last few decades. The continual challenge regarding most of these policy instruments is to keep them up-to-date and enforced while increases occur in the consumption of conventional or unconventional fuels.

All of these environmental issues need to be fully considered in assessing the real costs of different energy options. Further, agencies, other stakeholders, and funders concerned with environmental impacts must enhance their readiness for new challenges that are likely to emerge in the future regarding systems that make use of fossil fuels. These new challenges include the capture and storage of CO_2; potentially increased use of coal for coal-to-liquid fuel or coal-to-natural-gas production; shale oil and tar sands development; LNG safety; and water use.

A regulatory structure must be developed during the 2010–2020 period to enable large-scale deployment of the CCS necessary for continued use of fossil fuels. Pertinent issues include CO_2 pipeline-transport safety and land use, stability and leakage from underground carbon storage, and public acceptance of such storage.

Increased use of coal will intensify concerns about environmental and safety aspects of extraction as well as about pollutant emissions arising from power generation. Oil shale and tar sands production will also result in extraction issues, along with those pertaining to water availability and CO_2 production. Expansion of LNG imports may raise concerns about the potential coastal-area impacts of LNG storage facilities and their vulnerability to terrorist attacks, and the impacts of pipeline-capacity enlargements in some regions may raise concern as well. In general, increased fossil-fuel use for electricity generation will add to power plants' already substantial requirements for fresh water. In addition, there will be greater impacts on water quality, aquatic life, and surrounding ecosystems. Finally, although technologies exist to achieve high levels of control for most of the conventional pollutants produced in coal-to-liquid or gas-to-liquid fuel plants, performance standards relating to CCS will need to be written during the 2010–2020 period.

NUCLEAR ENERGY

Energy companies in the United States are expressing increased interest in constructing new nuclear power plants. Reasons cited include the need for additional baseload generating capacity; growing concerns about greenhouse gas emissions from fossil-fuel plants; volatility in natural gas prices; and favorable experience with existing nuclear plants, including ongoing improvements in reliability and safety.[14] No major R&D is needed for an expansion of U.S. nuclear power through 2020 and, likely, through 2035.

Nonetheless, the high cost of construction of new nuclear plants is a major concern, and the experience with the handful of new plants that could be built before 2020 will be critical to assess the future viability of the nuclear option. If

---

[14]The $18.5 billion in loan guarantees for new nuclear plants arising from the Energy Policy Act of 2005 may also contribute to this interest.

these plants are not built on time and on budget, or if the electricity produced is not cost competitive, few additional new plants are likely to follow, at least for a while.

Technologies

The nuclear plants now in place in the United States were built with technology developed in the 1960s and 1970s. In the intervening decades, ways to make better use of existing plants have been developed, along with new technologies that improve safety and security, decrease costs, and reduce the amount of generated waste—especially high-level waste. These technological innovations, now available or under development, include the following:

- *Improvements to existing plants.* The trend of technical and operational improvements in nuclear technology that has developed over the past few decades is expected to continue. Incremental improvements to the 104 currently operating U.S. nuclear plants have enabled them to produce more power over their operating lifetimes. Modifying existing plants to increase power output, referred to as "uprates," is considerably less costly than adding new capacity, and additional power uprates are expected in the future. In fact, nearly as much new nuclear capacity could be added in this way before 2020 as could be produced during that period by building new plants. Additionally, most currently operating nuclear power plants have received or are expecting to receive 20-year operating-license extensions, which will allow them to operate for a total of 60 years; discussions have recently commenced about extending licenses an additional 20 years (for a total of 80 years). Also, the periods when plants are off line have been reduced and can be further reduced. Average plant capacity factors have grown from 66 percent in 1990 to 91.8 percent in 2007, primarily through shortened refueling outages and improved maintenance, thereby greatly improving the plants' economic performance.
- *Evolutionary nuclear plants.* New plants constructed before 2020 will be based on modifications of existing plant designs, using technologies that are largely ready for deployment now.
- *Alternative nuclear plants.* Alternative designs in two broad categories are being developed or improved: thermal neutron reactor designs (all

current U.S. reactors are thermal) and fast neutron reactor designs. Thermal neutron reactor designs include plants that operate at higher temperatures, thereby offering process heat (which could be used, for example, for producing hydrogen) in addition to electricity production. Fast neutron reactor designs include plants intended to destroy undesirable isotopes associated with much of the long-lived radioactive waste burden in used fuel, and, in some cases, to breed additional fuel. These plants could reduce the volume of and the heat emitted by long-lived nuclear waste that must go to a repository for disposal.[15] Much R&D will be needed before any of these alternative reactor types can be expected to make significant contributions to the U.S. energy supply.

- *Alternative fuel cycles.* The United States currently employs a once-through nuclear fuel cycle in which used fuel is disposed of after removal from the reactor. In contrast, alternative (closed) nuclear fuel cycles involve the reprocessing of used fuel to produce new fuel. In principle, these alternative fuel cycles could extend fuel supplies and reduce the amount of long-lived nuclear waste requiring disposal. The reprocessing technology in common use today, called plutonium and uranium extraction (PUREX), is associated with an increased risk of nuclear weapons proliferation, as well as an increased risk of theft or diversion of nuclear materials,[16] because it yields a separated stream of plutonium. A modified version of PUREX that keeps uranium with the plutonium could result in modestly reduced proliferation risks relative to PUREX and could be deployed after 2020. Other alternatives are being investigated, but they are unlikely to be ready for commercial deployment before 2035. R&D is still needed on fuel design, separation processes, fuel fabrication, and fuel qualification, as well as on the associated alternative reactors.

---

[15]For about the first century, the major challenges for managing high-level waste are the heat and radioactivity emitted by short-lived fission products. If a closed fuel cycle is implemented, these fission products will likely need to be removed from the waste and dealt with separately to achieve a significant reduction in the number of repositories needed.

[16]The United States is a nuclear weapons state and the primary proliferation risk applies to the use of such technologies in countries that are not nuclear weapons states. There is also concern about the theft of weapons-usable materials from reprocessing, wherever it takes place. The risk of proliferation is a controversial subject, and there are differing points of view about how it should affect technology trajectories within the United States.

Deployment Potential

As many as five to nine new nuclear plants could be built in the United States by 2020; however, in light of the long lead times expected for construction, the first one is unlikely to be operating before 2015. These new plants will have evolutionary designs that are similar to existing power plants. Combining new power plants with increased capacity obtained by uprating currently operating plants, a 12–20 percent increase in U.S. nuclear capacity is possible by 2020.

After 2020, the potential magnitude of nuclear power's contribution to the U.S. energy supply is uncertain. The operating licenses of existing plants will begin to expire in 2028, and the plants will have to be shut down if license extensions (to 80-year total operating lifetimes) are not obtained; under these circumstances, about 24 percent of the current U.S. nuclear capacity would be retired by 2035. Because of the long construction times, many companies will need to decide soon whether to replace retiring plants with new nuclear plants. As noted previously, the major barrier to new construction is financial; thus, companies will need to know whether evolutionary plants can be built on budget and on schedule. One important purpose of providing federal loan guarantees is to acquire experience with a few early plants that will guide these decisions.[17] This experience will affect the U.S. electricity portfolio up to and after 2035.

The scale of new nuclear deployment after 2020 will depend on the performance of plants built during the next decade. If the first handful of new plants (say, five) to be constructed in the United States meet cost, schedule, and performance targets, many more plants could be deployed after 2020. Construction of as many as three plants per year could take place up to 2025, and as many as five

---

[17]The statute authorizes DOE to provide guarantees for loans covering up to 80 percent of the total project cost. When the government provides a guarantee for 100 percent of the debt instrument, the standard government loan-guarantee rules require that the government itself allocate and provide the capital for the investment (through the Department of the Treasury's Federal Financing Bank [FFB]), which is then repaid by the entity receiving the guarantee over the period of the loan. If an entity other than the FFB provides the loan, there is no federal money that changes hands at the outset. The program is intended to be revenue neutral to the government; that is, the company benefiting from the guarantee is required to pay a fee to cover the risk of failure to repay the loan, as well as the administrative costs. DOE is authorized to provide $18.5 billion in loan guarantees for nuclear power facilities, but it is not yet clear whether this allocation will be sufficient for the four to five plants the committee judges will be needed to demonstrate whether new nuclear plants can be built on schedule and on budget. DOE has found it difficult to implement the program, in part because of the challenge associated with estimating the appropriate fee.

plants per year could be constructed between 2026 and 2035. This could grow to 5–10 plants per year after 2035 if there is sufficient demand. However, if the first new plants do not meet their targets, few others are likely to follow, at least for a while.

Costs

The committee estimates that the LCOE at the busbar from new evolutionary nuclear plants could range from 8¢/kWh to 13¢/kWh (see Figure 2.10). Existing federal incentives—including loan guarantees such as those of the Energy Policy Act of 2005—could reduce the LCOE to about 6–8¢/kWh for plants that receive them. These levelized costs are higher than the current average cost of wholesale electricity, but they are likely to be comparable to future costs of electricity from other sources, particularly if fossil-fuel plants are required to store CO_2 or pay a carbon fee. The LCOEs for improvements to existing plants are from one-tenth to one-third those of new plants. The possible LCOEs from advanced plant designs and alternative fuel cycles are highly uncertain at this time. However, these costs are likely to be higher than the LCOEs from current designs using the once-through cycle, although cost advantages from reductions in long-lived high-level waste could offset some of these differences.

Barriers to Deployment

The potential barriers to the deployment of new nuclear plants are several:

- *Economics.* The high cost of new plants, with the resulting financial risk, is the most significant barrier to new deployment. Nuclear power plants have low operating costs per unit of electricity generation, but they incur high capital costs that present a financing challenge for generating companies, particularly given the long lead times for construction and the possibility of expensive delays.
- *Regulatory processes.* The U.S. Nuclear Regulatory Commission (USNRC) is implementing a revised licensing process that allows for reactor design certification, early site permits, and combined construction and operating licenses. Nevertheless, in light of the surge in recent applications, bottlenecks and delays could occur in the near term.
- *Public concerns.* Public opinion about nuclear power has improved in recent years, at least in part because of the safe and reliable perfor-

mance of existing plants, but it would likely become more negative if safety or security problems arose. The absence of a policy decision regarding the disposal of long-lived nuclear wastes, while not technically an impediment to the expansion of nuclear power, is still a public concern.[18] New reactor construction has been barred in 13 states as a result, although several of these states are reconsidering their bans.

- *Shortages of personnel and equipment.* These current shortages could limit construction during the next decade. The market should respond, however, and over time, the shortages should disappear.

Impacts

The impacts of an increased use of nuclear power include the following:

- *Diversity of supply.* Barring a crash program, renewable-energy sources and fossil fuels with CCS are unlikely to be able to provide all of the U.S. electricity demand projected for 2035, even with gains in energy efficiency. Future deployment of nuclear plants would help to ensure a diversity of sources for electric supply—at present, they provide a significant proportion (about 19 percent). Thus, they could serve as an insurance policy for the United States, which would be particularly needed if carbon constraints were applied.

- *Environmental quality.* A major factor in favor of expanding nuclear power is the potential for reduction in greenhouse gas emissions. Avoided CO_2 emissions could reach 150 million tonnes per year by 2020 and 2.4 billion tonnes per year by 2050 under the maximum nuclear power deployment rate discussed in this report.[19] However, an environmental challenge is presented by the disposal of the result-

---

[18]The USNRC previously determined that the used fuel could be safely stored without significant environmental impacts for at least 30 years beyond the licensed life of operation of a reactor, at or away from the reactor site, and that there was reasonable assurance that a disposal site would be available by 2025 (10 CFR 51.23). The USNRC is now revisiting this determination and has proposed to find that used fuel can be stored safely and without significant environmental impacts until a disposal facility can reasonably be expected to be available (73 Fed. Reg. 59,547 [Oct. 9, 2008]).

[19]This calculation assumes that nuclear plants replace traditional baseload coal plants emitting 1000 tonnes of CO_2 equivalent per gigawatt-hour and that nuclear plants emit 24–55 tonnes of CO_2 equivalent per gigawatt-hour on a life-cycle basis.

ing radioactive waste, particularly used fuel. The one site previously envisioned for such disposal—Yucca Mountain, Nevada—would not be ready until after 2020, if at all. And the prospects for the Yucca Mountain repository are substantially diminished by the declared intent of the Obama administration not to pursue this disposal site. Nonetheless, the safe and secure on-site or interim storage of used fuel for many decades—until a location for a permanent disposal location is agreed upon—is technically and economically feasible.

- *Safety and security.* Accidents or terrorist attacks involving nuclear reactors or used fuel storage could result in the release of radioactive material. Measures have been taken in recent years to reduce the likelihood and consequences of such events for existing plants, and evolutionary and advanced designs have features that further enhance safety and security.

- *Adequacy of resources.* The estimated supply of uranium is sufficient to support a doubling of current world nuclear power capacity through the end of this century.

ELECTRICITY TRANSMISSION AND DISTRIBUTION

The U.S. electric power transmission and distribution (T&D) systems—the vital link between generating stations and customers—are in urgent need of expansion and upgrading. Growing loads and aging equipment are stressing the system and increasing the risk of widespread blackouts.

Adding transmission lines and replacing vintage equipment currently in operation would solve this problem. But with an investment only modestly greater, new technology could be incorporated that would have many additional advantages. Among the benefits of modern T&D systems are the following:

- *Superior economics.* By improving the reliability of power delivery, enabling the growth of wholesale power markets, optimizing assets (reducing the need for new generating stations and transmission lines), and providing price signals to customers.

- *Better security.* By improving resilience against major outages and speeding restoration after a system failure.

- *Environmental quality.* In particular by accommodating a large fraction of generation from renewable-energy sources.

Technologies

Technologies used to modernize the T&D systems must be implemented systematically and nationwide, particularly with respect to the transmission system, to achieve maximum benefit. R&D will be important for reducing costs and improving performance, but except in a few cases, breakthroughs are not needed. In fact, most of the technologies already exist and could be deployed now.

Included among these key modernizing technologies are the following:

- *Advanced equipment and components.* Power electronics and high-voltage AC and DC lines offer the potential for long-distance transmission and grid operation that are more efficient. Power electronics both for transmission (Flexible Alternating Current Transmission System—FACTS) and distribution (Custom Power) currently exist and have been deployed in limited applications. Corresponding higher-voltage long-distance lines and substations could be deployed by 2020. High-voltage DC systems can be more economical than AC under some conditions, especially when lines must be underground or underwater, and several DC lines are already in operation. Cost-effective electric storage would be valuable in smoothing power disruptions, preventing cascading blackouts, and accommodating intermittent renewable-energy sources. Some storage technologies (e.g., compressed air energy storage and perhaps advanced batteries) will be ready for deployment before 2020, but significant development is still needed.
- *Measurements, communications, and control.* Modern T&D systems will have the ability to gather, process, and convey data on the state of the system far more effectively than can be done at present. Sampling voltage, frequency, and other important factors many times per second will give operators a much clearer picture of changes in the system and enhance their ability to control it. Most of the necessary technologies are available and have been installed to a limited degree. The communications and control software needed to take full advantage of these technologies could use further development but should be ready by

2020. The costs of installation of the technologies and development of the required software will be significant, however, and the monitoring, sensing, and communications technologies for distribution systems differ from those for transmission systems. Nevertheless, full deployment of modern T&D systems could be achieved by 2030.

- *Improved decision support tools.* The data that a modern grid collects and analyzes can assist operators in deciding when action must be taken, but only if the data are presented in timely and useful forms. During disruptions, split-second decision making may be necessary to prevent cascading failures. Improved decision support tools (IDST) will provide grid visualization to help operators understand the problem and the options available to resolve it. In addition, IDST can strengthen longer-term planning by identifying potential vulnerabilities and solutions. These technologies could be developed by 2020 and continually improved afterward.
- *Integrating technologies.* The technologies discussed in this section can achieve their maximum benefit only through integrated deployment, which poses the primary challenge to creating modern T&D systems. Even though many of these technologies are available now, continued R&D will be important for improvements and cost reductions.

Costs

Modernization and the necessary expansion of T&D systems could be completed in the next 20 years. The total costs are estimated to be about $225 billion for the transmission system and $640 billion for the distribution system. Expansion alone without modernization would cost $175 billion and $470 billion, respectively. Such estimates are complicated by the expansive and interconnected nature of the system and the difficulty of determining development costs, particularly for software.

Barriers to Deployment

Significant barriers hinder the development of modern T&D systems. First, even though most of the necessary technologies are now available, many are expensive and present some performance risk. Second, in the short term it is more costly to develop modern T&D systems than to just expand current systems, and utilities tend to be risk averse; many consumers are more interested in low rates than in reliability of service. And third, legislative and regulatory changes are needed to provide utilities and customers with adequate incentives to invest in modernization. Shortages of trained personnel and equipment could also be a barrier to T&D systems modernization, especially over the near term.

A clear vision for the modern grid is tantamount to providing an environment where utilities, regulators, and the public can understand the benefits and accept the costs, especially as the ownership, management, and regulation of the T&D systems are highly fragmented and collaboration will thus be required. Moreover, investments will be needed in locations and jurisdictions that do not directly benefit—e.g., areas that must be crossed by transmission lines to link generation and load centers. Such a vision would also provide a road map for integrating modernization of the various parts of the enormously complicated transmission system. It might also help expedite the construction of new transmission lines that are now subject to long delays. Clear metrics that measure benefits and progress, as well as the costs of *not* following this path, should be part of the strategy. In contrast, distribution systems can be modernized on a regional level, and some elements, such as smart meters, are appearing already.

Impacts

Modern T&D systems will provide substantial economic benefits by correcting the inefficiency and congestion of the current system and by reducing the number and length of power disruptions. Some estimates are that benefits will exceed costs by four to one. In addition, expanded capacity and improved information flows will raise the efficiency of the electricity markets. Modern T&D systems will be less vulnerable to potential disruptions because of their greater controllability and higher penetration of distributed generation, but the overlay of computer-driven communications and control will make cybersecurity an integral part of modernization. Environmental benefits from modern T&D systems will result from the greater penetration of large-scale intermittent renewable sources and of distributed and self-generation sources; better accommodation of demand-response technologies and electric vehicles; and improved efficiency. Finally, modern systems will be safer because improved monitoring and decision making allow for quicker identification of hazardous conditions, and less maintenance will be required.

REFERENCES

AEP (American Electric Power). 2007. Interstate transmission vision for wind integration. AEP white paper. Columbus, Ohio.

DOE (Department of Energy). 2008. 20 Percent Wind Energy by 2030—Increasing Wind Energy's Contribution to U.S. Electricity Supply. Washington, D.C.: U.S. Department of Energy, Energy Efficiency and Renewable Energy.

EIA (Energy Information Administration). 2008. Annual Energy Outlook 2008. DOE/EIA-0383(2008). Washington, D.C.: U.S. Department of Energy, Energy Information Administration.

EIA. 2009. Annual Energy Outlook 2009. DOE/EIA-0383(2009). Washington, D.C.: U.S. Department of Energy, Energy Information Administration.

JCSP (Joint Coordinated System Plan). 2009. Joint Coordinated System Plan 2008. Available at www.jcspstudy.org.

NAS-NAE-NRC (National Academy of Sciences-National Academy of Engineering-National Research Council). 2009a. Electricity from Renewable Resources: Status, Prospects, and Impediments. Washington, D.C.: The National Academies Press.

NAS-NAE-NRC. 2009b. Liquid Transportation Fuels from Coal and Biomass: Technological Status, Costs, and Environmental Impacts. Washington, D.C.: The National Academies Press.

NAS-NAE-NRC. 2009c. Real Prospects for Energy Efficiency in the United States. Washington, D.C.: The National Academies Press.

NRC (National Research Council). 2004. The Hydrogen Economy: Opportunities, Costs, Barriers, and R&D Needs. Washington, D.C.: The National Academies Press.

NRC. 2008. Transitions to Alternative Transportation Technologies—A Focus on Hydrogen. Washington, D.C.: The National Academies Press.

ANNEX 3.A: METHODS AND ASSUMPTIONS

This annex provides a description of some of the key methods and assumptions that were used to develop the energy supply, savings, and cost estimates made in this report. More detailed explanations of these methods and assumptions can be found in Chapters 4–9 of Part 2.

Energy Supply and Cost Estimates

The methodologies and assumptions used to develop the energy supply and cost estimates in this report are shown in Table 3.A.1. Each row in the table is described in the bulleted list that follows:

- *Reference scenario.* The statement of task for this study (see Box 1.1) called for the development of a reference scenario "that reflects a projection of current economic, technology cost and performance, and policy parameters into the future." This reference scenario is the "base case" for comparison with the AEF Committee's energy savings and supply estimates resulting from the accelerated deployment of technology. The committee adopted the Energy Information Administration's reference case as the reference scenario for this study (see Box 2.1). The reference case for 2007 (EIA, 2008) was used for all but one of the energy supply assessments. The exception was renewable energy, which used the reference case for 2008 (EIA, 2009) because it contained estimates of capital costs for renewable energy technologies that the committee judged to be more realistic than the EIA (2008) estimates.
- *Source of cost estimates* and *models used to obtain estimates* describe the methodologies that were used by the AEF Committee to estimate energy supply costs—either the levelized cost of electricity (LCOE; see Box 2.3) or the costs of liquid fuels. Committee-derived model estimates (i.e., developed by the committee itself or for the committee by consultants) were used for the costs of fossil, nuclear, and alternative liquid fuel technologies. The fossil- and alternative-liquid fuel cost estimates were developed using a common set of models and assumptions (see Box 7.2 in Chapter 7). The nuclear energy cost estimates were developed using a different but comparable set of models and assumptions (see Box 8.4 in Chapter 8). The renewable energy cost estimates were

developed through a critical review of published studies that employed a range of models and assumptions; two examples are shown in the table. The AEF Committee used expert judgment in selecting the estimates from these studies that it considered to be reliable.

- *Cost estimate limitations* are key knowledge gaps and uncertainties that could affect the accuracy of the cost estimates. These limitations arise primarily from technology immaturity or a lack of experience with deploying technologies at commercial scales. One would expect these uncertainties to be reduced as technologies mature and deployment experience is gained.

- *Plant maturity.* The costs of initial deployments of a new technology, sometimes referred to as first plant costs, are generally higher than the costs of deployments of mature proven technologies, sometimes referred to as Nth plant costs. The cost estimates presented in this report reflect the AEF Committee's judgments about the state of technology maturity in 2020. The committee presents first plant cost estimates for immature technologies, Nth plant costs for mature technologies (e.g., pulverized coal plants), and intermediate plant costs for technologies that are still maturing (e.g., IGCC, liquid fuels production). In some cases, cost contingencies were added for immature technologies to bring them closer to Nth plant estimates.

- *Plant size* is the nameplate capacity of the energy supply plant assumed in the cost estimates. The AEF Committee selected plant sizes that it deemed to be typical of each technology class.

- *Plant life* is the time over which the energy supply plant is assumed to generate electricity or liquid fuels. The AEF Committee generally followed industry convention in selecting plant lives for each technology class. In some cases, the plant lives selected were less than the lives of current generating assets (e.g., pulverized coal plants).

- *Feedstock and fuel costs* are the costs for the feedstocks and fuels that are used to produce electricity and liquid fuels. The fuel costs used in this report were selected by the committee based on examinations of historical costs, recent costs, and cost trends. In some cases, ranges of costs were used in the estimates. There are no fuel costs for some renewable energy supplies (e.g., solar and wind).

- *CO_2 prices* represent potential future costs to operators for emitting CO_2 to the atmosphere from energy production. A base-case CO_2 price of $0 per tonne was assumed for all of the energy supply cost estimates presented in this report; prices of $50 and $100 per tonne were also considered in the fossil energy and alternative liquid fuels estimates in order to assess the sensitivity of energy supply costs to CO_2 prices for a future in which climate change is taken seriously.

- *Financing period* is the length of time that capital borrowed for constructing the energy supply plant would be financed. The financing periods used in this report reflect current industry practices, which vary across technology classes.

- *Debt/equity* indicates the ratio of borrowed capital to equity capital in financing the construction of the energy supply plant. The ratios used in this report reflect current industry practices, which vary across technology classes. In some cases ranges were used.

- *Before-tax discount rate* was used to convert future energy supply costs into present values. The ratios used in this report reflect standard industry practice.

- *Overnight costs* represent the present-value costs, paid as a lump sum, for building an energy supply plant. The overnight costs do not include any costs associated with the acquisition of capital, the acquisition of land on which the plant would be built, or site improvements such as new or upgraded transmission equipment. In some cases, overnight costs are given as ranges. For the fossil-fuel estimates, however, 10 percent of the capital costs were added to account for owners' costs.

- *Source of supply estimates* describe the methodologies that were used by the AEF Committee to estimate the supply of electricity and liquid fuels. Many factors can affect deployment rates of a technology beyond its readiness for deployment. Consequently, it was not possible to develop a single methodology for estimating deployment rates for all of the energy supply technologies considered in this report. The committee's estimates of deployment rates were instead based on expert judgment informed by historical rates of technology deployments or by current deployment trends. The supply estimates represent new electricity or liquid fuel supplies and do not account for possible future supply reductions arising from retirements of existing assets.

- *Build time* is the estimated time required to construct a new energy supply plant. This estimate represents actual construction time; it does not include the time required to acquire a site, to design the plant, and to obtain any needed licenses, permits, or other approvals. The build times used in this report reflect current industry practices, which vary across technology classes.

- *Capacity factor* is the ratio (expressed as a percent) of the energy output of a plant over its lifetime to the energy that could be produced by that plant if it was operated at its nameplate capacity. Some capacity factors are expressed as ranges. The capacity factors used in this report reflect current experience and projected future improvements, both of which vary across technology classes.

- *Near-term build-rate limitations* identifies important factors that could limit the rates of plant deployments between 2009 and 2020. These limitations arise from a lack of experience in deploying new technologies (e.g., CCS), bottlenecks in obtaining critical plant components (e.g., large forgings for nuclear plants), and reduced availabilities of other materials and personnel. Most of these bottlenecks are expected to be temporary and should not present major impediments to deployment after 2020.

- *Resource limitations* are factors that could restrict the supply of energy obtained from the deployment of existing and new technologies. These limitations relate mainly to the availability of feedstocks and fuels that are needed to operate the energy supply plants.

TABLE 3.A.1 Sources and Key Assumptions Used to Develop Cost and Energy Supply
Estimates in This Report

	Fossil-Fuel Energy (Chapter 7)	Nuclear Energy (Chapter 8)	Renewable Energy (Chapter 6)
Reference scenario	EIA (2008)	EIA (2008)	EIA (2009)
COST ESTIMATES: SOURCES AND KEY ASSUMPTIONS			
Source of cost estimates	Committee-derived model estimates	Committee-derived model estimates	Critical assessment of the literature[a]
Models used to obtain estimates	NETL (2007) and Princeton Environmental Institute[b]	• Keystone (2007) model for LCOE[c] • Monte Carlo for sensitivity analysis	• NEMS model for EIA (2009) cost estimates • MERGE model for EPRI (2007) cost estimates • Other literature estimates are not model based
Cost estimate limitations	• IGCC, USPC, and CCS technologies are not yet mature and have not been deployed • Geologic storage of CO_2 has not been demonstrated on a commercial scale	Evolutionary nuclear technologies are mature but plants have not yet been deployed in the United States.	Solar technologies are undergoing rapid technological improvements that could bring down future costs.
Plant maturity	• Nth plant for pulverized coal • 3 percent premium on capital costs added for IGCC, PC-CCS, and IGCC-CCS to account for immaturity of technologies • 20 percent premium on CCS capital costs added for CCS 2020 estimates to account for immaturity of technologies	Nth plant	Nth plant
Plant size	500 MW (coal and gas)	1.35 GW, based on weighted average of current plant license applications	Variable

Alternative Transportation Fuels (Chapter 5)

Cellulosic Ethanol	Coal to Liquid	Coal + Biomass to Liquid
EIA (2008)	EIA (2008)	EIA (2008)
Committee-derived model estimates	Committee-derived model estimates	Committee-derived model estimates
See NAS-NAE-NRC (2009), Appendix I	Princeton Environmental Institute[b]	Princeton Environmental Institute[b]
Cellulosic technologies are not yet mature and have not been deployed	Geologic storage of CO_2 has not been demonstrated on a commercial scale	Geologic storage of CO_2 has not been demonstrated on a commercial scale
• Intermediate plant • No capital cost contingency included in estimate for CCS	• Intermediate plant • No capital cost contingency included in estimate for CCS	• Intermediate plant • No capital cost contingency included in estimate for CCS
4,000 bbl/d	50,000 bbl/d	10,000 bbl/d

continued

TABLE 3.A.1 Continued

	Fossil-Fuel Energy (Chapter 7)	Nuclear Energy (Chapter 8)	Renewable Energy (Chapter 6)
Reference scenario	EIA (2008)	EIA (2008)	EIA (2009)
COST ESTIMATES: SOURCES AND KEY ASSUMPTIONS			
Plant life (yr)	20	40	Variable
Feedstock and fuel costs	Coal: $1.71/GJ ($46/tonne) Gas: $6/GJ, $16/GJ	Average: 1.25¢/kWh Range: 0.8–1.7¢/kWh	Biomass: $15–35/MWh Others: $0
CO₂ prices ($/tonne)	0, 50, 100	0	0
Financing period (yr)	20	Average: 40 Range: 30–50	Variable
Debt/equity	55/45	• IPP: Average 60/40 Range: 50/50 to 70/30 • IOU: Average 50/50 Range: 45/55 to 55/45 • Also considered: 80/20 for IPP and IOU with federal loan guarantees	Variable
Before-tax discount rate (percent/yr)	7	• IOU: 6.9 • IPP: 7.7	Variable
Overnight costs (Millions of 2007$/kW) (Millions of 2007$/bbl)	• PC: 1625 • PC+CCS: 2961 • IGCC: 1865 • IGCC+CCS: 2466 • NGCC: 572 • NGCC+CCS: 1209 • –20%/+30% uncertainty	Average: 4500 Range: 3000–6000	• Biopower: 3390 • Traditional geothermal: 1585 • CSP: 2860–4130 • PV: 2547–5185 • Onshore wind: 916–1896 • Offshore wind: 2232–3552

ELECTRICITY OR LIQUID FUELS SUPPLY ESTIMATES: SOURCES AND KEY ASSUMPTIONS

Source of supply estimates	Committee-generated, based on historical build rates of plants in the United States	Committee-generated, based on historical build rates of plants in the United States	Committee-generated, based on an examination of natural resource base and other factors[d]

CO_2

Alternative Transportation Fuels (Chapter 5)

Cellulosic Ethanol	Coal to Liquid	Coal + Biomass to Liquid
EIA (2008)	EIA (2008)	EIA (2008)
20	20	20
$111/tonne dry biomass	$46/tonne coal	$46/tonne coal
		$111/tonne dry biomass
0, 50	0, 50	0, 50
20	20	20
70/30	55/45	55/45
7	7	7
349	4000–5000 (with CCS)	1340 (with CCS)
	(0.08–0.09/bbl per day)	(0.134/bbl per day)
Committee-generated, based partly on corn-ethanol plant build rates in the United States[e]	Committee-generated, based on historical build rates of plants in the United States	Committee-generated, based partly on corn-ethanol plant build rates in the United States[f]

continued

TABLE 3.A.1 Continued

	Fossil-Fuel Energy (Chapter 7)	Nuclear Energy (Chapter 8)	Renewable Energy (Chapter 6)
Reference scenario	EIA (2008)	EIA (2008)	EIA (2009)

ELECTRICITY OR LIQUID FUELS SUPPLY ESTIMATES: SOURCES AND KEY ASSUMPTIONS

	Fossil-Fuel Energy (Chapter 7)	Nuclear Energy (Chapter 8)	Renewable Energy (Chapter 6)
Build time (yr)[g]	3[b]	Average: 5.5 Range: 4–7	• 1–2 for solar and wind • Longer for biopower and hydrothermal
Capacity factor (percent)	85	Average: 90 Range: 75–95	• Biopower: 83–85 • Traditional geothermal: 90 • CSP: 31–65 • PV: 21–32 • Wind: 32.5–52
Near-term build-rate limitations	Learning curve for CCS slows build rate before 2025	Build rates slowed before 2020 by: • Time to acquire license and construct plants • Lack of domestic experience • Potential bottlenecks in obtaining plant components	Barriers to reach 20 percent renewables generation: • Availability of raw materials • Manufacturing capacity • Availability of personnel
Resource limitations	Historical resources limits considered	None	None for wind and solar; limited resource bases for biomass, traditional hydropower, hydrokinetic, and traditional geothermal

Note: CCS = carbon capture and storage; CSP = concentrating solar power (i.e., solar thermal); IGCC = integrated gasification combined cycle; IOU = investor-owned utility; IPP = independent power producer; MERGE = Model for Evaluating Regional and Global Effects [of greenhouse gases]; NEMS = National Energy Modeling System; NGCC = natural gas combined cycle; PC = pulverized coal; PV = photovoltaics; USPC = ultrasupercritical pulverized coal.
[a]The following studies were used to "bookend" the renewable energy cost estimates: ASES (2007), EIA (2008, 2009), EPRI (2007), and NREL (2007).
[b]See Kreutz et al. (2008) and Larson et al. (2008).
[c]This model was run using committee-developed assumptions as described in Chapter 8 in Part 2 of this report.

Alternative Transportation Fuels (Chapter 5)

Cellulosic Ethanol	Coal to Liquid	Coal + Biomass to Liquid
EIA (2008)	EIA (2008)	EIA (2008)
1	3	3
90	90	90
None	None	None
Biomass availability	Coal extraction rates	Biomass availability

[d]These additional factors included manufacturing and materials constraints, employment and capital requirements, and necessary deployment rates. The committee also considered current growth rates of renewables technologies and historical build rates of other types of plants.

[e]The committee assumed twice the capacity achieved for corn grain ethanol.

[f]The committee assumed a build-out rate slightly slower than that for corn grain ethanol because of issues involving accessing sites with about 1.0 million tonnes of biomass per year and a similar availability of coal.

[g]Estimates do not include the time required for permitting and other approvals.

[h]This estimate does not account for differences in complexity of different types of coal and natural gas plants.

Energy Savings and Cost Estimates

The methodologies and assumptions used to develop the energy savings and cost estimates are provided in Table 3.A.2. Each row in the table is described in the following bulleted list:

- *Reference scenarios.* The reference case for 2006 (EIA 2007) was used for the buildings and industrial sector estimates, but these were adjusted in some cases to reflect the 2007 reference case provided in EIA (2008). The transportation estimates were based on a committee-derived, no-change baseline.
- *Source of cost estimates* describes the methodologies that were used to estimate energy savings costs. As shown in the table, these estimates were derived from critical assessments of the literature.
- *Source of savings estimates* describes the methodologies that were used to estimate energy savings. As shown in the table, these estimates were derived from critical assessments of the literature and, for buildings and transportation, committee-derived analyses.
- *Key cost-effectiveness criteria* describes the criteria that were used to determine which energy savings were cost-effective. Different criteria were used in the buildings, transportation, and industrial sectors, as described in the table.
- *Technology lifetimes* are average useful lifetimes of the technologies used to obtain energy savings. These estimates are highly technology specific.
- *Before-tax discount rate* was used to convert future energy supply costs into present values. The ratios used in this report reflect standard industry practice.
- *Other considerations* describe other factors that were considered in developing the energy-savings cost and supply estimates.

TABLE 3.A.2 Sources and Key Assumptions Used to Develop Energy Savings and Cost Estimates

	Buildings Sector	Transportation Sector	Industry Sector[a]
Reference scenario	EIA (2007, 2008)	Developed by committee[b]	EIA (2007, 2008)
Source of cost estimates	Critical assessment of the literature	Critical assessment of the literature	Critical assessment of the literature
Source of savings estimates	Critical assessment of the literature on individual technologies and committee-derived conservation supply-curve analysis	• Critical assessment of the literature on specific technologies • For light-duty vehicles (LDVs), committee-derived illustrative scenario analysis of overall savings in fuel consumption	Critical assessment of the literature on industry-wide savings, industry-specific savings, and savings from specific crosscutting technologies
Key cost-effectiveness criteria	Levelized cost of energy savings is less than the average national electricity and natural gas prices	Recovery of discounted costs of energy savings over the life of the vehicle	Energy savings provide an internal rate of return on investment of at least 10 percent or exceed the company's cost of capital by a risk premium
Technology lifetimes	Technology specific	Average vehicle lifetime	Technology specific
Before-tax discount rate (percent/yr)	7	7	15
Other considerations	Assessment accounts for stock turnover in buildings and equipment	For LDVs, assessment considers how the distribution of specific vehicle types in the new-vehicle fleet affects the on-the-road fleet	Assessment of savings in specific industries used to confirm industry-wide estimates

[a]Manufacturing only.
[b]This is a "no-change" baseline in which, beyond 2020 (when Energy Independence and Security Act targets are met), any efficiency improvements are fully offset by increases in vehicle performance, size, and weight.

References for Annex 3.A

ASES (American Solar Energy Society). 2007. Tracking Climate Change in the U.S.: Potential Carbon Emissions Reductions from Energy Efficiency and Renewable Energy by 2030. Washington, D.C.

EIA (Energy Information Administration). 2007. Annual Energy Outlook 2007. DOE/EIA-0383(2007). Washington, D.C.: U.S. Department of Energy, Energy Information Administration.

EIA. 2008. Annual Energy Outlook 2008. DOE/EIA-0383(2008). Washington, D.C.: U.S. Department of Energy, Energy Information Administration.

EIA. 2009. Annual Energy Outlook 2009. DOE/EIA-0383(2009). Washington, D.C.: U.S. Department of Energy, Energy Information Administration.

EPRI (Electric Power Research Institute). 2007. The Power to Reduce CO_2 Emissions: The Full Portfolio. Palo Alto, Calif.

Keystone Center. 2007. Nuclear Power Joint Fact-Finding. Keystone, Colo.

Kreutz, T.G., E.D. Larson, G. Liu, and R.H. Williams. 2008. Fischer-Tropsch fuels from coal and biomass. In 25th Annual International Pittsburgh Coal Conference. Pittsburgh, Pa.

Larson, E.D., G. Fiorese, G. Liu, R.H. Williams, T.G. Kreutz, and S. Consonni. 2008. Coproduction of synthetic fuels and electricity from coal + biomass with zero carbon emissions: An Illinois case study. In 9th International Conference on Greenhouse Gas Control Technologies. Washington, D.C.

NAS-NAE-NRC (National Academy of Sciences-National Academy of Engineering-National Research Council). 2009. Liquid Transportation Fuels from Coal and Biomass: Technological Status, Costs, and Environmental Impacts. Washington, D.C.: The National Academies Press.

NETL (National Energy Technology Laboratory). 2007. Cost and Performance Baseline for Fossil Energy Plants. DOE/NETL-2007/1281, Revision 1, August. U.S. Department of Energy, National Energy Technology Laboratory.

NREL (National Renewable Energy Laboratory). 2007. Projected Benefits of Federal Energy Efficiency and Renewable Energy Programs. NREL/TP-640-4137. Golden, Colo. March.

PART 2

Part 2 of *America's Energy Future: Technology and Transformation* contains six chapters and supporting annexes that provide detailed assessments of the following energy supply and end-use technologies:

- *Energy efficiency* in the buildings, transportation, and industrial sectors (Chapter 4)
- Production and use of *alternative transportation fuels*, in particular biofuels as well as fuels derived from converting coal, or mixtures of coal and biomass, into liquids (Chapter 5)
- Production of electricity from *renewable energy* sources such as wind, solar, and geothermal, as well as hydropower and biopower (Chapter 6)
- Domestic *fossil-fuel energy,* particularly as coupled with technologies that would capture and safely store CO_2 (Chapter 7)
- Production of electricity from *nuclear energy* (Chapter 8)
- *Electrical transmission and distribution* systems that reliably accommodate intermittent energy supplies such as solar and wind and sophisticated demand-side energy efficiency technologies (Chapter 9).

The chapters on energy efficiency (Chapter 4), alternative transportation fuels (Chapter 5), and renewable energy (Chapter 6) were derived from three National Academies reports that were published as part of the America's Energy Future (AEF) Phase I project:

- *Real Prospects for Energy Efficiency in the United States* (available at www.nap.edu/catalog.php?record_id=12621)
- *Liquid Transportation Fuels from Coal and Biomass: Technological Status, Costs, and Environmental Impacts* (available at www.nap.edu/catalog.php?record_id=12620)
- *Electricity from Renewable Resources: Status, Prospects, and Impediments* (available at www.nap.edu/catalog.php?record_id=12619)

The chapters and supporting annexes in Part 2 of this report provide the AEF Committee's detailed technical assessments of the energy-supply and end-use technologies that it judged were most likely to have meaningful impacts on the U.S. energy system during the three time intervals considered in this study: 2009–2020, 2020–2035, and 2035–2050. The assessments were used to inform the committee's judgments about what *could* happen as a result of accelerated deployments of existing and new technologies. They are not forecasts of what *will* happen, however. As is noted in Chapter 1, the potential energy supply (or savings) and cost estimates presented in this report were developed independently for each class of technologies. The AEF Committee did not conduct an integrated assessment of these technologies to understand, for example, how policies, regulations, and market competition would affect energy savings, supplies, and costs. Predicting the nature and impacts of such policies and regulations on investments in particular energy-supply and end-use technologies and their deployment is well beyond the scope of this study. *Consequently, the estimates provided in these chapters should not be viewed as predictions.*

4 Energy Efficiency

The United States is the world's largest consumer of energy. In 2006, it was responsible for some 20 percent of global primary energy consumption, while its closest competitor, China, used 15 percent (IEA, 2009). But given the energy-security concerns over oil imports, recent volatility in energy prices, and the greenhouse gas emissions associated with energy consumption, using energy more efficiently has become an important priority. Fortunately, the potential for higher energy efficiency[1] is great.

This chapter focuses on the technologies that could increase energy efficiency over the next decade. It describes their state of development, the potential for their use, and their associated performance, costs, and environmental impacts. For these technologies to make a difference, however, they will have to be widely adopted. Hence, the chapter also addresses the sometimes formidable barriers to achieving such market penetration (see Box 4.1 for examples) and the experience that has been gained with policies and programs aimed at overcoming these barriers.

In fact, continued technological advances make energy efficiency a dynamic resource. When new efficient or otherwise advanced technologies reach the market, they hold the potential for reducing the then current level of energy use or moderating its growth. This chapter reviews some of these advanced technologies—some of which could become available and cost-effective in the

---

[1]The terms "energy efficiency" and "energy conservation" are often used interchangeably, but even though both can save energy, they refer to different concepts. Improving energy efficiency involves accomplishing an objective, such as heating a room to a certain temperature, while using less energy. Energy conservation involves doing something differently and can involve lifestyle changes—e.g., lowering the thermostat. This chapter primarily discusses energy efficiency.

BOX 4.1 *Why Energy Efficiency Opportunities Aren't More Attractive to Consumers and Businesses*

Why don't consumers and businesses take greater advantage of cost-effective energy efficiency opportunities? If so much energy can be saved, why doesn't everyone do it, especially when the cost savings over time tend to well outweigh the initial costs?

The answer is complex, as there is no one reason for this seeming behavior gap. Each of this chapter's sector discussions, as well as the policy discussion at the end of the chapter, identify factors—commonly called barriers—that impede the full uptake of energy efficiency technologies and measures. They fall into several categories, but the following examples illustrate how some of them affect decisions:

- Cost savings may not be the only factor influencing a decision to invest in an energy efficiency measure. For example, consumers purchase vehicles based on many factors, such as size, performance, and interior space, in addition to fuel economy. In reality, fuel economy may not come into the picture at all.
- Although energy and cost savings might be achievable with only a low first cost (investment), such savings may be a small-enough part of the family or company budget that they are not really relevant to economic decisions.
- The up-front financial investment might be small, but substantial investments of time and effort may be required to find and study information about potential energy-saving technologies, measures, and actions.
- It is well established that purchasers tend to focus much more on first costs than on life-cycle costs when making investments. This behavior is no different when it comes to energy efficiency. There is also the phenomenon of risk aversion—new products may be unfamiliar or not work as expected. The default behavior is often simply the status quo. Knowing this, producers may never design and develop energy-efficient products.
- Some of the behavior gap can be attributed to economic structural issues. For example, landlords of rental residential buildings are not motivated to pay for

---

2020–3035 timeframe and beyond—and the research and development (R&D) needed to support their development.

ENERGY USE IN THE UNITED STATES AND THE POTENTIAL FOR IMPROVED ENERGY EFFICIENCY

In 2008, the United States used 99.4 quadrillion Btu (quads) of primary energy (see Figure 4.1). About 31 percent of this total was consumed in industry,

technologies that are more efficient when their tenants pay the utility bills. And builders whose incentive is to minimize the cost of new homes may not offer highly efficient appliances that increase purchase prices but save buyers money over time.

- Other factors may involve retailers of equipment and appliances. If there is low demand for efficient products, retailers may not stock them. Even purchasers who might be motivated to search elsewhere for an efficient product may have to deal with limited choices in the event of an emergency purchase, such as when a refrigerator fails.
- Other reasons for the behavior gap are the subject of much social science research. They involve factors such as habits in purchasing or use, which can be very difficult to change. Some apparent consumer preferences—typically learned from parents, neighbors, and friends—may change very slowly, if at all.
- Energy-savings investments by businesses and industries are not always seen as beneficial. If energy accounts for only a small part of total costs, or if the available capital is limited, other investments may be preferred—e.g., in reducing other costs, improving products, or developing new ones. If the consequences of a new-product or production-method failure are large, this in itself can maintain the status quo.
- Firms may not be aware of the potential savings achievable by replacing equipment, such as older motors, with more efficient or variable-speed versions. When motors, large or small, are used throughout a facility, the savings from upgrading them can be substantial.
- Energy efficiency investments by companies are made in the context of complex business cultures. "Champions," or commitment at the highest levels, may be required.

More details on how barriers such as these play out in the buildings, transportation, and industrial sectors are given later in this chapter.

28 percent in transportation activities, and about 41 percent in the myriad activities and services associated with residential and commercial buildings. Figure 4.2 provides more detail, breaking out energy consumption by source and sector and also defining "primary" energy.

Energy use in the United States has grown steadily since 1949, with the exception of a dip in the mid-1970s during the oil crisis. Energy consumption today is double what it was in 1963 and 40 percent higher than it was in 1975 (the low point following the oil crisis). But there has also been progress in increasing the efficiency of energy use. The nation's energy use per dollar of gross domes-

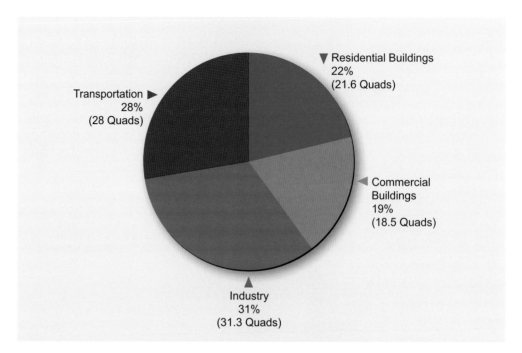

FIGURE 4.1 *Total U.S. energy use by sector, 2008 (quadrillion Btu, or quads).*
Notes: For each sector, "total energy use" is direct (primary) fuel use plus purchased elec-
tricity plus apportioned electricity-system losses. Economy wide, total U.S. primary energy
use in 2008 was 99.4 quads.
Source: EIA, 2009a, as updated by EIA, 2009b.

tic product (GDP) has been cut in half since 1973, with about 70 percent of that
decline resulting from improvements in energy efficiency (IEA, 2004). Nevertheless, the absolute amount of energy used continues to rise.

Yet the potential for higher energy efficiency is large, as illustrated by two
points. First, despite the impressive gains made by the United States over the last
30 years, almost all other developed nations use less energy per capita and less
energy per dollar of GDP (see Table 4.1 and Figure 1.5 in Chapter 1). Denmark's
levels of usage, for example, are about half on both measures. While there are
structural variations that account for part of this gap, some 50 percent of it results
from differences in energy efficiency (Weber, 2009).

The second point is that a greater number of energy-efficient and cost-
effective technologies are available today to supply such services as lighting, heating, cooling, refrigeration, transport, and computing—all of which are needed
throughout the economy and constitute the underlying driver of the demand
for energy. Hundreds of realistic and demonstrated technologies, some already

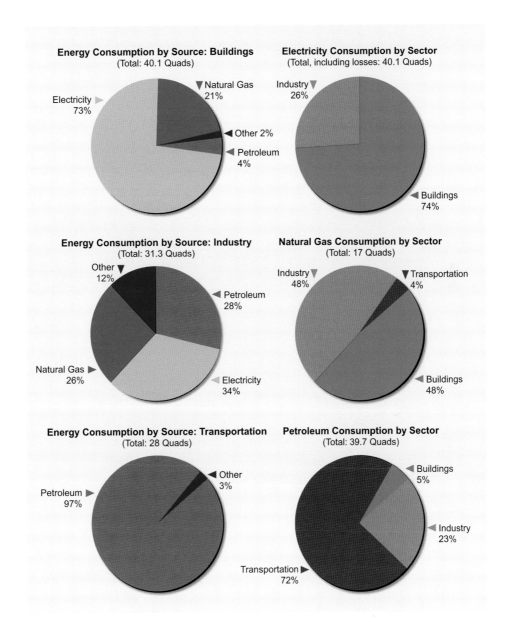

FIGURE 4.2 *U.S. energy consumption by source and end-use sector, 2008 (quads).
Note: Does not include consumption in the electric power sector. Electricity includes
delivered electricity as well as the allocated losses incurred in the generation, transmission, and distribution of electricity. "Delivered" energy refers to the electricity delivered
to a site plus the fuels used directly on site (e.g., natural gas for heating water). This
measure does not account for the losses incurred in generating, transmitting, and distributing the electricity. Delivered energy plus these losses is referred to as "primary" energy.
Source: EIA, 2009a, as updated by EIA, 2009b.*

TABLE 4.1 Energy Use in 2006, per Capita and per Dollar of GDP (2000 Dollars)

	Million Btu per Person	Btu per Dollar of GDP
Denmark	161	4971
Germany	178	7260
Japan	179	4467
France	181	7767
United States	335	8841

Source: DOE, 2006b.

commercial and some just beginning to enter the market, can provide these services more efficiently than is the case today, and they can collectively save large amounts of energy.

This chapter documents the AEF Committee's review of the performance, costs, and environmental impacts—primarily greenhouse gas emissions[2]—of energy-efficient technologies and processes that are currently ready for implementation; technologies that need some further development; and scientific concepts that promise major efficiency improvements in the future. The assessment followed the traditional organization of energy use into three sectors: buildings (both residential and commercial), transportation, and industry. Further, each was considered over three timeframes—the present to 2020, 2020 to 2035, and 2035 to 2050. The first period received major attention because so many cost-effective technologies are ready for implementation today or will be ready within a few years.

The committee examined the available energy efficiency literature and performed additional analyses with primary data. The committee was able to estimate energy efficiency supply curves for electricity and natural gas in the residential and commercial sectors, showing the amount of energy that could be saved over a range of costs. In the transportation sector, the committee focused on alternative technologies that could power the nation's cars and light trucks. By estimating the costs and energy savings associated with each technology as R&D improved

---

[2]Although greenhouse gas emissions are the primary environmental impact considered here, it should be noted that the evaluation of a specific application of a technology or measure should consider any other effects, including local effects, on the environment and natural resources.

it over time, and the timeframes in which specific technologies can be expected to penetrate the market, the committee was able to develop illustrative scenarios of how total energy consumption could evolve. Confronted with myriad, diverse manufacturing industries, the committee focused on the five most energy-intensive industries. The committee examined other technologies, although in less detail.

For each sector, comparisons were made to a baseline, or business-as-usual, case in order to derive the potential for energy savings. For the buildings and industrial sectors, this was the reference-case scenario of the U.S. Energy Information Administration's (EIA's) Annual Energy Outlook 2007 or 2008 (EIA, 2007a, 2008161). For the transportation sector, a committee-directed baseline was derived. In all cases, though, the study estimates the level of energy-efficiency improvement beyond the baseline or reference case. More details can be found in the report titled *Real Prospects for Energy Efficiency in the United States* (NAS-NAE-NRC, 2009).

ENERGY EFFICIENCY IN RESIDENTIAL AND COMMERCIAL BUILDINGS

Energy Use in U.S. Buildings

In 2006, the United States had approximately 81 million single-family homes, 25 million multifamily housing units, 7 million mobile homes, and 75 billion square feet of floor space contained within 5 million commercial buildings (EIA, 2008). The building stock is long-lived; homes last 100 years or more, commercial buildings often last 50 years or more, and appliances used in buildings last 10 to 20 years. In 2008, residential and commercial buildings accounted for 73 percent of total electricity use in the United States and 40 percent of total primary energy use (Figures 4.1 and 4.2).

Use of delivered energy in the residential sector increased by 15 percent from 1975 to 2005, and in the commercial sector it grew by 50 percent. Meanwhile, primary energy grew by 46 percent and 90 percent, respectively, in the residential and commercial sectors. Despite these increases, energy "intensity"—energy use per unit of service or activity—decreased over that time span.

In the residential sector, on-site energy intensity, measured as energy use per household, fell by about 33 percent during 1978–2001, while primary energy use per household declined by 20 percent. In the commercial sector, on-site energy intensity, measured as energy use per square foot of floor area, dropped by about

20 percent during 1979–2003, while primary energy use per square foot decreased by 6 percent. The difference between on-site and primary-energy-use growth rates was due to growing electrification, which engendered sizable generating, transmitting, and distributing losses.

Factors that have affected energy use in buildings over the last several decades include increased electrification, population shifts to milder climates, growing penetration of appliances and electronics, larger home sizes, smaller households, growing household incomes, and dramatic improvements in the energy efficiency of appliances and other equipment. The last item is a key factor in the decline in energy intensity of buildings over the past 30 years. For example, the average electricity use of new refrigerators sold in 2007 was 71 percent less than that of new refrigerators sold in 1977 (AHAM, 2008), despite their becoming larger and having more features.

Significant energy efficiency gains have also been made in lighting. Sales and use of compact fluorescent lamps, which consume about 75 percent less electricity per unit of light output than incandescent lamps consume, have greatly increased in the past decade. In commercial buildings, energy-efficient fluorescent lighting fixtures containing T8 fluorescent lamps and high-frequency electronic lamp ballasts use 15–30 percent less energy per unit of light output than do older fixtures with T12 lamps and electromagnetic ballasts. These devices also have been used increasingly in recent years, as periodic surveys by the EIA attest. However, a large fraction of commercial buildings still have not embraced common energy efficiency measures such as energy management and control systems.

The adoption of ENERGY STAR®-labeled products has also grown substantially in recent years. For example, the construction and certification of ENERGY STAR® new homes increased from about 57,000 in 2001 to 189,000 in 2006, or 11.4 percent of all new homes built that year.

Energy Efficiency Improvement in Buildings

Many studies, whether on the local, regional, national, or global levels, have estimated the potential for improved energy efficiency in buildings.[3] For the most part, these efforts evaluate the quantity of savings that could be realistically

---

[3]Citations to these studies are given in NAS-NAE-NRC (2009).

achieved as a function of the cost of the saved energy, and they generally show consistent findings despite differences in assumptions and approaches.

Across the two building sectors, the studies demonstrate a median technical potential for improved energy efficiency of 33 percent for electricity (32 percent for the residential sector and 36 percent for the commercial sector) and 40 percent for natural gas (48 percent for residential and 20 percent for commercial), after accounting for energy prices and implementation barriers. The median cost-effective and achievable potential is 24 percent for electricity (26 percent for residential and 22 percent for commercial). For natural gas, this measure is 9 percent (9 percent for residential and 8 percent for commercial), but it could increase considerably as gas prices rise or could decrease as gas prices fall.

These studies have limitations, however, and care must be taken in their use. The question, How much efficiency is available at what price? is not well framed because "available" is ambiguous for several reasons. Among them are the timeframe over which the potential applies, the level of incentive required, and the motivation of society. In addition, the studies can underestimate the potential because of biases that might, among other things, exclude new and emerging technologies, hold technology static, or fail to consider nonenergy benefits. Conversely, the studies may overestimate savings by being excessively optimistic about energy efficiency potential.

Nevertheless, the potential for cost-effective energy efficiency improvements in buildings is large. And the prospects for savings will grow as new technologies become available, existing technologies are refined, and energy efficiency measures begin to be implemented in an integrated manner—often, with synergistic effects (such as those that can result from a whole-building approach to building design).

Approaches to Understanding Efficiency Potential

Analysts have developed a variety of ways to investigate the technologies and design principles that could make buildings more efficient. The two most important are the integrated approach and the technology-by-technology approach.

Integrated Approach

An integrated (also known as a whole-building or system-wide) approach to improving energy efficiency considers the energy consumption, and the set of improvements that could save energy, for entire buildings. It accounts for the ability to reduce energy use through design considerations—such as incorporation of day lighting or reorientation or strategic placement of equipment to reduce heating and cooling loads—as well as through high-efficiency systems and equipment.

For residential buildings, a whole-house approach using a cost-effectiveness criterion can result in savings of 50 percent or more in heating and cooling and 30–40 percent reductions in total energy use. This conclusion is supported by the fact that more than 8,000 single-family households applied for the federal tax credit for 50 percent savings during its first year of availability. For 2008, the number of qualifying homes grew to more than 23,000, about 4.6 percent of all homes built. There are examples in Europe of new residences that have achieved even lower levels of energy consumption.[4]

For commercial buildings, several studies have reviewed the small but growing number of structures that have achieved 50 percent reductions in the energy needed for heating, cooling, and water heating. Most of these buildings have relied on

- High-efficiency electrical lighting systems, which use state-of-the-art lamps, ballasts, and luminaires (complete lighting fixtures)
- Luminaires chosen to provide the desired amount of lighting in the right places, coupled with the use of natural day lighting and associated controls that limit electrical lighting correspondingly
- Fenestration (window) systems that reduce heat gains while providing daylight
- Heating, ventilation, and air-conditioning controls that provide effective operation of the system during part-load conditions.

A few low-energy buildings have also made use of such on-site generation options as combined heat and power (CHP) systems[5] or solar photovoltaic (PV) systems.[6]

This whole-building approach is usually applied to new buildings, but in some cases it can be used to identify the potential for system-wide savings in existing buildings.

---

[4]See, for example, www.businessweek.com/globalbiz/content/apr2007/gb20070413_167016.htm.

[5]Combined heat and power (CHP) units transform a fuel (generally natural gas) into electricity and then use the remaining heat for applications such as space and hot-water heating or industrial and commercial processes.

[6]See the "Getting to Fifty" website, www.newbuildings.org/gtf.

End-Use and Technology Approach

Some integrated approaches—for example, strategic placement of ductwork—are most easily applied to new buildings. A second approach, useful for existing buildings, relies on the one-by-one review of major categories of energy use and consideration of the types of efficiency measures and technologies that could be applied to them. For example, efficiency in the provision of space heating and cooling could be raised by upgrading furnaces, using variable-speed motors, reducing leakage, increasing insulation, and applying other measures, most of which could be incorporated into existing buildings.

This technology-by-technology approach can be carried out on as detailed or disaggregated a level as desired. A drawback is that it misses the kinds of integrated measures that can be identified with the whole-building approach.

Potential for Efficiency Improvement: Conservation Supply Curves

Developing conservation supply curves, which have been used widely in analyses of energy use in buildings to display the results of technology-by-technology or measure-by-measure assessments, involves evaluating a comprehensive list of measures that could be taken and ranking them in order of the cost of conserved energy (CCE).[7] Each measure is evaluated not in isolation but in the context of the measures that have already been taken. Most of the studies reviewed for this report relied on the technology-by-technology approach to develop supply curves for both residential and commercial buildings. To reconcile the results across studies, this report integrates and updates these data to produce new conservation supply curves that can be applied at the national level.

The reference-case scenario of the EIA's Annual Energy Outlook 2007 (EIA, 2007a) is used as the baseline for this analysis,[8] which mostly involves technolo-

---

[7]As explained at the beginning of this chapter, the terms "energy efficiency" and "energy conservation" are often used interchangeably, but even though both can save energy, they refer to different concepts. This chapter discusses energy efficiency. However, the traditional term for a graph of the amount of energy that can be saved through energy efficiency measures at different prices is "conservation supply curve." The cost of these measures has traditionally been referred to as the cost of conserved energy. The traditional terminology has been retained in this section, but the fact that the curves refer to energy efficiency improvements should be kept in mind.

[8]The reference case of the Annual Energy Outlook 2008, which is used in some other parts of this report, has slightly different assumptions from those in the AEO 2007 reference case (e.g., slower growth in the housing stock). But because of other factors embedded in the assessment

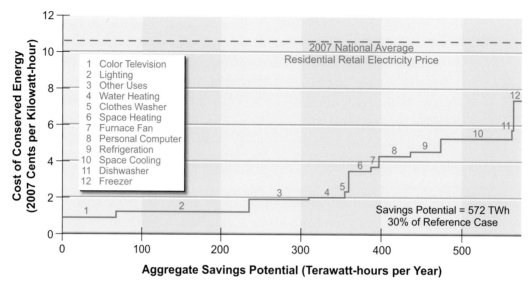

FIGURE 4.3 *Residential electricity savings potential, 2030.*
Source: Brown et al., 2008.

gies that are widely available in the market and well proven. Embedded in the scenario is the assumption that energy efficiency will improve over time in response to market forces as well as to codes and standards. Under these conditions, residential electricity use is projected to increase by an average of 1.4 percent per year, and commercial electricity use by 1.9 percent per year, over the period from 2006 to 2030. The potential for energy savings that is computed here is above and beyond that embodied in the baseline used for the analysis.

Figures 4.3 through 4.6 are the supply curves developed for this study. They illustrate the potential for energy efficiency improvements over the period 2010–2030 in the residential and commercial sectors for electricity and natural gas. The x-axis shows the total reduction in 2030 energy consumption, while the y-axis shows the CCE in fuel-specific units. Each step on the curve represents the total savings for a given end-use for all the cost-effective efficiency measures analyzed to that point. These plots are referred to as *supply* curves because they indicate how much energy savings is available for a given cost, with the CCE calculated as

here, and in the AEO 2008, the overall findings do not change. More details can be found in NAS-NAE-NRC (2009).

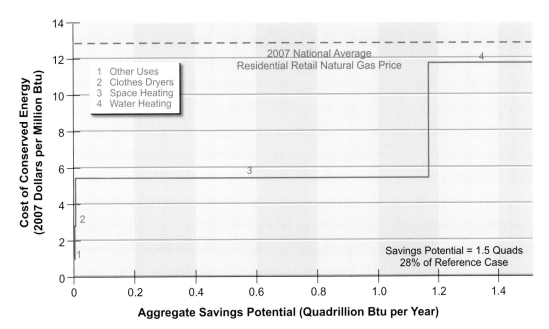

FIGURE 4.4 *Residential natural gas savings potential, 2030.*
Source: Brown et al., 2008.

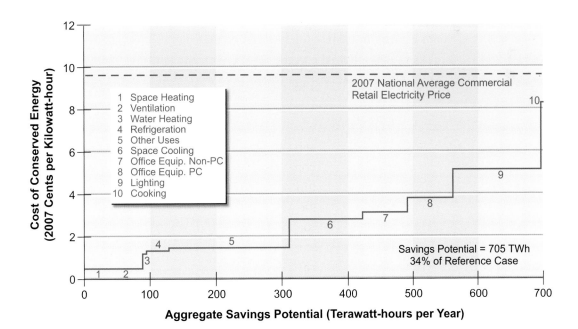

FIGURE 4.5 *Commercial electricity savings potential, 2030.*
Source: Brown et al., 2008.

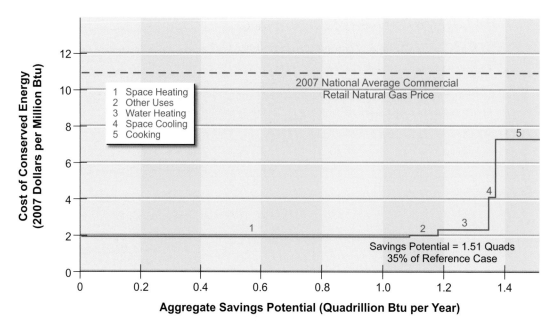

FIGURE 4.6 *Commercial natural gas savings potential, 2030.*
Source: Brown et al., 2008.

the weighted average of savings for all the measures in that end-use cluster. (The CCE here is given in terms of delivered energy.)

Table 4.2 compares the average CCE with the national average retail energy price. The analysis indicates that the projected baseline energy use in 2030 can be reduced by about 30–35 percent at a cost less than current average retail energy prices. That the results show an average CCE well below the retail energy prices in all areas means that adopting efficiency measures is very cost-effective for households and businesses: the average CCE for electricity-savings measures is only about one-quarter of the average retail electricity price. Of course, factors such as local energy prices and weather will influence cost-effectiveness in any particular location.

For the entire buildings sector, the cumulative capital investment needed between 2010 and 2030 is estimated to be about $440 billion to achieve annual energy-bill savings in 2030 of nearly $170 billion. These savings result in an average simple payback period of 2½ years, or savings over the life of the measures that are nearly 3.5 times larger than the investment required. (This analysis considers only the cost [or incremental cost] of the efficiency measures to purchasers; it does not include the costs of policies and programs aimed at increasing the adoption of the measures.)

TABLE 4.2 Comparison of Average Cost of Conserved Energy
in Residential and Commercial Buildings with National Average
Retail Energy Price, 2007

Sector and Energy Type	Average Cost of Conserved Energy	National Average Retail Energy Price
Residential buildings		
Electricity	2.7¢/kWh	10.6¢/kWh
Natural gas	$6.9/million Btu	$12.7/million Btu
Commercial buildings		
Electricity	2.7¢/kWh	9.7¢/kWh
Natural gas	$2.5/million Btu	$11.0/million Btu

Note: Estimates are in 2007 dollars.
Source: Brown et al., 2008.

Advanced Technologies and Integrated Approaches

Advanced technologies for buildings promise large additional increases in efficiency. Most are available today but await further development and cost reduction. They include discrete technologies such as solid-state lighting, advanced windows, and high-efficiency air-conditioning equipment as well as the full integration of technologies into new and highly efficient buildings, both residential and commercial. These technologies demonstrate that energy efficiency is a dynamic resource—new and improved alternatives now under development will reach the marketplace in the future, thereby increasing the potential for energy efficiency and energy savings. (This review does not include many promising advanced technologies related to building materials, design, and appliances.)

Solid-State Lighting

Unlike compact fluorescent lights (CFLs), light-emitting diodes (LEDs) are point light sources that require no warm-up period, do not contain mercury, and are readily dimmable. They are increasingly available today. The best white LEDs are now more efficient than fluorescent lamps are. White LEDs are expected to reach 150 lumens per watt, more than twice the efficiency of CFLs (Craford, 2008).

Cost is the primary issue with LEDs, but it is decreasing rapidly, and this trend is expected to continue. For example, a 1000-lumen LED source that costs around $25 (wholesale) today is projected by the U.S. Department of Energy (DOE) to cost $2 in 2015 (DOE, 2008).

Highly Integrated Cooling Systems

Cooling is one of the largest users of energy in residential and commercial buildings, responsible for about 10 percent of total U.S. electricity consumption and 25–30 percent of total peak electricity demand (DOE, 2007b; Koomey and Brown, 2002). Cooling demand could be reduced or eliminated entirely in some climates by combining existing measures—such as highly efficient building envelopes, shading, reflective surfaces and roofs, natural ventilation, and thermal storage—with emerging technologies that are designed to supplement or replace vapor compression-based cooling with low-energy, thermally driven cooling approaches. These technologies include indirect and indirect-direct evaporative cooling, solar-thermal cooling systems, and thermally activated desiccants. Each option has already been used commercially as an individual component, but further R&D and commercial demonstration projects are needed to develop the technologies as integrated systems boasting optimized performance.

Technologies to Reduce Energy Consumption in Home Electronics

Consumer products dealing with information processing are responsible for about 13 percent of residential electricity use (Roth and McKenney, 2007; EIA, 2008). Numerous efficiency improvements have already been incorporated into them, and many observe ENERGY STAR® specifications, but energy use continues to increase in a few important products such as flat-panel televisions and set-top boxes. Strategies to reduce the energy used by consumer electronics include improving power supplies; designing smaller and more efficient circuitry; incorporating automatic power-down features, allowing products on a network to operate at a low-power sleep level without losing network connectivity; and designing "power strips" to manage energy consumption in clusters of products. These strategies are already evident in a few products but have had minimal impact on energy use to date. The efficiency of the products themselves can be expected to increase as the devices become ever smaller, and with the development of new materials and other advances that can reduce energy demands.

Advanced Window Technologies

Windows are responsible for about 2.7 quads of energy use annually in homes and about 1.5 quads in the commercial sector—primarily from the losses through energy leakage that they allow—and they affect another 1.0 quad of potential lighting energy savings (Apte and Arasteh, 2006). Advances have largely been made by reducing the heat-transfer coefficient (U-value) of windows through the use of low-emissivity (low-E) coatings and by reducing the solar-heat-gain coefficient (SHGC) via spectrally selective low-E coatings. The U-value is the primary determinant of winter heat loss; the SHGC is the primary determinant of summer cooling loads.

Two new window-technology advances, now available in niche markets, could have far-reaching implications if they became mainstream products and systems. The first is highly insulating "superwindows" that achieve U-values in the range of 0.1–0.2, as compared with a typical U-value of 0.5 for double-glazing and 0.35–0.4 for the ENERGY STAR® windows currently being sold in cold climates. The second advance is a new generation of dynamic products that reduce cooling loads and electric illumination when daylighting is available in commercial buildings. The full penetration of these advanced technologies into the building stock, which could take decades, might shift the role of windows in buildings to being approximately "energy neutral."

Low-Energy and Zero-Net-Energy New Homes

It is possible to construct homes that combine high levels of energy efficiency—in the building envelope, heating and cooling systems, and appliances—with passive and active solar features in order to approach zero-net-energy consumption.[9]

The whole-building approach described earlier is being used by the DOE to reach a zero-net-energy consumption goal. For example, two highly instrumented homes were built with the same floor plan in Lakeland, Florida, in 1998 (Parker et al., 2000). Over 1 year, the control home used 22,600 kWh of electricity. During that same year, the experimental home used only 6,960 kWh, a reduction of 70 percent. The experimental home also had solar PV production of 6,180 kWh. When this production is accounted for, the experimental home's net energy use was only 780 kWh, about a 97 percent reduction relative to the control home.[10]

---

[9]A home with zero-net-energy consumption may at times produce more energy than it consumes (for example, through PV panels on the roof) and at other times consume more energy than it produces.

[10]The full cost of providing backup power in the electrical supply system for those times when the PV array is not generating electricity would need to be included in any analysis of the overall systems cost of such low-energy homes.

Low-Energy New Commercial Buildings

The best-performing commercial buildings in the country achieve energy-use reductions of 50 percent or more below standard practice by successfully integrating multiple "state-of-the-shelf" technologies[11] (Turner and Frankel, 2008).[12] This represents a huge opportunity for improved energy performance using existing and available technologies (Griffith et al., 2007). To achieve such high performance requires that integration, interaction, quality control, and monitoring be employed throughout the design, construction, and operation of the building.

Barriers to Improving Energy Efficiency in Buildings

Numerous barriers hinder the adoption of energy-efficient technologies in buildings. They vary in their relevance to particular situations and in the difficulty of their being overcome. Many of these barriers also apply to other sectors.

Environmental and social costs are usually not reflected in the price of energy. The price does include costs associated with meeting environmental standards, but other adverse environmental impacts, such as emissions of mercury or carbon dioxide, land disruption, and legal water contamination, are not factored in. Also, the costs incurred by society in defending sources and protecting shipments of oil and other energy imports are not included in energy prices. As a result, more fossil energy is used than would otherwise be the case.

Various types of *fiscal policies* may discourage investment in energy efficiency. For example, capital investments in commercial buildings must be depreciated over more than 30 years, while energy purchases can be fully deducted from taxable income the year they occur (Brown, 2001).

Some *regulatory policies* also discourage investment in energy efficiency. In particular, policies that allow public utilities to increase their profits by selling more electricity or natural gas are disincentives to effective utility energy efficiency programs (Carter, 2001). Many utilities also have applied tariffs and interconnection standards that discourage end users from adopting energy-efficient CHP

---

[11]"State-of-the-shelf" technologies represent the state-of-the-art selection of technologies that are widely available (on the shelf) today.

[12]See also www.gettingtofifty.org for a searchable database of information about projects with energy performance targets that are 50 percent beyond the ASHRAE (American Society of Heating, Refrigerating and Air-Conditioning Engineers) standard 90.1-2001 (NBI, 2008).

systems (Brooks et al., 2006a,b). Variabilities in the stringency and enforcement of building energy codes across states and localities constitute another barrier to energy efficiency in buildings (Brown et al., 2007).

Misplaced incentives, also known as split incentives or principal-agent problems, exist in numerous situations. The most visible example is in rental markets, where building owners are responsible for investment decisions but tenants pay the energy bills. Studies have revealed lower levels of energy efficiency in U.S. dwellings occupied by renters compared to those occupied by owners.

Misplaced incentives are found in new-construction markets as well, where decisions about building design and features are also made by people who are not responsible for paying the energy bills. Architects, builders, and contractors have an incentive to minimize first cost in order to win bids and maximize their profits (Koomey, 1990; Brown et al., 2007). Moreover, commercial leases are often structured so that the landlord allocates energy costs to tenants based on the amount of square footage leased rather than on the amount of energy used (Lovins, 1992).

It can take many years to *inform and educate* a large majority of households and businesses about energy efficiency options. For example, after nearly 8 years of active promotion and incentives for the use of compact fluorescent lamps, nearly one-third of households surveyed in the Pacific Northwest in late 2004 were still unaware of them (Rasmussen et al., 2005).

Lack of information is an even greater problem and a harder one to fix for individual end-uses. For example, when a tenant of a commercial building buys office equipment, its electricity usage is not individually metered. And not a single end-use in homes is ever metered separately. Householders have no direct information as to whether their computer or video game box or hair dryer is a big energy user or a trivial one.

Businesses tend to pay limited attention to energy use and energy-savings opportunities if energy costs are a small fraction of the total cost of owning or operating the business or factory—or if energy efficiency is not viewed as a priority by company management. As a result, many businesses limit energy efficiency investments to projects with payback periods of no more than 2 or 3 years (DeCanio, 1993; Geller, 2003).

Some highly efficient appliances or other energy efficiency measures are relatively new and still not widely available in the marketplace or not well supported by product providers (Hall et al., 2005). And some very effective energy efficiency services, such as duct testing and sealing, and recommissioning of existing buildings, are not widely available in many parts of the country.[13]

Many households have *limited resources and limited access to credit*, which restricts their ability to invest in energy efficiency measures. In addition, some businesses (particularly small ones) have insufficient capital or borrowing ability.

Drivers for Improving Energy Efficiency in Buildings

Numerous factors—rising fuel and electricity costs, growing environmental awareness, increasing interest by consumers in cutting greenhouse gas emissions, the expanding number of "green" buildings, and corporate environmental initiatives—can help to overcome the barriers just described. Likewise, many energy efficiency measures provide nonenergy benefits that encourage their adoption. In addition, public policies—including building energy codes, appliance efficiency standards, and state and utility efficiency programs—are stimulating greater adoption of efficiency measures. See the section titled "Energy Efficiency Policies and Programs: Experience and Lessons Learned" (later in this chapter) for a review of experiences with some of these policies and programs.

Findings: Buildings

Studies taking several different approaches are consistent in finding the potential for large, cost-effective energy savings in buildings. Median predictions of achievable and cost-effective savings are 1.2 percent per year for electricity and 0.5 percent per year for natural gas, amounting to a 25–30 percent energy savings for the buildings sector as a whole over 20–25 years. The committee's analysis suggests that baseline energy use can be reduced by 30–35 percent by 2030 at a cost less than current retail energy prices. If these savings were to be achieved, they would hold energy use in this sector about constant, in contrast to the current trend of continuing growth.

The full deployment of cost-effective energy-efficient technologies in buildings alone could eliminate the need to add to electricity generation capacity. Since the estimated electricity savings in buildings exceeds the EIA forecast for new net generation, implementing these efficiency measures would mean that no new

---

[13]It should be noted, however, that some energy efficiency measures, such as insulation, compact fluorescent lamps, or ENERGY STAR® appliances, are readily available.

generation would be required except to address regional supply imbalances, to replace obsolete generation assets, or to substitute more environmentally benign generation sources. This conclusion assumes that electricity demand does not grow faster than the EIA forecast or that new demand does not require more generation capacity (e.g., electric vehicles are charged at night).

Studies of energy efficiency potential are subject to a number of limitations and biases. Factors such as not accounting for new and emerging energy efficiency technologies can lead these studies to underestimate energy-savings potential, particularly in the midterm and the long term. On the other hand, some previous studies have been overly optimistic about the cost and performance of certain efficiency measures, thereby overestimating energy-savings potential, particularly in the short term.

Many advanced technologies, including LED lights, innovative window systems, new types of integrated cooling systems, and power-saving electronic devices, are either commercially available already or likely to become available within the next decade. Their availability will further increase the energy-savings potential in buildings. In addition, new homes and commercial buildings with low overall energy use have been demonstrated throughout the country. With appropriate policies and programs, they could become the norm in new construction.

There are substantial barriers to widespread energy efficiency improvements in buildings, but a number of factors are counteracting these barriers. Drivers of increased energy efficiency include rising energy prices; growing concern about global climate change and the resulting willingness of consumers and businesses to take action to reduce emissions; more consumers moving toward "green buildings"; and growing recognition of the significant nonenergy benefits offered by energy efficiency measures.

ENERGY EFFICIENCY IN TRANSPORTATION

Energy Use in the U.S. Transportation Sector

The U.S. transportation sector is varied and complex, consisting of vast networks of land, air, and marine vehicles that are owned and operated by combinations of public and private entities. As a whole, the sector's activities use about 28 percent of the nation's primary energy supply (see Figure 4.1), including more than 70 percent of all the petroleum. U.S. transportation is almost entirely dependent on petroleum, 56 percent of which was imported in 2008. Transportation also has environmental impacts. For example, it is responsible for about a third of all U.S. CO_2 emissions arising from energy use, as well as for significant fractions of other air pollutants.

Passenger transport is dominated by personal automobiles and aviation.[14] (Mass transit and scheduled intercity rail and bus services play important roles in some regions, but overall, they account for a modest proportion of total passenger-miles.) On the *freight* side, trucking dominates both with respect to tons and value of shipments.[15] Thus, highway travel, for passengers and freight alike, is the preponderant mode of transportation in the United States, accounting for about 75 percent of all transportation energy use. Consequently, efficiency gains in highway vehicles will have the greatest effect on the transportation sector's total consumption of energy.[16]

The motivators for energy efficiency in commercial transportation differ from those for private transportation. Lifetime operating cost, and thus energy efficiency, is important to companies supplying passenger and freight transportation. The commercial transportation sector is so highly competitive that even small cost differentials among firms can have major impacts on their profitability and growth. In contrast, consumer purchases of motor vehicles are influenced by many factors, including vehicle comfort, style, and operating performance. Historically, vehicle fuel efficiency has not been a major factor in consumer decisions. In addition, although there are many reasons for consumer choice of vehicles, from 1980 until recently U.S. gasoline prices had been falling (after accounting for inflation), which encouraged consumers to buy (and manufacturers to offer) larger, more powerful, and heavier vehicles.

Transportation energy consumption is also influenced by the physical networks of infrastructure through which vehicles move; by the logistic, institutional,

---

[14]Bureau of Transportation Statistics, National Transportation Statistics. Available at www.bts.gov/publications/national_transportation_statistics/.

[15]Ibid.

[16]Nonetheless, other modes, such as mass transit, intercity rail, and water, have important roles to play in bringing about more energy-efficient passenger and freight transportation, particularly if traffic is shifted to them from the more energy-intensive highway and aviation modes.

commercial, and economic considerations that determine the types of vehicles selected and how they are used; and by the performance of the infrastructure itself (e.g., in managing congestion). Energy use in air transportation, for example, is influenced by air-traffic management, and energy use in freight transport is affected by the possibilities (or lack thereof) for intermodal transfers. The functioning of the underlying physical and economic systems, in other words, can foster—or in some cases hinder—energy efficiency.

Potential for Energy Efficiency Improvements in Passenger Transportation

Automobiles account for the vast majority of local and medium-distance passenger-trips[17] (those under 800 miles); airlines dominate for longer trips.

Light-Duty Vehicles

Globally, the major motivators for energy efficiency in light-duty vehicles (LDVs) are fuel prices, vehicle fuel-economy regulation, personal preferences, and environmental concerns. In Europe, a long history of elevated fuel taxes has been a major reason that motorists there have put a high priority on fuel efficiency when purchasing automobiles. In the United States, the corporate average fuel-economy (CAFE) standards have been the main impetus for boosting vehicle efficiency. Falling real fuel prices from 1980 to 2005, however, encouraged consumers to purchase larger, more powerful, and heavier vehicles rather than to seek greater fuel economy. However, during periods of high fuel prices (such as those prevailing in mid-2008), U.S. consumers have demonstrated more interest in fuel economy.

Today, the average fuel economy of new vehicles sold in the United States is about 25 mpg (new cars average 27.5 mpg compared with 22.3 mpg for light trucks). The U.S. Energy Independence and Security Act of 2007 (EISA 2007; P.L. 110-140) requires that CAFE standards be set for LDVs for model years 2011 through 2020. This provision aims to ensure that, by 2020, the industry-wide CAFE for all new passenger cars and light trucks combined will be at least 35 mpg[18]—a 40 percent increase over today's average of 25 mpg.

While fuel economy in the United States has not improved for almost 30

---

[17]One passenger taking one trip, regardless of trip length, is referred to as a passenger-trip.

[18]The Obama administration has recently proposed that these requirements, specified by Subtitle A of EISA 2007 (P.L. 110-140), be accelerated.

years, vehicle fuel efficiency *has* improved. But the efficiency gains have been off-set by increases in vehicle size and performance (Lutsey and Sperling, 2005; An and DeCicco, 2007).[19]

Current technologies offer many fuel-economy improvements, which become increasingly attractive as fuel prices rise. Opportunities through 2020 will apply primarily to today's vehicle fleet of spark-ignition (SI) engines, compression-ignition (CI) diesel engines, and hybrid-electric vehicles (HEVs), fueled with petroleum, biofuels, or other nonpetroleum hydrocarbon fuels. Annual incremental improvements to engines, transmissions, and nonpropulsion systems are expected to continue. During the subsequent decade, plug-in hybrid-electric vehicles (PHEVs), using electricity plus any of the above-mentioned fuels, may become a significant part of new vehicle sales. Longer-term, substantial sales of hydrogen fuel-cell vehicles (FCVs) and battery-electric vehicles (BEVs) are possible. What follows are summaries of the possible improvements in efficiency that can be expected from new technologies for LDVs.

Engine Improvements in Light-Duty Vehicles

- *Gasoline spark-ignition engines.* Technologies that improve the efficiency of gasoline SI engines, such as variable valve timing, cylinder deactivation, direct injection, and turbocharging with engine downsizing, could be deployed in large numbers over the next decade. Many of these are already being produced in low volumes. They have the potential to reduce fuel consumption[20] in new vehicles, on average, by about 10–15 percent in the near term (through 2020) and an additional 15–20 percent over the longer term (15–20 years). It is expected that turbocharged but downsized gasoline engines will steadily replace a significant fraction of naturally aspirated (non-turbocharged) engines.

[19]"Fuel efficiency" relates to the amount of useful work derived from the combustion of fuel. Increased fuel efficiency can be used to improve fuel economy (vehicle-miles traveled on a gallon of fuel, for example) or to permit increases in vehicle size and performance without degrading fuel economy.

[20]As used here, "fuel consumption" is the inverse of fuel economy—that is, the amount of fuel consumed in traveling 1 mile (or some other distance).

- *Diesel compression-ignition engines.* Diesel CI engines offer about a 20–25 percent fuel consumption benefit over gasoline SI engines (when adjusted for the energy density of diesel fuel). There are opportunities for further efficiency improvements that could reduce the fuel consumption of new diesel-engine vehicles relative to current diesel vehicles by about 10 percent by 2020 and an additional 10–15 percent by 2030. New technologies are emerging for after-treatment that reduce emissions of particulate matter and nitrogen oxides to levels comparable to those of SI engines. The primary challenges for diesel engines in the United States are the added costs and fuel penalties (of about 3–6 percent) associated with those after-treatment systems (Bandivadekar et al., 2008; Johnson, 2008, 2009; Ricardo, Inc., 2008).

- *Gasoline hybrid-electric vehicles.* HEVs combine an internal-combustion engine (ICE) with a battery-electric motor/generator system. Their primary efficiency benefits derive from smaller engines, regenerative braking, elimination of idling, and optimization of engine operating conditions. Hybrid vehicles span a range of technologies and fuel-economy levels. Diesel HEVs are also under development.

- *Plug-in hybrid-electric vehicles.* PHEVs have larger batteries than regular hybrids do, and they can be recharged from an external source of electricity. They also require a larger electric motor and higher-capacity power electronics. Hybrid vehicles, including PHEVs, are designed to allow all-electric operation powered by the battery. The driving range with all-electric power depends on factors such as the size of the battery, the weight of the vehicle, and the driving cycle. Unlike a hybrid-electric vehicle, a PHEV's external power connection can recharge the battery when the vehicle is at rest and plugged in; the internal combustion engine can also recharge the battery, provide power to the wheels, or both, extending vehicle range. The capacity of the battery and the distance of the trip(s) determine gasoline savings; current hybrids in commercial production have a range of less than 10 miles on all-electric power. The vast majority of U.S. vehicles are driven less than 40–60 miles per day in normal operation. Thus, a battery that can power the vehicle for 40 to 60 miles could substantially reduce petroleum consumption for this duty cycle. Commercial PHEVs with a variety of ranges on all-electric power will likely be introduced to the U.S. market over the next 5 years. However, success for a mass-market vehicle with relatively long driving range on battery power alone (e.g., 40 miles or more) will require development of a low-cost, lightweight battery that can store the needed electricity and last for 10 years or more (Box 4.2).

- *Battery-electric vehicles.* Successful development and deployment of PHEVs using advanced battery technology might lead to a battery suitable for BEVs (see Box 4.2). Although several models of BEVs are being introduced into the market today in low volumes, in the foreseeable future the only commercially viable BEVs may be small cars with modest performance expectations, such as "city BEVs."

- *Hydrogen fuel-cell vehicles.* Several scientific, engineering, and business challenges must be met before hydrogen FCVs can be successfully commercialized.[21] The principal challenges are to increase the durability and lower the costs of fuel cells, achieve cost-effective storage of hydrogen in fueling stations and on board vehicles, and deploy a hydrogen supply and fueling infrastructure with low greenhouse gas emissions. These vehicles offer tremendous potential for reductions in oil imports and CO_2 emissions in the long term (beyond 2035) but little opportunity for impact before 2020 because of the time required to address the technical and cost challenges and, subsequently, to achieve high-volume production.

Transmission Improvements in Light-Duty Vehicles

Automatic-transmission efficiency is likely to improve in the near term to midterm through increasing the number of gears and reducing losses in bearings, gears, sealing elements, and hydraulic systems. Seven- and eight-speed transmissions may become standard in the midterm. A continuously variable transmission (CVT) would in principle allow an engine to operate near its maximum efficiency, but its estimated actual efficiency improvement is lower than that expected for six- or seven-speed transmissions.

---

[21]See, for example, NRC, 2004, 2008a,b; Crabtree et al., 2004.

Nonpropulsion System Improvements in Light-Duty Vehicles

Improvements to nonpropulsion systems can involve better tires with lower rolling resistance, body designs that reduce aerodynamic drag, and reductions in vehicle weight. Weight reduction can be achieved by using lightweight materials, by redesigning vehicles, and by reducing vehicle size. A 10 percent reduction in vehicle weight can reduce fuel consumption by 5–7 percent, when accompanied by appropriate engine downsizing at constant performance (Bandivadekar et al., 2008).

Summary of Potential Improvements and Costs for Light-Duty Vehicles

Table 4.3 shows plausible reductions in fuel consumption and CO_2 emissions stemming from evolutionary improvements in LDVs as well as the use of new vehicle types. Evolutionary improvements could reduce the fuel consumption of gasoline ICE vehicles by up to 35 percent over the next 25 years. While diesel engines will also improve, the gap between gasoline and diesel fuel consumption is likely to narrow. Hybrid vehicles—both HEVs and PHEVs—could deliver deeper reductions in fuel consumption, although they would still depend on gasoline or other liquid fuels. Vehicles powered by batteries and hydrogen fuel cells need not depend on hydrocarbon fuels; if they were to run on electricity or hydrogen, they could have zero tailpipe emissions of CO_2 and other pollutants. If the electricity or hydrogen were generated without CO_2 emissions, they would have the potential to reduce total life-cycle CO_2 emissions dramatically.

Table 4.4 shows the approximate incremental retail price of different vehicle systems (including the costs of emission-control systems), as compared with a baseline 2005 gasoline-fueled ICE vehicle. The estimates shown in Table 4.4, when combined with the estimates of fuel-consumption reductions shown in Table 4.3, indicate that from the driver's perspective evolutionary improvements in gasoline ICE vehicles are likely to prove the most cost-effective choice for reducing petroleum consumption and CO_2 emissions. Given that these vehicles will be sold in large quantities in the near term, it is critical that efficiency improvements in these vehicles not be offset by increased power and weight. While the current hybrids appear less competitive than a comparable diesel vehicle, they are likely to become more competitive over time, in part because hybrids can deliver greater absolute emission reduction than diesel vehicles can.

PHEVs, BEVs, and FCVs appear to be more costly alternatives for reducing petroleum consumption and CO_2 emissions. Among these three technologies,

BOX 4.2 *Status of Advanced Battery Technology*

Lead acid batteries were invented in the 19th century and are still the standard battery technology in vehicles today. The GM EV1, a production battery-electric vehicle (BEV), used this battery technology as recently as 1999, and then transitioned to the nickel-metal hydride (NiMH) battery.

The next generation of batteries, based on lithium-ion chemistry, is widely deployed in consumer electronic devices. Of course, the power and energy storage requirements of these devices are much smaller than those of electric vehicles.

Hybrid-electric vehicles (HEVs) require batteries with high power (commonly stated in units of watts per kilogram). Plug-in HEVs (PHEVs) and BEVs require significant energy storage (along with sufficient power). Today's batteries have an energy storage capacity of 150–200 Wh/kg. A typical vehicle consumes approximately 0.25 kWh per mile in all-electric mode. Typical electric motors that can propel a vehicle require power ranging between 50 and 150 kW.

Chemistries

Table 4.2.1 summarizes the promising advanced battery chemistries and their performance characteristics. Significant amounts of research and development are being devoted to promising new versions of the chemistries of cathode materials, anode materials, and electrolytes, as well as to manufacturing processes.

TABLE 4.2.1 Lithium-ion Battery Cathode Chemistries

	Lithium Cobalt Oxide	Lithium Manganese Spinel	Lithium Nickel Manganese Cobalt	Lithium Iron Phosphate
Automotive status	Limited auto applications (due to safety concerns)	Pilot	Pilot	Pilot
Energy density	High	Low	High	Moderate
Power	Moderate	High	Moderate	High
Safety	Poor	Good	Poor	Very good
Cost	High	Low	High	High
Low-temperature performance	Moderate	High	Moderate	Low
Life	Long	Moderate	Long	Long

Source: Adapted from Alamgir and Sastry, 2008.

Performance and Cost Targets

The U.S. Advanced Battery Consortium (USABC) has established a set of long-term performance goals for electrochemical energy storage devices:

- The target for PHEV batteries is an energy storage capacity of 11.6 kWh with an energy density of 100 Wh/kg and a unit cost of stored energy of $35/kWh.
- The target for BEV batteries is an energy storage capacity of 40 kWh with an energy density of 200 Wh/kg and a unit cost of stored energy of $100/kWh.

In addition, goals were established for battery life in terms of the number of 80 percent discharge cycles. Meeting these goals is likely to be required for widespread commercialization of electrically powered vehicles.

Lithium-ion batteries currently lead in energy density (Wh/kg) metric and have an average annual improvement rate of 3.7 percent. Lead-acid batteries lead in the cost of stored energy ($/kWh) at $50/kWh and have an average annual reduction rate of around 3 percent. However, lead-acid batteries are unable to satisfy the battery life requirements for PHEVs and BEVs. Today's lithium-ion batteries that have the cycle life desired for automotive applications cost between $500/kWh and $1000/kWh.

The cost target (in $/kWh) is currently viewed as the greatest challenge for lithium-ion battery technology.

Industry Developments

The lithium-ion consumer electronics market is currently at around 2 billion units annually. The volume of lithium-ion batteries in automotive applications, however, is very small. Frost & Sullivan (2008) predict a 19.6 percent compound annual growth rate for shipments of HEV batteries, as well as a smaller but rapidly growing market for PHEV and BEV batteries.

An auto battery alliance has been promoted by the U.S. Department of Energy's Argonne National Laboratory and includes 3M, ActaCell, All Cell Technologies, Altair Nanotechnologies, EaglePicher, EnerSys, Envia Systems, FMC, Johnson Controls-Saft, MicroSun, Mobius Power, SiLyte, Superior Graphite, and Townsend Advanced Energy.

All major vehicle manufacturers have partnered with major battery manufacturers: Ford with Johnson Controls-Saft, General Motors with LG Chem, Chrysler with General Electric, Toyota with Panasonic/Sanyo, Nissan with NEC via the Automotive Energy Supply joint venture, and Honda with GS Yuasa.

Specialists anticipate that it may be 10 to 20 years before advanced battery technology can reach the USABC performance and cost targets.

TABLE 4.3 Potential Reductions in Petroleum Use and Greenhouse Gas Emissions from Vehicle Efficiency Improvements over the Next 25 Years

Propulsion System	Vehicle Petroleum Consumption (gasoline equivalent)[a]		Greenhouse Gas Emissions[a]	
	Relative to Current Gasoline ICE	Relative to 2035 Gasoline ICE	Relative to Current Gasoline ICE	Relative to 2035 Gasoline ICE
Current gasoline	1	—	1	—
Current turbocharged gasoline	0.9	—	0.9	—
Current diesel	0.8	—	0.8	—
Current hybrid	0.75	—	0.75	—
2035 gasoline	0.65	1	0.65	1
2035 turbocharged gasoline	0.57	0.89	0.57	0.88
2035 diesel	0.55	0.85	0.55	0.85
2035 HEV	0.4	0.6	0.4	0.6
2035 PHEV	0.2	0.3	0.35–0.45	0.55–0.7
2035 BEV[b]	—	—	0.35–0.5	0.55–0.8
2035 HFCV[b]	—	—	0.3–0.4	0.45–0.6

Note: These estimates assume that vehicle performance and size (acceleration and power-to-weight ratio) are kept constant at today's levels. BEV, battery-electric vehicle; HEV, hybrid electric vehicle; HFCV, hydrogen fuel-cell vehicle; ICE, internal-combustion engine; PHEV, plug-in hybrid vehicle.

[a]Greenhouse gas emissions from the electricity used in 2035 PHEVs, 2035 BEVs, and 2035 HFCVs are estimated from the projected U.S. average electricity grid mix in 2035. Greenhouse gas emissions from hydrogen production are estimated for hydrogen made from natural gas.

[b]The metric "vehicle petroleum consumption" is not applicable to vehicles powered by batteries and hydrogen fuel cells. Estimated greenhouse gas emissions are those resulting from production of the needed electricity and hydrogen.

Source: Bandivadekar et al., 2008.

PHEVs are likely to become more widely available in the near term to midterm, whereas BEVs and FCVs are high-volume alternatives for the midterm to long term.

Deployment of Light-Duty Vehicle Technologies

To have a significant effect on fuel use in the vehicle fleet and on associated CO_2 emissions, advanced-technology vehicles must garner a sizable market share. Generally, however, a decade or more elapses in developing a technology to a stage that it can be deployed, introduced on a commercial vehicle, and then achieve significant sales. There are also technical constraints on the speed with which the

TABLE 4.4 Estimated Additional Cost to Purchaser of Advanced Vehicles Relative to Baseline 2005 Average Gasoline Vehicle

Propulsion System	Additional Retail Price (2007 dollars)	
	Car	Light Truck
Current gasoline	0	0
Current diesel	1,700	2,100
Current hybrid	4,900	6,300
2035 gasoline	2,000	2,400
2035 diesel	3,600	4,500
2035 hybrid	4,500	5,500
2035 PHEV	7,800	10,500
2035 BEV	16,000	24,000
2035 HFCV	7,300	10,000

Note: Cost and price estimates depend on many assumptions and are subject to great uncertainty. For example, different companies may subsidize new vehicles and technologies with different strategies in mind. Costs listed are additional costs only, relative to baseline average new car and light truck purchase prices (in 2007 dollars) that were calculated as follows:
  —Average new car: $14,000 production cost × 1.4 (a representative retail price equivalent factor) = an average purchase price of $19,600.
  —Average new light truck: $15,000 × 1.4 = $21,000.
These are not meant to represent current average costs. Rather, they are the costs used in this analysis. Details on how the costs were estimated can be found in NAS-NAE-NRC (2009).
  For the purpose of these estimates, the PHEV all-electric driving range is 30 miles; the BEV driving range is 200 miles. Advanced battery and fuel-cell system prices are based on target battery and fuel-cell costs from current development programs.
Source: Bandivadekar et al., 2008.

market shares of advanced technologies can grow, such as the need for break-throughs in battery performance and for a hydrogen-distribution infrastructure.

Table 4.5 shows the AEF Committee's judgment, based on the constraints just outlined, of the extent to which these advanced vehicle technologies could plausibly penetrate the new LDV market in the United States. (Note that Table 4.5 is not intended to imply that all these technologies would necessarily be deployed together.) The estimates are intended as illustrations of achievable deployment levels, based on historical case studies of comparable technology changes; these estimates suggest that relative annual increases of 8–10 percent in the deployment rate are plausible. With changes in the factors that affect vehicle attributes or purchases, such as stricter regulatory standards or high fuel prices, the timeline for reaching these market shares could be shortened.

TABLE 4.5 Plausible Share of Advanced Light-Duty Vehicles in the New-Vehicle Market by 2020 and 2035 (percent)

Propulsion System	2020	2035
Turbocharged gasoline SI vehicles	10–15	25–35
Diesel vehicles	6–12	10–20
Gasoline hybrid vehicles	10–15	15–40
PHEV	1–3	7–15
HFCV	0–1	3–6
BEV	0–2	3–10

Note: The percentage of hydrogen fuel-cell vehicles being "plausible" is in contrast to the percentages reported in NRC (2008a), which represent "maximum practical" shares.

Savings in Total Fleet Fuel Consumption from Deployment of Light-Duty Vehicles

As noted previously, the Energy Independence and Security Act of 2007 (EISA 2007) requires CAFE standards to be set for LDVs through 2020 in order to ensure that the industry-wide average fuel economy by that time is at least 35 mpg. This would be a 40 percent increase over today's average of 25 mpg.[22]

The AEF Committee examined two scenarios to explore how the deployment of the advanced technologies listed in Table 4.3, together with vehicle-efficiency improvements (such as reductions in vehicle weight, aerodynamic drag, and tire rolling resistance), could reduce the petroleum consumption of the LDV fleet in the United States. These scenarios, based on the methodology described in Bandivadekar et al. (2008), are not predictions of what the LDV fleet will be like in the future. Instead, they are intended as illustrative examples of the degree of change to the LDV fleet that will be necessary to improve fleet average fuel economy. The two scenarios—termed "optimistic" and "conservative"—are described below.

- *Optimistic scenario.* The new CAFE target of 35 mpg for LDVs is met in 2020. This improvement rate is then extrapolated out through 2035. Under this scenario, 75 percent of the improvement is used to reduce actual fuel consumption; the remaining 25 percent is offset by increases

[22]As noted previously, the Obama administration recently announced new policies that will accelerate the implementation of these fuel-economy standards.

in vehicle size, weight, and performance. The resulting new LDV fuel economy in 2035 is double today's value.

- *Conservative scenario.* The new CAFE target is met 5 years later, in 2025. This improvement rate is then extrapolated out through 2035. Under this scenario, only half of the improvement is used to reduce actual fuel consumption; the remaining half is offset by increases in vehicle size, weight, and performance. The resulting new LDV fuel economy in 2035 is 62 percent above today's value.

Both scenarios are compared with a "no-change" baseline that corresponds roughly to meeting the EISA target for 2020. The baseline also includes some growth in overall fleet size and miles driven, but no resulting change in fuel consumption. This is because the baseline extrapolates the history of the past 20 years, during which time power train efficiency improvements essentially offset any negative effects on fuel consumption from increasing vehicle performance, size, and weight.

Based on the estimated fuel consumption characteristics of individual vehicle types, shown in Table 4.3, and the fleet efficiency improvements represented in the scenarios, Table 4.6 shows examples of the sales mixes and weight reduction that would be required to meet the CAFE targets and to meet the scenario assumptions beyond 2020. Figure 4.7 shows, for the two scenarios, the corresponding annual gasoline consumption of the U.S. in-use LDV fleet from the present out to 2035. Table 4.7 shows the cumulative fleet-wide fuel savings, as compared with the no-change baseline. These savings can be substantial so long as the proposed fuel-economy standards are met and the rate of improvement is sustained. Table 4.8 gives the corresponding annual fuel savings from the no-change baseline in 2020 and 2035.

Air Transportation

Air transportation represents almost half of nonhighway transportation energy use (personal and freight), or about 10 percent of total transportation energy consumption. Fuel expenditures are the largest operating cost for most airlines, thereby driving their investment decisions toward higher energy efficiency. For example, Boeing's and Airbus's newest generation of airliners, the Boeing 787 Dreamliner and 747-8, and the Airbus 350XWB, attain a 15–20 percent improvement in fuel efficiency over the aircraft they replace. The new aircraft all employ

TABLE 4.6 Illustrative Vehicle Sales Mix Scenarios

	Percent Emphasis on Reducing Fuel Consumption[a]	Percent Light Trucks vs. Cars	Percent Vehicle Weight Reduction	Market Share by Power Train (percent)						Percent Fuel Efficiency Increase from Today
				Naturally Aspirated SI	Turbo SI	Diesel	Hybrid	Plug-in Hybrid	Total Advanced Power Train	
Optimistic										
2020	75	40	17	52	26	7	15	0	48	+38
2035	75	30	25	36	26	9	20	9	64	+100
Conservative										
2025	50	40	17	55	24	7	14	0	45	+38
2035	50	40	20	49	21	7	16	7	51	+62

Note: Assumed average new-vehicle weight (cars and light trucks) currently is 1900 kg (4180 lb). Thus, average weight reductions of 700–1050 lb per vehicle would be required. Neither of these scenarios includes BEVs or FCVs.
[a]The amount of the efficiency improvement that is dedicated to reducing fuel consumption (i.e., that is not offset by increases in vehicle power, size, and weight).

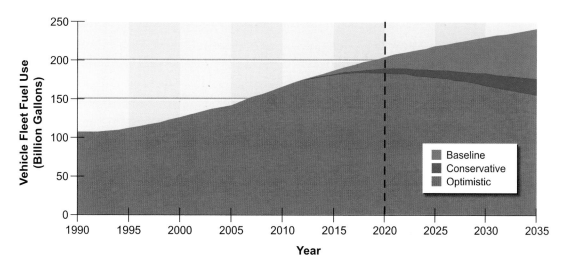

FIGURE 4.7 *Vehicle fleet fuel use for the no-change baseline and the optimistic and conservative scenarios described in the text.*

TABLE 4.7 Cumulative Fuel Savings from the Baseline Shown in Figure 4.7

	Today through 2020 (billion gallons)	2020 through 2035 (billion gallons)
Optimistic scenario	86	834
Conservative scenario	64	631

Note: The no-change baseline assumes constant sales mix by power train, constant ratio of light trucks versus cars, 0.8 percent compounded annual growth in new-vehicle sales, and 0.1–0.5 percent increase in vehicle travel.

TABLE 4.8 Annual Fuel Savings in 2020 and 2035 from the No-Change Baseline Shown in Figure 4.7

	2020 (billion gallons/year)	2035 (billion gallons/year)
Optimistic scenario	21	86
Conservative scenario	16	66

Note: The no-change baseline assumes no change in average new-vehicle fuel consumption, a constant ratio of light trucks versus cars, and a 0.8 percent compounded annual growth in new-vehicle sales. It also assumes that growth in vehicle travel slows from 0.5 percent to 0.1 percent per year over 25 years, and that any efficiency improvements are fully offset by increases in vehicle performance, size, and weight.

weight-reducing carbon composite structural materials and less energy-intensive electric systems.

Because the DOE expects air travel to grow about 3 percent per year over the next several decades, efforts to reduce aviation-fuel consumption face a stiff challenge. The expected efficiency improvement of 1–2 percent per year (Lee et al., 2004) will not be enough to offset the expected growth in demand.

Potential Energy Efficiency Improvements in Freight Transportation

The movement of freight represents about 6–7 percent of the U.S. GDP. Given that the EIA expects freight transport to continue to increase by 2 percent per year over the next two to three decades, energy use in the freight sector could grow by more than 40 percent by 2030.

Truck Transport

Pressure to reduce fuel costs has led truck manufacturers to make continuous progress in raising engine efficiency. Technological improvements have included more sophisticated fuel-injection systems, enhanced combustion, higher cylinder pressure (due to increased turbocharging), and automated manual-transmission systems. Technologies on the horizon include continuously variable transmissions, power-shift transmissions, and hybrid-electric systems that could be used to modulate auxiliaries (pumping, fans, compressors, air-conditioning, and power steering) and reduce idling. Auxiliary-power units with greater efficiency could increase fuel economy, as could use of utility-supplied electricity when parked at truck stops. Reduced idling is especially desirable in urban-duty cycles and in sleeper cabs, where idling alone can account for 10 percent of the vehicle's fuel use.

Air, Rail, and Waterborne Freight Transport

A small proportion (less than 1 percent by weight) of total freight shipments is transported by air. Potential gains in efficiency would stem from the same improvements made to passenger aircraft.

Rail accounts for 2 percent of transportation energy use in the United States but about 10 percent of all freight shipments by weight. Freight railroads are nearly all diesel powered, unlike the mostly electrified rail systems of Europe and Japan. Per ton-mile, rail is 10 times more efficient than trucking is. Still, improvements in railroad technology would offer modest opportunities for gains in U.S.

transportation-energy efficiency. Advances in railroad operation could contribute to improved efficiency as well.

Shifting some freight from trucks to rail could save fuel. Candidates for diversion include trailers and containers carrying commodities that are not time sensitive and are being transported more than 500 miles.

The main fuels used in waterborne shipping—by ocean, inland, coastal, and Great Lakes routes—are diesel fuel (about 70 percent) and heavy fuel oil (30 percent). Waterborne freight accounts for about 3.5 percent of all shipments by weight.

Measured in tonnage, the oceangoing segment accounts for about half the freight moved on water into or within the United States. In terms of energy consumed per ton-mile, ocean shipping is highly efficient, as the vessels carry large payloads over long distances. Gains in energy efficiency are still possible, however. For example, one study estimates that improvements of 20–30 percent could be achieved in ocean shipping by 2020 (Kromer and Heywood, 2008). Speed reduction offers the greatest potential, although there are strong incentives to move shipments rapidly.

Potential System-Level Improvements in Transportation

Transitions in transportation systems—such as expanded use of rail for freight or passenger service—provide opportunities to boost overall energy efficiency. Such changes are usually costly and complicated, however, and are often driven by factors other than energy efficiency (such as productivity). Nevertheless, energy costs can play a motivating role.

The freight sector offers examples. The use of shipping containers has stimulated intermodal transfers among trucks, rail, ships, and even cargo airplanes, leading to dramatic productivity improvements, although gains in overall energy efficiency are less clear. Rail is much more energy-efficient than trucking is; thus, enhancing the quality of rail services and facilitating intermodal transfers should lead to significant gains in freight-transport energy efficiency.

In passenger transport, the opportunities for systemic approaches to improve energy efficiency may be even greater. Some studies have suggested that greater use of advanced information and communication technologies—"intelligent" transportation systems that electronically link vehicles to one another and to the infrastructure—could enable better traffic management. These and other studies have also examined the potential for reducing vehicle use by enhancing collective modes of travel, substituting information and communication technologies for travel, enhancing nonmotorized travel, and reorganizing land use to achieve higher population densities.

In order for a diversified and efficient system to evolve, two types of policy changes are needed: better land-use management and greater use of pricing. The former would have to be implemented on a substantial scale to have a significant effect on fuel consumption and greenhouse gas emissions, and the timeframes of such deployments would span several decades, but the long-term implications could be enormous. The net effect of a concerted effort to internalize congestion-related and environmental externalities into prices could also be large, especially in reducing the numbers of single-occupant vehicles on the roads and encouraging the use of small and efficient city cars for local travel.

Barriers to Improving Energy Efficiency in Transportation

Numerous factors hinder the improvement of energy efficiency and the reduction of fuel consumption in passenger and freight transportation. Some of the most important are noted in the following list:

- In the United States, many factors—including a century of falling energy prices and rising incomes, together with personal preferences and some government policies—have contributed to decentralized land-use patterns and a transportation-intensive economy.
- Low-priced energy has led to consumer purchasing behavior, vehicle designs, and operating decisions that emphasize convenience, style, and speed over fuel economy in automobiles and light trucks. Changing these preferences, which have been developed and reinforced over decades, will not be easy.
- The primary barriers to realizing greater energy efficiency in the transportation sector are the expectations of individuals and companies about future energy prices, fuel availability, and government policies. Although there is an extensive menu of existing technologies for saving energy in transportation, before decision makers decide to invest in these technologies they must be convinced that energy-price increases (or other factors that influence market demand) will persist.
- Even if sufficient demand exists for certain vehicle technologies, there must be the capacity to supply them at the needed scale—vehicle manu-

facturers and their suppliers must have been able to make adequate capital investments for bringing new production capacity on line. Typical product-development times for individual automotive products are 3–5 years, but to deploy a new-vehicle technology across all product platforms and vehicle classes usually takes more than a decade, unless mandated by law.

- Even when new or improved vehicle technologies are available on the market, barriers to purchasing them include high initial cost, safety concerns, reliability and durability issues, and lack of awareness. Reaching a substantial fraction of vehicle sales usually takes more than a decade unless mandated by law or made possible by clear consumer demand for the new or improved technology.

Findings: Transportation

In the transportation sector, the potential for reducing fuel consumption resides both in increasing the efficiency with which liquid fuels (especially petroleum) are used and in shifting some of the vehicle fleet's energy demand to electricity. The greenhouse gas emissions and other environmental effects of such a shift depend on how the electricity (or hydrogen, if fuel-cell vehicles are used) is generated.

An extensive menu of technologies exists today for increasing energy efficiency in transportation. Even so, improving new-vehicle fuel economy substantially is a challenging task. A continued decrease in fuel consumption (and associated greenhouse gas emissions) beyond 2020, when the EISA standards must be met, will require that the historic emphasis on ever-increasing vehicle power and size virtually be abandoned.

In the near term, reductions in fuel consumption will come predominantly from improved gasoline and diesel engines, superior transmissions, and reduced vehicle weight and drag. Evolutionary improvements in gasoline internal-combustion engine vehicles are likely to prove the most cost-effective choice for reducing petroleum consumption in the 2009–2020 timeframe. Gasoline-electric hybrids will play an increasingly important role as their production volumes increase and their cost, relative to conventional vehicles, decreases. Meeting the EISA standards is likely to require that, over the next decade or two, an ever-larger fraction of the new-vehicle fleet be hybrids or plug-in hybrids.

Beyond 2020, continuing reductions in fuel consumption are possible. Plausible efficiency improvements and weight reductions in LDVs, alongside more extensive use of hybrid and plug-in hybrid (and possibly battery-electric) vehicles, could reduce transportation fuel consumption beyond 2020 to below the levels implied by EISA 2007 so long as a plausible rate of improvement can be sustained. An especially important R&D focus here will be developing marketable vehicles that use electricity, which will require improving the performance and reducing the cost of high-energy-storage batteries.

A parallel long-term prospect is fuel cells, with hydrogen as the energy carrier. But major improvements, especially in reducing costs, are needed if this option is to be attractive. Widespread implementation requires significant investment in low-emissions hydrogen supply and in ensuring efficient distribution systems. Onboard hydrogen storage is another key R&D issue. Because establishing a new propulsion-system technology and new fuel infrastructure on a large scale is a formidable task, significant deployment is unlikely before 2035.

Reduced energy use in freight transportation can occur both by improving vehicle efficiency and by streamlining freight-system logistics and infrastructure. Reductions of 10–20 percent in the fuel economy of heavy- and medium-duty vehicles appear feasible over a decade or so. Meanwhile, a broad examination is needed of the potential for further reductions in energy consumption stemming from improved freight-system effectiveness.

Most transportation-related energy efficiency studies and proposals have focused on the considerable gains that could be achieved with improved vehicles rather than on changing the transportation system as a whole. This emphasis is appropriate, given the potential impact of such gains. But major improvements will also come from a broader as well as deeper understanding of transportation-system issues for all transportation modes. The potential overall impact of systemic changes, such as densifying and reorganizing land uses and enhancing collective modes of travel, needs further exploration and quantification. Developing better tools for analyzing and forecasting the effects of different policies and investments on entire transportation systems is therefore an important task.

ENERGY EFFICIENCY IN INDUSTRY

U.S. industrial energy efficiency has improved over the past several decades in response to volatile fossil-fuel prices, fuel shortages, and technological advances, but improving the energy efficiency of the nation's industrial sector even further is essential for maintaining its viability in an increasingly competitive world. Yet there still remain opportunities to incorporate cost-effective, energy-efficient technologies, processes, and practices into U.S. manufacturing. This section describes the progress made to date and the magnitude of the remaining opportunities, stemming both from broader use of current best practices and from a range of possible advances enabled by future innovations.

Energy Use in U.S. Industry

The U.S. industrial sector is composed of an exceptionally diverse set of businesses and products with a broad range of prospects for energy efficiency. While industry used more than 50 percent of the worldwide delivered energy in 2005, in the United States in 2008 industry's share was only 31 percent (Figure 4.1), reflecting the high energy intensity of the domestic housing and transportation sectors as well as the net import into the United States of products containing embodied energy. U.S. industrial energy use is substantial: 31.3 quads of primary energy in 2008 (almost a third of the national total) at a cost of $205 billion. In 2006, about 7 quads of this total was dedicated to nonfuel needs, such as petroleum feedstocks for petrochemicals and coke used in the production of steel (DOE, 2009). Industries in the United States use more energy than those of any other G8 nation and about half of the total energy used by China.

The average annual growth of energy use in the U.S. industrial sector is projected to be 0.3 percent out to 2030. Industry's CO_2 emissions are projected to increase more slowly, at 0.2 percent annually (EIA, 2008). These low growth rates are due partly to the presumed updating with more energy-efficient technologies and practices in industry. They also reflect the restructuring of the economy away from energy-intensive manufacturing and toward service- and information-based activities.

The most energy-intensive industries are metals (iron, steel, and aluminum), petroleum refining, basic chemicals and intermediate products, glass, pulp and paper, and nonmetallic mineral products such as cement, lime, limestone, and soda ash. Less energy-intensive industries include the manufacture or assembly of automobiles, appliances, electronics, textiles, food and beverages, and other products.

Petroleum and natural gas are the two fuels most commonly consumed by the industrial sector. In 2002, they accounted for about 44 percent and 40 percent, respectively, of the sector's primary energy use. While petroleum use in the industrial sector increased by some 24 percent from 1985 to 2002, coal consumption dropped by 27 percent. The use of renewable energy has exhibited a fluctuating pattern over the years, totaling about 1.4 quads in 1978, rising to about 1.9 quads in 1985, and then retreating to about 1.68 quads in 2002. Energy use in the manufacturing industries continues to be significantly higher than in the non-manufacturing industries, which include agriculture, forestry, fisheries, mining, and construction.

Energy-Intensity Trends and Comparisons

Between 1985 and 2004, real GDP in the U.S. industrial sector increased by nearly 45 percent, while total energy use was virtually unchanged; this led to a decrease in energy intensity by nearly a third. However, this apparent improvement in energy intensity was due primarily to a change in the mix of products manufactured in the United States rather than to improvements in energy efficiency. The share of industrial GDP accounted for by such energy-intensive industries as petroleum refining and paper manufacturing had declined and was replaced by relatively non-energy-intensive sectors such as computers and electronics. In general, industries in most industrialized countries are more energy-efficient than their counterparts in the United States.[23] The differences in energy use among countries stem from multiple sources, including natural resource endowments, energy-pricing policies, and the average age of industrial plants.

The Potential for Energy Efficiency Improvement in Industry

This section briefly reviews two studies that attempted to assess the potential for cost-effective energy efficiency improvements across the U.S. industrial sector. Many other studies of energy efficiency potential either examine individual industrial subsectors, such as aluminum, chemicals, or paper manufacturing, or they focus on the potential impacts of specific technologies (such as membranes or combined heat and power) or a family of technologies (e.g., sensors and controls or fabrication and materials). Such cross-sectional studies are also reviewed here

---

[23]More details are available in NAS-NAE-NRC (2009).

for major energy-consuming industries and for important crosscutting technologies and processes. Because they do not treat the industrial sector comprehensively, however, these less-comprehensive studies cannot enable a sector-wide estimation of economic energy efficiency potential, though they do provide valuable benchmarking for the two comprehensive studies discussed below. State-level and international assessments of industrial energy efficiency potential are drawn on as well in the following paragraphs.

Two major studies have estimated the potential for energy savings in U.S. industry by 2020 at 14–17 percent (IWG, 2000; McKinsey and Company, 2007). These savings would occur largely through the deployment of technologies reported as being attractive on a commercial basis—e.g., with internal rates of return (IRRs) of at least 10 percent or that exceed a firm's cost of capital by a risk premium.

In the first major study—the comprehensive and extensively reviewed *Scenarios for a Clean Energy Future* (the CEF study; IWG, 2000)—a portfolio of advanced policies was estimated to reduce energy consumption in the industrial sector by 16.6 percent relative to a business-as-usual forecast, and at no net cost to the economy (IWG, 2000; Brown et al., 2001; Worrell and Price, 2001). The business-as-usual forecast used in the CEF study was that of the *Annual Energy Outlook 1999*, which estimated that the industrial sector would require 41.2 quads of energy in 2020. The most recent EIA projection (*Annual Energy Outlook 2008* [EIA, 2008]), however, forecast industrial sector consumption of energy at only 34.3 quads in 2020. Scaling the 16.6 percent savings estimate to this lower level suggests a savings of 5.7 quads. In a separate assessment, the CEF study concluded that new CHP systems could reduce industrial-sector energy requirements by another 2.0 quads in 2020 (IWG, 2000; Lemar, 2001), for a total savings of 7.7 quads.

The second and more recent major study, documented in McKinsey and Company (2007), concurred that U.S. industries have a significant opportunity for energy efficiency gains. Financially attractive investments (defined as those with an IRR of 10 percent or greater) were estimated to offer 3.9 quads in energy-usage reduction in 2020, as compared with a business-as-usual forecast. Projects with lower returns, with a positive IRR below 10 percent, were estimated to reduce energy use by 1.4 quads in 2020.

Table 4.9 summarizes the studies' estimates of energy-savings potential for different industrial subsectors. The CEF study foresaw a large potential for energy savings in pulp and paper manufacturing (6.3 percent), iron and steel (15.4 per-

TABLE 4.9 Estimated Economic Potential for Energy Efficiency Improvements in Industry in the Year 2020: Sector-Wide and for Selected Subsectors and Technologies

	CEF Study (IWG, 2000) Scaled to EIA (2008) (quads)	McKinsey and Company (2007) (quads)	Other U.S. Studies (quads)	Global Estimates from IEA (2007) (percent improvement)
Selected subsectors				
Petroleum refining	N/A	0.3	0.61–1.21 to 1.4–3.28[a]	13–16
Pulp and paper	0.14[b]	0.6	0.37–0.85[c,d]	15–18
Iron and steel	0.21[e]	0.3	0.76[f]	9–18
Cement	0.08[g]	0.1	0.01–0.22[h,d]	28–33
Chemical manufacturing	N/A	0.3	0.19[i] to 1.1[j]	13–16
Combined heat and power	2.0	0.7	6.8[k,d]	
Total industrial sector	7.7 (22.4%)	4.9 (14.3%)		18–26

[a]Based on range of 10–20 percent savings (LBNL, 2005) to 23–54 percent savings (DOE, 2006a) from a baseline of 6.075 quads.

[b]6.3 percent of the 2.311 quads of energy consumption forecast for the paper industry in 2020 by EIA (2008).

[c]Based on 16 percent savings (Martin et al., 2000) and 25.9 percent savings (DOE, 2006c) from a baseline of 3.3 quads.

[d]Values updated from prepublication draft.

[e]15.4 percent of the 1.364 quads of energy consumption forecast for the iron and steel industry in 2020 by EIA (2008).

[f] Based on 40 percent savings (AISI, 2005) from a baseline of 1.9 quads.

[g]19.1 percent of the 0.431 quads of energy consumption forecast for the cement industry in 2020 by EIA (2008).

[h] Based on a range of 34.8–71.9 percent savings (Worrell and Galitsky, 2004) for the 0.431 quads of energy consumption forecast for the cement industry in 2020 by EIA (2008).

[i]NREL, 2002.

[j]DOE, 2007c.

[k]Bailey and Worrell, 2005.

cent), and cement (19.1 percent) (IWG, 2000, Table 5.8; Worrell et al. 2001). On a segment-by-segment basis, McKinsey and Company (2007) concluded that the largest untapped opportunities for U.S. industrial energy efficiency savings reside in pulp and paper and in iron and steel. The CEF study's estimates for iron and steel and for cement were similar to those of the McKinsey and Company study, but its estimate for pulp and paper was significantly lower.

Business-as-Usual Efficiency Improvement

The McKinsey and Company (2007) analysis assumes that a significant amount of energy efficiency improvement, resulting from capital-stock turnover of out-dated technologies and from cost reductions and performance improvements that happen through economies of scale and the advance of science and technology, is included in the business-as-usual forecast. Thus, the level of energy efficiency improvement anticipated in the year 2020 relative to today could far exceed 6.3 quads. Prime targets are waste-heat recovery and improved energy-management and integration practices. These are the kinds of cost-saving potential that EIA assumes will be absorbed in the business-as-usual case. Thus, relative to today's energy efficiency practices, industrial energy efficiency improvements in 2020 could save considerably more energy than the 4.9 quads estimated by McKinsey and Company (2007) if the "naturally occurring" efficiency improvements relative to today's technology were added on.

Looking beyond 2020, a wide array of advanced industrial technologies could make significant contributions to reducing industrial energy consumption and CO_2 emissions. Possible revolutionary changes include novel heat and power sources, as well as innovative processes for new products that take advantage of developments in nanotechnology and micro-manufacturing. Examples include the microwave processing of materials and nanoceramic coatings, which show great potential for boosting the efficiency of industrial processes. In addition, advances in resource recovery and utilization—e.g., aluminum recycling—could reduce the energy intensity of U.S. industry. Many of these approaches provide other benefits as well, such as improved productivity and reduced pollution.

The Role of Innovation

Most of the discussion in this chapter focuses on new technology that lowers industry's energy use. In some cases, energy savings of greater importance come from adapting the new technologies, such as fuel cells for CHP production, used in other sectors. This role of industry in the development of emerging technologies suggests even greater energy savings than might be apparent from looking at industry's own energy-use patterns alone. Companies are adopting a much broader view of their energy and environmental responsibilities by, for example, addressing the sustainability of their products and services together with those of their suppliers.

Major Energy-Consuming Industries

Chemical Manufacturing and Petroleum Refining

The chemical industry manufactures an extensive array of organic and inorganic chemicals and materials. Feedstocks include hydrocarbons from petroleum refining, mined chemicals and minerals, and even animal and plant products such as fats, seed oils, sugars, and timber. For energy sources, the chemical industry uses petroleum-based fuels, natural gas, coal, electricity, and, to a lesser but growing extent, biomass (DOE, 2000, 2006b). Most of the larger chemical companies that have a major presence in the consumer marketplace are R&D oriented because of the continual need to generate new and improved products, enhance quality and yields, and conform to environmental regulations.

The petroleum-refining industry (DOE, 2007c) is similar to the chemical industry in its use of energy sources and process equipment, though its normal output is narrower—a range of refined hydrocarbon products, made in high volume, for the transportation industry. Many refining companies do have a bulk-chemical arm, however, to manufacture a limited spectrum of high-volume organic chemicals and bulk polymer intermediates that are natural extensions of their refining operations. Petroleum companies vary in research intensiveness; generally, they are less dependent on finding new products and processes than is the chemical industry.

Despite some differences, the chemical and petroleum-refining industries have many similarities in raw materials, energy sources, and process technologies. Together, they have the opportunity to achieve significant energy efficiency improvements. Technologies in common include high-temperature reactors, distillation columns for liquid-mixture separation, gas-separation technologies, corrosion-resistant metal and ceramic-lined reactors, sophisticated process-control hardware and software, and many others.

Benchmarking data indicate that most petroleum refineries can economically improve energy efficiency by 10–20 percent (LBNL, 2005); analyses of individual refining processes estimate energy savings ranging from 23 to 54 percent (DOE, 2006a). Major areas for energy efficiency improvement are utilities (e.g., steam generation and distribution, power generation, and compressors), process optimization, heat exchangers, and motors and motor applications.

Pulp and Paper

Pulp and paper production, which makes up a majority of the U.S. forest products enterprise, consumes about 2.4 quads of energy annually. Drying and the recovery of chemicals, the most energy-intensive parts of the papermaking process, rely on steam, which is also used for pulp digesting. Electricity is required to run equipment such as pumps and fans and to light and cool buildings.

Several energy-efficient methods of drying have been developed, many of which are cost-effective today. One of them involves using waste heat from other processes, such as power generation and ethanol production, as the energy source for evaporation (Thorp and Murdoch-Thorp, 2008). Advanced water-removal technologies can also reduce energy use substantially in drying and concentration processes (DOE, 2005). More generally, membrane and advanced filtration methods could effect significant reductions in the total energy consumption of the pulp and paper industry (ORNL and BCA, Inc., 2005). High-efficiency pulping technology that redirects green liquor to pretreat pulp and reduce lime-kiln and digester load is another energy-saving method for this industry (DOE, 2005). Modern lime kilns are available with external dryer systems and modern internals, product coolers, and electrostatic precipitators (DOE, 2006c).

Estimates of the cost-effective energy efficiency potential of the pulp and paper industry in 2020 range from a low of 16 percent (Martin et al., 2000) to a high of 25.9 percent (DOE, 2006c), yielding a range of energy savings of 0.37–0.61 quads by the year 2020. Additional savings are possible from the use of CHP technologies.

Iron and Steel

The iron and steel industry consumes 1.4–1.9 quads per year (EIA, 2007a). In 2006, approximately 43 percent of raw steel was produced by integrated steelmakers—i.e., blast furnace and basic oxygen furnace (BOF) operation—and 57 percent by electric arc furnace (EAF) operation. Energy intensities for the two production methods vary substantially, reflecting the fact that BOFs produce new steel mainly from iron ore whereas EAFs use mainly scrap steel. To produce hot rolled steel from iron ore takes almost five times the energy per ton as making the same product from scrap steel.

In 2006, yield losses totaled 8 million tons. Losses occur in many different operations and appear as "home" scrap and waste oxides. Integrated producers also lose a small percentage of coal and coke. The steel industry consumes about

18.1 million Btu per ton of product, which is 22 percent more than the practical minimum energy consumption of about 14 million Btu per ton (AISI, 2003). These energy losses are a result of the production energy embedded in yield losses and process inefficiencies.

Energy consumption per ton of steel has decreased 27 percent since 1990, while CO_2 emissions fell by 16 percent. For 2002–2005, energy intensity per ton of steel decreased by 12 percent. Energy-use improvements from avoiding yield losses would contribute another 20 percent in savings.

In 2005, the American Iron and Steel Institute announced a goal of using 40 percent less energy per ton of steel in 2025 than what steelmakers were using in 2003 (AISI, 2005); this goal will require the development and implementation of "transformational technologies." The most promising opportunities include EAF melting advances, BOF slag-heat recovery, integration of refining functions, heat capture from EAF waste gas, and increased direct carbon injection. The majority of these technologies may be available before 2020, assuming continued R&D. Several revolutionary new steelmaking technologies and concepts—the use of hydrogen as an iron-ore reductant or furnace fuel, for example, or electrolytic or biometallurgy-based iron and steel production—could be ready in the 2020–2035 timeframe.

Cement

The cement industry accounts for 5 percent of the energy used in the U.S. manufacturing sector, or 1.3 quads (DOE, 2002), and about 9 percent of global industrial energy use (IEA, 2007). The industry is also responsible for about 5 percent of worldwide anthropogenic CO_2 emissions and for 2 percent in the United States (Worrell et al., 2001; Worrell and Galitsky, 2004).

Cement plants increase in efficiency with size, and advanced dry-kiln processes are substantially more efficient than older wet-kiln processes. In the United States, energy consumption varies from 6.2 million Btu per ton of clinker (the cement precursor produced from limestone and other chemicals in cement kilns) for smaller wet plants to 3.8 million Btu per ton of clinker for dry preheater-precalciner kilns (van Oss, 2005). Coal is the chief fuel consumed in U.S. plants, although they utilize an increasing proportion of waste materials, used tires, and petroleum coke.

Energy use also varies with the process and characteristics of a plant, but in general, about 90 percent of the energy use, and all of the fuel use, occurs in the manufacture of clinker in the kiln. Half of the CO_2 generated in cement manufacturing derives from energy use and the other half from the chemical process that converts limestone to lime, the key ingredient of clinker.

While most major energy savings in cement processes require a major upgrade to an advanced dry-kiln process, other technologies that incrementally improve energy efficiency include advanced control systems, combustion refinements, indirect firing, and optimization of components such as the heat shell. Opportunities vary with specific plants; however, the combination of these activities appears to yield energy savings on the order of 10 percent.

The most attractive available energy efficiency technologies, with potential energy savings of 10–20 percent, derive from changing the chemistry of cement to reduce the need for calcination. Blended cements include higher proportions of other cementitious materials, such as fly ash. Steel slag, which is already calcined, is an alternative to limestone for the production of clinker. Technologies that allow production of cement with a lower per-ton share of clinker thus yield multiple benefits: savings in fuel consumption and reductions in greenhouse gas emissions by a factor of two or more above what is associated with energy efficiency alone.

Advanced technologies with a potential to further improve energy efficiency and emissions include fluidized bed kilns, advanced comminution processes, and the substitution of mineral polymers for clinker. A Battelle (2002) study concluded that non-limestone-based binders may yield a reduction of 30 percent in CO_2 emissions. Additional advanced approaches to reducing CO_2 emissions are hybrid cement-energy plants, currently under investigation in the United States, and the incorporation of carbon capture and storage.

Crosscutting Technologies for Energy Efficiency in Industry

Several illustrations of technologies and approaches that could improve industrial energy efficiency are given in this section. Some have already been introduced but could have much greater application, while others are still in the development stage.

- *Combined-heat-and-power* units transform a fuel (generally, natural gas) into electricity and then use the residual heat for space and hot-water heating or for industrial and commercial processes. Estimates of the economic energy-savings potential of CHP nationwide range from 0.7 quads (McKinsey and Company, 2007) to 2.0 quads (IWG, 2000; Lemar, 2001).

- The general class of *separation processes* is one of the most attractive targets for improving energy efficiency in industry, as some separation processes have thermal efficiencies as low as 6 percent. Much of industrial separation now is done by distillation, especially in the petroleum and chemical industries, and technologies are available that could significantly reduce the energy required for this process. However, attention has recently begun to shift to the development of separation processes based on membranes and other porous materials, which could reduce industry's energy intensity and total energy use.

- The industrial sector needs *advanced materials* that resist corrosion, degradation, and deformation at high temperatures and pressures. Relatively low-energy-intensity materials with particular properties and potential uses, such as composites and nanomaterials for structural applications, could substitute for energy-intensive materials such as steel.

- Numerous technologies and practices are available that could optimize and improve *steam heating and process heating* in industrial facilities. Some changes could take place immediately and could reduce natural gas consumption. Even further efficiency gains could result from R&D, particularly on ultrahigh-efficiency boilers. Such boilers, which employ a combination of advanced technologies, could offer considerable efficiency gains over today's state-of-the-art boilers.

- *Electric motors* make up the largest single category of electricity end-use in the U.S. economy. They also offer considerable opportunity for electricity savings, especially in the industrial sector. Based on an inventory of motor systems conducted in 1998 (Xenergy, Inc., 1998), it is estimated that industrial motor energy use could be reduced by 11–18 percent if facility managers undertook all cost-effective applications of proven energy efficiency technologies and practices. Specifically, implementation of all established motor-system energy efficiency technologies and practices that meet reasonable investment criteria could yield annual energy savings of 75–122 billion kWh. A next generation of motor and drive improvements is on the horizon, including motors with high-temperature superconducting materials that could extend savings much further.

- New *fabrication processes* could improve yields per unit energy cost for multiple elements of the manufacturing supply chain, reduce waste, and lower air- and water-pollutant emissions. New processes include net and near-net design and manufacturing; advanced casting, forming, joining, and assembly; engineering of functional materials and coatings; and nanomanufacturing, which would enable the mass production and application of nanoscale materials, structures, devices, and systems.

- *Sensor development* for energy efficiency in the United States is being led by the DOE's Office of Energy Efficiency and Renewable Energy, much of it in collaboration with industrial firms. The developers' approach essentially entails data gathering by automated monitoring, automated data analysis, automated feedback and control, and effective communication among the components. Sensors include inferential controls, real-time and nondestructive sensing and monitoring, wireless technology, and distributed intelligence.

- *Remanufacturing* of used products for resale is gaining recognition as a potentially profitable and resource-efficient business opportunity. Examples include the remaking of automobile pumps and photocopiers. Relative to making a new product from scratch, remanufacturing appears to offer substantial energy efficiency benefits because of the energy saved directly in the production process and indirectly by forgoing the use of many of the raw materials. However, a thorough assessment would require analysis of the options for collecting used products, remanufacturing them, and redistributing them (Savaskan et al., 2004).

Summary of Potential Energy Savings in Industry

Table 4.10 summarizes the potential energy savings stemming from energy efficiency improvements in industry. As shown, if the full potential were realized, industrial energy consumption in 2020 could fall to 14–22 percent below its projected level.

Barriers to Improving Energy Efficiency in Industry

Economic, managerial, and political barriers such as those described below can inhibit the broad deployment of otherwise available technologies:

- *Technical risks of adopting a new technology.* Uncertainties about a technology's benefits and impacts, particularly on existing production lines, can be significant. Such perceived risks result in longer and larger-

TABLE 4.10 Energy Use in Industry and Estimated Energy Savings in 2020 Due to Energy Efficiency Improvements

Industry	Energy Use (quads)			Savings over Business as Usual in 2020[a,b,c] (quads)
	2007	Business-as-Usual Projection (EIA, 2008)		
		2020	2030	
Petroleum refining	4.39[d]	6.07	7.27	0.3–3.28[d]
Iron and steel	1.38	1.36	1.29	0.21–0.76[d]
Cement	0.44	0.43	0.41	0.1–0.22[d]
Bulk chemicals	6.85	6.08	5.60	0.19–1.1[d]
Pulp and paper	2.15	2.31	2.49	0.14–0.85[d]
Total savings—all industries (including those not shown)				4.9–7.7[e] 14–22%

[a]The savings cited are the same as those listed in Table 4.9, which provides references.
[b]Based on a review of studies for specific energy-using industries, both for industrial combined heat and power (CHP) and for industry as a whole.
[c]Savings shown are for cost-effective technologies, defined as those providing an internal rate of return of at least 10 percent or exceeding a company's cost of capital by a risk premium.
[d]Values updated from prepublication draft of this report.
[e]Includes 0.7–2.0 quads from CHP systems.
Source: Compiled from the sources given in Table 4.9.

scale field testing of new technologies, more stringent investment criteria, and a slower pace of technology diffusion.

- *Relatively high costs.* Because new technologies often have longer payback periods than does energy-efficient traditional equipment, they represent a more serious financial risk, given uncertainty about future energy prices.

- *External benefits and impacts that can be difficult to value quantitatively.* Industrial plant managers are thus often inhibited from investing in greenhouse gas mitigation and other pollution-abatement efforts. Companies generally do so only when compelled by law or when they expect to be rewarded with lower energy costs, lower raw material costs, or other economic advantages. Moreover, firms may be reluctant to develop new technologies for reducing emissions without an assured market for their innovations.

- *Distorted price signals* skew the demand for electricity in today's retail markets. While time-of-use (TOU) pricing is available for many major industrial customers, electricity rates generally do not reflect the real-time costs of electricity production, which can vary by a factor of 10 over a single day. Most customers in traditionally regulated markets buy electricity under time-invariant prices that are set months or years ahead of actual use. As a result, current market structures actually block price signals from reaching customers (Cowart, 2001), who are thus rendered unable to respond. By contrast, broadly applied TOU pricing would encourage industrial customers to use energy more efficiently during high-price periods. According to Goldman (2006), 2700 commercial and industrial customers were enrolled in TOU programs in 2003, representing 11,000 MW. Three programs in the Southeast (TVA, Duke Power, and Georgia Power) accounted for 80 percent of these participants, most of which used large amounts of energy. Thus there would appear to be considerable room for expanding TOU programs to other regions and to smaller enterprises.

- *Lack of specialized knowledge.* Industrial managers can be overwhelmed by the numerous technologies and programs that tout energy efficiency, especially in the absence of in-house energy experts. Managers may find it risky to rely on third-party information to guide investments, given that energy consulting firms, for example, often lack the industry-specific knowledge to provide accurate energy and operational cost assessments.

- *Incomplete or imperfect information* is a barrier to the diffusion of energy-efficient industrial technologies and practices, such as those involving CHP systems, materials substitution, recycling, and changes in manufacture and design. This barrier is exacerbated by the *high transaction costs for obtaining reliable information* (Worrell and Biermans, 2005). Researching new energy-efficient industrial technologies consumes precious time and resources, especially for small firms, and many industries prefer to expend their human and financial capital on other investment priorities. In some cases, industrial managers are simply unaware of energy efficiency opportunities and low-cost ways to implement them.

  This barrier is made more onerous by the limited governmental collection and analysis of data on energy use in the industrial sector. Consider, for example, the Manufacturing Energy Consumption Survey, a widely used publication that is published every 4 years by the DOE's Energy Information Administration (EIA). In it, one can find the fuel breakdown of the petroleum industry, but there is no estimation of how much energy is used in distillation columns or other separations. On the other hand, for other sectors, annual reports contain substantially more detailed statistics than those available for manufacturing. More frequent and comprehensive collection and publication of such data and analysis are needed.

- Investments in industrial energy efficiency technologies are hindered by *market risks caused by uncertainty* about future electricity prices, natural gas prices, and unpredictable long-term product demand. Additionally, industrial end-use energy efficiency faces *unfavorable fiscal policies*. Because tax credits designed to encourage technology adoption are limited by alternative-minimum-tax rules, tax-credit ceilings, and limited tax-credit carryover to following years, qualified companies are often prevented from utilizing tax credits to their full potential. Similarly, outdated tax-depreciation rules that require firms to depreciate energy efficiency investments over a longer period of time than other investments can distort the efficiency investment options' cost-effectiveness (Brown and Chandler, 2008).

- *Capital market barriers.* Although, in theory, firms might be expected to borrow capital any time a profitable investment opportunity presents itself, in practice they often ration capital—that is, firms impose internal limits on capital investment. The result is that mandatory investments (e.g., those required by environmental or health regulations) and those that are most central to the firms' product line often are made first. Moreover, projects to increase capacity or bring new products to market typically have priority over energy-cost-cutting investments; the former have a greater return on investment or are otherwise more important to the firm. Firms wishing to make energy efficiency investments may face problems raising capital—for example, when the technology involved is new to the market in question, even if it is well-demonstrated elsewhere.

- *Regulatory barriers* can also inhibit energy-saving improvements. For example, the Environmental Protection Agency's New Source Review

(NSR) program tends to hinder energy efficiency improvements at industrial facilities. As part of the 1977 Clean Air Act Amendments, Congress established the NSR program and modified it in the 1990 Amendments, but old coal plants and industrial facilities were exempted from the New Source Performance Standards (NSPSs) to be set. NSPSs are intended to promote use of the best air-pollution control technologies, taking into account the costs of such technologies, their energy requirements, and any non-air-quality-related health and environmental impacts. However, investment in an upgrade could trigger an NSR, and the threat of such a review has prevented many upgrades from occurring.

Drivers for Improving Energy Efficiency in Industry

Helping to overcome the barriers to improving energy efficiency in industry is a set of motivators that include the following:

- *Rising energy prices and fuel/electricity availability.* Rapid increases in fuel prices command management's attention. To remain competitive, industry must find ways to reduce costs, and higher energy costs can make efficiency investments more beneficial.
- *Air quality.* Many states are allowing industry to use energy efficiency to qualify for NO_x and SO_2 offsets in non-attainment areas. Increasingly stringent ambient air-quality standards, together with cap and trade markets, have resulted in rising prices for NO_x and SO_2 allowances. The high costs of these allowances provide incentives to reduce energy use.
- *Demand charges and demand-response incentives.* Demand charges to industrial and commercial customers—based on their peak electricity demand—can be greater than the payments for the consumed energy itself. These charges provide strong incentives for a plant to manage its electricity usage to avoid peaks and to shift power use from periods of peak prices.
- *Collateral benefits.* An efficiently run plant, in terms of both energy use and other factors, will likely also have excellent product quality, high labor productivity, reliable production schedules, and an enviable safety record.

- *Corporate sustainability.* Voluntarily reducing greenhouse gas emissions can help boost shareholder and investor confidence, encourage favorable future legislation, improve access to new markets, lower insurance costs, avoid liability, and enhance competitiveness.

Findings: Industry

Independent studies using different approaches agree that the potential for improved energy efficiency in industry is large. Of the 34.3 quads of energy forecast to be consumed in 2020 by U.S. industry (EIA, 2008), 14–22 percent could be saved through cost-effective energy efficiency improvements (those with an internal rate of return of at least 10 percent or that exceed a company's cost of capital by a risk premium). These innovations would save 4.9–7.7 quads annually.

Comparisons of the energy content of manufactured products across countries underscore the potential for U.S. industry to reduce its energy intensity. Japan and Korea, for instance, have particularly low levels of industrial energy intensity. Care is needed, however, to avoid unrealistic assessments. The savings potentials of existing industrial plants in the United States cannot easily be derived from comparing them with new state-of-the-art facilities in rapidly growing economies.

Additional efficiency investments could become attractive through accelerated energy research, development, and demonstration. Enabling and crosscutting technologies—such as advanced sensors and controls, microwave processing of materials, nanoceramic coatings, and high-temperature membrane separation—could provide efficiency gains in many industries as well as throughout the energy system. For example, these innovations could apply to vehicles, feedstock conversion, and electricity transmission and distribution.

Energy-intensive industries such as aluminum, steel, and chemicals have devoted considerable resources to increasing their energy efficiency. For many other industries, energy represents no more than 10 percent of costs and is not a priority. Energy efficiency objectives compete for human and capital resources with other goals, including increased production, introduction of new products, and compliance with environmental, safety, and health requirements. Outdated depreciation capital schedules, backup fees for CHP systems, and other policies also hamper energy efficiency investment.

More detailed data, collected more frequently, are needed to better assess the status of energy efficiency efforts in industry and their prospects. In order to achieve this goal, proprietary concerns will have to be addressed.

ENERGY EFFICIENCY POLICIES AND PROGRAMS: EXPERIENCE AND LESSONS LEARNED

Although policy recommendations are beyond the scope of this study, policy actions will doubtless be an integral part of the nation's efforts to transform the ways in which Americans use energy. To inform the policy debate, the AEF Committee reviewed some experiences with—and, just as important, lessons learned from—the use of policies and programs to influence energy use in the United States. This brief review concentrates on federal actions, but it also covers state policy initiatives as well as some programs that have been adopted by electric utilities. Among the important initiatives at the state level, the most successful and interesting are those in California and New York.

Barriers to Adoption of Energy-Efficient Technologies

There is no single market for energy efficiency. Instead, there are hundreds of end-uses, thousands of intermediaries, and millions of consumers (Golove and Eto, 1996). The preceding sections have identified some specific factors that hinder the adoption of energy-efficient technologies and practices by these consumers—individuals, organizations, and businesses—in each of the three end-use sectors. Summarized in the following list, the barriers include:

- *Limited supply and availability* of some energy efficiency measures, such as newer products manufactured on a limited scale or not yet widely marketed;
- *Lack of information, or incomplete information*, on energy efficiency options for businesses, households, and other venues;
- *Lack of funds* to invest in energy efficiency measures, often resulting from constraints imposed within the financial system rather than from the financial inability of the would-be user to raise capital;
- *Fiscal or regulatory policies* that discourage energy efficiency investments, often inadvertently;

- *Decision making that does not consider or value energy efficiency;*
- *Perceived risks* associated with the performance of relatively new energy efficiency measures;
- *Energy prices that do not reflect the full costs imposed* on society by energy production and consumption (i.e., that insufficiently account for externalities);
- *Human and psychological factors,* such as risk aversion, loss aversion, and status-quo bias.

The AEF Committee reviewed policies and programs at the federal and state levels that have attempted to overcome or compensate for these barriers so as to reduce energy use. These approaches are discussed below.

Federal, State, and Utility Policies and Programs

Certain policies and programs have played important roles in reducing energy use and energy intensity in the United States. For example, over the past 30 years the federal government has devoted billions of dollars to energy efficiency R&D. It has also adopted a number of laws—notably during the 1975–1980 period—that stimulated educational efforts, created financial incentives, and authorized the setting of efficiency standards. More recent legislation has established minimum efficiency standards on a wide range of household appliances and commercial/industrial equipment, as well as tax incentives to motivate commercialization and adoption of highly efficient products and buildings. In addition, many states have implemented building energy codes, utility-based energy efficiency programs, and other policies to complement the federal initiatives.

This review does not consider energy taxes that have been enacted over the past 30 years because increases have been very modest. The federal tax on gasoline, for example, was increased incrementally from 4¢/gal in 1973 to a total of 18.4¢/gal in 1993, but it has not been increased since then. Corrected for inflation, the gasoline tax in 2006 was only 26 percent greater than its value in 1973.

Vehicle Efficiency Standards

The United States adopted energy efficiency standards for cars and light trucks, known as CAFE standards, in 1975. These standards played a leading role in the near-doubling of the average new-car fuel economy and the 55 percent increase in the fuel economy of light trucks from 1975 to 1988 (Greene, 1998). Unfortunately, the trend reversed between that period and 2006–2007, attributable mainly to the shift from cars toward less-efficient sport utility vehicles, pickup trucks, and minivans (EPA, 2007a). This shift to less-efficient vehicles, together with a greater number of vehicle-miles driven, resulted in a 31 percent increase in U.S. gasoline consumption during 1986–2006 (EIA, 2007b).

EISA included the first significant advance in fuel-economy standards in more than 30 years. Assuming that these standards are met, the average fuel economy of cars and light trucks combined will reach at least 35 mpg in 2020, a 40 percent increase.[24] It is estimated that the new CAFE standards will save 1 million barrels per day of gasoline by 2020 and 2.4 million barrels per day by 2030 (ACEEE, 2007). These estimates account for the "rebound effect," that is, the increase in travel demand due to the reduction in the cost per mile driven as vehicle fuel economy improves. This effect is generally thought to be real but small (Greene, 1998; NRC, 2002; Small and Van Dender, 2007).

Appliance Efficiency Standards

Appliance efficiency standards, first enacted by California, New York, Massachusetts, and Florida during the late 1970s and early 1980s, were followed by national standards in 1987. These standards led to dramatic improvements in the energy efficiency of new refrigerators, air conditioners, clothes washers, and other appliances. For example, the combination of state and federal standards resulted in a 70 percent reduction in the average electricity use of new refrigerators sold in the United States from 1972 to 2001 (Geller, 2003).

In 1992, minimum efficiency standards were extended to motors, heating and cooling equipment used in commercial buildings, and some types of lighting products. In 2005, standards were adopted for a variety of "second-tier" products, among them torchiere light fixtures, commercial clothes washers, exit signs, distribution transformers, ice makers, and traffic signals. With the addition of these products, national minimum efficiency standards were in place for better than 40 different types of products.

National appliance efficiency standards saved an estimated 88 terawatt-hours (TWh) of electricity in 2000, or 2.5 percent of national electricity use that year (Nadel, 2002); based both on the time required to turn over the appliance stock

---

[24]As noted previously, the Obama administration recently announced new policies that will accelerate the implementation of these fuel-economy standards.

and on the new and updated standards adopted for a number of products since 2000, the energy savings are expected to grow to about 268 TWh (6.9 percent) in 2010 and to 394 TWh (9.1 percent) by 2020 (Nadel et al., 2006). These projections are underestimates, as they consider only the savings from standards adopted as of 2007. Federal law requires dozens of additional standards to take effect before 2020, and some states are setting standards for appliances not covered by the federal standards.

Additional appliance efficiency standards were included in the 2007 federal energy legislation. Most noteworthy are efficiency standards for general service lamps, standards that will make it illegal to sell ordinary incandescent lamps after the standards take effect. In phase one, which takes effect in three stages during 2012–2014, manufacturers will be able to produce and sell improved incandescent lamps as well as CFLs and LED lamps that meet the efficacy requirements, namely, the minimum lumens of light output per watt of power consumption. In phase two, which takes effect in 2020, only CFLs and LED lamps will qualify unless manufacturers are able to roughly triple the efficacy of incandescent lamps. It is estimated that these new standards will save 59 TWh per year by 2020, additive to the savings from standards for other products (ACEEE, 2007).

Building Energy Codes

Most state and local authorities have adopted mandatory energy codes for new houses and commercial buildings, often following models such as the International Energy Conservation Code, although some state or local codes are more stringent. Building energy codes for new homes and commercial structures built during the 1990s are estimated to have reduced U.S. energy use by 0.54 quads in 2000. The DOE estimates that if all states adopted the model commercial building energy code approved in 1999 by the American Society of Heating, Refrigerating and Air-Conditioning Engineers, owners and occupants would save about 0.8 quads over 10 years (DOE, 2007a). Building energy codes are enforced at the local level, however, and there is evidence that enforcement and compliance are weak in many jurisdictions.

Research, Development, and Demonstration

The DOE spent more than $7 billion (in 1999 dollars) on energy efficiency research, development, and demonstration (RD&D) programs during 1978–2000 (NRC, 2001). The resulting efforts contributed to the evolution and commercialization of high-efficiency appliances, electronic lighting ballasts, and low-emissivity windows, which provided net economic benefits to the buildings sector far in excess of the RD&D costs.

In contrast to the outcomes of its buildings technology program, the DOE's transportation technology RD&D program has had little effect on the vehicle marketplace to date. During the 1980s and 1990s, the DOE chose to focus on advanced engines and power systems for which the technological problems could not ultimately be solved or that evoked little industry or customer interest. The more recent emphasis on hybrid and fuel-cell technologies, implemented through government-industry RD&D partnerships, shows greater promise (NRC, 2008a,b). This experience demonstrates that RD&D projects should be carefully selected and designed, taking into account the technological, institutional, and market barriers involved.

The DOE operates a number of programs to promote greater energy efficiency in industry, including RD&D on advanced technologies as well as deployment programs. These efforts are estimated to have saved about 3 quads of energy cumulatively and about 0.4 quads in 2005 alone (DOE, 2007b).

Federal Incentives and Grants

Federal tax credits of 15 percent for households and 10 percent for businesses were created in the late 1970s and early 1980s to stimulate investment in energy efficiency measures. Subsequent studies, however, were unable to show that the tax credits had expanded purchases of the technologies or measures involved (Clinton et al., 1986; OTA, 1992). This failure has been attributed to flaws in the design of the programs, notably that the incentives were too low, that they were based on cost rather than performance, and that they applied exclusively to commonplace energy efficiency measures such as home insulation and weather stripping. After costing the U.S. Treasury around $10 billion, these unsuccessful tax incentives were discontinued in 1985 (OTA, 1992).

Based in part on this experience, new tax credits were enacted in 2005 for innovative energy efficiency measures that included hybrid, fuel-cell and advanced diesel vehicles, highly efficient new homes and commercial buildings, and very efficient appliances. These tax credits were intended to support the commercialization and market development of these innovative technologies, but not necessarily to save a significant amount of energy. In addition, a 10 percent tax credit of up to $500 was adopted for energy retrofits to the building envelope of existing homes. Other than the tax credits for advanced vehicles, these new tax credits expired at the end of 2007, although most were extended as part of the financial rescue legislation enacted in October 2008. It is too early to evaluate the impact of the 2005 tax credits.

State and Utility Programs[25]

Joint actions of states and electric utilities have played a major role in advancing energy efficiency. Many state utility regulatory commissions or legislatures require electric utilities to operate energy efficiency programs, also known as demand-side management (DSM). Most of these programs are funded through a small surcharge on electricity sales. In some states, utilities are allowed to earn more profit on their energy efficiency programs than on building new power plants or other sources of energy supplies, thereby reducing or removing the utilities' financial disincentives to promote energy savings.

Overall, state/utility energy efficiency programs reduced electricity use in 2004 by about 74 TWh, or 2 percent of electricity sales nationwide (York and Kushler, 2006). But certain states stood out. California, Connecticut, Minnesota, Vermont, and Washington reduced electricity use in 2004 by 7–9 percent. Further, energy savings have risen since 2004 because overall DSM funding has increased. Assuming typical energy-savings rates, national savings reached approximately 90 TWh in 2006.

Promoting Combined Heat and Power Systems

Policy initiatives have also improved the efficiency of energy conversion and supply, specifically by expanding the use of combined heat and power (CHP), also known as cogeneration. Installed CHP reached 82 GW, at a total of more than 2800 sites, by 2004 (Hedman, 2005). It is estimated that the use of CHP systems resulted in total energy savings of about 2.8 quads in 2006, with perhaps 60 percent attributable to the Public Utilities Regulatory Policies Act (PURPA) of 1978 and other policy initiatives (Elliott and Spurr, 1999).

Consumer Education, Training, and Technical Assistance

Complementing the minimum efficiency standards and financial incentives just discussed, the ENERGY STAR® product-labeling program informs U.S. consumers

---

[25]This section summarizes a more extensive discussion in NAS-NAE-NRC (2009).

of the most efficient products in the marketplace at any given time. The ENERGY STAR® label helps consumers by reducing uncertainties about energy performance and lowering transaction costs for obtaining such information. The ENERGY STAR® label applies to a wide range of products, including personal computers and other types of office equipment, kitchen and laundry appliances, air conditioners and furnaces, windows, commercial appliances, and lighting devices. Whole structures—energy-efficient commercial buildings and new homes—also can qualify for the ENERGY STAR® label.

The ENERGY STAR® program in aggregate is estimated to have saved about 175 TWh of electricity in 2006 (EPA, 2007b). The program has achieved the most energy savings in the areas of commercial building improvements and personal computers, monitors, and other types of office equipment. The ENERGY STAR® program continues to develop criteria and adopt labeling for additional products, for instance, televisions and water heaters.

Summary of Estimated Savings from Policies and Programs

Table 4.11 provides estimates of the annual energy savings resulting from most of the policies and programs addressed in this chapter. In some cases (e.g., for CAFE standards and PURPA), the savings reflect expert judgments of the relative importance of the policies and market forces. The total energy savings from the nine policies and programs listed in Table 4.11, about 13.3 quads per year, was equivalent to 13-plus percent of national energy use in 2007. This level of savings is greater than the energy supplied by nuclear power and hydroelectric power combined. It is also more than five times the increase in the supply of renewable energy in the United States between 1973 and 2006.

It should be noted, however, that these policies and programs provided only a moderate amount of the total energy savings associated with the 50 percent decline in national energy intensity during 1973–2007. Increasing energy prices, ongoing technological change, and structural change have also contributed to the steep decline in energy intensity in the past 35 years.

Comparing energy savings across the various policies and programs listed in Table 4.11, regulatory initiatives such as the CAFE standards, appliance efficiency standards, and PURPA provided the greatest amount of energy savings. It should be recognized that some energy efficiency policy initiatives, such as RD&D efforts in the buildings sector, are not included in Table 4.11 in order to avoid double counting of savings.

TABLE 4.11 Estimates of Energy Savings from Major Energy Efficiency Policies and Programs

Policy or Program	Electricity Savings (TWh/yr)	Primary Energy Savings (Quads/yr)	Year	Source
CAFE vehicle efficiency standards	—	4.80	2006	NRC, 2002[a]
Appliance efficiency standards	196	2.58	2006	Nadel et al., 2006[b]
PURPA and other CHP initiatives	—	1.62	2006	Shipley et al., 2008[c]
ENERGY STAR® labeling and promotion	132	1.52	2006	EPA, 2007b[d]
Building energy codes	—	1.08	2006	Nadel, 2004[e]
Utility and state end-use efficiency programs	90	1.06	2006	York and Kushler, 2006[f]
DOE industrial efficiency programs	—	0.40	2005	DOE, 2007b
Weatherization assistance program	—	0.14	2006	DOE, 2006d[g]
Federal energy management program	—	0.11	2005	FEMP, 2006[h]
Total	—	13.31	—	—

Note: Estimates are based on the sources shown, augmented or modified as indicated.

[a]Extrapolation to 2006 of fuel savings estimated by NRC (2002), and assuming that 75 percent of the energy savings from vehicle efficiency improvements are due to the CAFE standards.

[b]Extrapolates between savings estimates by ACEEE for 2000 and 2010.

[c]Assumes that 85 percent of the energy savings from all CHP systems installed in 2006 was due to PURPA and other policy initiatives.

[d]Assumes 75 percent of energy savings estimated by U.S. EPA in order to avoid double counting savings with utility and state programs.

[e]Increases energy savings estimate for new buildings constructed during 1990–1999 from Nadel (2004) by 100 percent to account for the impact of codes prior to 1990 and post–1999.

[f]Extrapolates 2004 national electricity savings estimate to 2006 based on national DSM budget estimates for 2005 and 2006.

[g]Assumes 5.6 million weatherized households and average energy savings of 25 million Btu/yr per household, from Berry and Schweitzer (2003).

[h]Based on the reported reduction in energy use per square foot of floor area during 1985–2005 and actual primary energy use in federal buildings as of 2005 (i.e., excluding energy use by transport vehicles and equipment).

Experience in California and New York

This section describes the experience of two large states that have put many energy efficiency programs in place, predominantly for electricity, and have collected extensive data on the results. Both states have achieved electricity consumption per capita that is about 40 percent below the national average. Figure 4.8 illustrates electricity use per capita from 1960 to 2006 in California, New York, and the United States as a whole.

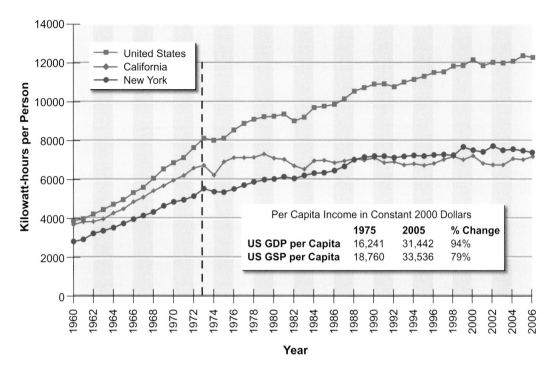

FIGURE 4.8 *Per capita electricity consumption in California, New York, and the United States, 1990–2006 (not including on-site generation).*
Note: GDP = gross domestic product; GSP = gross state product.
Source: Energy Information Administration, State Energy Consumption, Price, and Expenditure Estimates, available at www.eia.doe.gov/emeu/states/_seds.html.

California

As shown in Figure 4.8, California maintained nearly flat per capita electricity consumption from 1975 to the present. Per capita use in California is currently about 40 percent less than in the United States as a whole, even though the two were nearly equal in the 1960s.

There are many regulatory, demographic, and climatic reasons why the per capita electricity use in California has differed from that of the rest of the United States. In addition, the relatively high price of electricity in California has helped to lower demand. Approximately 25 percent of the differential, however, appears to be related to California's policies designed to reduce electricity use (Sudarshan and Sweeney, 2008). They include tiered electricity prices, utility-based incentives and other DSM programs, and codes and standards for energy use in buildings.

California first enacted efficiency standards for major types of appliances, as well as for new residential and commercial buildings, in the mid-1970s. These standards have been updated many times since then and have been extended to additional appliances.

California also adopted a number of policies intended to stimulate utility energy efficiency programs. They included the decoupling of utility profits from sales, the inclusion of efficiency as part of integrated resource planning, and the creation of performance incentives to meet or exceed efficiency targets. California's investor-owned utilities spent in excess of $600 million per year to promote more efficient electricity use by their customers as of 2007.[26] They can now earn a profit on these expenditures through the performance-based incentive program.

The combination of appliance standards, building energy codes, and utility efficiency incentives has resulted in considerable electricity savings in California. It is estimated that these initiatives have saved a total of some 40,000 GWh per year as of 2003, equivalent to about 15 percent of actual electricity use in the state that year (CEC, 2007).

New York

New York State has a long history of implementing policy actions to encourage more efficient use of energy across all sectors. They have included adoption and continual updating of building codes and appliance standards, for example, and well-funded research and development programs. Consequently, New York has maintained a relatively flat level of total energy use per capita (about 36 percent lower than the national average in 2005) for the past 30 years (see Figure 4.8).

New York's energy efficiency programs targeting energy consumers are designed to promote behavioral changes that favor adoption of a greater number of energy efficiency technologies, appliances, and services. Programs directed at electric utilities include the implementation of utility-run DSM efforts and a revenue-decoupling mechanism to allow utilities to recover revenues lost from reductions in energy demand due to efficiency measures. As a result of its energy efficiency initiatives since 1990, New York has lowered its annual electricity use by nearly 12,000 GWh, or about 8 percent (New York Energy $mart Annual Evaluation and Status Report, 2008).

---

[26]These utilities provide service to about 75 percent of the state's population. The remainder is served by municipal utilities and other public agencies.

Lessons Learned

The experiences of these two states in particular show that well-designed policies can overcome barriers to the use of energy-efficient technologies and can result in substantial energy savings. This is clear from the estimates in Table 4.11.

Minimum efficiency standards can be a very effective strategy for stimulating energy efficiency improvements on a large scale, especially if the standards are periodically updated. Such standards should not only be technically and economically feasible but also provide manufacturers with enough lead time to phase out production of nonqualifying products in an orderly manner.

Government-funded RD&D has contributed to the development and commercialization of a number of important energy efficiency technologies. While technological advancement is always a central objective of such grants, experience demonstrates that more attention should be devoted in the future to commercialization and market development. Also, a prudent RD&D portfolio should include high-risk but potentially high-payoff projects as well as those involving incremental improvements and lower risk (NRC, 2001).

Financial incentives, including those provided by utilities, can increase the adoption of energy efficiency measures. But these incentives should be carefully designed so as to avoid costly efforts that have little or no incremental impact on the marketplace.

Information dissemination, education, and training can raise awareness of energy efficiency measures and improve know-how with respect to energy management, including the successful implementation of building energy codes.

In general, energy efficiency policies and programs work best if they are integrated into market-transformation strategies that address the range of barriers present in a particular locale (Geller and Nadel, 1994). In the appliance market, for example, government-funded RD&D helps to nurture and commercialize new technologies; product labeling educates consumers; efficiency standards eliminate inefficient products from the marketplace; and incentives offered by some utilities and states encourage consumers to purchase products that are significantly more efficient than what the minimum standards specify.

Energy efficiency policies should be kept in place for a decade or more in order to ensure an orderly development of markets. Meanwhile, policies such as efficiency standards and targets, product labeling, and financial incentives should be periodically revised, as past successes and disappointments have shown. Dynamic policies steadily improved residential appliance efficiency, while stagnant policies failed to achieve continuing efficiency improvements in cars and light trucks during the 1990s and early part of this decade.

GENERAL FINDINGS: REAL PROSPECTS FOR ENERGY EFFICIENCY IN THE UNITED STATES

Energy efficiency technology for the buildings, transportation, and industry sectors exists today, or is expected to be developed in the normal course of business, that could save about 30 percent of the energy used in the U.S. economy by 2030. If energy prices remain high enough to motivate investment in energy efficiency, or if public policies have the same effect, energy use could be lower than business-as-usual projections by 15–17 quads by 2020 and by 32–35 quads by 2030. These energy efficiency improvements would save money as well as energy.

There are formidable barriers to improving energy efficiency. Overcoming them will require significant public and private support, as well as sustained resourcefulness. The experiences of states provide valuable lessons for national, state, and local policy makers in the leadership skills required and in the policies and programs that are most effective.

Particular attention must be paid to buildings, infrastructure, and other long-lived assets. Once an asset is installed, it embodies a level of energy use that is difficult to modify. Thus, it is important to take advantage of windows of opportunity for putting efficient technologies and systems in place.

REFERENCES

ACEEE (American Council for an Energy-Efficient Economy). 2007. Energy Bill Savings Estimates as Passed by the Senate. Washington, D.C.: ACEEE. Available at http://www.aceee.org/energy/national/EnergyBillSavings12-14.pdf.

AHAM (Association of Home Appliance Manufacturers). 2008. Data compiled by the Association of Home Appliance Manufacturers. Washington, D.C. www.aham.org.

AISI (American Iron and Steel Institute). 2003. Steel Industry Technology Roadmap: Barriers and Pathways for Yield Improvements. Prepared by Energetics, Inc. for the AISI. Washington, D.C.: AISI. October.

AISI. 2005. Saving One Barrel of Oil per Ton. Washington, D.C.: AISI. October.

Alamgir, M., and A.M. Sastry. 2008. Efficient batteries for transportation applications. SAE paper 2008-21-0017. SAE Convergence, Detroit, Mich. October.

An, F., and J. DeCicco. 2007. Trends in Technical Efficiency Trade-offs for the U.S. Light Vehicle Fleet. Technical Paper Series No. 2007-01-1325. Society of Automotive Engineers. April.

Apte, J., and D. Arasteh. 2006. Window-Related Energy Consumption in the US Residential and Commercial Building Stock. LBNL-60146. Berkeley, Calif.: Lawrence Berkeley National Laboratory. Available at http://gaia.lbl.gov/btech/papers/60146.pdf.

Bailey, O., and E. Worrell. 2005. Clean Energy Technologies: A Preliminary Inventory of the Potential for Electricity Generation. Report LBNL-57451. Berkeley, Calif.: Lawrence Berkeley National Laboratory. September.

Bandivadekar, A., K. Bodek, L. Cheah, C. Evans, T. Groode, J. Heywood, E. Kasseris, K. Kromer, and M. Weiss. 2008. On the Road in 2035: Reducing Transportation's Petroleum Consumption and GHG Emissions. Laboratory for Energy and the Environment Report, Massachusetts Institute of Technology.

Battelle (Battelle Memorial Institute). 2002. Toward a Sustainable Cement Industry: Climate Change. Substudy 8 of an independent study commissioned by the World Business Council for Sustainable Development. Columbus, Ohio: Battelle Memorial Institute.

Berry, Linda, and Martin Schweitzer. 2003. Metaevaluation of National Weatherization Assistance Program Based on State Studies, 1993–2002. ORNL/CON-488. Oak Ridge, Tenn.: Oak Ridge National Laboratory. February.

Brooks, S., B. Elswick, and R. Neal Elliott. 2006a. Combined Heat and Power: Connecting the Gap Between Markets and Utility Interconnection and Tariff Practices (Part I). American Council for an Energy-Efficient Economy (ACEEE) Technical Report IE062. Washington, D.C.: ACEEE.

Brooks, S., M. Eldridge, and R. Neal Elliott. 2006b. Combined Heat and Power: Connecting the Gap Between Markets and Utility Interconnection and Tariff Practices (Part II). American Council for an Energy-Efficient Economy (ACEEE) Technical Report IE063. Washington, D.C.: ACEEE.

Brown, M.A. 2001. Market failures and barriers as a basis for clean energy policies. Energy Policy 29:1197-1207.

Brown, M.A., and S. Chandler. 2008. Governing confusion: How statutes, fiscal policy, and regulations impede clean energy technologies. Stanford Law and Policy Review 19:427-509.

Brown, M.A., J. Chandler, M.V. Lapsa, and B.K. Sovacool. 2007. Carbon Lock-In: Barriers to Deploying Climate Change Mitigation Technologies. ORNL/TM-2007/124. Oak Ridge, Tenn.: Oak Ridge National Laboratory.

Brown, M.A., M. Levine, W. Short, and J. Koomey. 2001. Scenarios for a clean energy future. Energy Policy 33:1179-1196.

Brown, R., S. Borgeson, J. Koomey, and P. Biemayer. 2008. U.S. Building-Sector Energy Efficiency Potential. Berkeley, Calif.: Lawrence Berkeley National Laboratory.

Carter, S. 2001. Breaking the Consumption Habit: Ratemaking for Efficient Resource Decisions. The Electricity Journal 14:66-74.

CEC (California Energy Commission). 2007. California Energy Demand: 2008–2018. Staff Revised Forecast, FINAL Staff Forecast, Second Edition. CEC-200-2007-015-SF2, Table 4.2. Sacramento, Calif. November.

Clinton, J., H. Geller, and E. Hirst. 1986. Review of government and utility energy conservation programs. Annual Review of Energy 11:95-142.

Cowart, R. 2001. Efficient Reliability: The Critical Role of Demand-Side Resources in Power Systems and Markets. Report to the National Association of Regulatory Utility Commissioners by the Regulatory Assistance Project. Available at http://www.raponline.org/Pubs/General/EffReli.pdf. Accessed January 11, 2009.

Crabtree, G., M. Dresselhaus, and M. Buchanan. 2004. The hydrogen economy. Physics Today 57(12):39-44.

Craford, M.G. 2008. High power LEDs for solid state lighting: Status, trends, and challenges. Presented at First International Conference on White LEDs and Solid State Lighting, Tokyo, November 26–30, 2007.

DeCanio, S.J. 1993. Barriers within firms to energy-efficient investments. Energy Policy 21(9):906-914.

DOE (U.S. Department of Energy). 2000. Energy and Environmental Profile of the U.S. Chemical Industry. Energetics Incorporated. Available at http://www1.eere.energy.gov/industry/chemicals/tools_profile.html.

DOE. 2002. 2002 Manufacturing Energy Consumption Survey. Table 2a. Available at www.eia.doe.gov/emeu/efficiency/mecs_trend_mecs_trend9802.html.

DOE. 2005. U.S. Climate Change Technology Program Technology Options for the Near and Long Term. Climate Change Technology Program. Available at www.climatetechnology.gov. Accessed April 29, 2009.

DOE. 2006a. Energy Bandwidth for Petroleum Refining Processes. Energetics, Inc. October. Available at www1.eere.energy.gov/industry/petroleum_refining/bandwidth.html. Accessed April 29, 2009.

DOE. 2006b. 2006 Manufacturing Energy Consumption Survey. Available at www.eia.doe.gov/emeu/mecs/mecs2006/2006tables.html.

DOE. 2006c. Pulp and Paper Industry Energy Bandwidth Study. Prepared by Jacobs Engineering Group and the Institute of Paper Science and Technology for the American Institute of Chemical Engineers and DOE. Washington, D.C.: DOE. August. Available at http://www1.eere.energy.gov/industry/forest/bandwidth.html.

DOE. 2006d. Weatherization Assistance Program Fact Sheet. Washington, D.C.: DOE Office of Energy Efficiency and Renewable Energy, Weatherization Assistance Program. Available at http://www1.eere.energy.gov/office_eere/pdfs/weatherization_fs.pdf.

DOE. 2007a. Energy Department Determines That Model Commercial Building Code Will Save Energy and Benefit Consumers. Office of Energy Efficiency and Renewable Energy, Industrial Technologies Program. Available at www.energycodes.gov/implement/determinations_com_news.stm. Accessed April 29, 2009.

DOE. 2007b. Impacts. Industrial Technologies Program: Summary of Program Results for CY 2005. Office of Energy Efficiency and Renewable Energy, Industrial Technologies Program. Washington, D.C.

DOE. 2007c. Energy and Environmental Profile of the U.S. Petroleum Refining Industry. Prepared by Energetics, Inc., for the U.S. Department of Energy, Industrial Technologies Program. Washington, D.C. November. Available at http://www1.eere.energy.gov/industry/petroleum_refining/analysis.html.

DOE. 2008. Solid State Lighting Research and Development Multi-Year Program Plan FY'09–FY'13. Available at www1.eere.energy.gov/buildings/ssl/.

DOE. 2009. 2006 Manufacturing Energy Consumption Survey. Washington, D.C. Available at http://www.eia.doe.gov/emeu/mecs/contents.html.

EIA (Energy Information Administration). 2007a. Annual Energy Outlook 2007. DOE/EIA-0383(2007). Washington, D.C.: U.S. Department of Energy, Energy Information Administration.

EIA. 2007b. Annual Energy Review 2006. DOE/EIA-0384(2006). Washington, D.C.: U.S. Department of Energy, Energy Information Administration.

EIA. 2008. Annual Energy Outlook 2008. DOE/EIA-0383(2008). Washington, D.C.: U.S. Department of Energy, Energy Information Administration.

EIA. 2009a. Annual Energy Review 2008. DOE/EIA-0384(2008). Washington, D.C.: U.S. Department of Energy, Energy Information Administration.

EIA. 2009b. Monthly Energy Review (July). Washington, D.C.: U.S. Department of Energy, Energy Information Administration.

Elliott, R.N., and M. Spurr. 1999. Combined Heat and Power: Capturing Wasted Energy. Washington, D.C.: American Council for an Energy Efficient Economy.

EPA (U.S. Environmental Protection Agency). 2007a. Light-Duty Automotive Technology and Fuel Economy Trends: 1975 Through 2007. EPA420-R-07-008. Office of Transportation and Air Quality.

EPA. 2007b. ENERGY STAR® and Other Climate Protection Partnership Programs 2006 Annual Report. Office of Air and Radiation.

FEMP (Federal Energy Management Program). 2006. Annual Report to Congress on Federal Government Energy Management and Conservation Programs, Fiscal Year 2005. U.S. Department of Energy. Washington, D.C.

Frost & Sullivan. 2008. World Hybrid/Electric Vehicle Battery Markets. MC1812769. June 26.

Geller, H. 2003. Energy Revolution: Policies for a Sustainable Future. Washington, D.C.: Island Press.

Geller, H., and S. Nadel. 1994. Market transformation strategies to promote end-use efficiency. Annual Review of Energy and the Environment 19:301-346.

Goldman, C. 2006. Utility experience with real-time pricing. Workshop on Smart Meters and Time-Based Rates, March 15, Lawrence Berkeley National Laboratory.

Golove, W.H., and J.H. Eto. 1996. Market Barriers to Energy Efficiency: A Critical Reappraisal of the Rationale for Public Policies to Promote Energy Efficiency. LBL-38059. Berkeley, Calif.: Lawrence Berkeley National Laboratory.

Greene, D.L. 1998. Why CAFE worked. Energy Policy 26:595-614.

Griffith, B., N. Long, P. Torcellini, D. Crawley, and J. Ryan. 2007. Energy efficiency improvements that use the best available technologies and practices and integrated whole-building design approaches can, on average, reduce energy consumption by 43%. In: Assessment of the Technical Potential for Achieving Net Zero Energy Buildings in the Commercial Sector. NREL/TP-550-41957. Golden, Colo.: National Renewable Energy Laboratory. Available at http://www.nrel.gov/docs/fy08osti/41957.pdf.

Hall, N., C. Best, J. Roth, P. Jacobs, and L. Lutzenhiser. 2005. Assessing markets to design programs that more fully attack key market barriers and take advantage of market opportunities—Why do we continue to miss so many opportunities? Pp. 697-713 in Reducing Uncertainty Through Evaluation. Proceedings of the International Energy Program Evaluation Conference, August, Brooklyn, N.Y.

Hedman, B. 2005. CHP market status. Presentation at the Gulf Coast Roadmapping Workshop, April 26. Available at files.harc.edu/Sites/GulfCoastCHP/News/RoadmapWorkshop2005/GulfCoastCHPOverview.pdf. Accessed April 29, 2009.

IEA (International Energy Agency). 2004. Oil Crises and Climate Challenges: 30 Years of Energy Use in IEA Countries. Paris: Organisation for Economic Co-operation and Development (OECD)/IEA.

IEA. 2007. Tracking Industrial Energy Efficiency and CO_2 Emissions. Paris: IEA.

IEA. 2009. Energy Balances by Country/Region. Paris: OECD/IEA. Database available at http://www.iea.org/Textbase/stats/prodresult.asp?PRODUCT=Balances. Accessed June 25, 2009.

IWG (Interlaboratory Working Group on Energy-Efficient and Clean-Energy Technologies). 2000. Scenarios for a Clean Energy Future. ORNL/CON-476 and LBNL-44029. Oak Ridge National Laboratory and Lawrence Berkeley National Laboratory.

Johnson, T. 2008. Diesel emission control technology in review. SAE Technical Paper 2008-01-0069. Warrendale, Pa.: Society of Automotive Engineers. April.

Johnson, T. 2009. Diesel emission control in review. SAE Technical Paper 2009-01-0121. Warrendale, Pa.: Society of Automotive Engineers. April.

Koomey, J. 1990. Energy Efficiency Choices in New Office Buildings: An Investigation of Market Failures and Corrective Policies. PhD Thesis. Energy and Resources Group, University of California, Berkeley. Available at enduse.lbl.gov/Info/JGKdissert.pdf. Accessed April 29, 2009.

Koomey, J., and R.E. Brown. 2002. The Role of Building Technologies in Reducing and Controlling Peak Electricity Demand. LBNL-49947. Berkeley, Calif.: Lawrence Berkeley National Laboratory. September.

Kromer, M., and J. Heywood. 2008. A comparative assessment of electric propulsion systems in the 2030 U.S. light-duty vehicle fleet. SAE Technical Paper 2008-01-0459. Warrendale, Pa.: Society of Automotive Engineers.

LBNL (Lawrence Berkeley National Laboratory). 2005. Energy Efficiency Improvement and Cost Saving Opportunities for Petroleum Refineries: An ENERGY STAR® Guide for Energy and Plant Managers. Lawrence Berkeley National Laboratory.

Lee, J.J., S.P. Lukachko, and I.A. Waitz. 2004. Aircraft and Energy Use. Encyclopedia of Energy. Volume 1. Elsevier.

Lemar, P.L. 2001. The potential impact of policies to promote combined heat and power in U.S. industry. Energy Policy 29:1243-1254.

Lovins, A. 1992. Energy-Efficient Buildings: Institutional Barriers and Opportunities. Boulder, Colo.: E-Source.

Lutsey, N., and D. Sperling. 2005. Energy efficiency, fuel economy, and policy implications. Transportation Research Record 1941:8-25.

Martin, N., N. Anglani, D. Einstein, M. Khrushch, E. Worrell, and L.K. Price. 2000. Opportunities to Improve Energy Efficiency and Reduce Greenhouse Gas Emissions in the U.S. Pulp and Paper Industry. LBNL-46141. Berkeley, Calif.: Lawrence Berkeley National Laboratory.

McKinsey and Company. 2007. The Untapped Energy Efficiency Opportunity in the U.S. Industrial Sector. Report prepared for the U.S. Department of Energy, Office of Energy Efficiency and Renewable Energy.

Nadel, S. 2002. Appliance and equipment efficiency standards. Annual Review of Energy and the Environment 27:159-192.

Nadel, S. 2004. Supplementary Information on Energy Efficiency for the National Commission on Energy Policy. Washington, D.C.: American Council for an Energy Efficient Economy.

Nadel, S., A. de Laski, M. Eldridge, and J. Kleisch. 2006. Leading the Way: Continued Opportunities for New State Appliance and Equipment Standards. Report ASAP-6/ACEEE-A062. (Updated from and supersedes report ACEEE-051.) Washington, D.C.: American Council for an Energy-Efficient Economy (ACEEE) and Boston, Mass.: Appliance Standards Awareness Project. March.

NAS-NAE-NRC (National Academy of Sciences-National Academy of Engineering-National Research Council). 2009. Real Prospects for Energy Efficiency in the United States. Washington, D.C.: The National Academies Press.

NBI (New Buildings Institute). 2008. Getting to Fifty. White Salmon, Wash.: NBI. Available at http://www.newbuildings.org/gtf.

New York Energy $mart Annual Evaluation and Status Report. 2008. Available at http://www.nyserda.org/pdfs/Combined%20Report.

NRC (National Research Council). 2001. Energy Research at DOE: Was it Worth It? Washington, D.C.: National Academy Press.

NRC. 2002. Effectiveness and Impact of Corporate Average Fuel Economy (CAFE) Standards. Washington D.C.: The National Academies Press.

NRC. 2004. The Hydrogen Economy: Opportunities, Costs, and R&D Needs. Washington, D.C.: The National Academies Press.

NRC. 2008a. Transitions to Alternative Transportation Technologies: A Focus on Hydrogen. Washington D.C.: The National Academies Press.

NRC. 2008b. Review of the Research Program of the FreedomCar and Fuel Partnership: Second Report. Washington, D.C.: The National Academies Press.

NREL (National Renewable Energy Laboratory). 2002. Chemical Industry of the Future: Resources and Tools for Energy Efficiency and Cost Reduction Now. DOE/GO-102002-1529; NREL/CD-840-30969. October. Available at http://www.nrel.gov/docs/fy03osti/30969.pdf.

ORNL (Oak Ridge National Laboratory) and BCA, Inc. 2005. Materials for Separation Technologies: Energy and Emission Reduction Opportunities. Available at http://www1.eere.energy.gov/industry/imf/pdfs/separationsreport.pdf. Accessed April 29, 2009.

OTA (Office of Technology Assessment). 1992. Building Energy Efficiency. Washington, D.C.: U.S. Congress, OTA.

Parker, D.S., J.P. Dunlop, S.F. Barkaszi, J.R. Sherwin, M.T. Anello, and J.K. Sonne. 2000. Towards zero energy demand: Evaluation of super efficiency housing with photovoltaic power for new residential housing. Pp. 1.207-1.223 in Proceedings of the 2000 ACEEE Summer Study on Energy Efficiency in Buildings. Washington, D.C.: American Council for an Energy-Efficient Economy.

Rasmussen, T., V. Goepfrich, and K. Horkitz. 2005. Drivers of CFL purchase behavior and satisfaction: What makes a consumer buy and keep buying? In Reducing Uncertainty Through Evaluation: Proceedings of the International Energy Program Evaluation Conference. August. Brooklyn, N.Y., pp. 897-910.

Ricardo, Inc. 2008. A Study of Potential Effectiveness of Carbon Dioxide Reducing Vehicle Technologies. Prepared for EPA Office of Transportation and Air Quality. January. Available at http://www.epa.gov/OMS/technology/420r08004a.pdf.

Roth, K., and K. McKenney. 2007. Energy Consumption by Consumer Electronics in U.S. Residences. Report No. D5525. Cambridge, Mass.: TIAX, LLC.

Savaskan, R.C., S. Bhattacharya, and L.N. Van Wassenhoveet. 2004. Closed-loop supply chain models with product remanufacturing. Management Science 50(February):239-252.

Shipley, A., A. Hampson, B. Hedman, P. Garland, and P. Bautista. 2008. Combined Heat and Power: Effective Energy Solutions for a Sustainable Future. ORNL/TM-2008/224. Oak Ridge, Tenn.: Oak Ridge National Laboratory.

Small, K., and K. Van Dender. 2007. Fuel Efficiency and Motor Vehicle Travel: The Declining Rebound Effect. Working Paper 05-06-03. Irvine, Calif.: Department of Economics, University of California.

Sudarshan, A., and J. Sweeney. 2008. Deconstructing the "Rosenfeld Curve." Working paper, draft. Palo Alto, Calif.: Stanford University. June 1. Available at http://peec.stanford.edu/modeling/research/Deconstructing_the_Rosenfeld_Curve.php.

Thorp, B.A., and L.D. Murdock-Thorp. 2008. Compelling case for integrated biorefineries. Paper presented at the 2008 PAPERCON Conference, May 4-7, 2008, Dallas, Tex. Norcross, Ga.: Technical Association of the Pulp and Paper Industry.

Turner, C., and M. Frankel. 2008. Energy Performance of LEED for New Construction Buildings. White Salmon, Wash.: New Buildings Institute. Available at http://www.newbuildings.org/downloads/Energy_Performance_of_LEED-NC_Buildings-Final_3-4-08b.pdf.

van Oss, H.G. 2005. Background Facts and Issues Concerning Cement and Cement Data. Open-File Report 2005-1152. Washington, D.C.: U.S. Department of the Interior, U.S. Geological Survey.

Weber, C.L. 2009. Measuring structural change and energy use: Decomposition of the U.S. economy from 1997 to 2002. Energy Policy 37(April):1561-1570.

Worrell, E., and G. Biermans. 2005. Move over! Stock turnover, retrofit, and industrial energy efficiency. Energy Policy 33:949-962.

Worrell, E., and C. Galitsky. 2004. Energy Efficiency Improvement and Cost Saving Opportunities for Cement Making—An ENERGY STAR® Guide for Energy and Plant Managers. LBNL-54036. Lawrence Berkeley National Laboratory.

Worrell, E., and L. Price. 2001. Policy scenarios for energy efficiency improvement in industry. Energy Policy 29(14):1223-1241.

Worrell, E., L. Price, N. Martin, C. Hendriks, and L.O. Meida. 2001. Carbon dioxide emissions from the global cement industry. Annual Review of Energy and the Environment 26:303-329.

Xenergy, Inc. 1998. United States Industrial Electric Motor Systems Market Opportunities Assessment. Prepared for the U.S. Department of Energy, Office of Energy Efficiency and Renewable Energy. Available at http://www1.eere.energy.gov/industry/bestpractices/pdfs/mtrmkt.pdf.

York, D., and M. Kushler. 2006. A nationwide assessment of utility sector energy efficiency spending, savings, and integration with utility system resource acquisition. In Proceedings of the 2006 ACEEE Summer Study on Energy Efficiency in Buildings. Washington, D.C.: American Council for an Energy-Efficient Economy.

5 Alternative Transportation Fuels

The U.S. transportation sector relies almost exclusively on oil. Because domestic sources are unable to supply sufficient oil to satisfy the demands of the transportation and petrochemical industry sectors, the United States currently imports about 56 percent of its petroleum supply. Volatile crude oil prices and tight global supplies, coupled with fears of oil production peaking in the next 10–20 years, further aggravate concerns over oil dependence. The other key issue is greenhouse gas emissions from the transportation sector, which contribute one-third of the country's total emissions. These issues have motivated the search for alternative domestic sources of liquid fuels that also have significantly lower greenhouse gas emissions.

CONVERSION OF COAL AND BIOMASS TO LIQUID FUELS

Coal and biomass are in abundant supply in the United States, and they can be converted to liquid fuels for use in existing and future vehicles with internal-combustion and hybrid engines. Thus, they could be attractive candidates for providing non-oil-based liquid fuels to the U.S. transportation system. There are important questions, however, about the economic viability, carbon impact, and technology status of these options.

While coal liquefaction is potentially a major source of alternative liquid transportation fuels, the technology is capital intensive. Moreover, on a life-cycle basis,[1] coal liquefaction yields about twice the greenhouse gas emissions produced by petroleum-based gasoline when the carbon dioxide (CO_2) is vented to the atmosphere. Capturing this CO_2 and geologically storing it underground—a process frequently referred to as carbon capture and storage, or CCS—is therefore a requirement for production of coal-based liquid fuels in a carbon-constrained world. However, the viability of CCS, its costs, and its safety could pose a barrier to commercialization.

Biomass is a renewable resource that, if properly produced and converted, can yield biofuels with lower greenhouse gas emissions than petroleum-based gasoline yields. However, biomass production on fertile land already cleared might displace food, feed, or fiber production; moreover, if ecosystems were cleared to produce biomass for biofuels, the accompanying releases of greenhouse gases could negate for decades to centuries any greenhouse gas benefits from the biofuels (Fargione et al., 2008). Thus, there are questions about using biomass for fuel without seriously competing with other crops and without causing adverse environmental impacts.

This chapter assesses the potential for using coal and biomass to produce liquid fuels in the United States; provides consistent analyses of technologies for the production of alternative liquid transportation fuels; and discusses the potential for use of coal and biomass to substantially reduce U.S. dependence on conventional crude oil and also reduce greenhouse gas emissions in the transportation sector. Quantities in this chapter are expressed in the standard units commonly used by biomass producers. Greenhouse gas emissions, however, are expressed in tonnes of CO_2 equivalent, as in other chapters in this report. Details of the analyses and numerical estimates presented in this chapter can be found in the America's Energy Future panel report *Liquid Transportation Fuels from Coal and Biomass: Technological Status, Costs, and Environmental Impacts* (NAS-NAE-NRC, 2009).

---

[1]Life-cycle analyses include the "well-to-wheel," "mine-to-wheel," or "field-to-wheel" estimates of total greenhouse gas emissions—for example, from the time that the resource for the fuel is obtained from the oil well (in the case of petroleum-based gasoline) or from the coal mine (in the case of coal-to-liquid fuel) to the time that the fuel is combusted. In the case of biomass, the life-cycle analysis starts during the growth of biomass in the field and continues to the time that the fuel is combusted. Greenhouse gas emissions as a result of indirect land-use change, however, are not included in the estimates of greenhouse gas life-cycle emissions presented in this report.

FEEDSTOCK SUPPLY

Biomass Supply and Cost

While it is important that both the development of feedstocks for biofuels and the expansion of biofuel use in the transportation sector be achieved in a socially, economically, and environmentally sustainable manner, the social, economic, and environmental effects of domestic biofuels production have so far been mixed. In 2007, the United States consumed about 6.8 billion gallons of ethanol, made mostly from corn grain, and 491 million gallons of biodiesel, made mostly from soybean (EIA, 2008b), for a combined total of less than 3 percent of the U.S. transportation-fuel consumption. Diverting corn, soybean, or other food crops to biofuel production induces competition among food, feed, and fuel uses. Moreover, both for corn grain ethanol and soybean biodiesel, the use of fossil fuels and other inputs are substantial, and greenhouse gas reductions compared to petroleum-based gasoline emissions are small at best (Farrell et al., 2006; Hill et al., 2006). Thus, the committee judges that corn grain ethanol and soybean biodiesel are merely intermediates in the transition from oil to cellulosic biofuels or other biomass-based liquid hydrocarbon transportation fuels (for example, biobutanol and algal biofuels).

Assuming that technologies for conversion will be commercially viable, liquid fuels made from lignocellulosic biomass[2] can offer major greenhouse gas reductions relative to petroleum-based fuels, as long as the biomass feedstock is a residual product of some forestry and farming operations or is grown on marginal lands that are not used for food and feed crop production. Therefore, the committee focused on the lignocellulosic resources available for producing biofuels, and it assessed the costs of different feedstocks of this type—corn stover, wheat and seed-grass straws, hay, dedicated fuel crops, woody biomass, animal manure, and municipal solid waste—delivered to a biorefinery for conversion. Societal needs were considered by examining recent analyses of trade-offs between land use for biofuel production and land use for growing food, feed, and fiber, as well as for ecosystem services.

---

[2]Lignocellulosic biomass refers to biomass made of cellulose, hemicellulose, and lignin. Cellulose is a complex carbohydrate that forms the cell walls of most plants. Hemicellulose is a matrix of polysaccharides present, along with cellulose, in almost all plant cell walls.

The committee estimated the amounts of cellulosic biomass that could be produced sustainably in the United States and result in fuels with significantly lower greenhouse gas emissions than petroleum produces. For the purpose of this study, the committee considered biomass to be produced in a sustainable manner if it met the following criteria: (1) croplands would not be diverted for biofuels (so that land would not be cleared elsewhere to grow the crops thus displaced); and (2) the growing and harvesting of cellulosic biomass would incur minimal adverse environmental impacts—such as erosion, excessive water use, and nutrient runoff—or even reduce them.

The committee estimated (1) that about 400 million dry tons (365 million dry tonnes) per year of biomass could potentially be made available for the production of liquid transportation fuels using technologies and management practices of 2008 and (2) that the cellulosic biomass supply could increase to about 550 million dry tons (500 million dry tonnes) each year by 2020 (Table 5.1). A key assumption in the committee's analysis was that 18 million acres of land currently enrolled in the Conservation Reserve Program (CRP) would be used to grow perennial grasses or other perennial crops for biofuel production, and that the acreage would increase to 24 million acres by 2020 as knowledge increased with time. Other key assumptions were that (1) harvesting methods would be developed for efficient collection of forestry or agricultural residues; (2) improved

TABLE 5.1 Estimated Amount of Lignocellulosic Feedstock That Could Be Produced Annually for Biofuel Using Technologies Available in 2008 and in 2020

Feedstock Type	Million Tons	
	With Technologies Available in 2008	With Technologies Available by 2020
Corn stover	76	112
Wheat and grass straw	15	18
Hay	15	18
Dedicated fuel crops	104[a]	164
Woody biomass	110	124
Animal manure	6	12
Municipal solid waste	90	100
Total	416	548

[a]CRP land has not been used for dedicated fuel crop production as of 2008. As an illustration, the committee assumed that two-thirds of the CRP land would be used for dedicated fuel production.

TABLE 5.2 Estimate of Biomass Suppliers' Willingness-to-Accept Price
(in 2007 Dollars) per Dry Ton of Delivered Cellulosic Material

| Biomass | Willingness-to-Accept Price (dollars per ton) | |
	Estimated in 2008	Projected in 2020
Corn stover	110	86
Switchgrass	151	118
Miscanthus	123	101
Prairie grasses	127	101
Woody biomass	85	72
Wheat straw	70	55

management practices and harvesting technology would raise agricultural crop yield; (3) yield increases would continue at the historic rates seen for corn, wheat, and hay; and (4) all cellulosic biomass estimated to be available for energy production would be used to make liquid fuels. The last assumption allowed the committee to estimate the potential amount of such fuel that could be produced.

Although the committee estimated that 550 million dry tons of cellulosic feedstock could be harvested or produced sustainably in 2020, those estimates are not predictions of what would be available for fuel production in 2020. The actual supply of biomass could be greater if existing croplands were used more efficiently (Heggenstaller et al., 2008) or if genetic improvements to dedicated fuel crops resulted in higher yields. But the supply could be lower if producers decided not to harvest agricultural residues or grow dedicated fuel crops on their CRP land.

The committee also estimated the costs of biomass delivered to a conversion plant (Table 5.2). In this analysis, the price that the farmer or supplier would be willing to accept was assumed to include land-rental cost; other forgone net returns from not selling or using the cellulosic material for feed or bedding; and all other costs incurred in sustainably producing, harvesting, storing, and transporting the biomass to the processing plant. The cost or feedstock price is the long-run equilibrium price that would induce suppliers to deliver biomass to the conversion plant. Because an established market for cellulosic biomass does not exist, the analysis relied on estimates obtained from the literature. The committee's estimates are higher than those of other published reports because transportation and land-rental costs are included.

The geographic distribution of biomass supply is an important factor in the

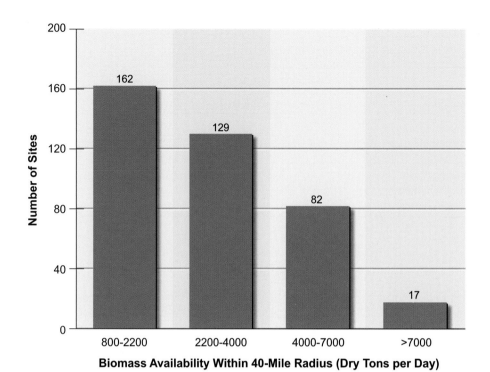

FIGURE 5.1 *The number of sites in the United States that can supply the indicated daily amounts of biomass from within a 40-mile radius of each site.*

development of the U.S. biofuels industry. For illustrative purposes, the committee estimated the quantities that could, for example, be available within a 40-mile radius (about a 50-mile driving distance) of fuels-conversion plants across the United States (Figure 5.1). With the exception of transport of woody material (primarily pulpwood), 40–50 miles has historically been the maximum distance considered economically feasible for biomass transport. An estimated 290 sites could supply from 1,500 up to 10,000 dry tons per day (from 0.5 million to 3.7 million dry tons per year) of biomass to conversion plants within a 40-mile radius. Notably, the wide geographic variation in potential biomass availability for processing plants affects their sizes. This variation suggests the potential to optimize each individual conversion plant to decrease costs and maximize environmental benefits and supply within a given region. Increasing the distance of delivery could result in larger conversion plants with lower fuel costs.

To help realize the committee's projected sustainable biomass supply, incentives could be provided to farmers and developers for using a systems approach to address biofuel production; soil, water, and air quality; carbon sequestration; wildlife habitat; and rural development in a comprehensive manner. Such incentives might encourage farmers, foresters, biomass aggregators, and biorefinery operators to work together to enhance technology development and ensure that best management practices were used for every combination of landscape and potential feedstock.

Findings: Biomass Supply and Cost

An estimated annual supply of 400 million dry tons of cellulosic biomass could be produced sustainably with technologies and management practices already available in 2008. The amount of biomass deliverable to conversion facilities could probably be increased to about 550 million dry tons by 2020. The committee judges that this quantity of biomass can be produced from dedicated energy crops, agricultural and forestry residues, and municipal solid wastes with minimal effects on U.S. food, feed, and fiber production and minimal adverse environmental effects.

Biomass availability could limit the size of a conversion facility and thereby influence the cost of fuel products from any facility that uses biomass irrespective of the conversion approach. Biomass is bulky and difficult to transport. The density of biomass growth will vary considerably from region to region in the United States, and the biomass supply available within 40 miles of a conversion plant will vary from less than 1,000 tons per day to 10,000 tons per day. Longer transportation distances could increase supply but would increase transportation costs and could magnify other logistical issues. The development of technologies that increase the density of biomass in the field, such as field-scale pyrolysis, could facilitate transportation of biomass to larger-scale regional conversion facilities.

Improvements in agricultural practices and in plant species and cultivars will be required to increase the sustainable production of cellulosic biomass and to achieve the full potential of biomass-based fuels. A sustained research and development (R&D) effort to increase productivity, improve stress tolerance, manage diseases and weeds, and improve the efficiency of nutrient use will help to improve biomass yields. Focused R&D programs supported by the federal government could provide the technical bases for improving agricultural practices and biomass growth to achieve the desired increase in sustainable production of cellulosic biomass. Attention could be directed toward plant breeding, agronomy, ecology, weed and pest science, disease management, hydrology, soil physics, agricultural engineering, economics, regional planning, field-to-wheel biofuel systems analysis, and related public policy.

Incentives and best agricultural practices will probably be needed to encourage sustainable production of biomass for production of biofuels. Producers need to grow biofuel feedstocks on degraded agricultural land to avoid direct and indirect competition with the food supply; they also need to minimize land-use practices that result in substantial net greenhouse gas emissions. For example, continuation of CRP payments for CRP lands when they are used to produce perennial grass and wood crops for biomass feedstock in an environmentally sustainable manner might be an incentive. A framework could be developed, with input from agronomists, ecologists, soil scientists, environmental scientists, and producers, to assess the effects of cellulosic-feedstock production on various environmental characteristics and natural resources. Such a framework would provide guidance to farmers on sustainable production of cellulosic feedstock and contribute to improvements in energy security and in the environmental sustainability of agriculture.

Coal Supply

Deployment of coal-to-liquid fuel technologies would require large quantities of coal and thus an expansion of the coal-mining industry. For example, because a plant producing 50,000 barrels per day (bbl/d) of liquid transportation fuels uses approximately 7 million tons of coal per annum, 100 such plants—producing 5 million bbl/d of liquid transportation fuels—would require about 700 million tons of coal per year, or a 70 percent increase in the nation's coal consumption. That would require major increases in coal-mining and transportation infrastructure, both in bringing coal from the mines to the plants and in bringing fuel from the plants to the market. These issues would represent major challenges, but they could be overcome. Thus, a key question is whether sufficient coal is available in the United States to support such increased consumption while also supplying other coal users, such as coal-fired electric power plants. In evaluating domestic coal resources, the National Research Council concluded:

> Despite significant uncertainties in existing reserve estimates, it is clear that there is sufficient coal at current rates of production to meet anticipated needs through 2030. Further into the future, there is probably sufficient coal to meet the nation's needs for more than 100 years at current rates of consumption. [However, a] combination of increased rates of production with more detailed reserve analyses that take into account location, quality,

recoverability, and transportation issues may substantially reduce the number of years of supply. Future policy will continue to be developed in the absence of accurate estimates until more detailed reserve analyses—which take into account the full suite of geographical, geological, economic, legal, and environmental characteristics—are completed. (NRC, 2007)

Recently, the Energy Information Administration estimated the proven U.S. coal reserves to be about 260 billion tons (EIA, 2009). A key conclusion of these two studies is that coal reserves in the United States are probably sufficient to meet the nation's needs for more than 100 years at current rates of consumption—and possibly even with increased rates of consumption. The primary issue is likely not to be reserves per se, however, but rather the increased mining of coal and the opening of many new mines. Increased mining would have numerous potential environmental impacts—and, possibly, heightened public opposition—which would need to be addressed in acceptable ways. Meanwhile, the cost of coal, which currently is low relative to the cost of biomass, would undoubtedly increase.

Finding: Coal Supply

Despite the vast coal resource in the United States, it is not a forgone conclusion that adequate coal will be mined and available to meet the needs of a growing coal-to-fuels industry and the needs of the power industry. The potential for a rapid expansion of the U.S. coal-supply industry would have to be analyzed by the U.S. coal industry, the U.S. Environmental Protection Agency, the U.S. Department of Energy, and the U.S. Department of Transportation so that the critical barriers to growth, environmental effects, and their effects on coal costs could be delineated. The analysis could include several scenarios, one of which would assume that the United States will move rapidly toward increasing use of coal-based liquid fuels for transportation to improve energy security. An improved understanding of the immediate and long-term environmental effects of increased mining, transportation, and use of coal would be an important goal of the analysis.

CONVERSION TECHNOLOGIES

Two key technologies, biochemical conversion and indirect liquefaction, are used for the conversion of biomass and coal into fuels, as illustrated in Figure 5.2.

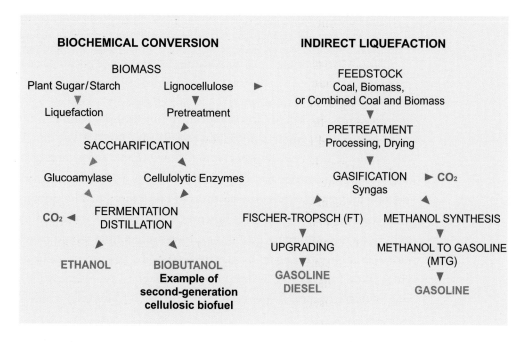

FIGURE 5.2 *Steps involved in the biochemical conversion of biomass and the thermo-chemical conversion (indirect route only) of coal, biomass, or combined coal and bio-mass into liquid transportation fuels.*

Biochemical conversion typically uses enzymes to transform starch (from grains) or lignocelluloses into sugars (saccharification), which are then converted into ethanol by microorganisms (fermentation). Thermochemical conversion includes indirect liquefaction, which uses heat and steam to convert biomass or coal into carbon monoxide and hydrogen (synthesis gas). The synthesis gas can then be catalytically converted into liquid fuels such as diesel and gasoline. The CO_2 from the fermentation process in biochemical conversion or from the offgas streams of the thermochemical processes can be captured and geologically stored. Direct liquefaction of coal (not shown in Figure 5.2), which involves adding hydrogen to slurried coal at high temperatures and pressures in the presence of suitable catalysts, represents another route from coal to liquid fuels, but it is less developed than is indirect liquefaction.

Biochemical Conversion

The biochemical conversion of starch (from grains) to ethanol, as depicted on the left side of Figure 5.2, has been commercially deployed. But while this process is important for stimulating public awareness and enhancing the industrial infrastructure for fuel ethanol, the committee considers grain-based ethanol to be a transition to cellulosic ethanol and other so-called advanced biofuels, because grain-based ethanol does not meet the sustainability criteria discussed above. The biomass supplies likely to be available by 2020 could technically be converted into ethanol by biochemical conversion, thereby displacing a significant proportion of petroleum-based gasoline and reducing greenhouse gas emissions, but the conversion technology has to be demonstrated first and developed into a commercially deployable state.

Over the next decade or two, cellulosic ethanol could be the main product of the biochemical conversion of biomass into fuels. Further research and development could also lead to commercial technologies that convert sugars into other biofuels such as butanol and alkanes, which have higher energy densities and could be distributed by means of the existing infrastructure. Although the committee focused on cellulosic ethanol as the most deployable technology over the next 10 years, it sees a long-term transition to conversion of cellulosic biomass to higher-energy alcohols or hydrocarbons—so-called advanced biofuels—as having significant long-term potential.

The challenge in biochemical conversion of biomass into fuels is to first break down the resistant structure of a plant's cell wall and then to break down the cellulose into five-carbon and six-carbon sugars fermentable by microorganisms; the effectiveness with which this sugar is generated is critical to economic biofuel production. The process for producing cellulosic ethanol, as shown in Figure 5.2, includes (1) preparation of the feedstock to achieve size reduction by grinding or other means; (2) pretreatment of the feedstock with steam, liquid hot water, or an acid or base to release cellulose from the lignin shield; (3) saccharification, by which cellulase hydrolyzes cellulose polymers into cellobiose (a disaccharide) and glucose (a monosaccharide), and hemicellulase breaks down hemicellulose into monosaccharides; (4) fermentation of the sugars into ethanol; and (5) distillation to separate the ethanol. The CO_2 generated by the conversion process and the combustion of the fuel is mostly offset by the CO_2 uptake during the growth of the biomass. The unconverted materials are burned in a boiler to generate steam for the distillation; some surplus electricity can thus be generated.

As of the end of 2008, no commercial-scale cellulosic ethanol plants were in operation. However, the U.S. Department of Energy (DOE) announced in February 2007 that it would invest up to $385 million for six biorefinery projects (two of them based on gasification) over 4 years to help bring cellulosic ethanol to market (DOE, 2007). When fully operational, the total production of these six plants would be 8000 bbl/d. In addition, a number of companies are actively pursuing commercialization of cellulosic ethanol plants. The corresponding technologies will continue to evolve over the next 5–10 years as challenges are overcome and experience is gained in the first technology-demonstration and commercial-demonstration plants. As a result, the committee expects deployable and commercialized technology to be in place by 2020 if technology-demonstration plants continue to be built, despite the current economic crisis, and if they are rapidly followed by commercial-demonstration plants.

The committee developed a model, in collaboration with the Massachusetts Institute of Technology, that estimated costs and CO_2 emissions for converting the biomass feedstocks just discussed into ethanol via biochemical pathways. The model included the effects of enzyme cost (10–40 cents per gallon [¢/gal]),[3] feedstock composition, solids loading (18–25 percent), and plants size (40 and 100 million gallons per year, corresponding to daily feed rates of 1400 and 3500 dry tons, respectively). The analysis also included the effects of pretreatment, hydrolysis, and fermentation yields. Three scenarios (representing low, medium, and high levels of improvements) were developed, in the form of process-cost estimates, representing current technology for the biochemical conversion of cellulosic feedstocks, reasonable evolutionary advancement of the technology, and the most optimistic advancement of the technology. (See NAS-NAE-NRC, 2009, for details on the analyses and results.)

The committee judges that the reasonable-improvement scenario best represents where the technology will be for 2020 deployment, and that the major-improvement scenario shows the considerable potential likely to remain. Results of the modeling for the woody biomass poplar, as an illustration of how technology improvements and the size of the ethanol plant could affect costs, are given in Table 5.3. The current costs of production are estimated for a biorefinery with a production capacity of 40 million gallons of ethanol per year; the committee accounted for the costs of production by 2020 by assuming reasonable technological advancements between now and then for the same-size plant. The estimated cost of production in 2020 at a biorefinery with a production capacity of 100 million gallons of ethanol per year is also shown to illustrate the economy of scale.

Table 5.3 shows that the cost of biomass (listed as "raw material-dependent

---

[3]Enzyme companies project enzyme costs to be about 40–50 cents by 2010 and about 20–30 cents by 2020 (Jensen, 2008). The cost of cellulase per gallon of ethanol produced in the Nth plant is modeled to be about 10–20 cents (DOE/EERE, 2007).

TABLE 5.3 Comparison of Costs (in 2007 Dollars) for Three Scenarios That Represent Low, Medium, and High Levels of Improvements in Technology and Process Efficiency in a Biorefinery Using the Woody Biomass Poplar

	Level of Improvement			Cost at Higher Capacity and Medium Improvement
	Poplar Low	Poplar Medium	Poplar High	
Plant capacity (million gallons)	40	40	40	100
Total capital ($ million)	223	194	174	349
Total capital ($ per annual gallon)	5.65	4.85	4.34	3.49
Total capital ($ per barrel per day)	87,000	75,000	67,000	61,000
Biomass used (dry tons)	593,000	514,000	461,000	1,286,000
Yield (gallons per ton)	67	78	87	78
Ethanol operating cost ($ per gallon)	1.95	1.40	0.90	1.30
Ethanol production cost ($ per gallon)	2.70	2.00	1.50	1.82
Facility-dependent fraction of cost (percent)	34	39	48	36
Raw material-dependent fraction of cost (percent)	57	51	40	57

fraction of cost") is a significant component of ethanol production costs. But with significant evolutionary improvement of the technology and scaling up of the operation, the process economics can be improved.

Ethanol has 66 percent as much energy as gasoline does. Ethanol is also hygroscopic and cannot be transported in existing fuel-infrastructure pipelines because of its affinity for water. It also is corrosive and can damage seals, gaskets, and other equipment and induce stress-corrosion cracking in high-stress areas. Ethanol is currently shipped by rail or barge. If ethanol is to be used in a fuel at concentrations higher than 20 percent ethanol (for example, in E85, which is a blend of 85 percent ethanol and 15 percent gasoline), the number of refueling stations will have to be increased. If ethanol is to replace a substantial volume of transportation gasoline, an expanded infrastructure will be required for its distribution. (The transport and distribution of synthetic diesel and gasoline produced from thermochemical conversion are less challenging because they are compatible with the existing infrastructure for petroleum-based fuels.)

Some key research, development, and demonstration challenges related to the ethanol-production process need to be overcome before the fuel's widespread commercialization can be achieved. These challenges are as follows: (1) improve the effectiveness of pretreatment in removing and hydrolyzing the hemicellulose, separating the cellulose from the lignin, and loosening the cellulose structure; (2) reduce the production costs of the enzymes for converting the cellulose to sugars; (3) reduce operating costs by developing more effective enzymes and more efficient microorganisms for converting the sugar products of biomass-deconstruction into biofuels; (4) demonstrate the biochemical-conversion technology on a commercial scale; and (5) begin to optimize capital costs and operating costs. The size of the biorefineries will likely be limited by the supply of biomass available from the surrounding regions. Such limitations could result in potential loss of the economies of scale that characterize large plants.

Findings: Biochemical Conversion

Process improvements in cellulosic-ethanol technology are expected to reduce the plant-related costs associated with ethanol production by up to 40 percent over the next 25 years. Over the next decade, process improvements and cost reductions are expected to come from evolutionary developments in technology, from learning gained through commercial experience and increases in the scale of operation, and from research and engineering in advanced chemical and biochemical catalysts that will enable their deployment on a large scale. Federal support for R&D programs is important for resolving the major technical challenges facing ethanol production from cellulosic biomass: pretreatment, suitable enzymes, tolerance to toxic compounds and products, solids loading, engineering microorganisms, and novel separations for ethanol and other biofuels. Designing the R&D programs with a long-term perspective could address current problems at a fundamental level and contribute to visible industrial goals. Furthermore, R&D programs that are closely coupled with pilot and commercial-scale demonstrations of cellulosic-ethanol plants could help resolve issues that arise during demonstrations.

Biochemical conversion processes, as configured in cellulosic-ethanol plants, produce a stream of relatively pure CO_2 from the fermenter that can be dried, compressed, and made ready for geologic storage or used in enhanced oil recovery with little additional cost. Geologic storage of the CO_2 from biochemical conversion of plant matter (such as cellulosic biomass) further reduces greenhouse gas life-cycle emissions from advanced biofuels, whose greenhouse gas life-cycle emissions would become highly negative. Because geologic storage of CO_2 from biochemical conversion of biomass to fuels could be important in reducing greenhouse gas emissions in the transportation sector, it could be evaluated and demonstrated in parallel with a program of geologic storage of CO_2 from coal-based fuels.

Future improvements in cellulosic technology that entail invention of biocatalysts and related biological processes could produce fuels that supplement ethanol production in the next 15 years. In addition to ethanol, advanced biofuels (such as lipids, higher alcohols, hydrocarbons, and other products that are easier to separate than ethanol) should be investigated because they could have higher energy content and would be less hygroscopic than ethanol and therefore could fit more compatibly into the current petroleum infrastructure than ethanol can. Large-scale commercial application of advances in biosciences (genomics, molecular biology, and genetics) and in biotechnologies to convert biomass directly to produce lipids, higher alcohols, and hydrocarbons fuels (that can be directly integrated into the existing transportation infrastructure) poses many challenges. These challenges will need to be resolved by R&D and demonstration if major advances in the production of alternative liquid fuels from renewable resources are to be realized. Research support from the federal government could help focus advances in bioengineering and the expanding biotechnologies on the development of advanced biofuels.

The need to expand the delivery infrastructure to meet a high volume of ethanol deployment could delay and limit the penetration of ethanol into the U.S. transportation-fuels market. Replacing a substantial proportion of transportation gasoline with ethanol will require a new infrastructure for ethanol's transport and distribution. Although the cost of delivery is a small fraction of the overall ethanol fuel cost, the logistics and capital requirements for widespread expansion could present many hurdles if they are not well planned.

A comprehensive study could be conducted jointly by the DOE and the biofuels industry to identify the infrastructure system requirements of, the research and development needs in, and the challenges facing the expanding biofuels industry. Such a study would consider the long-term potential of truck or barge delivery versus the potential of pipeline delivery that is needed to accommodate increasing volumes of ethanol, in addition to the timing and role of advanced biofuels that are compatible with the existing gasoline infrastructure.

Thermochemical Conversion

Indirect liquefaction converts coal, biomass, or mixtures of coal and biomass to liquid fuels by first gasifying the feedstocks to produce syngas, then cleaning it and adjusting its H_2-to-CO ratio (whereupon it is called synthesis gas) and catalytically converting the synthesis gas using Fischer-Tropsch (FT) technology into high cetane, clean diesel, and some naphtha (which can be upgraded to gasoline). The synthesis gas can also be converted into methanol using commercial technology, and methanol-to-gasoline (MTG) technology can then be used to produce high-octane gasoline from the methanol (Figure 5.2). These technologies can be integrated with those that compress the CO_2 emitted during production and store it underground—for example, in deep saline aquifers. Unlike ethanol, the gasoline and diesel produced via FT and MTG are fully compatible with the existing infrastructure and vehicle fleet.

Gasification has been used commercially worldwide for nearly a century by the chemical, refining, and fertilizer industries and for more than 10 years by the electric power industry. More than 420 gasifiers are currently in use in some 140 facilities worldwide, with 19 plants operating in the United States. Application to coal-to-liquid-fuel systems, and to combined coal-and-biomass gasification, will lead to further improvements in the technology so that it might become more robust and efficient by 2020. Gasification of biomass alone has been commercially demonstrated but requires added operational experience to render it more robust.

FT technology was first commercialized by the South African firm Sasol in the mid-1950s. Sasol now produces more than 165,000 bbl/d of transportation fuels from coal, and it has built large plants based on conversion of natural gas into synthesis gas, which is then converted into diesel and gasoline by FT. As with several other ready-to-deploy technologies, FT will likely undergo significant process improvements by 2020. For example, more robust and efficient technology for producing liquid transportation fuels, and significant catalyst improvements for coal applications, can be expected.

In technologies based on methanol synthesis, synthesis gas is converted to methanol using available commercial technology; plants as large as 6000 tons per day are currently operating. The methanol can be used directly or upgraded into high-octane gasoline using the proprietary MTG catalytic process developed by ExxonMobil and commercialized in New Zealand in the late 1980s.[4] Standard

---

[4]Some would place the option of methanol-to-olefins, gasoline, and diesel (commonly referred to as MOGD) on this list of technology options. Because of the lack of data and operating experience with that option, however, only the FT and MTG processes are described in this section.

MTG technology is considered by the committee to be commercially deployable today; a number of projects are in fact moving toward commercial deployment. Meanwhile, several variations on the technology, which could provide improvements, are ready for commercial demonstration.

While the technologies involved in thermochemical conversion of coal have all been commercialized and their operators have logged years of experience, geologic storage of CO_2 has not been adequately developed and demonstrated. For power generation from coal, most of the costs for CCS are in the CO_2 capture part of the process, and this technology has been demonstrated on a large scale. However, geologic storage of CO_2 in the subsurface has not been developed and demonstrated, except for use in enhanced oil recovery, and so there is insufficient confidence in its efficiency and long-term efficacy for commercial application at required scales. This is an important consideration for coal-to-liquid-fuels technology, as its CO_2 emissions are high because of the high carbon content of coal (about twice the carbon content of oil). Even with geologic storage of CO_2, the well-to-wheel emissions from coal-to-liquid fuels are about the same as those of gasoline because, as for any hydrocarbon fuel, CO_2 is released when the fuel is combusted in vehicles.

Inclusion of biomass in the feedstock with coal decreases the greenhouse gas life-cycle emissions because the biomass takes up atmospheric CO_2 during its growth. Thus, it is possible to optimize the biomass-plus-coal indirect liquefaction process to produce liquid fuels that have somewhat lower life-cycle greenhouse gas emissions than does gasoline, and even to make carbon-neutral liquid fuels if geologic storage of CO_2 is used. Although the notion of gasifying mixtures of coal and biomass to produce liquid fuels is relatively new and commercial experience is limited, several demonstration units are currently running in Europe. The committee judges that the technology for co-feeding biomass and coal is close to being ready for commercial deployment.

Gasifiers for biomass alone, designed around limited biomass availability, operate on a smaller scale than those for coal and thus will be more costly because of the diseconomies of scale of small plants. However, the fuels produced from such plants can have greenhouse gas life-cycle emissions that are close to zero without geologic storage of CO_2, and they can have highly negative carbon emissions if geologic storage of CO_2 is employed. The committee judges that stand-alone biomass gasification technology is probably 5–8 years away from commercial scale-up.

Working with the Princeton Environmental Institute, the committee analyzed the costs and CO_2 balances for thermochemical conversion of coal and biomass.

In these analyses, the viability of CCS was assumed to have been demonstrated by 2015 so that integrated coal-to-liquid fuel plants could start up by 2020. This assumption is ambitious, and focused and aggressive government action will be needed to make it happen. Four technologies, with and without CCS, were evaluated:

- A 50,000 bbl/d plant converting coal into diesel and naphtha using FT and then upgrading the naphtha to gasoline.
- A 50,000 bbl/d plant converting coal into gasoline using MTG.
- A 4,000 bbl/d plant converting biomass into diesel and naphtha using FT and then upgrading the naphtha to gasoline. The capacity of the plant is limited by the biomass supply of 4,000 dry tons per day.
- A 10,000 bbl/d plant converting biomass and coal into gasoline and diesel at a 40:60 ratio by feedstock energy (about 4,000 tons per day of biomass) using FT or MTG.

Some key results of the analysis are given in Table 5.4, and the complete results are contained in the report *Liquid Transportation Fuels from Coal and Biomass: Technological Status, Costs, and Environmental Impacts* (NAS-NAE-NRC, 2009). Details of models can be found in Kreutz et al. (2008) and Larson et al. (2008).

Table 5.4 shows that a large-scale coal plant with a 50,000 bbl/d capacity could produce fuels at a cost of about $50–70/bbl of crude oil (or about $60–80/bbl of gasoline equivalent). However, without CCS, the plant's CO_2 emissions would be double those of petroleum-based gasoline on a life-cycle or well-to-wheels basis. Results with MTG are comparable. But even with CCS, both the FT and the MTG process produce low-cost fuels, and the CO_2 emissions are similar to those of petroleum gasoline.

The engineering cost of CCS is about $10–15 per tonne of CO_2 avoided. The coal-to-liquid plant configurations produce a concentrated stream of CO_2 as an integral part of the process, so CO_2 capture can be readily and more cheaply achievable than that, for example, in integrated gasification combined-cycle or pulverized-coal plants. The FT and MTG options without CCS are relevant if reduced CO_2 emissions are not desired and if energy supply and diversity of supply are the overriding societal issues. However, in a carbon-constrained world, there will be a drive to produce fuels with zero net CO_2 emissions. A plant that used combined coal and biomass as a feedstock with CCS could produce

TABLE 5.4 Fuel Costs and CO_2 Emissions for Thermochemical Conversion of Coal and Biomass

	Coal-to-Liquid FT	Coal-to-Liquid FT	Coal-to-Liquid MTG	Coal-and-Biomass-to-Liquid FT	Biomass-to-Liquid FT
	Without CCS	With CCS	With CCS	With CCS	With CCS
Inputs:					
Coal (tons per day as received)	26,700	26,700	23,200	3,030	0
Biomass (dry tons per day)	0	0	0	3,950	3,950
Biomass (mass %)	0	0	0	57	100
Biomass energy (%, low heating value)	0	0	0	42	100
Outputs:					
Gasoline (bbl/d)	21,290	21,290	50,000	4,260	
Diesel (bbl/d)	28,700	28,700	0	5,750	
Total liquid fuels (bbl/d)	50,000	50,000	50,000	10,000	4,410
Economic metrics:					
Specific total plant cost ($/bbl per day)	97,600	98,900	80,400	134,000	147,000
Total liquid fuels cost ($/gal of gasoline equivalent)	1.50	1.64	1.57	2.52	3.32
Break-even oil price ($/bbl)	56	68	51	103	139
Emissions relative to petroleum-derived fuels	2.18	1.03	1.17	−0.02	−1.35
Cost of avoided CO_2 ($/tonne)[a]	—	11	10	15	20

Note: CCS = carbon capture and storage; FT = Fischer-Tropsch; MTG = methanol-to-gasoline.
[a]Includes the costs of CO_2 transport and geologic storage and is expressed as dollars per tonne of CO_2 equivalent avoided.

10,000 bbl/d of fuels with close to zero CO_2 emissions. Note that the case shown is for FT, but the economics would look similar if MTG were used. FT primarily produces diesel; MTG produces gasoline. The economics show that the capital costs of coal-and-biomass-to-liquid fuel plants are higher than the costs of coal-to-liquid fuel plants.

The CO_2 emissions are near zero on a life-cycle basis because the biomass in the feedstock is a carbon sink, offsetting some of the coal carbon. The key assumption in this case is that biomass availability is limited to 4000 tons per day by regional harvesting and transportation considerations. In those sites where locally sustainable biomass densities are higher (see Figure 5.1), larger plants—perhaps as many as 100 nationwide—could be built at similar biomass-to-coal ratios and result in lower cost for carbon-neutral fuels. The last column of Table 5.4 shows the case for gasification with biomass as the single feedstock. The costs are high because of the small plant size, limited by feedstock availability. However, the life-cycle CO_2 emissions are net negative, which would be attractive if overall costs of production could be brought down.

The area of greatest uncertainty for conversion of coal and biomass into liquid fuels is the geologic storage of CO_2. As of late 2008, few commercial-scale geologic storage demonstrations had been carried out or were ongoing. Yet well-monitored and commercial-scale demonstrations are needed to gather data sufficient to assure industry and governments of the long-term viability, costs, and safety of geologic CO_2 storage and to develop procedures for site choice, permitting, operation, regulation, and closure. These objectives are particularly critical to the commercial success of thermochemical technology, which relies on the political and commercial acceptability of large-scale geologic storage of CO_2.

The potential costs of CCS of \$10–15 per tonne of CO_2 avoided are "bottom-up" estimates, based largely on engineering estimates of expenses for transport, land purchase, permitting, drilling, capital equipment, storage, well capping, and monitoring for an additional 50 years. However, uncertainty about the regulatory environment arising from concerns of the general public and policy makers has the potential to raise storage costs and slow commercialization of thermochemical fuel production technology. Ultimate requirements for design, monitoring, carbon-accounting procedures, and liability for long-term monitoring of geologically stored CO_2, as well as the associated regulatory frameworks, are dependent on future commercial-scale demonstrations of geologic storage of CO_2. These demonstrations will have to be pursued aggressively over the next few years if thermo-chemical conversion of biomass and coal with geologic storage of CO_2 is to be ready for commercial deployment in 2020 or sooner.

As a first step toward accelerating the commercial demonstration of coal-to-liquid and coal-and-biomass-to-liquid fuels technology and addressing the CO_2 storage issue, commercial-scale demonstration plants could serve as sources of CO_2 for geologic storage demonstration projects. So-called capture-ready plants that vented CO_2 would create liquid fuels with higher CO_2 emissions per unit of usable energy than petroleum-based fuels produce; commercialization of these plants would not be encouraged unless they were integrated with geologic storage of CO_2 at their start-up.

Direct liquefaction of coal—which involves relatively high temperature, high hydrogen pressure, and liquid-phase conversion of coal directly into liquid products—has a long history, as does the FT process. Direct liquefaction products generally are heavy liquids that require significant further upgrading into liquid transportation fuels. The technology is not ready for commercial deployment. Further, because of the absence of recent detailed design studies in the available literature, the committee's ability to estimate costs and performance is limited.

The three most significant R&D priorities for commercialization of thermochemical technologies are these:

- Immediate construction of a small number of commercial first-mover projects, combined with geologic storage of CO_2, that put the technology on the path toward reduced cost, improved performance, and robustness. These projects would have major R&D components that focus on solving problems identified in the operation of plants and on developing technology for specific improvements.
- R&D programs, associated with commercial-scale geologic CO_2 storage demonstrations, that involve detailed geologic analysis and a broad array of monitoring tools and techniques to provide the data and understanding upon which future commercial projects will depend.
- Research that determines the penalties associated with preprocessing of biomass, the choice of a best gasifier for a given biomass type, the technical problems with feeding biomass to high-pressure gasification systems, and the answers to related questions. Biomass gasification and combined biomass and coal gasification have potential CO_2-reduction benefits, but they can be brought to commercialization only if such practical issues are resolved.

Findings: Thermochemical Conversion

Technologies for the indirect liquefaction of coal to transportation fuels are commercially deployable today; without geologic storage of the CO_2 produced in the conversion, however, greenhouse gas life-cycle emissions will be about twice those of petroleum-based fuels. With geologic storage of CO_2, coal-to-liquid transportation fuels could have greenhouse gas life-cycle emissions equivalent to those of equivalent petroleum-derived fuels.

Technologies for the indirect liquefaction of coal to produce liquid transportation fuels with greenhouse gas life-cycle emissions equivalent to those of petroleum-

based fuels can be commercially deployed before 2020 only if several first-mover plants are started up soon and if the safety and long-term viability of geologic storage of CO_2 is demonstrated in the next 5–6 years.

Indirect liquefaction of combined coal and biomass to transportation fuels is close to being commercially deployable today. Coal can be combined with biomass at a ratio of 60:40 (on an energy basis) to produce liquid fuels that have greenhouse gas life-cycle emissions comparable to those of petroleum-based fuels if CCS is not implemented. With CCS, production of fuels from coal and biomass would have a carbon balance of about zero to slightly negative. A program of aggressive support for first-mover commercial plants that produce coal-to-liquid transportation fuels and coal-and-biomass-to-liquid transportation fuels with integrated geologic storage of CO_2 would have to be undertaken immediately if the United States were to address energy security with those fuels that have greenhouse gas emissions similar to or less than those of petroleum-based fuels. If decisions to proceed with commercial demonstrations are made soon so that the plants could start up in 4–5 years, and if CCS is demonstrated to be safe and viable, those technologies would be commercially deployable by 2020.

The technology for producing liquid transportation fuels from biomass or from combined biomass and coal via thermochemical conversion has been demonstrated but requires additional development to be ready for commercial deployment. For example, key technologies for biomass gasification would have to be demonstrated on an intermediate scale, alone and in combination with coal, to obtain the engineering and operating data required to design synthesis-gas-production units on a commercial scale.

Geologic storage of CO_2 on a commercial scale is critical for producing liquid transportation fuels from coal without a large adverse greenhouse gas impact. This is similar to the situation for producing power from coal. The operational procedures, monitoring, safety, and effectiveness of commercial-scale technology for geologic storage of CO_2 would have to be demonstrated in an aggressive program if geologic storage of CO_2 is to be ready for commercial deployment by 2020. Three to five commercial-scale demonstrations (each with about 1 million tonnes of CO_2 per year and operated for several years) would have to be set up within the next 3–5 years in areas with different geologic stroage media. The demonstrations would focus on the site choice, permitting, monitoring, operation, closure, and legal procedures needed to support the broad-scale application of the technology and would provide the needed engineering data and other information to determine the full costs of geologic storage of CO_2.

COSTS, CO_2 EMISSIONS, AND SUPPLY

This section compares the life-cycle costs, CO_2 emissions,[5] and potential supplies of the alternative liquid fuel options for technologies deployable by 2020. The result of its analyses is a supply curve of fuels that use biomass, coal, or combined biomass and coal as feedstocks.

It should be noted that the supply curve does not represent the actual amounts of fuels that would be commercially available in 2020. Those supplies could well be smaller because of critical lags—both in the decisions to construct new conversion plants and in the construction itself—as discussed in the deployment section that follows. In addition, some of the coal and biomass supplies that appear to be economical might not be made available for conversion to alternative fuels because of logistical, infrastructural, and organizational issues or because they have already been committed to electric power plants. The analyses show how the potential supply curve might change with alternative carbon dioxide prices and alternative capital costs.

As mentioned earlier, the committee worked with several research groups to develop the costs and CO_2 emissions of the individual conversion technologies and the cost of biomass. The analyses presented in this section use those inputs to derive life-cycle costs and CO_2 emissions for the alternative fuels.

To examine the potential supply of liquid transportation fuels from non-petroleum sources, the committee developed estimates of the unit costs and quantities of various biomass sources that could be made available. The committee's analysis was based on use of land that is currently not used for growing foods, although the committee cannot ensure that this land would not be used for food production in the future. The estimates of biomass supply were combined with estimates of supply of corn grain to satisfy the current legislative requirement to produce 15 billion gallons of ethanol per year. The analysis allowed the estimation

---

[5]This section only assesses CO_2 emissions because the committee was unable to determine changes in other greenhouse gases throughout the life cycle of fuel production. Such changes, however, are likely to be small.

of a supply curve for biomass that shows the quantities of biomass feedstocks that would potentially be available at various unit costs. *Coal was assumed to be available in sufficient quantities at a constant unit cost if used with biomass in thermochemical conversion processes.* Quantitative analyses were developed to compare alternative pathways to convert biomass, coal, or combined coal and biomass to liquid transportation fuels using thermochemical technologies. Biochemical technology that produced ethanol from biomass was also evaluated quantitatively on as consistent a basis as possible. Various combinations of biomass feedstocks could, in principle, be converted with either thermochemical or biochemical conversion processes.[6] However, rather than examining all possible combinations, the committee first examined the cost of and CO_2 emission associated with each of the various thermochemical and biochemical conversion processes by using a generic biomass feedstock with approximately a median cost and biochemical composition (the committee used *Miscanthus* in the analysis) and then examined the costs, supplies, and CO_2 emissions associated with one thermochemical conversion process and one biochemical conversion process that would use each of the different biomass feedstocks. The following assumptions underlie the analyses:

- All suitable CRP land is allocated to the growing of biomass for liquid fuels. Conversion plants that use biomass as a feedstock by itself or combined with coal (with 60 percent coal and 40 percent biomass on an energy basis) have the capacity of about 4000 dry tons of biomass per day.
- All product prices are free of government subsidies. The total cost of CO_2 avoided, which includes the costs of drying, compression, pipelining, and geologic storage of CO_2, is estimated to be in the range of $10–15 per tonne.
- If a carbon price is imposed, it applies to the entire life-cycle CO_2 net emissions—the balance of CO_2 removal from the atmosphere by plants, CO_2 released in the production of biomass, emissions from conversion of the feedstock to fuel, and emissions from combustion of the fuel. A process that removes more CO_2 from the atmosphere than it produces receives a net payment for CO_2.

---

[6]In addition, the committee included a biochemical conversion of corn grain to ethanol but did not focus the quantitative analysis on this process.

- No indirect greenhouse gas emissions result from land-use changes in the growing and harvesting of biomass.
- The price of subbitumimous Illinois #6 coal is $42 per dry ton.
- Electricity generated as a coproduct is valued at $80/MWh, absent any price placed on greenhouse gas emissions.
- The biomass and co-fed coal/biomass conversion plants are sized for biomass feed rates of approximately 4000 dry tons per day.
- The biomass feedstock is *Miscanthus*, a high-yield perennial grass costing $101 per dry ton.

Costs and CO_2 Emissions

The estimated 2020 supply function for biomass cost versus availability is shown in Figure 5.3. The costs of two of the feedstocks—corn grain and hay—are based on recent market prices. The corn price in particular is assumed to have dropped sharply from the 2008 high of $7.88 per bushel to $3.17 per bushel in 2020, corresponding to $130 per dry ton—a price more consistent with its historical levels. The price of hay is assumed to be $110 per dry ton, also similar to historical prices. The costs of most of the other feedstocks—corn stover, straw, high-yield grasses (such as *Miscanthus*), normal-yield grasses (such as native and mixed grasses and switchgrass), and woody biomass—are estimated from the growing, harvesting, transportation, and storage costs reported in the literature. Finally, the cost of using municipal solid wastes is based on a rough estimate of the costs of gathering, transporting, and storing them; although such costs can be highly variable, the committee assumes that they add up to $51 per dry ton.

The costs of producing alternative liquid fuels through the various pathways were estimated on the basis of the feedstock, capital, and operating costs, the conversion efficiencies, and the assumptions outlined above. Figure 5.4 shows the estimated gasoline-equivalent[7] costs of alternative liquid fuels, without a CO_2 price, produced from biomass, coal, or combined coal and biomass. Liquid fuels are produced using biochemical conversion—to make cellulosic ethanol from *Miscanthus*—or using thermochemical conversion via FT or MTG. For thermochemical conversion, FT and MTG are shown both with and without CCS. The cost of ethanol produced from corn grain is also included in Figure 5.4. For

[7]Costs per barrel of ethanol are divided by 0.67 to put ethanol costs on an energy-equivalent basis with gasoline. For FT liquids, the conversion factor is 1.0.

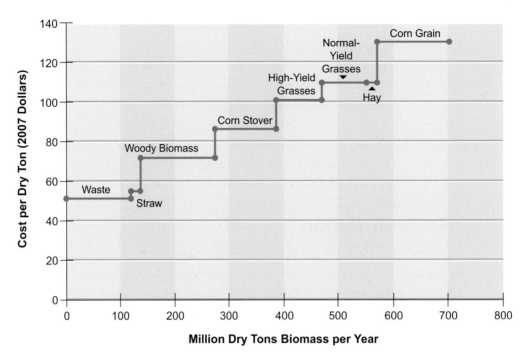

FIGURE 5.3 *Supply function for biomass feedstocks in 2020. High-yield grasses include* Miscanthus *and normal-yield grasses include switchgrass and prairie grasses.*

comparison, costs of gasoline are shown in Figure 5.4 for two different crude oil prices: $60/bbl and $100/bbl (that is, $73 and $113 per barrel of gasoline equivalent). Results are also shown in Table 5.5.

Figure 5.5 shows the net CO_2 emissions per barrel of gasoline equivalent produced by various production pathways. The CO_2 released during combustion of the fuel is similar among the options, with ethanol releasing less CO_2 than is released with either gasoline or synthetic diesel and gasoline. But a large variation in net releases results from the CO_2 taken out of the atmosphere when biomass is grown and from the significant differences in CO_2 released into the atmosphere during the conversion process. CO_2 emissions for corn grain ethanol are slightly lower than those of gasoline. In contrast, CO_2 emissions of cellulosic ethanol without CCS are close to zero.

Figure 5.4 shows that FT coal-to-liquid fuel products with and without geologic CO_2 storage are cost-competitive at gasoline-equivalent prices below $70/bbl (this represents equivalent crude-oil prices of about $55/bbl) and that prices for MTG are somewhat lower. Figure 5.5 shows that without CCS, both FT and MTG vent a large amount of CO_2—over twice that of petroleum gasoline on a

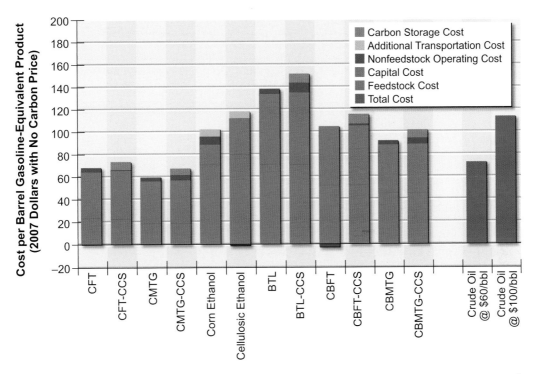

FIGURE 5.4 *Cost of alternative liquid fuels produced from coal, biomass, or coal and biomass with no carbon price.*
Note: BTL = biomass-to-liquid fuel; CBFT = coal-and-biomass-to-liquid fuel, Fischer-Tropsch; CBMTG = coal-and-biomass-to-liquid fuel, methanol-to-gasoline; CCS = carbon capture and storage; CFT = coal-to-liquid fuel, Fischer-Tropsch; CMTG = coal-to-liquid fuel, methanol-to-gasoline.

life-cycle basis. With CCS, the life-cycle CO_2 emissions from FT and MTG are about the same as those from petroleum gasoline.

The biochemical conversion of biomass produces fuels that are more expensive than coal-to-liquid fuels because the conversion plants are small and the feedstock is more expensive—biomass costs almost four times as much as coal on an energy-equivalent basis. The production cost of cellulosic ethanol is around $115/bbl on a gasoline-equivalent basis. The cost of thermochemical conversion of biomass, without coal, is higher than the cost of cellulosic ethanol on an energy-equivalent basis and with geologic storage has the potential for large negative net releases of CO_2; that is, the process involves a net removal of CO_2 from the atmosphere. For biomass-to-liquid and venting of CO_2, the estimated fuel cost is $140/bbl if electricity is sold back to the grid at $80/MWh; with geologic storage of CO_2, it is $150/bbl if electricity is sold back to the grid at $80/MWh. The

TABLE 5.5 Estimated Costs of Various Fuel Products With and Without a CO_2-Equivalent Price of $50 per Tonne[a]

Fuel Product	Cost Without CO_2-Equivalent Price ($/bbl gasoline equivalent)	Cost With CO_2-Equivalent Price of $50/Tonne ($/bbl gasoline equivalent)
Gasoline at crude oil price of $60/bbl	75	95
Gasoline at crude oil price of $100/bbl	115	135
Cellulosic ethanol	115	110
Biomass-to-liquid without CCS	140	130
Biomass-to-liquid with CCS	150	115
Coal-to-liquid without CCS	65	120
Coal-to-liquid with CCS	70	90
Coal-and-biomass-to-liquid without CCS	95	120
Coal-and-biomass-to-liquid with CCS	110	100

[a]Numbers are rounded to nearest $5. Estimated costs of fuel products for coal-to-liquids conversion represent the mean costs of fuels produced via FT and MTG.

results of the relatively small co-fed coal and biomass plant (total feed, 8000 tons per day) are particularly interesting. Fuels produced by that plant cost about $95/bbl on a gasoline-equivalent basis without CCS, and CO_2 atmospheric releases from plants with CCS are negative. Those results point to the importance of that option in the U.S. energy strategy.

The important influence of CO_2 price on fuel price is shown in Figure 5.6. In reading the graph, it is important to note that it shows the breakdown of all costs, including negative costs such as credit from electricity generation or carbon uptake. These negative costs must be subtracted from the positive ones in order to obtain the actual costs. For example, the cost of biomass-to-liquid fuel with CCS is $151/bbl – $37/bbl = $114/bbl. CO_2 emissions for corn grain ethanol are slightly lower than for gasoline. In contrast, CO_2 emissions of cellulosic ethanol without CCS are close to zero.

Figure 5.6 shows that a CO_2 price of $50 per tonne significantly increases the costs of the fossil-fuel options, including the costs of petroleum-based gasoline. The large amount of CO_2 vented in the coal-to-liquids process without CO_2 storage almost doubles the cost of product once a carbon price of $50 per tonne of CO_2 is imposed. The carbon price brings the cost of biochemical conversion options down to about $110/bbl.

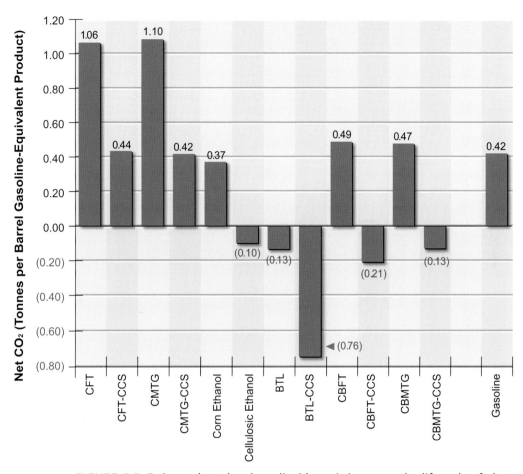

FIGURE 5.5 *Estimated total carbon dioxide emissions over the life cycle of alternative fuels production—from the mining and harvesting of resources to the conversion to and consumption of fuels.*
Note: BTL = biomass-to-liquid fuel; CBFT = coal-and-biomass-to-liquid fuel, Fischer-Tropsch; CBMTG = coal-and-biomass-to-liquid fuel, methanol-to-gasoline; CCS = carbon capture and storage; CFT = coal-to-liquid fuel, Fischer-Tropsch; CMTG = coal-to-liquid fuel, methanol-to-gasoline.

Inclusion of a carbon price does not increase the total costs of all thermochemical pathways. For example, thermochemical conversion of biomass costs about $150/bbl of gasoline equivalent with CCS, but with the carbon price and CCS, the produced fuels become competitive with petroleum-based fuels at about $115/bbl of gasoline equivalent ($100/bbl of crude oil equivalent). In general, if a pathway takes more CO_2 from the atmosphere than it releases in other parts of its

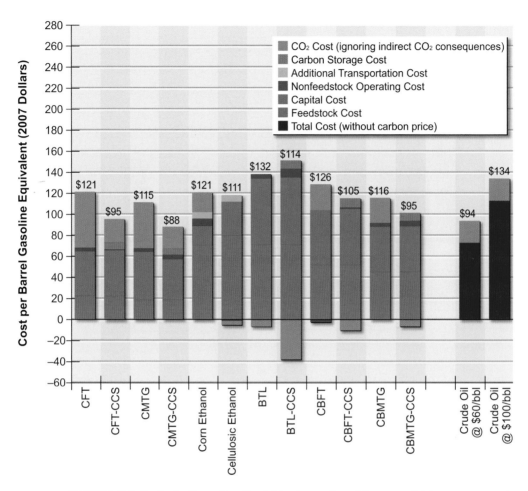

FIGURE 5.6 *Cost of alternative liquid fuels produced from coal, biomass, or coal and biomass with a $50/tonne CO₂ price. Negative cost elements must be subtracted from the positive elements; the number at the top of each bar indicates the net costs.*
Note: BTL = biomass-to-liquid fuel; CBFT = coal-and-biomass-to-liquid fuel, Fischer-Tropsch; CBMTG = coal-and-biomass-to-liquid fuel, methanol-to-gasoline; CCS = carbon capture and storage; CFT = coal-to-liquid fuel, Fischer-Tropsch; CMTG = coal-to-liquid fuel, methanol-to-gasoline.

life cycle, the inclusion of a carbon price reduces the pathway's total cost of producing liquid fuel. Note that these estimates are all based on costs for small gasification units operating at a feed rate of 4,000 dry tons per day. If larger units were deployed in regions where potential biomass availability is large—for example, 10,000 dry tons per day—the result could be significantly lower costs.

Costs and Supply

As previously noted, the cost estimates for biochemical conversion and thermo-chemical conversion are based on only one biomass feedstock, *Miscanthus*. Moreover, Figures 5.4 to 5.6 do not show how much fuel could be produced at the estimated costs. To provide a more complete picture of alternative liquid fuels, the supply function from Figure 5.3 for all biomass feedstocks has been combined with the conversion-cost estimates. (The potential supply of gasoline and diesel from coal-to-liquids technology is discussed in the section below titled "Deployment of Alternative Transportation Fuels.") The results are presented in Figures 5.7 and 5.8.

Figure 5.7 shows the potential gasoline-equivalent supply of ethanol from biochemical conversion of lignocellulosic biomass and corn grain, with technology deployable in 2020. The supply of grain ethanol satisfies the current legislative requirement to produce 15 billion gallons of ethanol per year in 2022. Figure 5.7

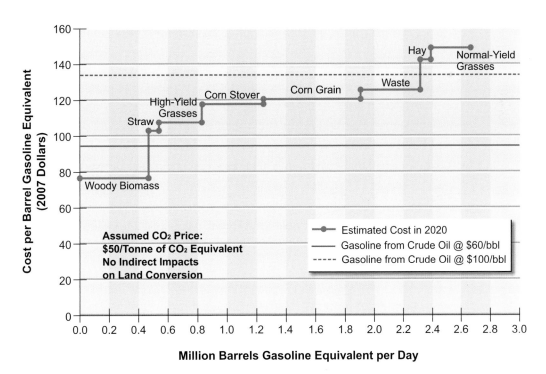

FIGURE 5.7 *Estimated supply of cellulosic ethanol plus corn grain ethanol at different price points in 2020. The red solid and dotted lines show, for comparison, the supply of crude oil at $60 and $100 per barrel.*

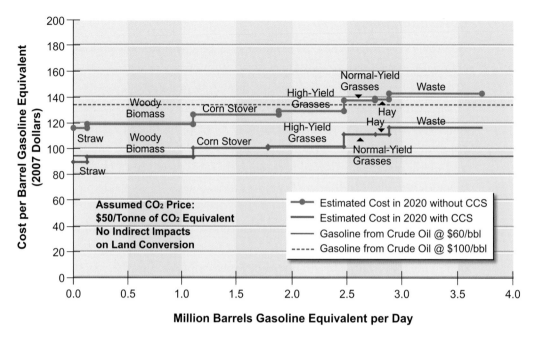

FIGURE 5.8 *Estimated supply of gasoline and diesel produced by thermochemical conversion via Fischer-Tropsch, with or without CCS, at different price points in 2020.*

shows potential supply and not the committee's projected penetration of cellulosic ethanol in 2020. This is because it does not incorporate lags in implementation of the technology that will result because of the time required to obtain permits for and build the infrastructure to produce and transport these alternative liquid fuels. The estimated supply of synthetic gasoline and diesel derived from coal and biomass as feedstocks is shown in Figure 5.8. Two different supply functions are plotted, one with CCS and the other without CCS. They show that if the CCS technologies are viable and a price of $50 per tonne of CO_2 is implemented, then for each feedstock it will be less costly to use CCS than to release the CO_2 into the atmosphere.

Either of the production processes underlying Figures 5.7 or 5.8 would use the same supplies of biomass. Therefore the quantities cannot be added. If all of the production (in addition to ethanol produced from corn grain) were based on cellulosic conversion, the quantities shown in Figure 5.7 would be applicable. If all production were based on thermochemical conversion co-fed with biomass and coal, then the quantities shown in Figure 5.8 would be applicable. Most likely, some of the production would be based on cellulosic processes and some based on thermochemical processes, so the actual potential supply function would lie between the two sets of supply functions shown in Figures 5.7 and 5.8. If corn grain ethanol (shown in Figure 5.7) has not been phased out by 2020, it would add about 0.67 million barrels per day of gasoline-equivalent production to the supply.

To put the results in perspective, the gasoline and diesel used by light-duty vehicles (LDVs) in the United States in 2008 was estimated to be about 9 million barrels of oil equivalent per day (1 bbl of crude oil produces about 0.85 bbl of gasoline equivalent). Total liquid fuels used in the United States was 21 million barrels per day, 14 million of which were used for transportation and 12 million of which were imported. Thus the 2 million barrels of gasoline equivalent of ethanol produced from cellulosic biomass and the 0.7 million barrels of gasoline equivalent of ethanol produced from corn grain have the potential to replace about 30 percent of the U.S. petroleum-based fuel consumed by LDVs, or almost 20 percent of all transportation fuels.

The potential supply of gasoline or diesel fuel from thermochemical conversion of a combination of biomass and coal (with CCS) is greater than with biochemical conversion of biomass alone. Moreover, the costs of thermochemical conversion of combined coal and biomass are lower than those of either biochemical or thermochemical conversion of biomass alone. The cost differences occur because coal is a lower-cost feedstock than is biomass. In addition, co-feeding coal and biomass allows a larger plant to be built and reduces capital costs per volume of product.

Using 60 percent coal and 40 percent biomass on an energy basis, almost 4 million barrels per day of gasoline equivalent—and thus of oil—can potentially be displaced from transportation. This would amount to 60 billion gallons of gasoline equivalent per year, or almost 45 percent of the gasoline and diesel used by LDVs in 2008. (The calculation assumes that all of the 550 million dry tons of cellulosic biomass sustainably grown for fuel will be used for coal-and-biomass-to-liquid fuel production. Thus the estimates represent the maximum potential supply.)

Findings: Costs and Supply

Alternative liquid transportation fuels from coal and biomass have the potential to play an important role in helping the United States to address issues of energy security, supply diversification, and greenhouse gas emissions with technologies that are commercially deployable by 2020.

- **With CO_2 emissions similar to those from petroleum-based fuels,** a substantial supply of alternative liquid transportation fuels can be produced with thermochemical conversion of coal with geologic storage of CO_2 at a gasoline-equivalent cost of $70/bbl.
- **With CO_2 emissions substantially lower than those from petroleum-based fuels,** up to 2 million barrels per day of gasoline-equivalent fuel can technically be produced with biochemical or thermochemical conversion of the estimated 550 million dry tons of biomass available in 2020 at a gasoline-equivalent cost of about $115–140/bbl. Up to 4 million barrels per day of gasoline-equivalent fuel can be technically produced if the same amount of biomass is combined with coal (60 percent coal and 40 percent biomass on an energy basis) at a gasoline-equivalent cost of about $95–110/bbl. However, the technically feasible supply does not equal the actual supply inasmuch as many factors influence the market penetration of fuels.

DEPLOYMENT OF ALTERNATIVE TRANSPORTATION FUELS

The discussion in this chapter thus far has addressed the potential supply of alternative transportation fuels from technologies ready to be commercially deployed by 2020; potential supply, however, does not translate into what will be *available* at that time. The rates at which alternative liquid fuels can penetrate the market will depend on many variables. In addition to technological readiness, they include such factors as oil price, carbon taxes, the construction environment, and labor availability. To illustrate the lag between the time when technology becomes commercially deployable and the time when significant market penetration of its product occurs, the committee developed a few plausible scenarios.

Cellulosic Ethanol

Regarding biochemical conversion to cellulosic ethanol, the committee took into account the current activities with demonstration plants, the announced commercial plants, the DOE road map, and the rate of construction of grain ethanol plants. It assumed that a capacity of 1 billion gallons per year of cellulosic ethanol would be achievable by 2015 and that the capacity build beyond 2015 would follow one of two scenarios. The first tracks the maximum capacity build experienced with grain ethanol (about a 25 percent yearly increase in capacity over a 6-year period); the second scenario is an aggressive capacity build rate that is approximately twice that achieved for grain ethanol. The two scenarios project 7–12 billion gallons of cellulosic ethanol per year (about 0.3–0.5 million barrels of gasoline equivalent per day) by 2020. Continued aggressive capacity build could conceivably achieve the Renewable Fuel Standard's[8] mandated capacity of 16 billion gallons of cellulosic ethanol per year by 2022, but this would be a stretch. Continued aggressive capacity build could yield 30 billion gallons of cellulosic ethanol per year by 2030 and up to 40 billion gallons per year of cellulosic ethanol by 2035. The latter would consume about 440 million dry tons of biomass annually and replace 1.7 million barrels per day of petroleum-based fuels.

Coal-to-Liquid Fuels with CCS

If commercial demonstrations of coal-to-liquids fuel production with CCS were begun immediately and CCS were proven viable and safe by 2015, commercial plants could be starting up before 2020. The subsequent growth rate could be about two to three plants per year. This scenario would reduce dependence on imported oil, but it would not reduce CO_2 emissions from transportation. At a build-out rate of two plants (at 50,000 bbl/d of fuel) per year, 2 million bbl/d of liquid fuels would be produced from 390 tons of coal annually by 2035, at a cost of about $200 billion. At a build-out rate of three plants per year, 3 million bbl/d of liquid fuels would be produced from 580 million tons of coal each year. The latter case would replace approximately one-third of the current U.S. oil use in light-duty transportation and increase U.S. coal production by 50 percent. At a build-out rate of three plants starting up per year, five to six plants would be under construction at any one time.

Coal-and-Biomass-to-Liquid Fuels

The technology for co-fed biomass and coal plants is close to being developed, and several commercial plants without CCS have in fact started to co-feed bio-

---

[8]The Renewable Fuel Standard (RFS) was created by the 2005 U.S. Energy Policy Act; the Energy Independence and Security Act of 2007 amended RFS to set forth "a phase-in for renewable fuel volumes beginning with 9 billion gallons in 2008 and ending at 36 billion gallons in 2022." The 36 billion gallons would include 16 billion gallons of cellulosic ethanol.

mass. Although this will allow them to acquire some operational experience and reduce cost, gaining experience with CCS in particular is critical, as it will probably be required. Because coal and biomass plants are much smaller than coal-to-liquid fuel plants (at 10,000 bbl/d of fuel, a coal-and-biomass-to-liquid fuel plant size is one-fifth the size of a coal-to-liquid fuel plant), biomass feed rates are similar to those of cellulosic ethanol plants. Thus, penetration rates should in principle closely follow the cellulosic plant build out. But most likely the coal and biomass build out will be much slower than the aggressive rate of building cellulosic plants just presented because of more complex plant design and the need to site the plants near *both* biomass and coal production.

Thus, the committee assumed that penetration rates of the coal-and-biomass-to-liquid fuel plants will be slightly less than that of the cellulosic ethanol build-out case that follows the experience of grain ethanol (which has experienced a 25 percent average annual growth rate). At a 20 percent average annual growth rate until 2035, when 280 plants would be in place, 2.5 million bbl/d of gasoline equivalent would be produced. This would consume about 300 million dry tons of biomass (less than the projected biomass availability) and about 250 million tons of coal per year. The analysis shows that capacity growth rates would have to exceed historical rates considerably if 550 million dry tons per year of biomass were to be converted to liquid fuels in 2030.

Findings: Coal-and-Biomass-to-Liquid Fuels

If commercial demonstration of cellulosic-ethanol plants is successful and commercial deployment begins in 2015, and if it is assumed that capacity will grow by 50 percent each year, cellulosic ethanol with low CO_2 life-cycle emissions can replace up to 0.5 million barrels of gasoline equivalent per day by 2020 and 1.7 million barrels per day by 2035.

If commercial demonstration of coal-and-biomass-to-liquid fuel plants with carbon capture and storage is successful and the first commercial plants start up in 2020, and if it is assumed that capacity will grow by 20 percent each year, coal-and-biomass-to-liquid fuels with low CO_2 life-cycle emissions can replace up to 2.5 million barrels of gasoline equivalent per day by 2035.

If commercial demonstration of coal-to-liquid fuel plants with carbon capture and storage is successful and the first commercial plants start up in 2020, and if it is assumed that capacity will grow by two to three plants each year, coal-to-liquid fuels with CO_2 life-cycle emissions similar to those of petroleum-based fuels can replace up to 3 million barrels of gasoline equivalent per day by 2035. That option would require an increase in U.S. coal production by 50 percent.

The deployment of alternative liquid transportation fuels aimed at diversifying the U.S. energy portfolio, improving energy security, and reducing environmental impacts by 2035 would require aggressive large-scale demonstration in the next few years and strategic planning to optimize the use of coal and biomass to produce fuels and to integrate them into the transportation system. Aggressive development and demonstration of cellulosic-biofuel and thermochemical-conversion technologies with CCS are necessary to advance these technologies and to address challenges identified in the commercial demonstration programs. Given the magnitude of U.S. liquid-fuel consumption (14 million barrels of crude oil per day in the transportation sector) and the scale of current petroleum imports (about 56 percent of the petroleum used in the United States is imported), a business-as-usual approach is insufficient to address the need to find alternative liquid transportation fuels, particularly because development and demonstration of technology, construction of plants, and implementation of infrastructure require 10–20 years per cycle. An assessment of the current government and industry programs would determine their adequacy to meet the commercialization timeline required to reduce U.S. oil use and CO_2 emissions over the next decade.

Developing detailed scenarios of market penetration rates of biofuels, coal-to-liquid fuels, and associated biomass and coal supply options would help clarify hurdles and challenges to achieving substantial effects on U.S. oil use and CO_2 emissions. Such analysis will provide policy makers and business leaders with the information needed to establish enduring policies and investment plans for accelerating the development and penetration of alternative-fuels technologies.

A potential optimal strategy for producing biofuels in the United States could be to locate thermochemical conversion plants that use coal and biomass as a combined feedstock in regions where biomass is abundant and locate biochemical-conversion plants in regions where biomass is less concentrated. Thermochemical plants require a larger capital investment per barrel of product than biochemical conversion plants require and thus benefit to a greater extent from economies of scale. This strategy could maximize the use of cellulosic biomass and minimize the costs of fuel products. An assessment of the spatial distribution of potential

U.S. biomass supply would allow determination of the optimal size of conversion plants for particular locations in relation to the road network and the costs and greenhouse gas effects of feedstock transport. The assessment could be conducted by the U.S. Department of Energy and the U.S. Department of Agriculture and the information could be combined with the logistics of coal delivery to such plants to develop an optimal strategy for using U.S. biomass and coal resources for producing sustainable biofuels.

ENVIRONMENTAL IMPACTS BEYOND GREENHOUSE GAS EMISSIONS

Biomass Supply

Although greenhouse gas emissions have been the central environmental focus regarding biomass production for alternative liquid fuels, other key effects must also be considered. On the one hand, lignocellulosic biomass feedstocks offer distinct advantages over food crop feedstocks with respect to water-use efficiency, nutrient and sediment loading in waterways, enhancement of soil fertility, emissions of criteria pollutants, and safeguarding habitat for wildlife and other species, especially those that provide biocontrol services for crop production. On the other hand, many of the traits of dedicated fuel crops have been shown to contribute to their invasiveness.

Biochemical Conversion

The biochemical conversion of cellulosic biomass into ethanol or other biofuels requires process water for cooling, heating, and mixing with reagents that are associated with hydrolysis and fermentation. The amount of water required is estimated at 2–6 gallons per gallon of ethanol produced; the lower levels would be approached if the plant's design included the recycling of process water. The processing of cellulosics into ethanol also results, in principle, in a residual water stream that needs to undergo wastewater treatment. However, an efficient process will ferment most of the feedstock's sugars into ethanol, leaving only low amounts of organic residuals.

Air emissions resulting from bioprocessing include CO_2, water vapor, and possibly sulfur or nitrogen. Fermentation processes release CO_2 as a result of microbial metabolism. Water vapor is also released, particularly if the lignin coproduct is dried prior to being shipped from the plant for use as boiler fuel at an off-site power-generation facility. The sulfur and nitrogen content of the fermentation residues is typically low, unless chemicals are used in the pretreatment in the biomass materials. Such chemicals, however, can be recovered.

Thermochemical Conversion

Coal-to-liquid fuel plants can be configured to minimize their impacts on the environment, given that the clean-coal technologies that have been developed for the electric power industry can also be used in coal-to-liquid applications. Coal-to-liquid fuel plants need to produce clean synthesis gas from coal using gasification and gas cleaning technologies. As a result, emissions of criteria pollutants and toxics such as sulfur oxides, nitrogen oxides, particulates, and mercury will be low.

The sulfur compounds in the coal are converted into elemental sulfur, which can be sold as a byproduct. The ammonia in the synthesis gas can either be recovered and sold as a fertilizer or sent to wastewater treatment, where it is absorbed by bacteria. All of the mercury, arsenic, and other heavy metals in the syngas are adsorbed on activated charcoal. The coal's mineral matter (ash), which is exposed to extremely high temperatures during gasification, becomes vitrified into slag. This slag is nonleachable and finds use in cement or concrete. Nitrogen-oxide emissions from existing conversion technologies are only about 3 parts per million.

Water usage in thermochemical conversion plants depends primarily on the water-use philosophy implicit in the plant design. For the conversion of coal and combined coal and biomass to transportation fuels with all water streams recycled or reused, the major consumptive use of water would generally be for cooling, hydrogen, and solids handling. If water availability were not limited—say, because of ready access to rivers—conventional forced or natural draft cooling towers would be used. In arid areas, where water is indeed limited, air-cooling would be used to the maximum degree possible. Depending on the degree of air-cooling, water consumption could range from about 1 to 8 barrels of water per barrel of product. For coal-to-liquid fuel plants, additional environmental impacts will be associated with the mining of coal, as discussed in the reports *Evolutionary and Revolutionary Technologies for Mining* (NRC, 2002) and *Coal Research and Development to Support National Energy Policy* (NRC, 2007).

BARRIERS TO DEPLOYMENT

Successful development of an industry to supply alternative liquid transportation fuels faces some technological and sociological challenges. These challenges are not trivial, but they can be successfully overcome.

Challenge 1

- Developing a systems approach through which farmers, biomass integrators, and those operating biofuel-conversion facilities can develop a well-organized and sustainable cellulosic-ethanol industry that will address multiple environmental concerns (for example, biofuel production; soil, water and air quality; carbon sequestration; wildlife habitat; rural development; and rural infrastructure) without creating unintended consequences through piecemeal development efforts.
- Determining the full greenhouse gas life-cycle signatures of various biofuel crops.
- Certifying the greenhouse gas benefits for different potential biofuel scenarios.

In other words, failure to link the critical environmental, economic, and social needs and address them as an integrated system could reduce the availability of biomass for conversion to levels significantly below the 550 million tons technically deployable in 2020.

Challenge 2

For the thermochemical conversion of coal, or of combined coal and biomass, to have any significant impact on reducing U.S. reliance on crude oil and on reducing CO_2 emissions over the next 20–30 years, CCS will have to be shown to be safe as well as economically and politically viable. The technological viability of CO_2 capture is already proven, although commercial-scale demonstration plants are now needed to quantify and improve costs and performance. Additional programs will be required to help resolve storage and regulatory issues associated with geologic CO_2 storage approaching a scale of gigatonnes per year. *In the analyses presented in this study, the viability of CCS was assumed to have been demonstrated by 2015 so that integrated coal-to-liquid fuel plants could start up by 2020. This assumption is ambitious and will require focused and aggressive government* action to make it happen. But uncertainties about the regulatory environment, especially those arising from concerns of the general public, have the potential to raise storage costs above those cited in this study. Meanwhile, ultimate requirements for selection, design, monitoring, carbon accounting procedures, liability, and associated regulatory frameworks have yet to be developed, creating the possibility of delay in initiating demonstration projects and, later, in licensing individual commercial projects. Large-scale demonstrations and establishment of procedures for operation and long-term monitoring of CCS have to be actively pursued in the next few years if thermochemical conversion of biomass and coal is to be ready for commercial deployment by 2020.

Challenge 3

Cellulosic ethanol is in the early stages of commercial development. A few commercial demonstration plants are expected to begin operations over the next several years, and most process improvements will likely come from evolutionary developments, knowledge gained through commercial experience, and increases in scale of operation. Incremental improvements of biochemical conversion technologies can be expected to reduce nonfeedstock process costs by up to 40 percent by 2030. It will take focused and sustained industry and government action to achieve those cost reductions, but some key technical challenges remain:

- Developing more efficient pretreatment to free up cellulose and hemicellulose and to enable more efficient downstream technology conversion. Improved pretreatment is not likely to reduce product cost substantially because the pretreatment cost is small relative to other costs.
- Creating better enzymes, not subject to end-product inhibition, for facilitating the conversion process.
- Maximizing solids loading in the reactors.
- Engineering organisms capable of fermenting the sugars in a toxic biomass hydrolysate and producing high concentrations of the final toxic product biofuel; improving microorganism tolerance to toxicity is a key issue.

Challenge 4

If ethanol is to be used in large quantities in LDVs, an expanded ethanol transportation and distribution infrastructure will be required. Because ethanol cannot be transported in pipelines used for petroleum transport, as discussed earlier in this chapter, it is currently transported by rail or barge. But if cellulosic biomass were dedicated to thermochemical conversion by FT or MTG, the resultant fuels would be chemically equivalent to conventional gasoline and diesel. They could thus be transported via existing pipelines, and the infrastructural challenge associated with ethanol would be minimized.

Challenge 5

The committee's analyses provide a snapshot of the potential costs of liquid fuels—from biomass by biochemical or thermochemical conversion, and from combined biomass and coal by thermochemical conversion. But the costs of fuels are dynamic, fluctuating as a result of externalities such as the costs of feedstocks, labor, and construction; the economic environment; and government policies. With the wide variation in most commodity prices, especially for oil, investors will need to have confidence that policies—including carbon caps, carbon price, mandated greenhouse gas reductions, or tariffs on imported oil—will ensure that alternative liquid transportation fuels can compete with fuels refined from crude oil. The price of carbon emissions, or the existence of fuel standards that require specified reductions in fuels' life-cycle greenhouse gas emissions, will affect the relative economic choices.

TECHNOLOGIES READY FOR DEPLOYMENT BEYOND 2020

Algal Biodiesel

Biodiesel refers to diesel fuel made by transesterifying oil from biological sources. A potential biodiesel feedstock that is not a commodity crop is algae, such as algal glycerolipids, which can be transesterified to produce fatty acid methyl (or ethyl) esters. Cellular lipids can also be converted via a catalytic hydrocracking process into a mixture of alkanes suitable for use as a jet fuel or gasoline ingredient; certain algae, such as *Botryococcus*, produce long-chain hydrocarbons that are potentially usable as a fuel after hydrocracking to reduce the chain length of the molecules. In most production schemes, the algal oil is extracted from the harvested algae.

Recent reevaluation suggests that current costs are well over $4/gal and that much more progress is needed if this technology is to have an impact in the foreseeable future (Pacheco, 2006). Many of the impediments are engineering challenges associated with how and where to grow the algae to achieve needed productivity. Because of the metabolic burden associated with the biosynthesis of high-energy lipids, production strains that accumulate high levels of oil tend to grow and reproduce more slowly than strains that do not. As a consequence, open cultures are subject to contamination by undesirable species unless the production strain is able to grow in specialized conditions that restrict the growth of those other species (for example, high-alkalinity environments).

Alternatively, production strains selected for high growth rates and high biomass yields without regard for oil content can often compete satisfactorily with contaminating strains, but the chemical composition of the algae would be better suited for anaerobic digestion than liquid fuel production. The use of closed photobioreactors can significantly lower the risk of culture contamination, although the capital costs of such systems are high.

Algal biodiesel has properties similar to those of biodiesel made from vegetable oil, except that algal biodiesel has better cold-weather properties. The energy content per gallon of biodiesel is about 93 percent that of petroleum-based diesel fuel, and it has a cetane number between 50 and 60, with 55 being typical. Also, it is somewhat more viscous. Biodiesel can be distributed by existing infrastructure and used in unmodified diesel-engine vehicles.

Biobutanol

Butanol is a four-carbon-atom alcohol—as opposed to ethanol, which is a two-carbon-atom alcohol. Biobutanol, the name given to butanol that has been made from biomass, is another potential entrant into the automotive biofuel market, and several technologies for producing it are in the R&D stage. The one receiving the most attention is the acetone-butanol-ethanol process. As currently envisioned, it involves the biochemical conversion of sugars or starches (from sugar beets, sugar cane, corn, wheat, or cassava) into biobutanol using a genetically engineered microorganism, *Clostridium beijernickii* BA101. The midterm goal is to start with cellulose, but that goal awaits the demonstration of economic success in converting cellulose and hemicelluloses into sugars.

Biobutanol has many attractive features as a fuel. Its energy content is close to that of gasoline, it has a low vapor pressure, it is not sensitive to water, it is less hazardous to handle and less flammable than gasoline is, and it has a slightly higher octane than gasoline has. Thus it is likely to be compatible with the existing distribution system and can substitute for gasoline directly. Its main drawback to date is the high cost of production. To reduce that cost and help initiate market entry, DuPont and BP have joined forces to retrofit an existing bioethanol plant to produce biobutanol using DuPont-modified biotechnology (Chase, 2006). Moreover, an improved next-generation bioengineered organism is projected to be available within the next few years.

Hydrocarbon Fuels from Biomass

The growing biofuel industry is based on well-established technology for producing ethanol via fermentation and distillation. This technology is energy-intensive, however, with approximately 60 percent of the product's fuel value consumed in these two processing steps (Katzen et al., 1981; Shapouri et al., 2002). In addition, fuel ethanol is expensive to distribute, as it cannot be added to gasoline prior to pipeline transport. At an estimated 13–18¢/gal, the cost of ethanol-fuel transportation is as much as six times that of transporting traditional petroleum-based fuels (GAO, 2007). Therefore approaches to developing hydrocarbon fuels produced directly from biomass, and that are analogous to fuels produced from petroleum, are being explored (Huber et al., 2006). Other proposed approaches include a hybrid hydrogen-carbon process for producing liquid hydrocarbons (Agrawal et al., 2007) and a catalytic strategy to produce dimethyl furan from carbohydrates (Román-Leshkov et al., 2007).

Gasoline Blend Stock

One approach produces straight-chain hydrocarbons, mostly hexane, via aqueous-phase hydrogenation of biomass-derived sugars followed by dehydration. The combination of reactions is exothermic and in theory could consume no net hydrogen. Because the reactants are dissolved in water, the hydrocarbons produced form a separate phase, and distillation is not required. This process, compared with the fermentation and distillation steps used in ethanol production, has the potential for higher energy efficiency and shorter residence times, but considerable development is required to confirm that this potential can be realized in a commercially viable process (Huber et al., 2005).

The product, consisting of linear hydrocarbons, can be isomerized in a conventional refining process to form branched hydrocarbons with higher octane, which are therefore more suitable for gasoline blending. Also, conventional refinery alkylation technology can be used to process the low-boiling straight-chain hydrocarbons to increase octane and boiling point to the extent needed for gasoline blending. Of course, if this production of hydrocarbons from biomass were widely commercialized, refining capability for isomerization and alkylation would likely have to be increased.

Diesel Fuel Components

Another approach to bio-hydrocarbon fuels being studied produces high-cetane diesel fuel material (Huber et al., 2005). In this process, sugars are first dehydrated and then hydrogenated to form cyclic oxygenated molecules that can undergo aldol condensation (self-addition) to form larger oxygenated molecules that remain soluble in water. The condensation products are then themselves hydrogenated and dehydrated to form mostly straight-chain hydrocarbons ranging from 7 to 15 carbon atoms per molecule. The final hydrogenation and dehydration reactions in this sequence are carried out in a four-phase reactor, with the phases being water with dissolved oxygenated hydrocarbon reactants, gaseous hydrogen, a solid catalyst, and hydrocarbons for reducing coke formation on the catalyst. The process can be modified to produce oxygenated compounds in the diesel-fuel boiling range that are soluble in the diesel fuel.

Status

Although the two processes described in this section have been shown to be feasible in the laboratory with pure feedstocks, much R&D remains before commercial applications can be undertaken. The concepts need to be tested using biomass-derived feedstocks with reactors that can be scaled for commercial operation. Based on work thus far, the keys to success in these processes appear to be the achievement of sufficient yield of the hydrocarbon product, development of high-activity catalysts with long-term stability, and minimization of coking reactions.

Bacteria- and Yeast-Based Direct Routes to Biofuels

With the rapid growth of synthetic biology and the enhanced ability to engineer organisms' metabolic pathways so as to produce specific chemical products, new approaches to renewable fuel production are emerging (Savage, 2007). They include using well-established recombinant DNA techniques to insert existing genes into microorganisms to make specific fuel precursors or even to directly synthesize hydrocarbon fuel components. Another approach involves redesigning genes, with computer assistance, to perform specific reactions and then synthesizing the desired genes for insertion into microbes. Yeasts can also be engineered to produce larger amounts of lipids, which with additional metabolic engineering can be converted to useful products—potentially, fuels. Using these techniques, it is possible that properly designed hydrocarbon products in either the diesel or the gasoline range would not require significant refining and could fit directly into the existing infrastructure.

Although none of these processes is approaching commercial production at this point, the level of activity and the current rate of progress could change that status in the not-too-distant future. Several companies are employing synthetic biology to create bacteria that produce increased amounts of fatty acids or other lipids that are then converted to hydrocarbons of virtually any length or structure desired. Moreover, the hydrocarbons phase-separate from the growth medium, thereby markedly reducing separation costs. The feedstock for the bacteria is renewable sugars, which can be obtained from sugar cane, grain, or cellulosic biomass (LS9, 2008). It is difficult to project the future of these and other nascent developments, but they deserve careful watching.

Technologies to Improve Biochemical Conversion

Significant advances are being made in the areas of genomics, molecular breeding, synthetic biology, and metabolic and bioprocess engineering that will likely enable innovation and advancement in the development of alternative transportation fuels. These and related technologies have the potential to greatly accelerate the creation of dedicated or dual-purpose energy crops as well as of microorganisms useful both for feedstock-conversion processes and biofuel production.

Genomics

The sequencing of full genomes continues to become faster and less costly, thus allowing energy crops such as tree species, perennial grasses, and nonedible oil seeds (castor and jatropha, for example) to be sequenced. The resulting data are extremely important for improving overall yields, for enabling improved nutrient and water utilization, and for understanding and manipulating biochemical pathways to enhance the production of desired products.

The sequencing data also have other uses. They can be used to target specific genes for downregulation by classical methods such as antisense and RNA interference, but also via complete inactivation using new and evolving procedures for homologous recombination-based gene disruption. In addition, rapid sequencing of breeding populations of energy crops can enable marker-assisted selection to accelerate the breeding of energy crops in ways previously not possible. And the rapid and inexpensive sequencing of fermentative and photosynthetic microorganisms in particular is redefining and shortening the timelines associated with strain-development programs for converting sugars, lignocellulosic materials, and CO_2 into alternative liquid fuels.

Strains generated through classical mutagenesis that have improved biocatalytic properties can now be analyzed at the molecular level to determine the specific genetic changes that result in the improved phenotype, allowing those changes to be implemented in other strains. In addition, "metagenome" sequence data, obtained by randomly sequencing DNA isolated from environmental samples, are providing vast numbers of new gene sequences that can be used to genetically engineer improved crops and microorganisms.

Synthetic Biology

Improved technologies for synthesizing megabase DNA molecules are being developed that will allow the introduction of entirely new biochemical pathways into energy crops and biofuel-producing microorganisms. These technologies could have a great impact on scientists' ability to generate plants and microorganisms with desired traits. For example, it is becoming conceivable that large portions of microorganisms' chromosomes, or even their complete chromosomes, can be replaced in ways that focus most of the cells' biochemical machinery on producing "next-generation" biofuel molecules boasting both cost and product advantages. Significant hurdles, however, could occur in maintaining the purity of such cultures and in dealing with mutants that gain competitive advantage by producing less of the desired chemicals.

Metabolic and Bioprocess Engineering

In addition to genetic manipulation, new bioengineering technologies are coming on line that will lower the cost of biofuel formation and recovery. While synthetic biology can now provide synthetic DNA for transferring heterologous genes into suitable host cells, metabolic engineering is the enabling technology for constructing functional and even optimal pathways for microbial fuel biosynthesis. This field has matured in only a few years and has an impressive record of accomplishments, many already in industrial practice (for example, biopolymers, alcohols,

1,3 propane-diol, oils, and hydrocarbons). Microbial strains have been developed that secrete hydrophobic fuels, similar to constituents of diesel and gasoline, into the culture medium. These fuels can be separated from the aqueous phase without distillation, thereby reducing the energy inputs and facilitating continuous production.

By taking a systems view, metabolic engineering has developed tools for overall biosystems optimization. They are now facilitating the construction of biosynthetic pathways and eliciting novel multigenic cellular properties of critical importance to biofuels production, such as tolerance to fuel toxicity. In the bioprocessing area, the successful development of membrane-based alcohol separations would greatly reduce energy costs from those of the typically used distillation process. Gas stripping, liquid-liquid extractions of secreted fuel molecules, and new adsorbent materials are also being developed that will allow continuous production modes for fermentation-based products. The photosynthetic production of biofuels—the development of low-cost photobioreactors and associated recovery systems for algal biofuel production—is another area of substantial interest that could have major benefits for overall-process economics.

OTHER TRANSPORTATION-FUEL OPTIONS READY FOR DEPLOYMENT BY 2020 AND 2035

So far in this chapter, the committe has focused strictly on certain liquid fuels and considered only biomass and coal as feedstocks, but in this section it explores the advantages and disadvantages of other known transportation-fuel options. The first to be considered is compressed natural gas (CNG). Thereafter, other liquid fuels that can be produced from syngas, including gas-to-liquid (GTL) diesel, dimethyl ether, and methanol, are described. Finally, the technology implications of using hydrogen in fuel-cell-powered vehicles for transportation are discussed.

The earlier sections discussed how coal, biomass, or combined coal-and-biomass gasification produces syngas, which can be converted to diesel and gasoline or to methanol, which can be converted to gasoline. Syngas can also be produced by reforming natural gas. Only if large supplies of inexpensive domestic natural gas were available—for example, from natural-gas hydrates—would the United States be likely to use natural gas as a feedstock for transportation-fuel production. Methanol can be produced from coal synthesis gas and used as a transportation fuel, but the committee judges that the best approach is to convert synthesis gas to methanol and use methanol-to-gasoline technology to produce gasoline, which fits directly into the existing U.S. fuel-delivery infrastructure. Hydrogen has the potential to reduce U.S. greenhouse gas emissions and oil use, as discussed in two National Research Council reports, *Transitions to Alternative Transportation Technologies—A Focus on Hydrogen* (NRC, 2008) and *The Hydrogen Economy: Opportunities, Costs, Barriers, and R&D Needs* (NRC, 2004). It is a long-term option, nonetheless.

Compressed Natural Gas

In 2007, the main uses for natural gas in the United States were electric power generation (30 percent) and industrial (29 percent), residential (20 percent), and commercial (13 percent) use. Only 0.1 percent was used in vehicles (EIA, 2008a). But natural gas is the cleanest and most efficient hydrocarbon fuel—it is environmentally superior to coal for electric power generation—and for similar reasons it could be a sound choice for transportation fuels.

Natural gas consumption levels in 2008 were satisfied mainly by domestic production (Chapter 7 on fossil fuels includes estimates of U.S. natural gas resources). However, a switch to natural gas for a large segment of U.S. transportation use would most likely trigger its increased importation. Even if natural gas were to be used for transportation rather than electricity generation, there is a potential to supply only about one-fifth to one-fourth of U.S. transportation needs from North American natural gas reserves, and only with investment in the distribution infrastructure. In any case, the technologies for producing transportation fuels from natural gas will be ready for deployment by 2020.

In 2008, there were more than 150,000 natural gas vehicles (NGVs) and 1,500 NGV fueling stations in the United States. Natural gas is sold in gasoline-equivalent gallons; each gasoline-equivalent gallon of natural gas has the same energy content (124,800 Btu) as a gallon of gasoline. NGVs are more expensive to purchase than are hybrid or gasoline vehicles. The Civic GX NGV has a manufacturer's suggested retail price of $24,590, compared to $22,600 for the company's hybrid sedan and $15,010 for its regular sedan (Rock, 2008).

Of all the fossil fuels, natural gas produces the least amount of CO_2 when burned because it contains the lowest carbon-to-hydrogen ratio. It also releases lesser amounts of criteria air pollutants. NGVs emit unburned methane, which has a higher greenhouse forcing potential than does CO_2, but this might be offset by the substantial reduction in CO_2 emissions. When compared with gasoline-powered vehicles, dedicated NGVs have lower exhaust emissions of carbon monoxide, nonmethane organic gases, nitrogen oxides, and carbon dioxide.

Natural-gas engines are more fuel efficient than gasoline engines are, and CNG in the past has had a low price (about 80 percent that of gasoline on a gasoline-equivalent gallon basis). Also, transport and distribution are relatively inexpensive because infrastructures already exist for delivery both to households and to industries (Yborra, 2006). Despite these advantages, however, NGVs still face many hurdles. The two main hurdles are insufficient numbers of refueling stations and inconvenient onboard CNG tanks, which take up most of the trunk space.

An NGV market can be analyzed using the vehicle-to-refueling-station index, or VRI, defined as the ratio of number of NGVs (in thousands) to the number of natural gas refueling stations. According to Yeh (2007), "Using techniques including consumer preference surveys and travel time/distance simulations, it has been found out that the sustainable growth of alternative fuel vehicles (AFVs) during the transition from initial market development to a mature market requires [that] the number of alternative-fuel refueling stations be a minimum of 10 to 20 percent of the number available for conventional gasoline stations." A thriving NGV market tends to have an index of 1; this gives rise to a problem: new stations are not being opened because of the lack of users, but few people use NGVs because of the lack of refueling stations.

A key disadvantage of NGVs is their limited range. While the average gasoline or diesel vehicle can go 400 miles on a tank full of fuel, the range of an NGV is only 100–150 miles, depending on the natural gas compression. Given this fact, together with the shortage of refueling stations, the current prevalent choice is to use a bi-fuel NGV that can run both on natural gas and on gasoline. The problems associated with bi-fuel engines include slightly less acceleration and about 10 percent power loss compared with a dedicated NGV, given that bi-fuel engines are not optimized to work on natural gas. Further, warranties on new gasoline vehicles are strongly reduced if they are converted into bi-fuel NGVs. But perhaps the most important barrier to NGVs could be the public perception that compressed natural gas is a dangerous "explosive" to have on board one's vehicle and that self-service refueling with a high-pressure gas may be too risky to offer to the general public.

About 22 percent of all new transit-bus orders are for natural-gas-powered vehicles. Therefore buses, together with corporate-fleet cars that stay in town, have been the main markets for NGVs. Both of these uses have occurred mainly in response to the Clean-Fuel Fleet Program set up by the U.S. Environmental Protection Agency to reduce air pollution.

Synthetic Diesel Fuel

The GTL process for producing synthetic diesel fuel is similar to the indirect lique-faction of coal. Instead of syngas production via the gasification of coal, however, syngas is produced by the steam reforming of natural gas. The synthesis gas can then be converted to an olefinic distillate, called synthol light oil, and wax using a catalytic modification of the FT process discussed earlier. The olefinic distillate and wax are hydrocracked to produce high-quality diesel, as well as naphtha and other streams that form the basis of specialty products such as synthetic lubricants.

Although it is technically difficult, the naphtha can also be upgraded to gasoline. Naphtha is an ideal feedstock for manufacture of chemical building blocks (for example, ethylene), and GTL diesel provides high-quality automotive fuel or blending stock (Johnson-Matthey, 2006) like coal-to-liquids technology. GTL is an option for producing diesel from "stranded" natural gas, such as that which exists in the Middle East and Russia. However, a couple of GTL plants would produce enough naphtha to swamp the chemical market for this material.

Hypothetically, there are several advantages to converting natural gas into GTL diesel rather than into CNG. All diesel vehicles can run on GTL diesel, which gives gas producers access to new market opportunities. Vehicle driving range for diesel is much higher than for compressed natural gas because of diesel's higher energy density. Engine efficiency and performance are not compromised by the adjustment to GTL diesel fuel. GTL diesel can be shipped in normal tankers and unloaded at ordinary ports (The Economist, 2006).

Currently, there are several commercial GTL plants. Sasol in Nigeria and Qatar, as well as Shell in Malaysia and Qatar, produce GTL diesel fuel; a number of companies, including World GTL and Conoco Phillips, have plans to build GTL plants in the next several years. Because the economics of GTL plants are very closely tied to the natural gas price, viability depends in large part on inexpensive stranded gas. GTL diesel is viewed mainly as an alternative to liquified natural gas for monetizing large natural gas accumulations such as the one in Qatar. The high cost to produce GTL diesel makes its development in the United States unlikely unless an abundant and inexpensive source of natural gas is found (for example, natural gas hydrates).

Methanol

Methanol, an alcohol, is a liquid fuel that can be used in internal-combustion engines to power vehicles. During the late 1980s, it was seen as a route to diversifying the fuels for the U.S. transportation system; natural gas from remote fields around the world would be converted into methanol and transported to the United States. This strategy was seen by energy planners as a way to convert what was, at that time, cheap remote natural gas (on the order of $1 per thousand cubic feet) into a marketable product. Currently, however, while methanol is produced primarily from natural gas, it is used principally as a commodity chemical.

Methanol has a higher octane rating than gasoline does and is therefore a suitable neat fuel for internal-combustion engines (for example, in racing cars). In practical terms, the penetration of methanol into a transportation system for LDVs that are fueled primarily by gasoline would require flexible-fuel vehicles that could run on a mixture of gasoline and methanol. Further, the use of a mixture of 85 percent methanol (M85) and gasoline would avoid the cold-start problem caused by methanol's low volatility. However, methanol has about half the energy density of gasoline, which affects the driving range that a vehicle can achieve on a full tank of the fuel.

Other drawbacks of methanol include its corrosive, hydrophilic, and toxic nature and its harmfulness to human health in particular if ingested, absorbed through the skin, or inhaled. Methanol could thus potentially create environmental, safety, health, and liability issues for fuel station owners. In addition, introducing a new fuel such as methanol on a large scale would require the construction of a new distribution system and the use of flexible-fuel vehicles that could run on a mixture of gasoline and methanol. One means of avoiding these infrastructural barriers would be to convert the methanol to gasoline using the MTG process.

Dimethyl Ether

Dimethyl ether (DME) is a liquid fuel with properties similar to that of liquefied petroleum gas (LPG). It produces lower CO and CO_2 emissions when burned, compared to gasoline and diesel, because of its modest carbon-to-hydrogen ratio. Because DME contains oxygen, it also requires a lower air-to-fuel ratio than do gasoline and diesel. DME has a thermal efficiency higher than that of diesel fuel (Kim et al., 2008), which could enable a higher-efficiency engine design. The presence of oxygen in the structure of DME also minimizes soot formation (Arcoumanis et al., 2008). Other exhaust emissions, such as unburned hydrocarbons, nitrogen oxides, and particulate matter, are also reduced. In fact, because DME meets and surpasses the California Air Resources Board emissions standards for automotive fuel, it is considered an ultraclean fuel.

At present, the preferred route and more cost-effective method for producing DME are through the dehydrogenation of methanol from synthesis gas, which is a mixture of CO and H_2. The basic steps for producing DME are as follows:

1.   Syngas production either by steam reforming of natural gas or by the partial oxidation of coal, oil residue, or biomass.
2.   Methanol synthesis using copper-based or zinc oxide catalysts.
3.   Methanol dehydrogenation to DME using a zeolite-based catalyst.

The produced DME fuel is not suitable for spark-ignition engines because of its high cetane number, but it can run a diesel engine with little modification. DME has properties similar to those of GTL diesel, including good cold-flow properties, low sulfate content, and low combustion noise (Yao et al., 2006; Arcoumanis et al., 2008; Kim et al., 2008).

The principal advantage of using DME as an automotive fuel is that it is clean burning and easy to handle and store. But as with other potential alternative fuels, the primary challenge facing the use of DME is the lack of an infrastructure for distribution. Other disadvantages include low viscosity, poor lubricity, a propensity to swell rubber and cause leaks, and lower heating value compared with conventional diesel.

Hydrogen

Hydrogen, like electricity, is an energy carrier that can be generated from a wide variety of sources, including nuclear energy, renewable energy, and fossil fuels. Hydrogen also can be made from water via the process of electrolysis, although this appears to be more expensive than reforming natural gas. Used in vehicles, both hydrogen and electricity make efficient use of energy compared with liquid-fuel options on a well-to-wheel basis. As generally envisioned, hydrogen would generate electricity in a fuel cell, and the vehicle would be powered by an electric motor.[9] Developments in battery technology that might make plug-in hybrid-

---

[9]Hydrogen also can be burned in an internal-combustion engine (ICE), but the overall efficiency is much lower than with a combination of fuel cells and a motor. It would be difficult to

electric and all-electric vehicles feasible will be discussed in several forthcoming National Research Council reports.

Hydrogen fuel-cell vehicle (HFCV) technology has progressed rapidly over the last several years, and large numbers of such vehicles could be introduced by 2015. Current HFCVs are very expensive because they are largely hand built. For example, in 2008, Honda released a small number of HFCVs named FCX Clarity which cost several hundred thousands of dollars to produce (Fackler, 2008). However, technological improvements and economies of scale brought about by mass production should greatly reduce costs.

This section provides a synopsis of the National Research Council report *Transitions to Alternative Transportation Technologies—A Focus on Hydrogen* (NRC, 2008), which concluded that the maximum practical number of HFCVs that could be operating in 2020 would be about 2 million, among 280 million LDVs in the United States. By about 2023, as costs of the vehicles and hydrogen drop, HFCVs could become competitive on a life-cycle basis. Their number could grow rapidly thereafter to about 25 million by 2030, and by 2050 they could account for more than 80 percent of new vehicles entering the U.S. LDV market. Those numbers are not predictions but rather a scenario-based estimate of the maximum penetration rate assuming that technical goals are met, that consumers readily accept HFCVs, and that policy instruments are in place to drive the introduction of hydrogen fuel and HFCVs through the market transition period.

The scenario would require that automobile manufacturers increase production of HFCVs even while they cost much more than conventional vehicles do and that investments be made to build and operate hydrogen fueling stations even while the market for hydrogen is very small. Substantial government actions and assistance would be needed to support such a transition to HFCVs by 2020 even with continued technical progress in fuel-cell and hydrogen-production technologies.

A large per-vehicle subsidy would be needed in the early years of the transition, but the number of vehicles per year would be low (Box 5.1) (NRC, 2008). Subsidies per vehicle would decline with fuel-cell costs, which are expected to drop rapidly with improved technology and economies of scale. By about 2025, an HFCV would cost only slightly more than an equivalent gasoline vehicle. Annual expenditures to support the commercial introduction of HFCVs would

store enough hydrogen on board to give an all-hydrogen ICE vehicle an acceptable range. The BMW hydrogen ICE also can use gasoline.

BOX 5.1 *Projected Costs of Implementing Hydrogen Fuel-Cell Vehicles*

According to a scenario developed in NRC (2008),

By 2023 (break-even year):
- The government would have spent about $55 billion, including
  —$40 billion for the incremental cost of HFCVs,
  —$8 billion for the initial deployment of hydrogen-supply infrastructure, and
  —$5 billion for research and development.
- About 5.6 million HFCVs would be operating.

By 2050:
- More than 200 million HFCVs would be operating, and there would be
  —180,000 hydrogen stations,
  —210 central hydrogen-production plants, and
  —80,000 miles of pipeline.
- Industry would have profitably spent about $400 billion on hydrogen infrastructure.

---

increase from about $3 billion in 2015 to $8 billion in 2023, at which point more than 1 million HFCVs could be joining the U.S. fleet annually. The cost of hydrogen also would drop rapidly, and because the HFCV would be more efficient it would cost less per mile to drive than would a gasoline vehicle in about 2020. Combining vehicle and driving costs suggests that the HFCV would have lower life-cycle costs starting in about 2023. After that, there would be a net payoff to the country, which cumulatively would balance the prior subsidies by about 2028.

Substantial and sustained R&D programs will be required to reduce the costs and improve the durability of fuel cells, develop new onboard hydrogen-storage technologies, and reduce hydrogen production costs. Needed R&D investments are shown in Box 5.1. These programs would have to continue after 2023 to reduce costs and to further improve performance, but the committee did not estimate the necessary funding.

The 2008 National Research Council study determined the consequent

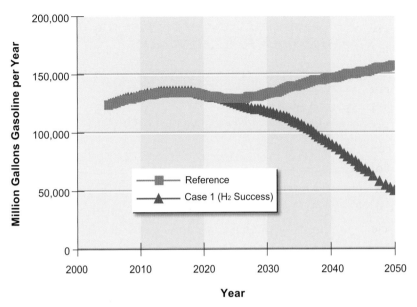

FIGURE 5.9 *Oil consumption with maximum practical penetration of HFCVs compared with reference case.*
Source: NRC, 2008.

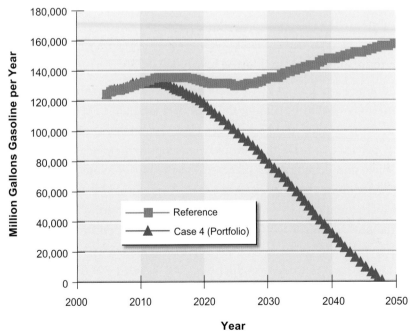

FIGURE 5.10 *Oil consumption for combined HFCVs, high-efficiency conventional vehicles, and biofuels compared with reference case.*
Source: NRC, 2008.

reductions in U.S. oil consumption and greenhouse gas emissions that could be expected in this scenario. HFCVs can yield large and sustained reductions in U.S. oil consumption and greenhouse gas emissions, but several decades will be needed to realize those potential long-term benefits. Figure 5.9 (on facing page) compares the oil consumption that would be required in this scenario with a reference case based on Energy Information Administration high oil-price projections, which include the recent increases in corporate average fuel economy standards. By 2050, HFCVs could reduce oil consumption by two-thirds. Greenhouse gas emissions would follow a similar trajectory if hydrogen produced from coal in large central stations were accompanied by carbon separation and sequestration.

The study then compared those reductions with the potential impact of alternative vehicle technologies (including conventional hybrid-electric vehicles) and biofuels oil consumption and greenhouse gas emissions. Over the next two decades, those approaches could deliver much greater reductions in U.S. oil use and greenhouse gas emissions than could HFCVs, but hydrogen offers greater longer-term potential. Thus, the greatest benefits will come from a portfolio of research and development in technologies that would allow the United States to nearly eliminate oil use in LDVs by 2050 (see Figure 5.10 on facing page). Achieving that goal would require substantial new energy-security and environmental-policy actions in addition to technological developments. Broad policies aimed at reducing oil use and greenhouse gas emissions will be useful, but they are unlikely to be adequate to facilitate the rapid introduction of HFCVs.

REFERENCES

Agrawal, R., N.R. Singh, F.H. Ribeiro, and W.N. Delgass. 2007. Sustainable fuel for the transportation sector. Proceedings of the National Academy of Sciences USA 104: 4828-4833.

Arcoumanis, C., C. Bae, R. Crookes, and E. Kinoshita. 2008. The potential of di-methyl ether (DME) as an alternative fuel for compression-ignition engines: A review. Fuel 87:1014-1030.

Chase, R. 2006. DuPont, BP join to make butanol: They say it outperforms ethanol as a fuel additive. USA Today, June 26.

DOE (U.S. Department of Energy). 2007. DOE selects six cellulosic ethanol plants for up to $385 million in federal funding. Available at http://www.energy.gov/print/4827.htm. Accessed October 16, 2008.

DOE/EERE (U.S. Department of Energy/Office of Energy Efficiency and Renewable Energy). 2007. Biomass Program 2007 Accomplishments Report. Available at www1. eere.energy.gov/biomass/pdfs/program_accomplishments_report.pdf. Accessed February 10, 2009.

Economist, The. 2006. Arabian alchemy. Vol. 379, Issue 8480, 00130613, June 3.

EIA (Energy Information Administration). 2008a. Natural gas consumption by end use. Available at http://tonto.eia.doe.gov/dnav/ng/ng_cons_sum_dcu_nus_m.htm. Accessed December 4, 2008.

EIA. 2008b. What are biofuels and how much do we use? Available at http://tonto.eia.doe. gov/energy_in_brief/biofuels_use.cfm. Accessed April 7, 2009.

EIA. 2009. Coal resources, current and back issues. Available at www.eia.doe.gov/cneaf/ coal/reserves/reserves.html#_ftp/. Accessed July 30, 2009.

Fackler, M. 2008. Latest Honda runs on hydrogen, not petroleum. New York Times, June 17, 2008. Available at www.nytimes.com/2008/06/17/business/worldbusiness/ 17fuelcell.html?_r=1&oref=slogin. Accessed April 7, 2009.

Fargione, J., J. Hill, D. Tilman, S. Polasky, and P. Hawthorne. 2008. Land clearing and the biofuel carbon debt. Science 319:1235-1238.

Farrell, A., R. Plevin, B. Turner, A. Jones, M. O'Hare, and D. Kammen. 2006. Ethanol can contribute to energy and environmental goals. Science 311:506-508.

GAO (U.S. Government Accountability Office). 2007. Biofuels: DOE Lacks a Strategic Approach to Coordinate Increasing Production with Infrastructure Development and Vehicle Needs. Washington, D.C.: GAO.

Heggenstaller, A.H., R.P. Anex, M. Liebman, D.N. Sundberg, and L.R. Gibson. 2008. Productivity and nutrient dynamics in bioenergy double-cropping systems. Agronomy Journal 100:1740-1748.

Hill, J., E. Nelson, D. Tilman, S. Polasky, and D. Tiffany. 2006. Environmental, economic, and energetic costs and benefits of biodiesel and ethanol biofuels. Proceedings of the National Academy of Sciences USA 103:11206-11210.

Huber, G. W., J.N. Chheda, C.J . Barrett, and J.A. Dumesic. 2005. Production of liquid alkanes by aqueous-phase processing of biomass-derived carbohydrates. Science 308:1446-1450.

Huber, G. W., S. Ibora, and A. Corma. 2006. Synthesis of transportation fuels from biomass: Chemistry, catalysts, and engineering. Chemical Review 106:4044-4098.

Jensen, T.H. 2008. Race for key biofuel breakthrough intensifies. Reuters, June 2, 2008. Available at www.reuters.com/article/reutersEdge/idUSL0248503420080602. Accessed April 7, 2009.

Johnson-Matthey. 2006. Reducing emissions through gas to liquids technology. Available at http://ect.jmcatalysts.com/pdfs/Reducingemissionart p2-3.pdf. Accessed October 20, 2008.

Katzen, R., W.R. Ackley, Jr., G.D. Moon, J.R. Messick, B.F. Brush, and K.F. Kaupisch. 1981. Low-energy distillation systems. In Fuels from Biomass and Wastes, D.L. Klass and G.H. Emert, eds. Ann Arbor, Mich.: Ann Arbor Science Publishers.

Kim, M.Y., S.H. Yoon, B.W. Ryu, and C.S. Lee. 2008. Combustion and emission characteristics of DME as an alternative fuel for compression ignition engines with a high pressure injection system. Fuel 87:2779-2786.

Kreutz, T.G., E.D. Larson, G. Liu, and R.H. Williams. 2008. Fischer-Tropsch fuels from coal and biomass. In 25th Annual International Pittsburgh Coal Conference, Pittsburgh, Pa.

Larson, E.D., G. Fiorese, G. Liu, R.H. Williams, T.G. Kreutz, and S. Consonni. 2008. Co-production of synthetic fuels and electricity from coal + biomass with zero carbon emissions: An Illinois case study. Energy Procedia 1:4371-4378.

LS9. 2008. Renewable Petroleum Technology. Available at www.ls9.com/technology/. Accessed May 1, 2008.

NAS-NAE-NRC (National Academy of Sciences-National Academy of Engineering-National Research Council). 2009. Liquid Transportation Fuels from Coal and Biomass: Technological Status, Costs, and Environmental Impacts. Washington, D.C.: The National Academies Press.

NRC (National Research Council). 2002. Evolutionary and Revolutionary Technologies for Mining. Washington, D.C.: The National Academies Press.

NRC. 2004. The Hydrogen Economy: Opportunities, Costs, Barriers, and R&D Needs. Washington, D.C.: The National Academies Press.

NRC. 2007. Coal Research and Development to Support National Energy Policy. Washington, D.C.: The National Academies Press.

NRC. 2008. Transitions to Alternative Transportation Technologies—A Focus on Hydrogen. Washington, D.C.: The National Academies Press.

Pacheco, M. 2006. Potential of biofuels to meet commercial and military needs. Presentation to the NRC Committee on Assessment of Resource Needs for Fuel Cell and Hydrogen Technologies.

Rock, B. 2008. An overview of 2007 American 2007 natural gas vehicles. Available at www.helium.com/items/451632-an-overview-of-2007-American-2007-natural-gas-vehicles. Accessed September 2, 2008.

Román-Leshkov, Y., C.J. Barrett, Z.Y. Liu, and J.A. Dumesic. 2007. Production of dimethylfuran for liquid fuels from biomass-derived carbohydrates. Nature 447:982-985.

Savage, N. 2007. Better biofuels. Technology Review 110:1.

Shapouri, H., J.A. Duffield, and M. Wang. 2002. The Energy Balance of Corn Ethanol: An Update. Agricultural Economic Report No. 814. U.S. Department of Agriculture, Office of the Chief Economist, Office of Energy Policy and New Uses.

Yao, M., Z. Chen, Z. Zheng, B. Zhang, and Y. Xing. 2006. Study on the controlling strategies of homogeneous charge compression ignition combustion with fuel of dimethyl ether and methanol. Fuel 85:2046-2056.

Yborra, S. 2006. Taking a second look at the natural gas vehicle. American Gas (Aug./Sept.):32-36.

Yeh, S. 2007. An empirical analysis on the adoption of alternative fuel vehicles: The case of natural gas vehicles. Energy Policy 35:5865-5875.

6

Renewable Energy

This chapter reviews the status of renewable resources as a source of usable energy. It describes the resource base, current renewables technologies, the prospects for technological advances, and related economic, environmental, and deployment issues. While the chapter's focus is on renewables for the generation of electricity, it also includes short discussions of nonelectrical applications. The use of biomass to produce alternative liquid transportation fuels is not covered in this chapter but rather in Chapter 5.

CURRENT STATUS OF RENEWABLE ELECTRICITY

Generation of Renewable Electricity in the United States

Renewables currently account for a small fraction of total electricity generation. According to the U.S. Energy Information Agency (EIA, 2007), conventional hydropower is the largest source of renewable electricity in the United States. Representing about 71 percent of the electric power derived from renewable sources, hydropower generated 6 percent of the electricity—almost 250,000 GWh out of a total of 4.2 million GWh—produced by the electric power sector in 2007.[1]

The nonhydropower sources of renewable electricity together contributed 2.5 percent of the 2007 total. Within this group, biomass electricity generation (called

---

[1]The electric power sector includes electricity utilities, independent power producers, and large commercial and industrial generators of electricity.

"biopower")[2] is the largest source, having produced 55,000 GWh in 2007. Wind power and geothermal supplied 32,000 GWh and 14,800 GWh, respectively, during that year. Except for wind power, none of these sources has grown much since 1990 in terms of either total electric power production or generation capacity.

The largest growth in the use of renewable resources for electricity generation is currently in wind power and, to a lesser extent, in solar power. Wind power technology, having matured over the last two decades, now accounts for an increasing fraction of total electricity generation in the United States. Though wind power in 2007 represented less than 1 percent, it grew at a 15.5 percent compounded annual rate over the 1990–2007 period and at a 25.6 percent compounded annual growth rate between 1997 and 2007. Wind power supplied almost 6,000 GWh more in 2007 than it had the year before. According to the American Wind Energy Association, an additional 8,300 MW of capacity was added in 2008 (AWEA, 2009a), representing an additional yearly generation of 25,000 GWh assuming a 35 percent capacity factor.[3] By the end of 2008, the overall economic downturn had caused financing for new wind power projects and orders for turbine components to slow, and layoffs began in the wind turbine manufacturing industry (AWEA, 2009a). Thus new capacity in 2009 recently looked to be considerably smaller than in 2008. However, AWEA (2009b) recently reported that 2.8 GW of new wind power generation capacity was installed in the first quarter of 2009. Further, analysis of the American Recovery and Reinvestment Act (ARRA) of 2009 shows that by 2012 wind power generation will more than double what it would have been without the ARRA (Chu, 2009).

Central-utility electricity generation from concentrating solar power (CSP) and photovoltaics (PV) combined was 600 GWh in 2007, just 0.01 percent of the U.S. total—a fraction that has been approximately constant since 1990. However, this estimate does not include contributions from residential and other small PV installations, which now account for the strongest growth in solar-derived electricity. Installations of solar PV in the United States have grown at a compounded annual growth rate of more than 40 percent from 2000 to 2005, with a genera-

---

[2]Biopower includes electricity generated from wood and wood wastes, municipal solid wastes, landfill gases, sludge wastes, and other biomass solids, liquids, and gases.

[3]The capacity factor is defined as the ratio (expressed as a percent) of the energy output of a plant to the energy that could be produced if the plant operated at its nameplate capacity.

tion capacity of almost 480 MW that, assuming a 15 percent capacity factor, produces approximately 630 GWh.

Current Policy Setting

At present, electricity generation from non-hydropower renewable sources is generally more expensive than from coal, natural gas, or nuclear power—the three leading U.S. options. Thus policies at the state and federal levels have provided the key incentives behind renewable sources' recent penetration gains.

One such policy is the renewables portfolio standard (RPS), which typically requires that a minimum percentage of the electricity produced or sold in a state be derived from some collection of eligible renewable technologies. Given that these RPSs have been developed at the state level, there are many different versions of them. The policies differ by the sources of renewables included (some states specify conventional hydropower or biopower); by the form, timeline, and stringency of the numerical goals; and by whether the goals include separate targets for particular renewable technologies. As of 2008, 27 states and the District of Columba had RPSs and another 6 states had related voluntary programs. Wiser and Barbose (2008) estimate that full compliance with these RPSs would result in an additional 60 GW of new renewables capacity by 2025. Assuming a 35 percent capacity factor, which means that the capacity produces electricity for approximately 3070 hours per year, an additional 180,000 GWh from renewable sources would be generated. This is compared to the estimated total of 4.2 million GWh generated in 2007.

Federal policies are also contributing to this era of strong growth in renewable-energy development. The major incentive, particularly for wind power, is the Federal Renewable Electricity Production Tax Credit (referred to simply as the PTC), which provides a $19 tax credit (adjusted for inflation) for every megawatt-hour (equivalent to 1.9¢/kWh) of electricity generated in the first 10 years of life of a private or investor-owned renewable electricity project brought on line through the end of 2008.[4] Congress most recently extended the PTC and expanded incentives for 1 year in the Emergency Economic Stabilization Act of 2008 and the ARRA of 2009. These two bills together extend the PTC for wind through 2012 and the PTC for municipal solid waste, qualified hydropower, biomass, geothermal, and marine and hydrokinetic renewable-energy facilities

---

[4]After adjusting for inflation, the current PTC is 2.1¢/kWh.

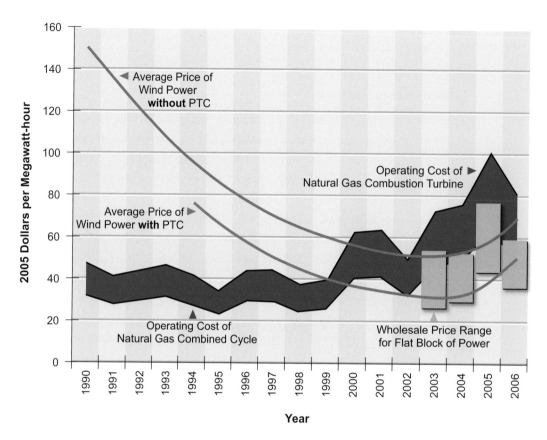

FIGURE 6.1 *Impacts of the PTC on the price of wind power compared to costs for natural-gas-fired electricity.*
Source: Wiser, 2008.

through 2013. Because of concerns that the current slowdown in business activity will reduce the capabilities of projects to raise investment capital, the ARRA allows owners of nonsolar renewable-energy facilities to elect a 30 percent investment tax credit rather than the PTC. Figure 6.1 shows the impact of the PTC on the price of wind power versus that of natural-gas-fired electricity, though it should be noted that other current electricity sources, such as coal, hydropower, and nuclear, have lower operating costs than do natural gas combined-cycle plants.

RESOURCE BASE

Size of Resource Base

The United States has significant renewable-energy resources. Indeed, taken collectively they are much larger than current or projected total domestic energy and electricity demands. But renewable resources are not evenly distributed spatially and temporally, and they tend to be diffuse compared to fossil and nuclear energy. Further, although the sheer size of the resource base is impressive, there are many technological, economic, and deployment-related constraints on using these sources on a large scale.

The United States has significant wind energy resources in particular; Figure 6.2 shows their distribution across the country. The total estimated electri-

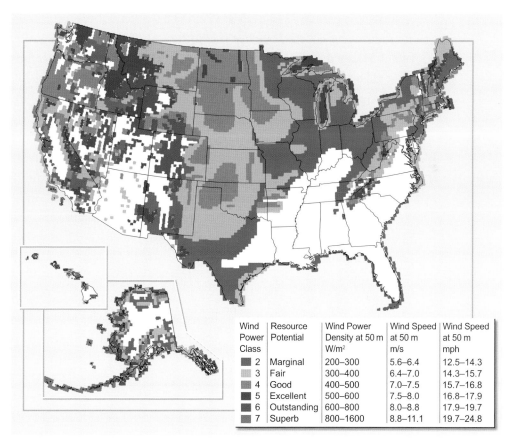

Wind Power Class	Resource Potential	Wind Power Density at 50 m W/m²	Wind Speed at 50 m m/s	Wind Speed at 50 m mph
2	Marginal	200–300	5.6–6.4	12.5–14.3
3	Fair	300–400	6.4–7.0	14.3–15.7
4	Good	400–500	7.0–7.5	15.7–16.8
5	Excellent	500–600	7.5–8.0	16.8–17.9
6	Outstanding	600–800	8.0–8.8	17.9–19.7
7	Superb	800–1600	8.8–11.1	19.7–24.8

FIGURE 6.2 *U.S wind resource map showing various wind power classes. Areas shown in white have class 1 wind resources.*
Source: NREL, 2007a.

cal energy potential for the continental U.S. wind resource in class 3 and higher wind-speed areas is 11 million GWh/yr (Elliott et al., 1991), greater than the 2007 electricity generation of about 4 million GWh. The 11 million GWh estimate was obtained from point-source measurements of the wind speed at a height of 50 meters (m); the actual value could differ substantially (Elliott et al., 1986). On the one hand, modern wind turbines can have hub heights of 80 m or greater, where larger wind energy resources are likely available. On the other hand, computer-model simulations of very-large-scale wind farm deployments have shown that an agglomeration of point-source wind speed data over large areas can significantly overestimate the actual wind energy resource (Baidya et al., 2004). Estimating the upper-bound limit for extraction of the resource at 20–25 percent of the energy in the wind field, and using the total domestic onshore wind electricity potential value of 11 million GWh, an upper bound for the annual extractable wind electric potential is perhaps 2–3 million GWh. This potential resource base is about half of the current electrical power use in the United States, and significant offshore wind energy resources also exist and increase the wind resource base considerably.

The solar energy resource also is very large indeed. Taking solar insolation to be a representative midlatitude, day/night average value of 230 W/m^2, in conjunction with the area of the continental United States of 8×10^{12} m^2, yields a yearly averaged and area averaged power-generation potential of 18.4 million GW. At 10 percent average conversion efficiency, this resource would therefore provide 1.6 billion GWh of electricity annually. For 10 percent conversion efficiency, coverage of 0.25 percent of the land of the continental United States would be required to generate the total 2007 domestic electrical generation value of 4 million GWh. However, the solar resource is very diffuse and, as shown in Figure 6.3, distributed unevenly across the country.

Additionally, the various technologies for tapping solar energy utilize different aspects of sunlight. Because CSP, for example, can exploit only the focusable direct-beam portion of sunlight, highly favored sites are located almost exclusively in the Southwest. Further, because CSP can use only the direct-beam portion of sunlight, energy input to the CSP plants falls to zero in the presence of clouds. However, most designs today decouple energy collection from the power cycle through the use of thermal storage, and thus the power output of the CSP plant will not immediately fall to zero in the presence of clouds. A recent analysis, which identified lands having high average insolation (>6.75 kW/m^2 per day) and excluded regions of such lands having a slope >1 percent or a small (<10 km^2) continuous area, estimated that CSP could deliver an average of

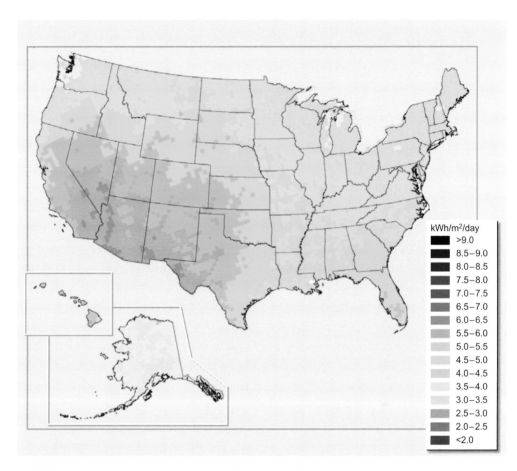

FIGURE 6.3 *Solar energy resources in the United States.*
Source: NREL, 2007b.

15–30 million GWh/yr of electrical energy, which is 4–7 times larger than the total U.S. supply (ASES, 2007).

Flat-plate PV arrays can be distributed more widely than concentrated solar power systems because flat-plate systems effectively utilize both the diffuse and the direct-beam components of sunlight. Analyses of the total rooftop area that would be suitable for installation of PV systems have produced resource estimates ranging from 0.9–1.5 million GWh/yr (ASES, 2007) to 13–17.5 million GWh/yr (Chaudhari et al., 2004). Only a fraction of rooftops and other lands can be developed economically at present for solar-based electricity generation, however; it is the economics of solar technologies, not the size of the potential resource, that significantly limit the ability of solar electricity alone to contribute substantially to electricity production.

There are two components to the geothermal resource base: hydrothermal (water heated by Earth) that exists down to a depth of about 3 km, and enhanced geothermal systems (EGS) associated with low-permeability or low-porosity heated rocks at depths down to 10 km. There is some potential for expanding electricity production from the hydrothermal resources and thus affecting regional electricity generation—for example, a regional study of known hydrothermal resources in the western states found that 13 GW of electric power capacity exists in identified resources within this region (WGA, 2006)—but in general, the resources are too small to have a major overall impact on total electricity generation in the United States.

It is the heat stored in the low-permeability and/or low-porosity rocks at great depths that represents the much larger resource base. As noted in a recent Massachusetts Institute of Technology study, a much larger potential for energy exists with EGS resources (MIT, 2006). The estimated geothermal resource below the continental United States, defined as the total amount of heat trapped to 10 km depth, has been estimated to be in excess of 1.3×10^{25} J (MIT, 2006). This is more than 130,000 times the total 2005 U.S. energy consumption of 1.00×10^{20} J. However, beyond the total amount of potentially available energy, the rate of extraction of this energy is especially critical in assessing the actual practical potential of this energy source. The mean geothermal heat flux at Earth's surface is on the order of 50 mW/m^2, and in many areas, the geothermal heat flux is significantly less than this value. Given that the electrical generation efficiency from use of this relatively low-temperature heat in a steam turbine is about 15 percent, the extractable and sustainable electrical power density from the geothermal resource is on the order of 10 mW/m^2. To provide substantial power, heat must be extracted at rates in excess of the natural geothermal heat flux (heat mining) in order to usefully tap sufficient geothermal resources. Indeed, the MIT report (2006) notes that some temperature drawdown should occur if EGS resources are to be used in their most efficient manner. The substantial technical challenges associated with tapping this resource are discussed later in this chapter.

Other renewable resources, including conventional hydropower, hydrokinetics (wave/tidal/current), and biomass, have significant resource bases, too. Because the conventional hydroelectric resource is generally accepted to be near its maximum utilization in the United States, further growth opportunities are relatively small. Regarding hydrokinetics, one study puts the size of the wave energy resource for the East and West Coasts at more than 0.5 million GWh/yr (EPRI, 2005). This study also estimates the wave energy base in Alaska to be 1.3 million GWh/yr, though it is unclear whether such a resource could be fully exploited. EPRI (2005) put the capacity of the tidal energy resource at a 152 MW annual average, which corresponds to an annualized electrical energy production of 1300 GWh/yr. The biomass resource base is discussed in Chapter 5.

Findings: Resource Base

Solar and wind renewable resources offer significantly larger total energy and power potential than do other domestic renewable resources. Solar energy is capable, in principle, of providing many times the total U.S. electricity consumption, even assuming low conversion efficiency. The land-based wind resource also is capable of making a substantial contribution to meeting current U.S. electricity demand without stressing the resource base. For these reasons, solar and wind resources are emphasized, but other non-hydroelectric renewables can make significant contributions to the electrical energy mix as well, at least in certain regions of the country. **However, renewable resources are not distributed uniformly.** Resources such as solar, wind, geothermal, tidal, wave, and biomass vary widely in space and time. **Thus, the potential to derive a given percentage of electricity from renewable resources will vary from location to location. Awareness of such factors is important in developing effective policies at the state and federal level to promote the use of renewable resources for generation of electricity.**

RENEWABLE TECHNOLOGIES

A renewable electricity-generation technology harnesses a naturally existing energy flux, such as wind, sun, or tides, and converts that flux into electricity. Such technologies range from well-established wind turbines to pilot-plant hydrokinetic systems to methods, such as those that exploit salinity and thermal ocean gradients, that are in the conceptualization or demonstration stages. Some of these technologies produce power intermittently (technologies that rely on wind and solar resources), whereas others are capable of producing baseload power (technologies that rely on hydropower, biomass, and geothermal resources). Though renewable-electricity technologies show much variability, they do have several shared characteristics: (1) the largest proportions of costs, external energy needs, and material inputs occur during the manufacturing and installation stages; (2) there are no associated fuel costs (except for biomass-fueled electricity generation); (3) opportunities for achieving economies of scale are greater at the manufacturing stage than at the generating site—larger-generation units do not necessarily reduce the average cost of electricity generation as much as they do for coal-fired or nuclear plants; and (4) renewable electricity technologies can be deployed in smaller increments and come on line more rapidly.

Technology Descriptions

Wind

Wind power uses a turbine and related components to convert the kinetic energy of moving air into electricity. A typical wind turbine assembly includes the rotor, controls, drive train (gearbox, generator, and power converter), other electronics (wiring, inverters, and controllers), and a tower. Each of these components has undergone significant development in the last 10 years, and the resulting modifications have been integrated into the latest turbine designs. Critical objectives for these and future improvements are to make it easier to integrate the wind power plants into the electrical system and to increase their capacity factors. Especially important has been the development of electronic controls that allow modern turbines to remain connected to the electricity grid during voltage disturbances and reduce the draw on the grid's reactive power resources. Advances in computerized controls will allow more aspects of the turbine to be monitored, resulting in more efficient use and the potential to better target and deploy technical upgrades.

Along with advances in electronics have come improvements in wind turbine structures, allowing turbine size and generating capacity to grow. Based on the fact that wind speed increases with height and that energy-capture ability depends on the turbine's rotor diameter, the most common turbines at present are three-bladed rotors with diameters of 70–80 m, mounted atop 60–80 m towers, that have a capacity of 1.5 MW. The rotor blade has gone through many generations of designs, using various types of materials and structures, to maximize its aerodynamic performance without compromising stability.

Wind power technologies are actively being deployed today, and no major technological breakthroughs are expected in the near future. However, evolutionary modifications in various turbine components are expected to bring 30–40 percent improvement in cost-effectiveness (cost per kilowatt-hour) over the next decade (Thresher et al., 2007). And while the turbine tower is not expected to get much taller, advances will likely occur in installing and maintaining these machines in difficult-to-reach locations. One possibility, for example, is self-erecting towers. In the future, turbine rotors will be made of advanced materials such as fiberglass, and they will have improved structural-aerodynamic designs, sophisticated controls, and higher speeds. By reducing the blade-soiling losses (e.g., through dust or insect buildup) and installing damage-tolerant sensors and robust control systems, reductions in energy loss and improvements in turbine availability can occur. In addition, drive trains will be modified to include fewer gear stages, medium- and low-speed generators, distributed gearbox topologies, permanent-magnet generators, and new circuit configurations. As shown in Table 6.1, these improvements will have significant impacts on annual wind energy production and capital costs over the next decade. It should be noted that future capital costs also will be greatly influenced by global supply and demand for wind turbines. Some of these issues are discussed in the section titled "Deployment Potential" later in this chapter, as well as in the report by the Panel on Electricity from Renewable Resources (NAS-NAE-NRC, 2009).

Along with improvements in onshore wind-turbine designs, offshore wind-turbine technologies will soon be actively enhanced to take advantage of the abundant U.S. offshore wind-energy resources. The technologies associated with offshore wind turbines will face fundamentally different challenges, however, attributable to the difficulties of building and operating turbines in the ocean and installing and maintaining transmission lines underwater.

Solar Photovoltaic Power

When sunlight strikes the surface of a PV cell, some of the light's photons are absorbed. This causes electrons to be released from the cell, which results in a current flow, namely, electricity. The two main PV technologies entail flat plates, which consist of crystalline silicon deposited on substrates, and concentrators, which typically involve lenses or reflectors that, together with tracking systems, focus the sunlight onto smaller and more efficient cells.

Silicon is used to form semiconductors in PV cells by taking advantage of the conductivity imparted when impurities ("doping" elements) are introduced. Because the efficiency of these crystalline PV modules is only 12–18 percent, further development is required—not only to increase efficiency but also to lower production costs (DOE, 2007a).

TABLE 6.1 Areas of Potential Technology Improvements for Wind Turbines

Technical Area	Potential Advances	Performance and Cost Increments (best/expected/least [percent])	
		Annual Energy Production	Turbine Capital Cost
Advanced tower concepts	• Taller towers in difficult locations • New materials or processes • Advanced structures/foundations • Self-erecting, initial, or for service	+11/+11/+11	+8/+12/+20
Advanced (enlarged) rotors	• Advanced materials • Improved structural-aerodynamic design • Active controls • Passive controls • Higher tip speed/lower noise levels	+35/+25/+10	−6/−3/+3
Reduced energy losses and improved availability	• Reduced blade-soiling losses • Damage-tolerant sensors • Robust control systems • Prognostic maintenance	+7/+5/0	0/0/0
Drive train (gear boxes, generators, and power electronics)	• Fewer gear stages or direct-drive • Medium/low speed generators • Distributed gearbox topologies • Permanent-magnet generators • Medium-voltage equipment • Advanced gear-tooth profiles • New circuit topologies • New semiconductor devices • New materials (gallium arsenide [GaAs], SiC)	+8/+4/0	−11/−6/+1
Manufacturing and learning curve	• Sustained incremental design and process improvements • Large-scale manufacturing • Reduced design loads	0/0/0	−27/−13/−3
Totals		+61/+45/+21	−36/−10/+21

Source: Thresher et al., 2007.

Recent increases in the cost of silicon have encouraged the development of alternative processes. Thin-film technologies have the potential for substantial cost advantages because of such factors as lower material use, fewer processing steps, and simpler manufacturing procedures for large-area modules. The most common materials used for thin films are amorphous silicon, cadmium telluride, and copper indium gallium diselenide and related alloys. Future directions for thin-film technologies include multi-junction assemblies aimed at significantly higher efficiencies, transparent and better-conducting oxide electrodes, and thin polycrystalline silicon films.

Another new technology, which takes advantage of photochemistry, is the dye-sensitized solar cell, in which organic dye molecules are adsorbed onto nanocrystalline titanium dioxide films (O'Regan and Grätzel, 1991). The dye molecules then absorb solar photons to create an excited molecular state that injects electrons into the film, the electrons are collected on a transparent electrode, and the dye is then reduced back to its initial state by accepting the electrons, which completes the circuit and generates electrical power in the external load. This type of solar cell is attractive because of its low cost and simplicity in manufacturing, but the device's efficiency and stability will need to be closely monitored before large-scale deployment is possible.

In organic solar cells, which also are in the early developmental stage, the sunlight creates an exciton, which separates into an electron on one side and a hole on the other side of a material interface within the device. This allows the cell to be thinner, which significantly reduces cost in at least four ways: inexpensive constituent elements (which do not require pure silicon), decreased material use, modest conversion efficiency, and high-volume production techniques. Some examples of organic solar cells include quantum dots embedded in an organic polymer, liquid crystal cells, and small-molecule chromophore cells.

Nanotechnology too could become a useful tool for making PV cells because it can tune the optical and electronic properties of the PV materials by precisely controlling their particle sizes and shapes. Nanoparticles produced by chemical solution methods may streamline the manufacturing process, but their long-term stability must be tested.

Solar PV technologies are at various stages of development. Silicon flat-plate PV cells are mature and are actively being deployed today. Reductions in the production costs of the cells and increases in efficiency and reliability will be needed, however, to make them more attractive to potential customers. Thin-film technologies, which have great potential to reduce module cost, are in a relatively immature stage compared to crystalline silicon PV; further research, testing, and deployment are required. Other competing technologies, such as dye-sensitized PV and nanoparticle PV, are at early stages of development; a great deal more effort will be required before commercialization is possible.

A critical uncertainty is whether the installed system cost of solar PV can be brought down to less than $1 per peak watt with at least a 10 percent module and system efficiency; this level is needed so that widespread deployment of PV may be promotable without subsidies. But reaching $1/W will require that the performance of cells and manufacturing processes be improved. It will also require that the balance-of-system costs (pertaining to installation, the inverter, cables, the support structures, grid hookups, and other components) be brought down significantly. For some systems, 50 percent or more of the total installed cost of a rooftop PV system is not in the module cost but in the balance-of-system costs. These costs must come down significantly through innovative system integration approaches or this aspect of a PV system will ultimately set a floor on the price of a full, installed PV system.

Concentrating Solar Power

CSP systems—sometimes referred to as solar thermal—employ optics to concentrate beam radiation, which is the portion of the solar spectrum that is not scattered by the atmosphere. The concentrated solar energy produces high-temperature heat, which can be used to generate electricity or to drive chemical reactions to produce fuels (syngas or hydrogen) (Fletcher, 2001). There are three types of CSP technologies—parabolic troughs, power towers, and dish-Stirling engine systems—which differ in their optical systems and receivers, where the concentrated solar radiation is absorbed and converted to heat.

The most mature technology is the parabolic trough combined with a conventional Rankine-cycle steam power plant. This technology uses concave, parabola-shaped mirrors to focus the directed beam radiation onto a linear receiver. In the trough systems, synthetic oil is circulated in tubes and then used to superheat steam, which drives a conventional turbine/generator. Parabolic trough plants can include short-term solar energy storage capabilities to extend generation for several hours. Also, many existing power plants using this technology have a backup fossil-fired capability for providing power during periods of low solar radiation and at night.

Power towers consist of numerous two-axis mirrors (heliostats) that track the sun and direct the incoming beam radiation to a receiver (located at the top of the tower). Dish technology uses a two-axis parabolic dish to concentrate solar energy into a cavity receiver, from which it is transferred to a heat engine/generator—specifically, a Stirling engine because of its high efficiency and power density (Mancini et al., 2003). In the future, some CSP plants will use molten-salt receivers, which have the advantage of short-term storage capability, thereby allowing the plant to generate electricity for a few hours after sunset. The molten-salt receiver was pioneered in the United States in the 1990s and is currently undergoing its initial commercial deployment in a small ($\approx$20 MW) plant in Spain.

Another evolving technology that relies on solar concentration is high-temperature chemical processing, in which the receiver placed at the focus of the concentrating reactor includes a chemical reactor. The main advantage of this system is its capability to provide long-term storage of intermittent solar energy (in the form of fuel or a commodity chemical). Also, a number of multiple-step cycles that could be used in parallel with CSP—including production of hydrogen using water as the feedstock, decarbonization of fossil fuels, production of metals (including aluminum), and processing and detoxification of waste—are being pursued by researchers (Fletcher, 2001; Perkens and Weimer, 2004; Steinfeld, 2005).

As discussed in the section titled "Costs" below in this chapter, concentrating-trough and power-tower systems are currently the lowest-cost utility-scale solar electricity technologies for the southwestern United States, as well as for other areas of the world with sufficient direct solar radiation. In the short term, costs will be driven down and uncertainties in performance reduced as designs are incrementally improved, as more systems are installed (the "learning curve"), and as economies of scale both for power plant sites and for manufacturing are realized. Over the medium term, advances in high-temperature and optical materials will be needed. For example, the development of less expensive yet durable optical materials—including selective surfaces for receivers in towers and dishes, transparent polymeric materials that are cheaper than glass, and reflective surfaces that prevent dust deposition—will help to further reduce cost, improve performance, and decrease water use. For the long term, thermochemical production of fuels using CSP is a promising mechanism for storing solar energy.

Geothermal

Geothermal electricity is currently produced by conventional power-generating technologies utilizing hydrothermal resources (hot water or steam) accessible within 3 km of Earth's surface. These plants operate 90–98 percent of the time, providing baseload electricity. Most hydrothermal plants are binary cycle, converting lower-temperature geothermal water (90–175°C) to electricity by routing it through a closed-loop heat exchanger, where a low-boiling-point hydrocarbon (such as isobutane or isopentane) is evaporated to drive a Rankine power cycle. The other possibilities are steam plants, which use steam directly from the source, or flash plants, which depressurize hot water from the source (175–300°C) to produce steam. These technologies are well developed but limited to particular geological areas—in the United States, the Southwest (WGA, 2006).

A more challenging possibility for tapping Earth's internal heat are enhanced geothermal systems. As discussed in the "Resource Base" section above in the chapter, this approach mines the heat stored deep below Earth's surface (typically beyond 3 km and down to about 10 km) in hot and low-permeability rock by artificially creating porous or fractured reservoirs. To access the stored thermal energy, the hot rock must first be "stimulated" by drilling a well to reach the rock, and high-pressure water is then used to form a fractured rock region. Injection and production (extraction) wells can then be drilled into the fractured region; water is circulated in the injection well to extract the stored heat.

A fractured EGS reservoir would cool significantly during heat-mining operations. MIT (2006) estimated that a normal project life would be some 20–30 years before reservoir temperatures fell by 10–15°C and abandonment occurred. The MIT study also found that production wells would need to be redrilled every 4–8 years during that project life.

EGS technology is not currently in large-scale operation. Significant challenges include a general lack of experience in drilling to depths approaching 10 km, even in oil and natural gas exploration, and the need to enhance heat-transfer performance for lower-temperature fluids in power production. Another challenge is to improve reservoir-stimulation techniques so that sufficient connectivity within the fractured rock can be achieved; in that way, the injection and production well system may realize commercially feasible and sustainable production rates (MIT, 2006). Genter et al. (2009) reviews the progress on the EGS plant at Soultz, France, a project that has been under way for more than two decades. Progress is ongoing at this site, with a recently added 1.5 MW power plant currently beginning a 3-year scientific and technical monitoring phase. EGS activity is also ongoing in the United States, Germany, and Australia.

Hydropower

Hydropower, whether through conventional hydroelectric technologies or emerging hydrokinetic technologies, captures the energy of moving water in order to generate electricity. Conventional hydroelectricity, which taps the energy in freshwater rivers, is the largest source of renewable electricity at present. Most hydroelectricity projects use a dam to back up and control the flow of water, a penstock to siphon water from the reservoir and direct it through a turbine, and a generator to convert the mechanical energy to electricity. The resources for conventional hydropower have been extensively exploited and the technologies required are fully developed. Currently, the major technological challenges relate to increasing the efficiency of existing facilities and mitigating any negative consequences, especially on anadromous fish (EPRI, 2007a). Additionally, there are pressures to return river systems back to free-running conditions; although this will not likely lead to the removal of any major hydropower dams, it could limit expansion of the resource.

Hydrokinetic technologies generate electricity from currents, tides, and ocean waves. Many pilot-scale projects for exploiting these sources currently exist, but only a few operate at a commercial scale worldwide. Tapping tidal, river, and ocean currents is done using a submerged turbine. Approaches for converting the energy in waves into electricity include floating and submerged designs that access the energy in the impacting wave directly or that utilize the hydraulic gradient between the top and bottom of a wave. Other approaches for generating electricity from the oceans include drawing on their thermal or salinity gradients. Ocean thermal energy conversion transforms solar radiation into electric power using the ocean's natural thermal differences to drive a power-producing cycle (SERI, 1989). Salinity-gradient power relies on the osmotic-pressure difference between freshwater and salt water (Jones and Finley, 2003). Further basic and applied research is needed before deployment of any of these approaches might be contemplated. As a result, the costs of their commercial-scale operation are unknown.

Biopower

Biopower is the generation of electricity by extracting the solar energy stored in biomass—that is, by burning it. Types of biomass for energy production fall into three categories: wood/plant waste; municipal solid waste/landfill gas; and "other" biomass, including agricultural by-products, biofuels, and selected waste products such as tires (Haq, 2001; EIA, 2007). There also is a recent increase in dedicated energy crops, but they are used mostly for liquid transportation fuels.

Given the wide variety of biomass resource types with diverse characteristics (e.g., with respect to moisture, energy, and ash content), numerous electricity-generation technologies are employed in biomass utilization. Despite the differences, several commonalities exist; indeed, production of electricity from biomass occurs in much the same manner as from fossil fuels. As with coal-fired power plants, the vast majority of biomass-fired power plants operate on a steam-Rankine cycle in which the fuel is directly combusted and the resulting heat is used to create high-pressure steam. The steam then serves as the working fluid to drive a generator for electricity production. With a gaseous fuel, a more efficient turbine engine (based on the gas-Brayton cycle) can be employed, in a manner similar to natural-gas-fired power plants. A gas reciprocating engine is also frequently used for installations of less than 5 MW, where a turbine would be too expensive.

A key difference between biomass power plants and those operating on coal is the size of the facilities. Wood-based biomass plants (accounting for about 80 percent of biomass electricity) rarely reach 50 MW in size, whereas conventional coal-fired plants typically range from 100 MW to 1500 MW. Similarly, power plants based on landfill gas—a methane-containing product of the anaerobic decomposition of solid waste—have capacities in the 0.5 MW to 5 MW range, whereas those operating on natural gas may be some 100 times larger, in the 50 MW to 500 MW range. The small sizes are partially due to the high cost of transporting low-energy-content biomass. A maximum 40-mile radius for the resource base is typical. And as a consequence of these sizes, biopower plants are typically less efficient than fossil fuel plants are; the cost of implementing high-efficiency technologies is not economically justified at small scales.

Beyond dedicated biomass power plants, biomass is also used to co-fire power plants that utilize coal as the primary fuel source. These plants exhibit high efficiencies, given their large sizes, with the added benefits of reduced CO_2, SO_x, and particulate emissions and of lower volumes of ash residue. However, co-firing may negatively affect the technologies for reducing nitric oxide emissions (Vredenbregt et al., 2003).

In the medium term, it is likely that new biopower capacity, if pursued, will incorporate a pretreatment step in which the biomass is converted to a gaseous or liquid fuel. This would be more suitable for power generation than the direct firing that is the norm today. In the fossil fuels chapter of this report (Chapter 7) is a discussion of a potential future technology for biopower—an oxygen-blown gasification-based combined-cycle power plant that would use switchgrass as a feedstock. As noted in that discussion, dedicating to this technology some 100 million dry tons per year of the potential 550 dry tons per year of biomass (the total that could be made available for the overall energy system) could provide an estimated 30 GW of biopower capacity. With a capacity factor of 85 percent, this could produce an additional 220,000 GWh/yr. However, in the absence of a program to grow dedicated energy crops, biomass from waste streams (forestry, agricultural, urban) is likely to grow but will probably remain a relatively small contributor to the nation's electricity supply.

Fundamental to biomass's future as an energy source is the competition between using it to generate electricity or to produce liquid transportation fuels. In particular, conversion from raw biomass into syngas or other fuels might render biomass more attractive for transportation applications. Indeed, the U.S. Department of Energy (DOE) has essentially stopped its biopower programs in favor of biofuels for transportation (Beaudry-Losique, 2007), given the legislative requirements to greatly expand the use of biofuels. In addition, the long-term potential of biomass is currently limited by the low conversion efficiency of the photosynthesis process.

For more details on the timing, costs, and impediments to the development and deployment of all of these technologies, the reader is referred to the report by the Panel on Electricity from Renewable Resources (NAS-NAE-NRC, 2009).

Technology Timeframes

To better assess the prospects of the various renewable electricity-generation technologies, the AEF Committee has distinguished between three timeframes: the present to the year 2020; the 2020–2035 period; and 2035–2050 and beyond.

For the first timeframe, there appear to be no technological constraints for the accelerated deployment of wind, solar PV, CSP, conventional geothermal, and biomass. The main barriers are cost competitiveness relative to fossil-based electricity (assuming no costs are assigned to carbon emissions or other currently unpriced externalities); markets that discourage renewable technologies from reaching scale and thus attaining economies of scale (especially with respect to manufacturing); and the lack of clear and sustained policies for achieving such goals.

Substantially expanded research and development will be needed to realize continued improvements and further cost reductions for these technologies. Such improvements, stimulated by appropriate policies, will greatly increase the competitiveness of electricity generation from renewable resources and thus enhance the technologies' levels of deployment. This achievement could, in turn, set the stage for an even greater and more cost-effective penetration in later time periods. With such accelerated deployment, it is reasonable to envision that nonhydroelectric renewables could collectively start to provide a material contribution to the nation's electricity generation within this first timeframe—reaching perhaps the 10 percent level or greater, with trends toward continued growth. Combined with hydropower, this would mean that renewable electricity could approach a total contribution of 20 percent of electricity by the year 2020.

It is clear that such an increase would require not only sufficient resources and technologies but also manufacturing, materials, and labor capabilities together with considerations of environmental impacts, siting issues, and systems integration. A critical aspect of an expansion of renewable electricity will be a corresponding expansion in transmission capacity and other improvements in the electricity transmission and distribution systems. Quantitative estimates of the deployment and installation requirements for wind power to meet this level of renewables penetration are provided later in this chapter and in the report by the Panel on Electricity from Renewable Resources (NAS-NAE-NRC, 2009). Sizing of transmission capacity solely for wind is an important issue because low capacity factors for wind mean that transmission build for wind power alone would not be utilized up to its capacity much of the time. This is the motivation for considering the colocation of other renewable and nonrenewable technologies with wind discussed later in this chapter.

In the second timeframe (2020–2035), assuming public- and private-sector research and development (R&D) efforts and sustained public policy incentives, it is reasonable to envision an even further acceleration in deployment of nonhydroelectric renewables, which could collectively provide 20 percent or more of domestic electricity generation. With conventional hydropower's contribution remaining at current levels, total electricity generation from renewables could account for more than 25 percent by 2035. The significant issues raised in reaching this level of renewables penetration are discussed later in this chapter and in NAS-NAE-NRC (2009).

In the third timeframe (2035–2050 and beyond), continued development of renewable electricity-generation technologies can be expected to provide lower costs and potentially to result in further increases in penetration. However, achieving a predominant (i.e., greater than 50 percent) level of renewable-electricity penetration, especially if the sources are predominately intermittent, will require scientific advances and dramatic changes in the ways we receive and use electricity. Beyond the continuing needs for improvements in cost, scalability, and performance of renewable electricity-generation technologies, some combination of intelligent two-way electric grids, cost-effective methods for large-scale and distributed storage (either direct electricity energy storage or generation of chemical fuels), widespread implementation of rapidly dispatchable fossil electricity technologies (for backup), and greatly improved technologies for cost-effective long-distance electricity transmission will also be required.

More Speculative Technologies

Given the long time periods covered in this report, it is conceivable that other technological approaches, in addition to those just discussed, will evolve. They include the possibility of high-altitude kites tethered to generators for producing wind-derived electricity, or satellites in space that would collect sunlight, convert the energy into a laser or microwave beam, and aim the beam at a terrestrial plant for generating electricity or producing hydrogen. Other approaches may be based on new processes developed from basic advances in fields such as bioengineering and nanotechnology. Additionally, new approaches may ensue from the application of systems engineering in a world that is highly constrained not only by energy supplies but also by water availability and waste-disposal options. For example, it is conceivable that neighborhood combined-energy-waste-and-water plants may be attractive for reducing pressures on the infrastructure and limiting the amounts of consumables entering the system. Such approaches could be of great benefit as the world's population becomes increasingly urbanized.

Findings: Technology

Wind, solar photovoltaics, concentrating solar power, conventional hydropower, hydrothermal, and biopower technologies are technically ready for accelerated deployment; both individually and collectively, these technologies could make significantly greater contributions to the U.S. electricity supply than they do today. Advances in the currently developed renewable electricity-generating technologies will be driven by incremental improvements in individual components, "learning curve" technology maturation, and achievement of economies of scale in commercial production.

Other technologies, including enhanced geothermal systems that mine the heat stored in deep low-permeability rock, as well as hydrokinetic systems that tap the energy in ocean tidal currents and waves, require further R&D before they can be considered viable entries into the marketplace. Thus these options could potentially be available in the second timeframe (2020–2035) or third timeframe (2035–2050 and beyond) if technological and economic conditions were favorable. Meanwhile, it is possible that plastic organic solar cells, dye-sensitized solar cells, and other new photovoltaic technologies could become commercially available during the second timeframe. In any case, basic and applied research efforts are expected to drive continued technological advances and cost reductions for all renewable electricity-generation technologies.

Because some of the technologies that tap renewable resources to produce electricity must operate under temporal and spatial constraints, special consideration of systems-integration and transmission issues will be needed in order for the penetration of renewable electricity to grow. Such considerations become especially important at sizable penetrations (greater than ~20 percent) of renewables in the domestic electricity generation mix. In the second timeframe in particular, a concurrent and unified overlaying of intelligent control and communications technologies (e.g., advanced sensors, smart meters, and improved software for forecasting and operations) would be required for assuring the viability and continued expansion of renewable electricity. Such improvements in the electricity transmission and distribution grid could enhance system integration and reliability, provide significant capacity and cost advantages, and reduce the need for backup power and energy storage.

COSTS

Given the experience with renewables over the past 20–30 years, it is clear that their economics have not been favorable. The economics of renewables is about profitability, and profitability depends on three drivers: (1) the market price of electricity; (2) the costs of renewables relative to those of other resources; and (3) policies designed to promote renewables or achieve environmental goals (particularly regarding climate) that raise the costs of using fossil fuels or subsidize the costs of renewables.

In order to enjoy greater market penetration in the future, renewables need to achieve cost reductions. And they must do so at a rate that is greater than the rate of cost improvement by technologies—including natural-gas- and coal-fired generation—that currently tend to set the market price of electricity. In some cases, these reductions might be the result of technological breakthroughs; in other cases, they could come from improvements in the manufacturing or in the operating performance of equipment (e.g., higher capacity factors for wind turbines).

Levelized Cost Estimates

In order to compare the costs of electricity-generation using different renewable technologies, both to each other and to the costs of electricity from fossil fuels and nuclear, a cost estimate is typically converted into a levelized cost of electricity (LCOE), which is expressed in dollars per megawatt-hour (or kilowatt-hour) of generation. A large portion of the cost of generating electricity, particularly for renewables, is the initial cost of the capital equipment and installation. Converting this large upfront cost to dollars per megawatt-hour requires making assumptions about the lifetime and capacity factor of the equipment as well as the discount rate and the timing of returns on capital. For intermittent technologies such as concentrating solar thermal, solar PV, and wind power, the capacity factor can vary considerably, depending on the quality of the resource (e.g., hours and intensity of sunlight or speed and constancy of wind), which varies by location. Table 6.2 compares LCOEs derived from various studies for four renewable electricity-generation technologies.[5] It also contains the cost breakdown used in the estimates where available. It compares cost estimates developed from information provided in the EIA AEO 2009 reference case forecast (EIA, 2008a) to estimates of the DOE's Office of Energy Efficiency and Renewable Energy Office (EERE) (NREL, 2007c), estimates of the Electric Power Research Institute (EPRI, 2007b), and inputs to the DOE's 20% Wind Energy study from Black & Veatch (DOE, 2008a; Black & Veatch, 2007). The committee has incorporated the more recent

[5]The 4.3 cents LCOE for onshore wind is a best-case wind site busbar cost that assumes a Class 7 wind resource and continued improvement in wind-turbine capacity factor raising it from the current value of almost 40 percent for the current vintage machines to 52 percent by 2020 by extrapolating the improvements over the period 1999–2006. As with any generation resource, transmission costs must be added in. Due to variation in wind resources, distance to existing transmission infrastructure, and integration costs, LCOEs for wind are site dependent. Additionally, the best-case solar PV cost assumes that overnight capital costs are reducd to $2547/kW by 2020.

TABLE 6.2 Electricity-Generating Costs for Renewable Technologies in 2020

Technology	Source	Case/ Scenario	Overnight Cost ($/kW)
Biopower			
Biopower	AEO 2009 (EIA, 2008a)	Reference	3390
Biopower–Stoker	EPRI (2007b)	Full Portfolio	
Biopower–Stoker	EPRI (2007b)	Limited Portfolio	
Biopower	ASES (2007)	WGA Biomass Task Force	
Geothermal			
Geothermal	AEO 2009 (EIA, 2008a)	Reference	1585
Concentrating Solar			
Concentrating solar	AEO 2009 (EIA, 2008a)	Reference	4130
Concentrating solar	EPRI (2007b)	Limited Portfolio	
Concentrating solar	EPRI (2007b)	Full Portfolio	
Photovoltaics			
Photovoltaics	EERE 2008 (NREL, 2007c)	Program	2547
Photovoltaics	EPRI (2007b)	Full Portfolio	
Photovoltaics	EPRI (2007b)	Limited Portfolio	
Photovoltaics	AEO 2009 (EIA, 2008a)	Reference	5185
Wind			
Onshore wind	AEO 2009 (EIA, 2008a)	Reference	1896
Onshore wind	EPRI (2007b)	Full Portfolio	
Onshore wind	EPRI (2007b)	Limited Portfolio	
Onshore wind	DOE (2008a), Black & Veatch (2007)	DOE *20% Wind Energy* study	1630
Offshore wind	AEO 2009 (EIA, 2008a)	Reference	3552
Offshore wind	DOE (2008a), Black & Veatch (2007)	DOE *20% Wind Energy* study	2232

Note: All cost estimates are in 2007 dollars.
[a]Cost estimate is for 2015.

Capacity Factor (percent)	Total Capital Cost ($/MWh)	Variable O&M/ Fuel Costs ($/MWh)	Fixed O&M ($/MWh)	Levelized Cost of Energy ($/kWh)
83	61.62	22.81	8.86	0.093
85				0.096
85				0.101
90				~0.080[a]
90	75.44	0.00	22.22	0.098
31	180.02	0.00	21.30	0.201
34				0.170
34				0.083[a]
21	135.81	0.00	5.59	0.141
				0.220
				0.260
22	292.84	0.00	6.21	0.299
35	81.38	0.00	9.95	0.091
42				0.078
33				0.097
38 to 52 (depending on wind class)	48.04–35.10	4.85	3.64–2.66	0.057–0.043
33	154.36	0.00	26.72	0.181
38 to 52 (depending on wind class)	64.1–46.29	4.87–3.52	4.62–3.34	0.074–0.053

estimates from AEO 2009 since they better reflect the recent increase in capital costs for electricity-generating technologies.

The snapshot of current costs presented in Table 6.2 is affected by a number of factors. One key factor is the assumed cost of capital equipment today and whether it reflects recent increases in material and labor costs for wind turbines, solar panel components, and construction in general. Another factor is technology learning and how it is likely to evolve over time (independent of scale of deployment). Details on how the individual renewable technologies might improve are included in the full report from the Panel on Electricity from Renewable Resources (NAS-NAE-NRC, 2009). Finally, projections that are more in the form of goals for renewable technology performance tend to be more optimistic than are projections based on some form of learning.

For example, though not shown in Table 6.2, estimates from the Solar Energy Industry Association (SEIA, 2004) for costs of solar technologies, and estimates from EERE for costs of wind technologies (NREL, 2007c), tend to be more optimistic than those from other sources; the SEIA and EERE estimates reflect how full funding of renewable-energy research at the DOE and elsewhere is expected to affect the future costs of renewable electricity generation. Thus these estimates tend to represent aspirations. (Detailed discussions of these factors, and on how differences in assumptions affect the various costs estimates, are provided in the full report of the Panel on Electricity from Renewable Resources [NAS-NAE-NRC, 2009].)

Taking all of these factors into consideration, the AEF Committee supported the idea that the upper-bound costs shown in these estimates were in line with the committee's estimate of a reasonable upper bound and that some of the lower-cost estimates, including ones not shown in Table 6.2, tended to represent aspirations.

Costs Beyond Electricity Generation Costs

The costs of purchasing, installing, and operating a specific power plant normally would not be the total costs to the system and to electricity consumers of deploying a new renewable generation facility. Costs that are normally missing from the traditional levelized cost measure are the costs of new infrastructure necessary to connect the renewable generator to the grid and to ensure continued quality of power supply. The report from the Panel on Electricity from Renewable Resources (NAS-NAE-NRC, 2009) reviews costs estimates for these components. The estimates focus almost exclusively on the costs of transmission and intermittency associated with wind power. For example, a recent report looked at cost estimates for 40 transmission projects for wind power (Mills and Wiser, 2009). The report found that the transmission costs associated with wind ranged from $0 to $1500/kW, and the majority were less than or equal to $500/kW, with a median of $300/kW. These numbers correspond to $0–79/MWh, with the majority below $25/MWh, and a median of $15/MWh. There have also been a large number of studies that look at intermittency costs, including DeCarolis and Keith (2004), EWEA (2005), Smith (2007), and DOE (2008b). Generation from intermittent resources such as wind requires other generation sources to provide backup when wind power is not available and to help track electricity loads, provide voltage support, and meet the needs for capacity reserves. Generally, these studies have found that, when the average cost of wind generation is about $80/MWh, the impact of grid integration costs is estimated to be less than 15 percent where wind produced 20 percent or less of total electricity generation.

Renewable Supply Curve: Example for Wind Power

Single national-average estimates of LCOEs fail to communicate how such costs vary with the quality of the resource. As shown in Figures 6.2 and 6.3, wind and solar intensities vary regionally, and other renewable resources show similar patterns. This phenomenon in turn affects capacity factor and LCOE. Figure 6.4, which plots LCOE as a function of the amount of wind capacity available, shows how the cost of energy increases from higher wind classes to lower wind classes and from onshore sites to offshore sites. It should be noted that Figure 6.4 does not reflect transmission costs or other factors that affect the cost of delivered wind energy. Figure 6.5 plots LCOE as a function of the amount of wind capacity available including transmission costs for both onshore and offshore sites. However, Figure 6.5 does not incorporate an estimate of integration costs, which would be dependent on a number of factors, including the current generation mix within each control area and regional power pool (Black & Veatch, 2007).

Findings: Costs

The evolution of the costs of renewables will depend both on technological advances and on stable and clear public policies that encourage greater penetration and accelerate production and deployment at scale. At present, onshore wind is an economically favored option relative to other nonhydroelectric renewables, and wind power could in principle scale to a significant penetration

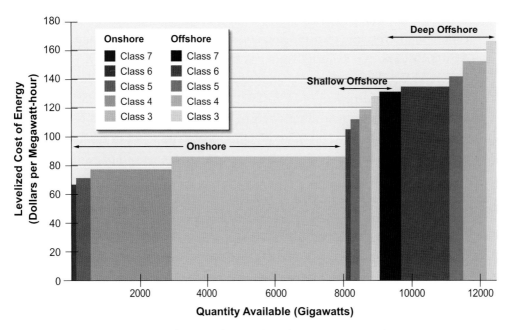

FIGURE 6.4 *Supply curve for wind without accounting for PTC, transmission, and integration cost.*
Sources: DOE, 2008a; Black & Veatch, 2007.

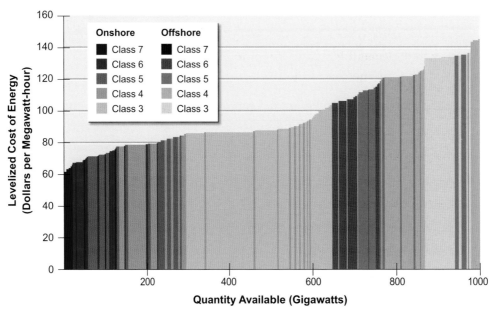

FIGURE 6.5 *Supply curve for wind accounting for transmission costs but no integration costs.*
Sources: DOE, 2008a; Black & Veatch, 2007.

of domestic electricity supply by 2020. Accordingly, wind power will continue to grow rapidly if recent policy initiatives, including the PTC, continue into the future.

Though AEO 2009 suggests that biopower will play an expanding role in meeting future RPS targets (EIA, 2008a), the degree of competition with and recent mandates for use of liquid biofuels for providing transportation fuel—and, of course, the use of biomass for food, feed, and fiber—all limit the prospects for greater use of biomass in the electricity market.

Supply-side utility-based renewable-electricity technologies, such as concentrated solar power, wind, and biomass, must compete on a cost basis with other technologies for utility electricity generation. But the future of distributed renewables generation, such as from residential PV, will depend more on policy, on how costs compare to the retail price of power delivered to residential or other customers, on whether prices fully reflect variations in cost over the course of the day, and on whether the full costs of using fossil-fuel generation—particularly their externalities, such as CO_2 emissions—are incorporated into prices.

Formulation of robust predictions about whether the price of electricity will meet or exceed the price required for renewable sources to be profitable and what their resulting level of penetration will be remain a difficult proposition. Furthermore, the profitability of renewables generation may be sensitive to investments in energy efficiency, especially if such efficiency improvements are sufficient to meet growth in demand for electricity or to lower the market-clearing price of electricity. If there are no changes in the financial operating environment for fossil fuels and other incumbent sources of electricity, then renewable electricity may be negatively impacted more than other electricity sources. However, renewable electricity is being driven at present by tax policies—in particular, the production tax credits—and by renewable portfolio standards.

ENVIRONMENTAL IMPACTS

The fact that renewable-energy sources have smaller environmental "footprints" than do fossil-fuel sources arguably provides the greatest impetus for making a transition. However, the use of renewables does not eliminate all environmental impacts, though the types and magnitudes differ substantially from those of fossil fuels and from one renewable technology to another.

Atmospheric Emissions and Land-Use Impacts

While most renewable technologies do not emit air pollutants during the generation of electricity,[6] they do emit them during manufacture or installation. Thus, an accurate accounting of emissions and other impacts must embrace the whole life cycle of a given technology.

The process of life-cycle analysis (LCA) aims to estimate the overall energy usage and environmental impacts of a given technology by considering all of its stages—from the extraction and refinement of raw materials to construction, use, and disposal. However, a major complication in comparing LCAs is that there is no set standard by which such analyses are carried out. While all LCAs attempt to cover technologies from "cradle to grave" in a systematic way, there is significant variability in the assumptions, boundaries, and methodologies used in these assessments. Further, the data are for normal operations and do not consider emissions or other impacts that result from start-ups, shutdowns, or accidents. Comparisons of LCA should therefore be made with caution; each LCA is only an approximation of a technology's actual set of impacts.

Not surprisingly, based on the range of estimates of carbon dioxide (CO_2)-equivalent (CO_2-eq)[7] emissions that appear in the literature, renewables have significantly lower CO_2-equivalent emissions than do coal, oil, and gas. Estimates for solar, wind, hydropower, and geothermal range from about 10 to 50 grams of CO_2-equivalent emissions per kilowatt-hour, similar to the numbers for nuclear power (Berry et al., 1998; Chataignere and Le Boulch, 2003; Denholm, 2004; Denholm and Kulcinski, 2003; European Commission, 1997a,b,c,d; Frankl et al., 2004; Fthenakis and Kim, 2007; Hondo, 2005; Mann and Spath, 1997; Meier, 2002; Odeh and Cockerill, 2008; Spath et al., 1999; Spath and Mann, 2000, 2004; Spitzley and Keoleian, 2005; Storm van Leeuwen and Smith, 2008; Vattenfall, 2004; White, 1998, 2006). Much higher values are found for fossil-fuel sources of electricity, upward of 500 g CO_2 eq per kilowatt-hour for natural gas and 1000 g CO_2 eq per kilowatt-hour for coal, though the relative advantage of renewables would be significantly reduced by adding carbon capture and storage (CCS) to coal and gas systems (if the commercial viability of CCS were demonstrated). As described in the report of the Panel on Electricity from Renewable

---

[6]Biomass and geothermal are notable exceptions.

[7]Equivalent carbon dioxide emissions (CO_2 equivalent) are the amount of greenhouse gas emissions expressed as carbon dioxide, taking into account the global warming potential of non–carbon dioxide greenhouse gases (e.g., methane, nitrous oxides).

Resources (NAS-NAE-NRC, 2009), renewable-electricity sources also have much lower emissions of other atmospheric pollutants, including sulfur oxides, nitrogen oxides, and lead.

On the other hand, because renewable-energy resources are more diffuse than fossil and nuclear energy resources, the land areas needed to collect renewable energy and convert it to electricity are, on an energy-equivalent basis, much larger than those of fossil and nuclear. Spitzley and Keoleian (2005) report land-use values of 9–14.3 m^2/MWh per year for solar, 69–94 m^2/MWh per year for wind, and 360–488 m^2/MWh per year for biopower. Spitzley and Keoleian (2005) included one LCA example of hydroelectric power with a land-use value of 122 m^2/MWh per year for a high-capacity, large-reservoir facility in the United States. This compares with estimates for natural gas, coal, and nuclear technologies of 0.45 m^2/MWh per year, 4.4–5.8 m^2/MWh per year, and 6.5 m^2/MWh per year, respectively. An exception to the statement that most renewable-electricity sources require large land areas is geothermal electricity, including both hydrothermal and EGS. Land requirements for geothermal energy are smaller than those for other renewables and are comparable, in fact, to those for nonrenewable sources of electricity (MIT, 2006). It should be noted, however, that in order to operate fossil-fuel and nuclear plants the fuel must first be extracted or mined. Most LCAs, including that of Spitzley and Keoleian (2005), do not account for that process in their assessment of land-use requirements.

Further, there is a wide spectrum of other environmental impacts of renewables that have not been addressed by LCA or analyzed in the studies cited previously. One impact often ascribed to wind power is bird and bat mortality. However, a recent NRC study concluded that although the impacts on bat populations were unclear, there was no evidence that bird fatalities caused by wind turbines had measurable effects on U.S. bird populations (NRC, 2007). Another consideration is that wind-farm deployments at levels needed to produce large amounts of power may extract a significant portion of the energy from the wind field. Continental-scale simulations indicate that high levels of wind power extraction could, to various degrees, affect regional weather as well as climate. In addition to limiting the efficiency of large-scale wind farms, model calculations suggest that the extraction of wind energy from very-large-scale wind farms could have some measurable effect on weather and climate at the local or even continental and global scales (Roy et al., 2004; Keith et al., 2004).

Other potential impacts from renewables include water diversion, water pollution, and land subsidence from geothermal drilling; ecosystem changes due to hydropower; and noise pollution and aesthetic degradation from wind power. In addition, CSP technologies, except for air-cooled dish-engine technologies, also require significant cooling water. While such impacts may not be widespread or overwhelming, they are often what raise the biggest concerns from local populations and regulators. This means that local and state authorities exercise significant authority over the siting and deployment of renewable-energy generating facilities. While this situation is not unique to renewable sources of electricity generation, there is a long, evolved permitting process that has been applied across the country in the case of traditional generating facilities such as coal and gas. For renewable technologies, on the other hand, the regulatory process is still in the developmental stage.

Findings: Impacts

Renewable electricity-generation technologies have inherently low life-cycle emissions of carbon dioxide and other atmospheric pollutants compared to those of fossil-fuel-based technologies. Most of the CO_2 emissions from renewable-electricity generation are incurred during the manufacturing and deployment stages. However, renewables emit much less CO_2 over their lifetimes than do fossil-fuel sources. Renewable electricity also has inherently low emissions of other regulated atmospheric pollutants, such as sulfur oxides and nitrogen oxides, and (except for biopower, and some geothermal and CSP technologies) it uses significantly less water than do nuclear, gas-fired, and coal-fired electricity technologies.

Because of the diffuse nature of the resources, the systems needed to collect renewable energy (e.g., wind turbines, solar panels, and concentrating systems) must be spread over large areas. These land-use impacts are mitigated to some extent by the low levels of atmospheric emissions and water use (except for biopower and some geothermal and CSP technologies), whose effects, if any, tend to remain localized. There also is the opportunity for some lands taken up by renewable technologies to be simultaneously used for other purposes. The land between turbines on large wind power installations may be used for agriculture, for example, and solar PV panels can be placed on roofs of residences and commercial and industrial establishments. At a high level of renewable technologies deployment, land-use and other local impacts would become quite important. The land-use impacts have caused, and will in the future cause, instances of local opposition to the siting of renewable electricity-generating facilities and associated transmissions lines. State and local government entities typically have primary jurisdiction over the local deployment of electricity generation, transmission, and distribution facilities. Significant increases in the deployment of renewable-electricity facilities will thus entail concomitant increases in the highly specific, administratively complex, environmental impact and siting review processes. Even though this situation is not unique to renewable electricity, a significant acceleration of its deployment will nevertheless require some level of coordination and standardization of siting and impact assessment processes.

DEPLOYMENT POTENTIAL

Renewable-energy technologies have the potential to grow into a significant segment of the electricity-supply mix—the necessary resources and technological capabilities are more than sufficient. The deployment of increased renewables capacity will also result in improved environmental quality. However, a number of challenges suggest that these outcomes are by no means assured. Adequate deployment capacities, successful renewable-electricity integration strategies, predictable policy conditions, acceptable financial risks, and access to capital all are needed to accelerate deployment.

Given the robust business and regulatory activities associated with the wind and solar energy industries, the examples discussed in this section—which illustrate the deployment issues that are associated, in varying degrees, with all renewable sources of electricity—are from those industries.

Deployment Capacity Considerations

Table 6.3 indicates the levels of 2006 employment associated with the renewables industry. These figures would increase, of course, if renewable electricity were to continue to grow at its present rate. There are constraints on such growth, however, which are related to restricted supplies of raw materials, limitations on manufacturing capacity, competition from other construction projects, and workforce shortages. These constraints have the potential to impede or even derail the large-scale deployment and integration of renewable electricity-generation technologies.

Consider, for example, scarcities of key raw materials. An illustration of the potential impacts of materials shortages for PV is given in Figure 6.6, which shows that recent shortages of polycrystalline silicon brought about by increased demand for PV have reversed the long-term decline in PV price. Originally, the

TABLE 6.3 Employment in the Renewable-Energy Sector in 2006

Industry Segment	Revenues/Budget ($ billions)	Employment
Wind	3.0	16,000
Photovoltaics	1.0	6,800
Solar thermal	0.1	800
Hydroelectric	4.0	8,000
Geothermal	2.0	9,000
Biopower	17.0	66,000
Federal government (including direct support contractors)	0.5	800
DOE laboratories (including direct support contractors)	1.8	3,600
State and local governments	0.9	2,500

Source: ASES, 2007.

primary use of polycrystalline silicon was for semiconductors, with PV manufacturers commanding a small fraction of silicon production and even using silicon recycled from the electronics industry. But now the PV industry has become the largest consumer of polycrystalline silicon. This has brought new entrants into polycrystalline manufacturing, including producers specifically oriented to PV as well as PV makers themselves that are trying to become more integrated into the supply chain (Flynn and Bradford, 2006). Despite these new entrants, there was still a shortage of polycrystalline silicon that continued to drive up the price for solar silicon PV modules through 2008, though this shortage was expected to subside by 2009. A recent report has confirmed this decrease in costs for solar PV, though the decline in price has been attributed to both increasing supplies and decreasing demands due to the global economic slowdown (Patel, 2009).

Consider also the limitations on production capacity for wind turbines. Because turbine manufacturers are still in the process of making the capital investments necessary to increase such capacity, developers face shortages of turbines as demand for wind power both in the United States and throughout the world continues to be strong (AWEA, 2008). Thus manufacturers are playing catch-up, with typical delays of 6 months or more from turbine order to delivery. Figure 6.7 shows that these shortages have spurred increases in installed wind power costs, largely due to increases in turbine prices. Other reasons for the turbine price

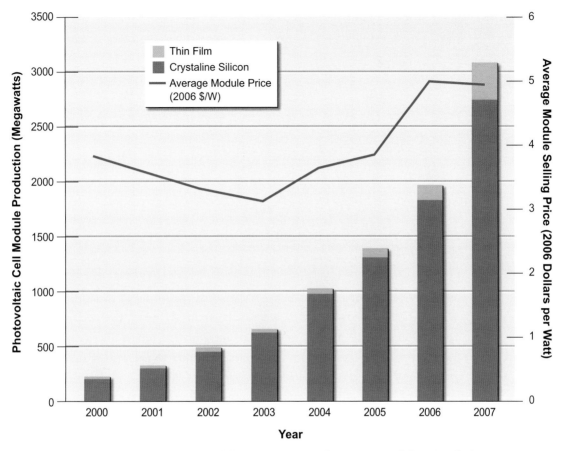

FIGURE 6.6 *Global PV module production and average module price during 2000–2007. Source: Courtesy of Paula Mints, Principal Analyst, Navigant Consulting PV Services Program.*

increases include higher costs for materials and energy inputs; component shortages; upscaling of turbine size and improvements in turbine design; the declining value of the U.S. dollar; and attempts to increase profitability in the wind turbine manufacturing industry (DOE, 2008b). Increases in wind power project costs continued through 2008; it is unclear how the economic downturn in 2009 will affect these costs.

Renewable Electricity Integration

Because renewable resources such as solar and wind have temporal (including short- and long-term) and spatial variability, they introduce intermittencies that

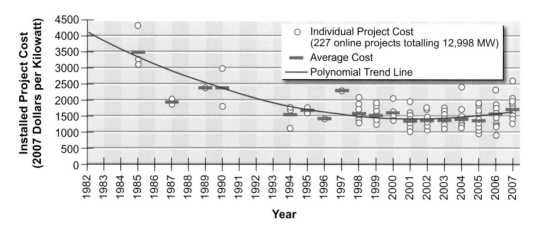

FIGURE 6.7 *Installed wind project costs over the past 25 years. Source: DOE, 2008b.*

must be accommodated in the electricity transmission and distribution systems. With the recent growth in renewables, particularly wind, the need to consider accommodating 20 percent or more of intermittent renewables brings into focus questions of how to maintain sufficient flexibility in the grid while accommodating these increases. Several reviews (e.g., Parsons et al., 2006; Holttinen et al., 2007; DOE, 2008a) analyze, summarize, and compare the various state and national studies on the integration of wind power in particular. These studies show that the integration of intermittent renewable electricity can be eased through modifications to the grid, including an increase in transmission capacity and installation of technologies to improve the grid's intelligence (thereby improving wind power forecasting capabilities). These studies also show that electricity storage is not needed for integrating intermittent renewable energy sources as long as they do not account for more than 20 percent of total electricity generation.

In the case of wind power, there is not always a clear match between the wind resource, available land, and where and when the power is needed. Capturing the full benefits of wind power will require the construction of transmission lines and improving wind power forecasting capabilities. Solar PV offers a way to bring on-site generation to the customer and modify a building's electric-load profile, as the implementation of sophisticated meters that allow time-of-day pricing and excess generation to be sold back to the grid enhances the value of the PV electricity generated. Further, co-siting of renewable-electricity generators (with other renewables electricity generation or conventional electricity generation technologies) or developing a geographically dispersed but interconnected resource base has the potential to smooth temporal variations of electricity generation associated with intermittent renewable resources and improve their integration into the electric system. A combination of intermittent sources backed by natural gas could make the combination of these sources dispatchable to the grid.

Market and Policy Factors

Continued deployment of renewable electricity-generation technologies requires concerted efforts to overcome market barriers and meet investment requirements. This includes navigating the "cash-flow valley of death," where a new business has insufficient ability to raise the capital required for a new technology (Murphy and Edwards, 2003). For renewable-energy technologies, this stage typically occurs during the transition from public-sector to private-sector funding.

Further, renewable electricity must participate in highly competitive electricity markets. But renewable electricity has some economic attributes that differ from those of basic electricity. One such characteristic is that the electricity created at the renewable facility and certain attributes associated with its production can be sold separately. That is, one customer can buy the kilowatt-hours of electricity, while another can buy the "renewableness" associated with the generation through tradable instruments known as renewable-energy certificates or credits. The ability to separately track and sell the electricity and its attributes has increased the number of ways that renewable electricity sales can ultimately occur, thereby expanding the opportunities for renewable-energy development.

Further, renewable-electricity markets operate within a web of interlocking, overlapping, and sometimes conflicting policy prescriptions and legal and regulatory structures. The key risks engendered by this pervasive regime relate to whether future policies will conform to reasonable expectations. For example, Figure 6.8 shows the impact that an off-and-on policy can have on wind power investment: the intermittency of the PTC for wind power generation has led to large fluctuations in demand for wind turbines and in annual installations of new wind power capacity. These fluctuations affect the whole wind power enterprise, from employment to manufacturing to investment.

Findings: Deployment

Increasing the deployment of renewable electricity-generation technologies will require consistent and long-term commitments from policy makers and the public to stimulate investment in business growth, enable market transformation, and

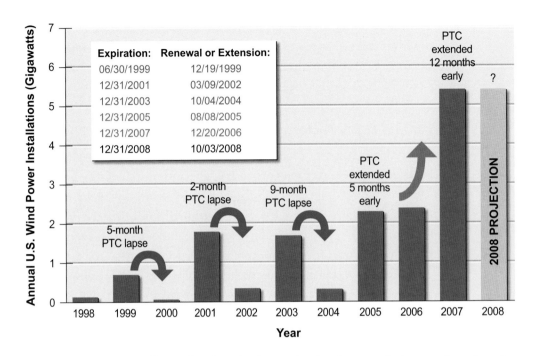

FIGURE 6.8 *Effects of PTC expiration and extension on wind power investment. Source: Wiser, 2008.*

support activities in market-enabling R&D. Public policies to encourage renewables will be more effective if they are predictable—as opposed to cycling on and off or having a highly uncertain future, which has been the case with the PTC. Comparisons of past forecasts of renewable-electricity penetration with actual data show that renewables technologies generally have met forecasts of cost reductions but have nevertheless fallen short of deployment projections.

Significant increases in renewable electricity-generation deployments are also dependent on concomitant improvements in several areas, including labor and workforce development, transmission and distribution grids, and the frameworks and regulations under which the electric systems are operated. One important element is that accommodation of the intermittent characteristics of wind and solar electricity into the overall system is critical for large-scale deployment. Advanced storage technologies will play an important role in supporting such integration beyond some initial level of renewables penetration (above about 20 percent). But in the case of wind power, the development of the resource is occurring long before storage technologies can help to make widespread deployment possible.

Storage is not a requirement for continued expansion of wind power development at present. In fact, it is the underutilization of transmission corridors—built to move wind-derived electricity to load centers—that represents an inefficient use of resources. But there is a stopgap approach that several interested parties are exploring to potentially address this underutilization issue. It is "co-siting," in either of two forms: (1) conventional dispatchable generating resources (natural-gas-fired combustion turbines or combined-cycle plants) paired with renewables; or (2) two or more renewable resources that interact synergistically in terms of transmission needs (e.g., solar power production during the day, wind power at night). Such a development strategy can rationalize the use of transmission resources and serve as an interim mechanism for increasing the value of renewables-generated electricity until advanced storage technologies are technically feasible and economical viable.

DEPLOYMENT SCENARIOS

Scenarios of how renewables might significantly increase their contribution to the electricity system provide a quantitative and conceptual framework to help describe and assess issues related to greatly increasing the scale of renewables deployment. Such scenarios are a primary way of quantifying the materials and manufacturing requirements, human and financial resources, and environmental impacts associated with greatly expanding renewables' electricity generation. In its report, the Panel on Electricity from Renewable Resources (NAS-NAE-NRC, 2009) considered scenarios recently developed by multiple sources. The scenarios typically incorporate computational models and other quantitative elements in an attempt to integrate environmental, technologic, economic, and deployment-related elements into an internally consistent assessment framework. These scenarios were chosen to represent aggressive but achievable rates of renewables deployment if considerable policy and financial resources were devoted to the effort. Other selection criteria used by the panel in choosing scenarios were whether they were developed with input from multiple stakeholder groups and underwent a peer review process. Thus the scenarios do not represent a simple extrapolation of historical growth rates but rather a more integrated perspective on the conditions needed to greatly scale up renewables deployment.

The panel considered two types of scenarios. The first involved increased penetration of a single resource, such as solar or wind. One prominent example is the 20% wind energy study (DOE, 2008a), which is described in more detail in the following subsection. Examples for solar include the DOE's Solar America Initiative (DOE, 2007b), the photovoltaics industry roadmap (SEIA, 2001, 2004), and the 10 percent solar study (Pernick and Wilder, 2008). These scenarios consider issues similar to those of the 20 percent wind power scenario, namely, the potential impacts of high renewables penetration on manufacturing, implementation, economics, and the environment. In addition, because of the higher costs associated with solar energy, all solar scenarios take into account the significant cost reductions that need to occur to make solar electricity widely competitive with other electricity sources.

The second type of scenario involved the interaction of renewables with other sources of electricity, other forms of energy, and end-use demands (U.S. Climate Change Science Program, 2007; EIA, 2008b). With the aid of long-term energy/economic models, these scenarios allow the potential influences of demographic, economic, and regulatory factors on renewable electricity to be assessed within a framework that considers how such factors interrelate with other sources of electricity and end-use energy demands.

Twenty Percent Wind Penetration Scenario

The most in-depth scenario describing increased renewables penetration into the electricity sector is the DOE's 20 percent wind penetration scenario (DOE, 2008a), which, as its name implies, examines the implications of increasing wind power's contribution to 20 percent of total U.S. electricity generation by 2030. The DOE developed this scenario in collaboration with the National Renewable Energy Laboratory (NREL), Lawrence Berkeley National Laboratory, and the American Wind Energy Association. The effort relied on contributions from more than 90 individuals, including stakeholders in the electric utility industry, wind power developers, engineering consultants, and staff members of environmental organizations. These individuals participated in every stage of the study, including its planning process, workshops, steering-group meetings, chapter writing, and review and oversight. The U.S. Office of Management and Budget (OMB) defined the report resulting from the effort as potentially "influential scientific information" disseminated by the agency, thereby requiring that the report be reviewed subject to the Information Quality Bulletin for Peer Review (OMB, 2004).

The DOE's 20 percent wind scenario includes an assessment of wind resources and available technologies; manufacturing, materials, and labor requirements; environmental effects and siting issues; transmission system integration;

and market requirements. All impacts of the scenario (such as costs and CO_2 emissions) are estimated through comparisons with a base case that assumed no new wind capacity additions after 2006—which is more pessimistic in terms of wind power than are the AEO 2008 base case (EIA, 2008c) or the early release AEO 2009 base case (EIA, 2008a). The scenario estimates that more than 300 GW of new wind power capacity would be needed to meet this goal, of which about 250 GW would be installed onshore and 50 GW offshore (DOE, 2008a). It also assumes that wind power capacity factors improve by 15 percent between 2005 and 2030.

Under the scenario, wind power would produce about 1.2 million GWh/yr out of a total electricity generation of 5.8 million GWh. Figure 6.9 shows the amounts of annual installed capacity needed to meet 300 GW by 2030, starting with the approximately 12 GW of total wind power capacity that was available in 2006; the scenario limited the annual capacity increase to 20 percent.

The scenario assumes that by 2018 the amount of annual installed capacity in the United States is in excess of 16 GW. This compares to a global wind-turbine

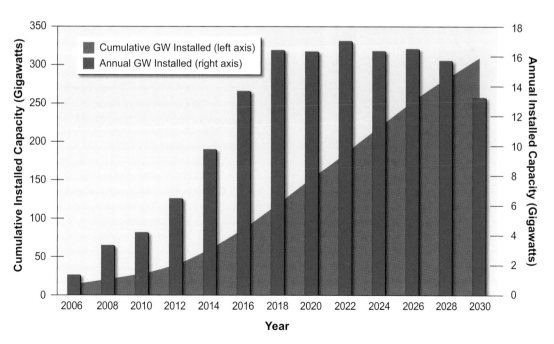

FIGURE 6.9 *Annual and cumulative generation needs to achieve 20 percent wind-generated electricity by 2030.*
Source: DOE, 2008a.

manufacturing output of about 20 GW in 2007, of which approximately 5 GW went to the U.S. market (DOE, 2008b). An additional 8.4 GW of capacity was added in the United States in 2008 and 2.8 GW of capacity in the first quarter of 2009, both of which exceeded the trajectory for the 20 percent wind scenario. Even assuming that growth outside the United States is more modest, this scenario would still require a continued large expansion of the manufacturing base. Additionally, the scenario assumes that wind turbines have an average life of 25 years; sustaining annual installations at approximately 16 GW/yr beyond 2030 would be needed to accommodate repowering of aging turbines and meeting increasing electricity demand to continue the 20 percent wind generation level (Laxson et al., 2006; DOE, 2008a).

Given these challenges, the DOE's 20 percent wind scenario considers the materials, capital, and employment requirements. Table 6.4 shows the level of raw materials that would be needed to meet the scenario. While some quantities are small relative to global production, Smith and Parsons (2007) conclude that supplying fiberglass, core materials (balsa and foam), and resins could be difficult, as could supplying a sufficient number of wind-turbine gearboxes. Assuming that the average-sized wind turbine would be in the 1–3 MW range, with some introduction of large 4–6 MW turbines, there would be a total of almost 100,000 wind turbines installed (Wiley, 2007; DOE, 2008a). This would mean that the average number of turbines installed would have to rise from its current level of 2000 per year to 7000 per year by 2017 (DOE, 2008a) and the average turbine size would be about 3 MW.

In 2030, the DOE's 20 percent wind scenario would require well over 140,000 direct manufacturing, construction, and operations jobs, according to

TABLE 6.4 Raw-Materials Requirements for the 20 Percent Wind Penetration Scenario, in Thousands of Tons per Year

Year	Concrete	Steel	Aluminum	Copper	Glass-Reinforced Plastic	Carbon Fiber Composite	Adhesive	Core
2010	6,800	460	4.6	7.4	30	2.2	5.6	1.8
2015	16,200	1,200	15	10	74	9	15	5
2020	37,000	2,600	30	20	162	20	34	11
2025	35,000	2,500	28	19	156	19	31	10
2030	34,000	2,300	26	18	152	18	30	10

Source: Wiley, 2007.

the DOE's Job and Economic Development model (Goldberg et al., 2004; Wiley, 2007; DOE, 2008a). This figure includes 20,000-plus jobs in manufacturing, almost 50,000 jobs in construction, and 75,000-plus jobs in operations (DOE, 2008a). (It should be noted that these job requirements are not the same as aggregate net new job creations, whose predictions would necessitate a thorough understanding of how expenditures in the wind power sector affect investments and employment in other sectors of the economy.)

The DOE's 20 percent wind scenario would have significant land-use and atmospheric-emissions impacts. The estimate land area needed would be 50,000 km^2, with only about 2–5 percent being for the turbines themselves. The rest of the land area would be devoted to spacing between turbines, and much of it might be available for agricultural uses. Atmospheric emissions of carbon dioxide and other pollutants would be significantly reduced if these initiatives displace electricity-generating plants that use fossil fuels. The scenario estimates that in replacing coal- and gas-fired electricity generation, wind power would reduce CO_2 emissions by 800 million tonnes per year in 2030 (about 25 percent of the estimated total CO_2 emissions from the electric sector in 2030). Increasing the amount of wind-generated electricity would also result in reductions of other atmospheric pollutants associated with fossil-fuel electricity generation. However, the impacts on NO_x and SO_x emissions are expected to be less than what would be estimated from simply assuming that electricity generation from fossil fuels is replaced with a non-carbon-emitting technology such as wind power. Emissions of both NO_x and SO_x are subject to caps on emissions; thus emissions reductions from wind-generated electricity may be reallocated to other plants. Other airborne toxics are also emitted from fossil-fuel electricity-generation technologies, especially coal-fired plants, that are not capped; thus they would be reduced in replacing fossil fuels with wind. Additionally, the 20 percent scenario would potentially reduce water consumption in the electric sector. The DOE report estimates that annual water consumption in the electric sector would be reduced by 17 percent (DOE, 2008a).

Participants in the DOE study, as well as other stakeholders, developed estimates of the costs required to reach 20 percent wind penetration by 2030. The study considered both the direct costs of installing the generating capacity and the costs of integrating this power into the electric system. Estimates of transmission and increased capital and operating and maintenance costs for offshore resources are included. Thus the NREL Wind Development System (WinDS) model, which simulated generation-capacity expansion in the U.S. electricity sector for wind and other technologies through 2030, has estimated that the 20 percent wind scenario would result in an increase in net present value (NPV) total electricity-sector costs of $43 billion (in 2006 dollars) over the no-new-wind case.

Overall, increases in wind power generation costs (capital and operation and maintenance expenses) are partially offset by lower capital, operation and maintenance, and fuel costs for other electricity sources (DOE, 2008a). The impact on the energy mix is largest for natural gas, with the 20 percent wind scenario displacing about 50 percent of electricity-utility natural gas consumption (DOE, 2008a). The need for imported liquefied natural gas would also be greatly reduced. However, maintaining electricity system reliability would require additional capacity from natural gas combustion turbines, which can respond quickly to low-wind conditions.

The total wind power capital cost under this scenario is $236 billion NPV, and the total operation and maintenance cost is $51 billion NPV. Though many studies have shown the feasibility of incorporating significant amounts of wind power into the electricity grid (Zavadil et al., 2004; GE Energy Consulting, 2005; DeMeo et al., 2005; UWIG, 2006; Parsons et al., 2006), integrating 20 percent wind power into the electricity system would require significant investments in the electric grid and other parts of the system. The DOE study estimates the cost of such an expansion of transmission capabilities at $23 billion, though it recognizes the significant barriers to installing new transmission in general. Separately, American Electric Power (AEP) developed a conceptual interstate transmission plan for integrating in excess of 300 GW from wind power and reducing existing transmission bottlenecks. AEP estimates that such a system would include 19,000 miles of new high-voltage (765-kV) transmission corridors and would require investments on the order of $60 billion (AEP, 2007).

The 20 percent wind scenario study has spurred related studies focused specifically on regional integration of a large fraction of intermittent renewable electricity. Ongoing large-scale studies on the regional integration of 20 percent and more of renewables into the electricity system include the Eastern Wind Integration and Transmission Study/Joint Coordinated System Planning Study (described at www.jcspstudy.org/) and the Western Wind and Solar Integration Study (JCSP, 2009; described at wind.nrel.gov/public/WWIS/). The JCSP study, as with the DOE's 20% wind energy study, included multiple stakeholders in a collaborative that held numerous public workshop meetings. JCSP (2009) looked at two scenarios: one a reference case with 5 percent market penetration by wind and the second with 20 percent wind. For the 5 percent wind scenario, the study estimated a need for 10,000 miles of new extra-high-voltage (EHV) transmission lines at an estimated cost of $50 billion. The 20 percent scenario assumes that almost 230 GW of new wind capacity and 36 GW of new baseload steam generation will be built by the year 2024. The projected transmission requirement for the 20 percent wind scenario was 15,000 miles of new EHV lines at an estimated cost of $80 billion.

In both cases, the additional transmission allowed renewable and baseload steam energy from the Midwest to be transmitted to a wider area. The study assumed that increased wind generation would primarily offset baseload steam production while requiring more production from fast-response, gas-fired combustion turbines. As with the DOE's 20 percent wind scenario, the JCSP study did not envision the need for electricity storage to be necessary for integrating 20 percent wind power into the study region.

Using Multiple Renewables to Reach 20 Percent Electricity Generation

The 20 percent wind scenario discussed in the previous section shows the potential for renewables to increase electricity generation and the scale and integration associated with rapid expansion of wind power only. In this section the committee describes a projection combining multiple renewable technologies that could meet the goal of providing 20 percent of total U.S. electricity generation by 2035 from new renewable-electricity generation.

Assuming the use of multiple renewable resources and technologies to reach the 20 percent goal might address some of the scale and integration issues associated with meeting this level of electricity generation with growth in a single renewable. Using an array of renewables could reduce the growth that would be required from individual sources and the scale-up challenges for manufacturers, materials, and human resources. Considering an array of renewables also might ease their integration into the electricity system, particularly for wind generation. Obtaining 20 percent of electricity generation from wind power as a single source would be a challenge, in that the 20 percent refers to an annual average, and wind power is intermittent. Because wind is not available all the time, it might have to represent much more than half the generation at times in order to reach the 20 percent annual average. Wind energy tends to be most abundant at night and in the spring and fall, when demand is low. Balancing wind with multiple renewable resources—including solar, which does not normally peak when wind does, and baseload power from geothermal and biomass—could mitigate the temporal variability in generation. Using multiple renewable resources would

TABLE 6.5 Capacity and Generation from Multiple Renewable Resources Sufficient to Meet 20 Percent of Estimated U.S. Electricity Demand in 2035

	Generating Capacity (GW)	Capacity Factor	Electricity Generation (GWh)
Wind	252	0.35	786,000
Solar	70	0.15	92,000
Biomass	13	0.90	102,000
Geothermal	13	0.90	102,000

Note: The estimate of total electricity generation for 2035 (5,386,000 GWh) comes from the AEO 2009 base-case estimate for 2030 (EIA, 2008a) projected out to 2035 using a 0.9 percent growth rate.

take advantage of the geographical variability in the resource base. Relying on multiple renewable resources would not eliminate the need to expand transmission capacity or to make other improvements in the electricity infrastructure to enable the integration of renewables, nor would it reduce the magnitude of costs. However, it can offer other attributes, such as providing baseload generation and combining different intermittent renewables to reduce the temporal variability in generation.

Table 6.5 lists a set of renewables that, under the projection described here, would reach 20 percent of electricity generation by 2035. Achieving that goal would depend on wind power capacity additions of 9.5 GW/yr, a slight increase over the 8.4 GW installed in 2008. Table 6.5 also shows solar growing to 70 GW by 2035, a smaller gain than those projected in the high market penetration solar scenarios described in the PV roadmap (SEIA, 2001, 2004), *Solar America Initiative* (DOE, 2007b), and Pernick and Wilder (2008). It assumes that an additional 13 GW would come from geothermal heat by 2035, which is consistent with the Western Governors' Association's estimated potential resource base in the western United States (WGA, 2006). It also assumes that an additional 13 GW would come from biomass, which would not greatly impact the biomass resource base needed to meet the alternative liquid fuels mandates. The mix of renewable resources shown in Table 6.5 is not presented as the optimal set to meet the target of obtaining 20 percent of total electricity generation from additions of renewable resources. This set is merely one mix that could be considered, given the available resource base, readiness of renewable-electricity technologies, and what might be practicable for an aggressive but achievable expansion of market penetration.

Findings: Scenarios

Understanding the scale of deployment necessary for renewables to make a material contribution to the U.S. electricity generation is critical. There are clearly sufficient resources and technologies, and some of the technologies currently are, or are close to being, economically competitive. There are also expanding manufacturing and deployment capabilities. **Over the first timeframe (the present to 2020), wind, solar photovoltaics, concentrating solar power, conventional hydropower, hydrothermal, and biopower technologies—which are technically ready (and, in some cases, close to being economically competitive) for accelerated deployment—could constitute a significantly greater contribution (up to about an additional 10 percent of electricity generation) to the U.S. electricity supply than they do today. In the second timeframe (2020–2035), it is reasonable to envision that further and accelerated deployment, along with public- and private-sector R&D efforts and supportive public policies (such as the Production Tax Credit), could result in nonhydroelectric renewables collectively providing 20 percent or more of domestic electricity generation.** This level of growth for renewables electricity is based on the objective of developing a scenario involving an accelerated deployment option as discussed in Chapter 2. However, simply continuing the level of deployment for wind that occurred in 2008 and assuming a capacity factor of 35 percent would have wind power contributing almost 14 percent of total projected electricity demand in 2035.[8]

But, as emphasized in the 20 percent wind scenario, greatly enhancing the penetration of renewable electricity will require large increases over current levels of manufacturing, employment, investment, and installation. The numbers from the 20 percent wind study demonstrate the potential challenges and opportunities: 100,000 wind turbines, up to $100 billion in additional capital costs and transmission upgrades, thousands of miles of new transmission lines, 100,000-plus jobs, and an 800-million-tonne annual reduction in CO_2 emissions. Increasing manufacturing and installation capacity, employment, and investment to the level required to meet this goal can be done, but the magnitude of the challenge emphasizes the need for a consistent long-term approach in the public policy arena. And although the technologies and resources are clearly sufficient to move beyond a 20 percent contribution to electricity generation from renewables, greater penetrations obviously carry with them larger deployment and integration issues.

---

[8]Total electricity generation in 2035 is based on the AEO 2009 base case (EIA, 2008a) that has been extrapolated out to 2035.

NON-ELECTRICITY RENEWABLE ENERGY

In addition to electricity generation, certain renewable resources—non-concentrating solar thermal technologies, low- and moderate-temperature geothermal, and biomass—can displace fossil fuels at the point of use, particularly in residential and commercial buildings and in light industry and agriculture. These resources are referred to as distributed renewables.

Solar Thermal

Applications for distributed solar thermal include water heating, space heating and cooling, and heat for industry and agriculture. Because the solar collector does not rely on concentrating the sun's energy and can use both direct and diffuse radiation, distributed solar thermal systems are applicable to the entire United States. However, solar insolation and costs vary across the United States.

The most prevalent and well-developed applications are for heating swimming pools and potable water (in homes and laundries), with performance standards overseen by the U.S. Solar Rating and Certification Corporation.[9] Systems include one or more collectors (which capture the sun's energy and convert it into usable heat), a distribution structure, and a thermal storage unit. Associated piping, heat exchangers, and storage tanks use technology found in conventional HVAC and water-heating systems.

The unique component is the solar collector. The flat-plate collector is most common in the United States, but the use of evacuated-tube collectors is growing rapidly. For heating swimming pools, which is the country's single largest application of solar thermal, the collector is an unglazed polymer absorber through which the pool water is circulated. Energy is delivered at moderate temperatures, usually less than 10°C above ambient. For domestic hot water, flat-plate collectors employ a copper plate absorber, usually coated with a wavelength-selective layer to reduce radiative losses. The absorber and tubes that contain the working fluid are mounted in an insulated box with a tempered glass cover. In evacuated-tube collectors, the absorber is mounted in an evacuated glass tube, and energy is delivered at temperatures up to 100°C.

Advances in manufacturing, materials, and industry standards have resulted in significant improvements since the 1980s in both the performance and the reli-

---

[9]See www.solar-rating.org/.

ability of solar thermal. Typically, systems reduce electricity or gas used for water heating by 40–70 percent. The number of buildings appropriate for solar thermal systems depends in general on population density and climate, but the actual market penetration depends on factors such as first and life-cycle cost, availability, customer preference, and local building codes and standards. The consensus is that the major obstacle to deployment is the relatively high initial purchase price, including installation, relative to electric and gas water heaters.

Projections of the potential for solar hot water to displace electricity and natural gas in residential and commercial buildings vary from 138 to 300 billion kWh (Davidson, 2005; Denholm, 2007). This would represent some 20–40 percent of all energy used for hot water heating in such buildings. An additional benefit would be reduced peak demand for electricity and gas, though this benefit would vary across the United States.[10] But despite the predicted benefits of solar water heating and the fact that life-cycle savings are reasonable, the technology accounts for a very small fraction (~3 percent or less) of the approximately 10 million gas and electric water heaters shipped each year.[11] The largest markets are China and European countries, many of which have aggressive financial incentives and public policies supporting distributed renewable technologies. In the United States, some (but not most) states include solar thermal in their RPS, and many have rebate programs for residential and commercial solar hot water. The Interstate Renewable Energy Council and the North Carolina Solar Energy Center maintain a database of state incentives for distributed renewable energy.[12]

The use of solar thermal systems to yield space heating and cooling in residential and commercial buildings could provide a greater reduction of fossil fuels than do water heaters, but at present these systems are largely an untapped opportunity. Recently there has been limited deployment of liquid-based solar collectors for radiant floor-heating systems and solar air heaters, but the challenge with these applications is the relatively large collector area required in the absence of storage. Solar cooling can be accomplished via absorption and desiccant cycles, but commercial systems are not widely available for residential use.

Future technological improvements potentially include cost-effective com-

---

[10]The U.S. Solar Rating and Certification Corporation estimates that, on average, for every solar water heater installed 0.5 kW of peak demand is deferred.

[11]U.S. Department of Energy, Office of Energy Efficiency and Renewable Energy. 2004 Buildings Energy Data Book, downloadable at buildingsdatabook.eren.doe.gov/.

[12]Database of State Incentives for Renewables and Efficiency, available at www.dsireusa.org/.

pact storage technologies, based on phase-change and thermochemical mechanisms (with higher storage density than water), and materials that replace the copper and low-iron glass used in today's collectors. Reductions in the cost of manufacture, materials, assembly, and shipping weight may be possible through a shift from metal/glass components to integrated systems, such as those associated with polymeric materials, that are manufactured using mass-production techniques.

Geothermal

Geothermal energy can be applied to a variety of end-uses, including agriculture (mainly greenhouse heating), aquaculture, industrial processes, and space heating and cooling of buildings.

Direct-use geothermal taps heated groundwater, without a heat pump or power plant, for the heating of facilities, with the technology generally involving resource temperatures between 38° and 150°C (Lindal, 1973). Current U.S. installed capacity of direct-use systems is 620 $MW_{thermal}(MW_t)$.[13] Municipalities and smaller communities provide district heating by circulating hot water from aquifers through a distribution pipeline to the points of use, though this application of geothermal energy remains modest, with systems in only seven states.[14] The barriers to increased penetration of direct geothermal heating and cooling systems are the high initial investment costs and the challenges associated with locating and developing appropriate sites. The resource for direct heating is richest in the western states.

Geothermal heat pumps have extended the use of geothermal energy into traditionally nongeothermal areas of the United States, mainly the midwestern and eastern states. A geothermal heat pump draws heat from the ground, groundwater, or surface water and discharges heat back to those media instead of into the air. The available land area and the soil and rock types at the installation site determine the best solution. Ground-coupled heat pumps are the most common type used. The efficiency of the heat pump is inversely proportional to the temperature difference between the conditioned space and the heat source or heat sink. As a consequence, heating and cooling efficiencies are improved because ground temperatures remain relatively constant throughout the year. The coefficient of perfor-

---

[13]Geo-Heat Center, Oregon Institute of Technology; see geoheat.oit.edu.
[14]See geoheat.oit.edu/directuse/district.htm.

Renewable Energy   *321*

mance is 3.4–4.4,[15] and the annual operating costs are 25–33 percent of comparable fossil-fuel heating costs.

The electric heat pump is standard off-the-shelf equipment, with minor modifications to handle the heat transfer from geothermal fluids or soil. The heat-pump equipment is located indoors, which reduces maintenance costs, and the more stable operating temperature and pressure of the compressor give it a longer life than in air heat pumps. The unique component is the heat-exchange interface with the soil or with the groundwater. Ground-coupled heat pumps use high-density polyethylene pipe buried either vertically or in horizontal trenches to exchange heat between a working fluid and the soil. Vertical loops cost more, but they provide access to more stable deep-soil temperatures and are the only option if land area is limited. Regulatory requirements for the bore holes vary across the country, with the primary regulatory issue being the potential for groundwater contamination. This problem is addressed by grouting the bore hole; the most commonly used material (bentonite-based grout) reduces heat transfer to the soil, but more conductive grouts such as cement mixtures and bentonite/sand mixtures provide superior performance.

Today, the United States has 700,000 installed units—with 8,400 MW$_t$ of capacity delivering about 7,200 GWh/yr[16]—and there are 1.5 million units worldwide. The rate of installation is estimated to be 10,000–50,000 units per year. One barrier to growth is the lack of sufficient infrastructure (i.e., trained designers and installers) and another is the high initial investment cost compared to conventional space-conditioning equipment. There are no major technical barriers to greater deployment.

Biomass

Burning wood to heat U.S. homes currently represents about 1 percent of fuel used for direct heating of buildings.[17] One-half to two-thirds of residential wood combustion in the United States occurs in wood stoves, as opposed to fireplaces (Fine et al., 2004). Solid fuels include conventional wood logs, which may or may not

---

[15]See www.eia.doe.gov/cneaf/solar.renewables/page/heatpumps/heatpumps.html, Tables 3.3 and 3.4. The coefficient of performance is the ratio of heat output per unit of energy input.

[16]Geo-Heat Center, Oregon Institute of Technology; see geoheat.oit.edu.

[17]U.S. Department of Energy, Office of Energy Efficiency and Renewable Energy. 2008 Buildings Energy Data Book, downloadable at buildingsdatabook.eere.energy.gov/. This figure does not include biomass that is used in electricity generation.

have been harvested sustainably, and pellets. Advanced biomass-fuel appliances use pellets, which are produced by compressing woody material that may include waste wood and sawdust, agricultural wastes, wastepaper, and other organic materials. Some pellet-fuel appliances can also burn corn kernels, nutshells, and wood chips. Pellet stoves use electricity to run fans, controls, and pellet feeders.

One of the concerns about solid-fuel combustion for home heating is air pollution. In areas where wood stoves are prevalent, wood smoke is a major source of fine particulates and gaseous pollutants, including nitrogen oxides, carbon monoxide, and organics. The mandatory smoke-emission limit set by the U.S. Environmental Protection Agency (EPA) for wood stoves is 7.5 grams of smoke per hour for noncatalytic stoves and 4.1 g/h for catalytic stoves.[18] Modern noncatalytic stoves have improved fireboxes to achieve high combustion efficiency. The most efficient wood-burning appliances also use catalytic converters to achieve nearly complete combustion of the feedstock and to reduce harmful emissions. Stoves are available with EPA-certified emissions as low as 1 g/h. Stoves require homeowner maintenance and catalyst replacement, however, to retain their high efficiencies and low emissions.

In summary, modern solid-fuel stoves are efficient and clean compared to the fireplaces of the past. The economics of using a stove to combust biomass products depends on the fuel being displaced and the distance from home to supplier.

CONCLUSION

A future characterized by a large penetration of renewable electricity represents a paradigm shift from the current electricity generation, transmission, and distribution system. There are many reasons why renewable electricity represents such a shift, including the spatial distribution and intermittency of some renewable resources, as well as issues related to greatly increasing the scale of deployment. Wind and solar—two renewable-energy resources with the potential for large near-term growth in deployment—are intermittent resources that have some of their resource bases located far from demand centers. The transformations required to incorporate a significant penetration of additional renewables include transformation in ancillary capabilities, especially the expansion of transmission

---

[18]See www.epa.gov/woodstoves/basic.html.

and backup power resources, and deployment of technologies that improve grid intelligence and provide greater system flexibility. Further, supplying renewable resources on a scale that would make a major contribution to U.S. electricity generation would require vast investment in and deployment of manufacturing and human resources, as well as additional capital costs relative to those associated with current generating technologies that have no controls on greenhouse gas emissions. The realization of such a future would require a predictable policy environment and sufficient financial resources. Nevertheless, the promise of renewable resources is that they offer significant potential for low-carbon generation of electricity from domestic sources of energy that are much less vulnerable to fuel cost increases than are other electricity sources. Overall success thus depends on having technology, capital, and policy working together to enable renewable-electricity technologies to become a major contributor to America's energy future.

REFERENCES

AEP (American Electric Power). 2007. Interstate transmission vision for wind integration. AEP white paper. Columbus, Ohio.

ASES (American Solar Energy Society). 2007. Tracking Climate Change in the U.S.: Potential Carbon Emissions Reductions from Energy Efficiency and Renewable Energy by 2030. Washington, D.C.

AWEA (American Wind Energy Association). 2008. 20 Percent Wind Energy Penetration in the United States: A Technical Analysis of the Energy Resource. Washington, D.C.

AWEA. 2009a. Wind energy grows by record 8,300 MW in 2008 (January 27, 2009, press release). Washington, D.C.

AWEA. 2009b. AWEA First Quarter 2009 Market Report. Washington, D.C.

Baidya, Roy S., S.W. Pacala, and R.L. Walko. 2004. Can large wind farms affect local meteorology? Journal of Geophysical Research 109:D19101.1-D19101.6.

Beaudry-Losique, J. 2007. Biomass R&D program and biomass-to-electricity. Presentation at the first meeting of the Panel on Electricity from Renewable Resources, Washington, D.C., September 18, 2007.

Berry, J.E., M.R. Holland, P.R. Watkiss, R. Boyd, and W. Stephenson. 1998. Power Generation and the Environment: A UK Perspective. European Commission. June.

Black & Veatch. 2007. 20 Percent Wind Energy Penetration in the United States: A Technical Analysis of the Energy Resource. Black & Veatch Project 144864. Prepared for the America Wind Energy Association, Walnut Creek, Calif.

Chataignere, A., and D. Le Boulch. 2003. Wind Turbine (WT) Systems. Final Report. ECLIPSE (Environmental and Ecological Life Cycle Inventories for Present and Future Power Systems in Europe). European Commission.

Chaudhari, M., L. Frantzis, and T.E. Holff. 2004. PV Grid Connected Market Potential Under a Cost Breakthrough Scenario. San Francisco, Calif.: The Energy Foundation and Navigant Consulting.

Chu, S. 2009. Department of Energy FY 2010 Budget—Transforming the Energy Economy through Science and Innovation. DOE Budget Rollout Presentation, Washington, D.C., May 7, 2009.

Davidson, J.H. 2005. Low-temperature solar thermal systems: An untapped energy resource in the United States. Journal of Solar Energy Engineering 127:305-306.

DeCarolis, J.F., and D.W. Keith. 2004. The economics of large-scale wind power in a carbon constrained world. Energy Policy 34:395-410.

DeMeo, E., W. Grant, M.R. Milligan, and M.J. Schuerger. 2005. Wind plant integration. IEEE Power Energy. Nov.-Dec.

Denholm, P.L. 2004. Environmental and Policy Analysis of Renewable Energy Enabling Technologies. Ph.D. dissertation. University of Wisconsin-Madison.

Denholm, P., and G. Kulcinski. 2003. Net Energy Balance and Greenhouse Gas Emissions from Renewable Energy Storage System. ECW Report Number 223-1. Energy Center of Wisconsin, June.

Denholm, P. 2007. The Technical Potential of Solar Water Heating to Reduce Fossil Fuel Use and Greenhouse Gas Emissions in the United States. Technical Report NREL. Golden, Colo.

DOE (U.S. Department of Energy). 2007a. National Solar Technology Roadmap: Wafer-Silicon PV. Energy Efficiency and Renewable Energy Office. Washington, D.C.

DOE. 2007b. Solar America Initiative: A Plan for the Integrated Research, Development, and Market Transformation of Solar Energy Technologies. Washington, D.C.

DOE. 2008a. 20% Wind Energy by 2030: Increasing Wind Energy's Contribution to U.S. Electricity Supply. Washington, D.C. Available at www1.eere.energy.gov/windand hydro/pdfs/41869.pdf. Accessed April 21, 2009.

DOE. 2008b. Annual Report on U.S. Wind Power Installation, Cost, and Performance Trends: 2007. Energy Efficiency and Renewable Energy Office. Washington, D.C.

EIA (Energy Information Administration). 2007. Annual Energy Review 2006. DOE/EIA-0384(2006). Washington, D.C.: U.S. Department of Energy, Energy Information Administration.

EIA. 2008a. Annual Energy Outlook 2009 Early Release. DOE/EIA-0383(2009). Washington, D.C.: U.S. Department of Energy, Energy Information Administration.

EIA. 2008b. Energy Market and Economic Impacts of S. 2191, the Lieberman-Warner Climate Security Act of 2007. Washington, D.C.: U.S. Department of Energy, Energy Information Administration.

EIA. 2008c. Annual Energy Outlook 2008. DOE/EIA-0383(2008). Washington, D.C.: U.S. Department of Energy, Energy Information Administration.

Elliott, D.L., C.G. Holladay, W.R. Brachet, H.P. Foote, and W.R. Sandusky. 1986. Wind Energy Resource Atlas of the United States. Washington, D.C.: National Renewable Energy Laboratory.

Elliott, D.L., L.L. Wendell, and G.L. Gower. 1991. An Assessment of the Available Windy Land Area and Wind Energy Potential in the Contiguous United States. Pacific Northwest Laboratory. Richland, Wash.

EPRI (Electric Power Research Institute). 2005. Final Summary Report: Offshore Wave Power Project. Palo Alto, Calif.

EPRI. 2007a. Assessment of Waterpower Potential and Development Needs. Palo Alto, Calif.

EPRI. 2007b. The Power to Reduce CO_2 Emissions: The Full Portfolio. Palo Alto, Calif.

European Commission. 1997a. External Costs of Electricity Generation in Greece. ExternE Project. Available at externe.jrc.es/reports.html.

European Commission. 1997b. ExternE National Implementation Denmark. Available at externe.jrc.es/reports.html.

European Commission. 1997c. ExternE National Implementation France. Available at externe.jrc.es/reports.html.

European Commission. 1997d. ExternE National Implementation Germany. Available at externe.jrc.es/reports.html.

EWEA (European Wind Energy Association). 2005. Large Scale Integration of Wind Energy in the European Power Supply: Analysis, Issues and Recommendations. Brussels, Belgium.

Fine, P.M., G.R. Cass, and B.R.T. Simoneit. 2004. Chemical characterization of fine particle emissions from the wood stove combustion of prevalant United States tree species. Environmental Engineering Science 21:705-721.

Fletcher, E.A. 2001. Solar thermal processing: A review. Journal of Solar Energy Engineering 123:63-74.

Flynn, H., and T. Bradford. 2006. Polysilicon: Supply, Demand, and Implications for the PV Industry. Cambridge, Mass.: Prometheus Institute.

Frankl, P., A. Corrado, and S. Lombardelli. 2004. Photovoltaic (PV) Systems. Final Report. ECLIPSE (Environmental and Ecological Life Cycle Inventories for Present and Future Power Systems in Europe). European Commission.

Fthenakis, V.M., and H.C. Kim. 2007. Greenhouse-gas emissions from solar electric and nuclear power: A life-cycle study. Energy Policy 35:2549-2557.

GE Energy Consulting. 2005. The Effects of Integrating Wind Power on Transmission System Planning, Reliability, and Operations. Prepared for New York State Energy Research and Development Authority, Albany, N.Y.

Genter, A., D. Fritsch, N. Cuenot, J. Baumgartner, and J. Graff. 2009. Overview of the current activities of the European EGS Soultz project: From exploration to electricity production. Proceedings of the Thirty-Fourth Workshop on Geothermal Reservoir Engineering, Stanford University, Stanford, Calif., February 9–11, 2009.

Goldberg, M., K. Sinclair, and M. Milligan. 2004. Job and economic development impact (JEDI) model: A user-friendly tool to calculate economic impacts from wind projects. Global Windpower Conference, Chicago, Ill.

Haq, Z. 2001. Biomass for Electricity Generation. Energy Information Agency, U.S. Department of Energy, Washington, D.C. Available at www.eia.doe.gov/oaif/analysispaper/biomass.

Holttinen, H., B. Lemstršm, P. Meibom, H. Bindner, A. Orths, F. van Hulle, C. Ensslin, A. Tiedemann, L. Hofmann, W. Winter, A. Tuohy, M. O'Malley, P. Smith, J. Pierik, J.O. Tande, A. Estanqueiro, J. Ricardo, E. Gomez, L. Sšder, G. Strbac, A. Shakoor, J.C. Smith, B. Parsons, M. Milligan, and Y. Wan. 2007. Design and operation of power systems with large amounts of wind power: State-of-the-art report. VTT Working Papers 82. VTT Technical Research Centre of Finland, October. Available at http://www.vtt.fi/inf/pdf/workingpapers/2007/W82.pdf.

Hondo, H. 2005. Life cycle GHG emission analysis of power generation systems: Japanese case. Energy 30:2042-2056.

JCSP (Joint Coordinated System Plan). 2009. Joint Coordinated System Plan 2008.

Jones, A.T., and W. Finley. 2003. Recent developments in salinity gradient power. Pp. 2284-2287 in OCEANS 2003: Celebrating the Past, Teaming Toward the Future. Columbia, Md.: Marine Technology Society.

Keith, D.W., J.F. DeCarolis, D.C. Denkenberger, D.H. Lenschow, S.L. Malyshev, S. Pacala, and P.J. Rasch. 2004. The influence of large-scale wind power on global climate. Proceedings of the National Academy of Sciences USA 101:16115-16120.

Laxson, A., M.M. Hand, and N. Blair. 2006. High Wind Penetration Impact on U.S. Wind Manufacturing Capacity and Critical Resources. National Renewable Energy Laboratory Technical Report NREL/TP-500-40482.

Lindal, B. 1973. Industrial and Other Uses of Geothermal Energy. Paris: UNESCO.

Mancini, T., P. Heller, B. Bulter, B. Osborn, S. Wolfgang, G. Vernon, R. Buck, R. Diver, C. Andraka, and J. Moreno. 2003. Dishing Stirling systems: An overview of development and status. Journal of Solar Energy Engineering 125:135-151.

Mann, M., and P. Spath. 1997. Life Cycle Assessment of a Biomass Gasification Combined-Cycle System. Golden, Colo.: National Renewable Energy Laboratory.

Meier, P. 2002. Life-Cycle Assessment of Electricity Generation Systems and Applications for Climate Change Policy Analysis. Ph.D. dissertation. University of Wisconsin-Madison.

Mills, A., and R. Wiser. 2009. The Cost of Transmission for Wind Energy: A Review of Transmission Planning Studies. Berkeley, Calif.: Lawrence Berkeley National Laboratory.

MIT (Massachusetts Institute of Technology). 2006. The Future of Geothermal Energy. Cambridge, Mass.

Murphy, L.M., and P.L. Edwards. 2003. Bridging the Valley of Death: Transitioning from Public to Private Sector Financing. Prepared for National Renewables Energy Laboratory, Task Order #7200.2050. Golden, Colo.

NAS-NAE-NRC (National Academy of Sciences-National Academy of Engineering-National Research Council). 2009. Electricity from Renewable Resources: Status, Impediments, and Prospects. Washington, D.C.: The National Academies Press.

NRC (National Research Council). 2007. Environmental Impacts of Wind-Energy Projects. Washington, D.C.: The National Academies Press.

NREL (National Renewable Energy Laboratory). 2007a. Wind Energy Resource Atlas of the United States. Golden, Colo.

NREL. 2007b. Very Large Scale Deployment of Grid Connected Solar PV in the United States. Golden, Colo.

NREL. 2007c. Projected Benefits of Federal Energy Efficiency and Renewable Energy Programs. NREL/TP-640-41347. Golden, Colo. March. See www1.eere.energy.gov/ba/pba/pdfs/41347.pdf. Accessed May 1, 2008.

Odeh, N.A., and T.T. Cockerill. 2008. Life cycle GHG assessment of fossil fuel power plants with carbon capture and storage. Energy Policy 38:367-380.

OMB (U.S. Office of Management and Budget). 2004. Final Information Quality Bulletin for Peer Review. December 15. Available at www.whitehouse.gov/omb/inforeg/peer2004/peer_bulletin.pdf. Accessed April 21, 2009.

O'Regan, B., and M. Grätzel. 1991. Low-cost, high-efficiency solar cell based on dye-sensitized colloidal TiO_2 films. Nature 353:737-740.

Parsons, B., M. Milligan, J.C. Smith, E. DeMeo, B. Oakleaf, K. Wolf, M. Schuerger, R. Zavadil, M. Ahlstrom, and D.Y. Nakafuji. 2006. Grid impacts of wind power variability: Recent assessments from a variety of utilities in the United States. Conference Paper NREL/CP 500-39955. Washington, D.C.: U.S. Department of Energy. July.

Patel, S. 2009. PV Sales in the U.S. soar as solar panel prices plummet. Power Magazine, March 1.

Perkens, C., and A.W. Weimer. 2004. Likely near-term solar-thermal water splitting technologies. International Journal of Hydrogen Energy 29:1587-1599.

Pernick, R., and C. Wilder. 2008. Utility Solar Assessment Study: Reaching Ten Percent by 2025. Clean Edge, Inc.

Roy, B.S., S.W. Pacala, and R.L. Walko. 2004. Can large wind farms affect local meteorology? Journal of Geophysical Research 109:D19101.

SEIA (Solar Energy Industries Association). 2001. Solar Electric Power: The U.S. Photovoltaic Industry Roadmap. Washington, D.C.

SEIA. 2004. Our Solar Power Future: The U.S. Photovoltaic Industry Roadmap Through 2030 and Beyond. Washington, D.C.

SERI (Solar Energy Research Institute). 1989. Ocean Thermal Energy Conversion: An Overview. SERI/SP-220-3024. Golden, Colo.

Smith, J.C. 2007. Integrating wind into the grid. Presentation to the Panel on Electricity from Renewable Resources, Washington, D.C., December 7.

Smith, J.C., and B. Parsons. 2007. What does 20 percent look like? IEEE Power and Energy 5:22-33.

Spath, P., and M. Mann. 2000. Life Cycle Assessment of a Natural Gas Combined-Cycle Power Generation System. Golden, Colo.: National Renewable Energy Laboratory.

Spath, P., and M. Mann. 2004. Biomass Power and Conventional Fossil Systems with and without CO_2 Sequestration: Comparing the Energy Balance, Greenhouse Gas Emissions, and Economics. Golden, Colo.: National Renewable Energy Laboratory.

Spath, P., M. Mann, and D. Kerr. 1999. Life Cycle Assessment of Coal-Fired Power Production. Golden, Colo.: National Renewable Energy Laboratory.

Spitzley, D., and G.A. Keoleian. 2005. Life Cycle Environmental and Economic Assessment of Willow Biomass Electricity: A Comparison with Other Renewable and Non-Renewable Sources. Report CSS04-05R. Center for Sustainable Systems, University of Michigan. March 2004 (revised February 10, 2005).

Steinfeld, A. 2005. Solar thermochemical production of hydrogen: A review. Solar Energy 78(5):603-615.

Storm van Leeuwen, J.W., and P. Smith. 2008. Nuclear Power: The Energy Balance. Available at www.stormsmith.nl/. Accessed April 21, 2009.

Thresher, R., M. Robinson, and P. Veers. 2007. To capture the wind. Power and Energy Magazine 5:34-46.

U.S. Climate Change Science Program. 2007. Scenarios of Greenhouse Gas Emissions and Atmospheric Concentrations; and Review of Integrated Scenario Development and Application. Washington, D.C.

UWIG (Utility Wind Integration Group). 2006. Grid Impacts of Wind Power Variability: Recent Assessments from a Variety of Utilities in the United States. Reston, Va. Available at http://www.uwig.org/opimpactsdocs.html.

Vattenfall, A.B. 2004. Nordic Countries Certified Environmental Product Declaration: EDP of Electricity. From Vattenfall's Nordic Hydropower. S-P-00088. February 2005.

Vredenbregt, L., R. Meijer, and J. Visser. 2003. The effect of co-firing 10 percent of secondary fuels on SCR catalyst deactivation. Proceedings of the Conference on Selective Catalytic Reduction and Non-Catalytic Reduction for NO$_x$ Control, October 29-30, 2003.

WGA (Western Governors' Association). 2006. Clean and Diversified Energy Initiative. Geothermal Task Force Report. Washington, D.C.

White, S. 1998. Net Energy Payback and CO$_2$ Emissions from Helium-3 Fusion and Wind Electrical Power Plants. UWFDM-1093. Ph.D. dissertation. Fusion Technology Institute, University of Wisconsin-Madison.

White, S. 2006. Net energy payback and CO$_2$ emissions from three midwestern wind farms: An update. Natural Resources Research 15:271-281.

Wiley, L. 2007. Utility scale wind turbine manufacturing requirements. Presentation at the National Wind Coordinating Collaborative's Wind Energy and Economic Development Forum, Lansing, Mich., April 24, 2007.

Wiser, R. 2008. The development, deployment, and policy context of renewable electricity: A focus on wind. Presentation at the fourth meeting of the Panel on Electricity from Renewable Resources, Washington, D.C., March 11, 2008.

Wiser, R., and G. Barbose. 2008. Renewables Portfolio Standards in the United States: A Status Report with Data through 2007. Berkeley, Calif.: Lawrence Berkeley National Laboratory.

Zavadil, R., J. King, L. Xiadong, M. Ahlstrom, B. Lee, D. Moon, C. Finley, L. Alnes, L. Jones, F. Hudry, M. Monstream, S. Lai, and J. Smith. 2004. Wind Integration Study Final Report. Xcel Energy and the Minnesota Department of Commerce, EnerNex Corporation and Wind Logics, Inc. Available at http://www.uwig.org/XcelMNDOCStudyReport.pdf.

7 Fossil-Fuel Energy

Total U.S. primary energy consumption in 2007 was about 100 quads, with fossil fuels—natural gas, petroleum, and coal—supplying about 85 percent, as shown in Table 7.1 (EIA, 2008a).[1] Liquid fuels (derived primarily from petroleum) were the main contributors, accounting for 40 percent of total consumption (see Figure 1.2 in Chapter 1). This fossil-fuel dominance has held steady for decades.

Even more striking, each of the fossil fuels accounts for a major segment of an important end-use market. Petroleum supplies 98 percent of the energy used in the transportation market, natural gas provides 74 percent of the nonelectric energy used in the residential and commercial market, and coal furnishes 52 percent of the energy used to generate electricity. Only in the electricity market, where nuclear and renewable energy sources account for 29 percent of the total energy supply, do serious competitors to fossil fuels exist.[2] Despite considerable efforts to expand biofuel production, for example, ethanol from corn provided only about 3 percent of the U.S. gasoline supply in 2005.

These distinctive structures exist because the attributes of liquid, gaseous, and solid fossil fuels closely match the needs of their respective end-use markets:

---

[1]Worldwide, the dominance of fossil fuels is little different; they provided 86 percent of world primary energy consumption in 2004.

[2]Oil and gas are the dominant suppliers of the industrial market, primarily for feedstocks in chemical production.

TABLE 7.1 U.S. Energy Consumption by Energy Source in 2007

Energy Source	Consumption (quadrillion Btu [percent])
Petroleum	39.77 [39]
Natural gas	23.63 [23]
Coal	22.75 [22]
Nuclear power	8.46 [8.3]
Hydropower	2.45 [2.4]
Biomass	3.60 [3.5]
Other renewable energy	0.77 [0.008]
Other	0.11 [0.001]
Total	101.55

Note: Numbers have been rounded.
Source: EIA, 2009a.

- Petroleum is easily stored and transported and has a relatively high energy density. These characteristics are well suited to the transportation market.
- Natural gas burns cleanly, is easily transported by pipeline, and can be stored in salt domes and old gas fields for peak use. As a result, it is a desirable fuel for the geographically distributed residential and commercial markets.
- Coal is abundant in the United States, is easily stored, and is less expensive, with lower price volatility than other fuels—attractive attributes for electricity generation.

Although the market-based reasons for using fossil fuels are thus very strong, U.S. reliance on this energy source carries some potentially adverse consequences. For one, reserves of petroleum—and, increasingly, of natural gas—are concentrated in only a few countries. In some cases, supplier nations have restricted supplies for nonmarket reasons. Moreover, such concentrations of production capacity, and the limited number of transportation routes from these facilities to their markets, create targets by which hostile states or nonstate actors may disrupt supplies. In either case, the security of petroleum and natural gas supplies is at risk, probably increasingly so.

A second concern is that the longer-term global demand for petroleum and natural gas is projected to grow faster than increases in production, resulting in tight market conditions and rising prices. The U.S. Energy Information Administration (EIA) and the International Energy Agency (IEA), along with other forecasters, do not anticipate that the factors underlying these market conditions will change anytime soon.[3] Under such conditions, maintaining significant spare production capacity is difficult.

From the point of view of net consuming nations, the resulting price increases could accelerate an economically disruptive wealth transfer from consumers to producers. While the dependence of the U.S. economy on oil has changed little in recent decades—in 1990, 39.7 percent of U.S. energy consumption was petroleum; in 2007, it was 39.2 percent—U.S. dependence on imports has doubled over this period.

Finally, fossil fuels pollute the atmosphere when burned, and they have other adverse environmental effects as well. While emissions of SO_x, NO_x, particulates, and other atmospheric contaminants have been reduced (albeit with an increase in solid, liquid, or recyclable wastes, including ash residuals), little has been done so far to address carbon dioxide (CO_2) emissions. U.S. energy use in 2007 was responsible for emissions of 6 billion tonnes of CO_2 (6 Gt CO_2). Of that amount, 43 percent came from petroleum, 36 percent from coal, and 21 percent from natural gas (EIA, 2008c). By market, the largest source was electric power generation (using coal and natural gas); it emitted some 2.4 Gt CO_2. Transportation, dominated by petroleum but also including some natural gas, accounted for 2 Gt CO_2. The remainder of the emissions resulted from industrial (1 Gt CO_2), residential (0.35 Gt CO_2), and commercial uses (0.25 Gt CO_2).[4] (See Figure 1.11 in Chapter 1.)

Thus the future of fossil fuels presents a serious dilemma for energy policy. On the one hand, because fossil fuels are well adapted to the needs of the market, a huge energy infrastructure has been put in place to take advantage of their value. The existing stocks of vehicles, home and business heating systems, and electric power stations were created with the expectation that petroleum, natural gas, and coal would be readily and reliably available. On the other hand, the

[3]For the latest IEA forecast, see *Energy Technology Perspectives 2008* (IEA, 2008a), p. 113ff. The downturn in the world economy apparent at the time of this writing will mitigate demand growth for a while, but the underlying determinants of demand remain in place.

[4]Note that while electric power is used in industrial, residential, and commercial settings, it is aggregated under electric power generation.

extraction and use of fossil fuels entail growing security, economic, and environmental risks. A crucial question, therefore, is whether this existing energy infrastructure can be supplied with liquid, gaseous, and solid fuels in the future at acceptable levels of such risks. If so, much of it can remain in place. If not, the embedded capital stock of technologies for energy production and use will need to change through a combination of market forces and policy choices.

Other chapters of this report discuss alternative pathways for providing the energy services that modern society demands. For example, the chapter on alternative transportation fuels (Chapter 5) provides an assessment of the technologies and environmental impacts of liquid fuels derived from biomass feedstocks, coal, or natural gas. This present chapter focuses on alternative ways of using fossil fuels to serve the existing energy-use infrastructure. Specifically, it explores:

- The extent to which the U.S. endowment of fossil fuels is limited in its ability to meet future needs for liquid, gaseous, and solid fuels by means of conventional pathways.
- New technologies that may become available for producing the desired form of fossil fuels. The focus in particular is on the generation of electricity from coal and natural gas with sharply reduced emissions of greenhouse gases, especially CO_2.
- Technologies and geologic settings suitable for the storage of CO_2 produced from electricity generation and other industrial processes.
- Environmental concerns that affect the future of fossil-fuel supply and use.

Given constraints on time and resources, the AEF Committee chose not to address issues relating to the current energy infrastructure, for example, the status of natural gas pipelines, oil refineries, rail and barge transportation for coal, and liquefied natural gas terminals.

OIL, GAS, AND COAL RESOURCES

Worldwide, the amount of oil, gas, and coal that can ultimately be produced is very large. Estimates of ultimately recoverable resources are uncertain, however, because they include not only those that are discovered though not yet economically or technically recoverable but also those that are yet to be discovered. Nevertheless, the potential is impressive. Roughly 3.3 trillion barrels of oil and 15,000 trillion cubic feet (Tcf) of natural gas are thought to be ultimately recoverable. By comparison, in 2006, world consumption of these resources was about 30 billion barrels of crude oil and 100 Tcf of gas. (See Tables 7.2, 7.3, and 7.4 for summaries of oil, gas, and coal statistics.)

Resources that are discovered, recoverable with current technology, commercially feasible, and remaining in the ground are classified as reserves. The size of

TABLE 7.2 Conventional Oil Resources, Reserves, and Production (billion barrels, variable years as noted)

	United States	World	U.S. Percent of World Total
Resources	430[a]	3345[b]	13.0
Reserves[c]	29	1390	2.1
Annual production	2.5/yr	29.8/yr	8.4
Annual consumption	7.5/yr	31.1/yr[d]	24.1

[a]DOE, 2006a, available at fossil.energy.gov/programs/oilgas/eor/Undeveloped_Domestic_Oil_Resources_Provi.html.
[b]NPC, 2007, p. 97.
[c]2007 data from British Petroleum, 2008.
[d]According to British Petroleum, 2008, discrepancies between world production and consumption "are accounted for by stock changes; consumption of nonpetroleum additives and substitute fuels; and unavoidable disparities in the definition, measurement, or conversion of oil supply and demand data."

TABLE 7.3 Natural Gas Resources, Reserves, and Production (trillion cubic feet, variable years as noted)

	United States	World	U.S. Percent of World Total
Resources	1,525[a]	15,401[b]	9.4
Reserves[c]	211	6,263	3.4
Annual production	19.3/yr	104.1/yr	18.5
Annual consumption[c]	23.1/yr	103.5/yr[d]	22.3

[a]PGC, 2006, available at www.mines.edu/research/pga/.
[b]NPC, 2007, p. 97.
[c]2007 data from British Petroleum, 2008.
[d]According to British Petroleum, 2008, discrepancies between world production and consumption are "due to variations in stocks at storage facilities and liquefaction plants, together with unavoidable disparities in the definition, measurement or conversion of gas supply and demand data."

TABLE 7.4 Coal Reserves and Production (million tonnes, variable years as noted)

	United States	World	U.S. Percent of World Total
Resources	3,968,000[a]	9,218,000[b]	43.0
Reserves[c]	242,721	847,488	28.6
Annual production[c]	1,039.2/yr	6,395.6/yr	16.2
Annual consumption[c]	1,015.3/yr	6,481.1/yr	15.7

[a]EIA, 1999.
[b]Hermann, 2006.
[c]2007 data from British Petroleum, 2008.

known reserves, while considerably smaller than the more speculative estimates of ultimately recoverable resources, is also large. British Petroleum has reported that proved reserves of oil in 2006 amounted to 1390 billion barrels and that proved natural gas reserves were 6263 Tcf (British Petroleum, 2007). World coal reserves were 900 billion tonnes, which is about 300 times the 2006 world coal consumption (British Petroleum, 2007).

Technology plays an important role in turning speculative resources into proved reserves. Sophisticated exploration and production methods for recovery of oil and natural gas are already commercially available, and the private sector is developing advanced versions of these techniques. The cumulative effect of continuing advances in exploration and production technology for oil and gas is that over the next 20 years much of the current resource base will become technically recoverable. (See Table 7.5 for a discussion of this technology.)

As noted previously, world reserves are annually producing about 30 billion barrels of oil and 104.1 Tcf of natural gas. The United States is the third-largest oil-producing country and the second-largest natural gas producer. Nevertheless, this country imports about 56 percent of its oil and about 14 percent of its natural gas.[5] Import dependence, especially for oil, creates serious economic and security risks, as global oil and gas supplies may be influenced by restrictions imposed by governments, by the actions of the Organization of the Petroleum Exporting Countries (OPEC), or by disruptions due to political instability or regional conflict. For this reason, the capacity to maintain or increase domestic production is

---

[5]Virtually all of the natural gas that the United States imports comes from Canada.

TABLE 7.5 Summary of Highly Significant Oil Exploration and Production Technologies

Technology	Timeframe	Discussion
Big increase in controlled reservoir contact	2015	Technologies allowing a continuing increase in the number of strategically placed horizontal wells will allow a much greater commercial access to reserves.
Horizontal, multilateral, and fishbone wells	2020	Multiply placed drainholes from a main wellbore will further extend commercial access to reserves.
Arthroscopic well construction	2025	The ability to place drain holes to within feet of every hydrocarbon molecule in the formation allows the ultimate recovery.
SWEEP (see, access, move)	2020	The combined technologies (including the four immediately below) allowing us to see, access, and move the hydrocarbons in the optimum way will bring a big increase to recoverable reserves.
Smart well (injection and production)	2015	The ability to control what fluids go where (at the wellbore).
Reservoir characterization and simulation	2015	Extending current technology to include simultaneous inversion of all measurements with a forward model.
Reservoir vision and management in real time	2020	Combining reserve scale measurements (pressure, seismic, electromagnetic, and gravity) in a joint inversion, with uncertainty and without bias.
Mission control for everything	2020	A full representation and control of the full system (subsurface and surface) allowing true optimization.
CO_2 flood mobility control	2020	Measurement and control of the CO_2 flood front is critical to successful implementation.
Artificial lift	2030	Produce only wanted fluids to surface.
Drilling efficiency	2015	A further extension of gains already made.
Steam-assisted gravity drainage (SAGD) or steam and alkaline-surfactant-polymers (ASPs)	2030	Technologies to perfect and optimize SAGD operations (including the use of ASPs) will be key to widespread economic exploitation of heavy oil.
Arctic subsea-to-beach technology	2020	Ice scouring of the seafloor surface presents a huge challenge to conventional approaches to subsea and subsea-to-beach operations.
Faster and more affordable, higher-definition 3D seismic	2015	Quicker, better, cheaper, could extend the already impressive "specialized" technology in universal use.

Source: NPC, 2007, Topic Paper 19, "Conventional Oil and Gas," Table V.1.

a major concern for energy policy. Technical, environmental, and economic uncertainties, however, constrain the pace at which domestic oil and gas production can or will be increased. Accordingly, the following sections focus on the ability of domestic oil, natural gas, and coal sources to maintain or increase production. Tables 7.2 and 7.3 summarize the current levels of resources, reserves, and production for domestic oil and natural gas, and Table 7.4 reports reserves and production for coal.

Oil

While Table 7.2 summarizes estimates of the quantities of various types of oil resources in the United States, Table 7.6 disaggregates them. "Proved reserves" in Table 7.6 are those that can reasonably be recovered at costs low enough to allow economic production of the resource. The remaining estimated resources listed are called "technically recoverable"—that is, they are generally expected to be recoverable using currently available technology, but without regard to economic viability. In some cases, the estimates are for oil that is yet to be discovered. These estimates are obviously less certain than for those resources already discovered.

Table 7.6 lists estimates of the range of costs that might be incurred to produce each of the resources. The wide ranges of estimated costs reflect considerable uncertainty; costs vary widely, depending on the location, size, and depth of the resource and on many other factors. Finally, Table 7.6 also estimates the time period in which a reasonable quantity of the resource might be available for use. Here again, there is considerable uncertainty because of costs and other limitations, such as access to drilling or mining and environmental impacts.

The resources listed in Table 7.6 for light oil enhanced oil recovery (EOR) are those that could be recovered primarily by CO_2 injection. Whereas conventional oil recovery processes (primary production under the natural pressure in the reservoir and water injection) typically recover about a third of the oil in place, this resource estimate is based on an assumption that total recovery in fields suited to CO_2 injection would reach 50 percent. The total amount recovered in some reasonable time period is likely to be lower than the total listed, however. Not all fields will be large enough to warrant the investment required, and sufficient CO_2 may not be available. Even so, the experience gained in operating CO_2 EOR projects in west Texas over the last three decades has advanced the technology significantly. EOR projects can now be undertaken with confidence that high-pressure injected CO_2 can displace oil efficiently in the zones that it invades.

TABLE 7.6 U.S. Oil Resources and Reserves

	Barrels (billion)	Estimated Cost Range ($/bbl)	Time Period for Significant Recovery
Oil Reserves (2007 annual U.S. production: 2.5 billion bbl[a])			
Conventional light oil proved reserves[b]	22	10–20	<2020
Natural gas liquid proved reserves[c]	8		<2020
Technically Recoverable Resources			
Light oil EOR[d]	90	20–45	<2020
Heavy oil EOR[b]	20	25–60	<2020
Residual zone EOR[c]	20	60–130	2020–2035
Undiscovered conventional (onshore)[b]	43	40–60	2010–2035
Undiscovered conventional (offshore)[b]	76	75–95	2020–2035
Undiscovered EOR (onshore)[b]	22	50–75	>2035
Undiscovered EOR (offshore)[b]	38	105–145	>2035
Reserve growth (conventional recovery)[b]	71	10–20	<2020
Reserve growth (EOR)[b]	40	20–45	2020–2035
Tar sands[b]	10	40–95	>2035
Oil shales[e]	500	40–95	>2035

[a]British Petroleum, 2008.
[b]DOE, 2006a.
[c]British Petroleum, 2008.
[d]DOE, 2006b.
[e]Bartis et al., 2005.

An extensive infrastructure of pipelines in west Texas delivers CO_2 to numerous oil fields. Much of that CO_2 is transported by pipeline from natural CO_2 sources in Colorado and New Mexico, though there are also significant EOR projects in west Texas, Wyoming, and Colorado that make use of CO_2 separated from natural gas (instead of venting it to the atmosphere). The pipeline infrastructure demonstrates CO_2 transport technology that would be needed to support large-scale geologic storage of CO_2. These projects also allow assessment of whether injected CO_2 has been retained in the subsurface (Klusman, 2003). For example, measurements of CO_2 seepage at the surface above the Rangely Field in Colorado indicate that the rate of CO_2 escape from the storage formation is very low (less than 170 tonnes per year over an area of 72 km^2). Currently, CO_2 injection for EOR is limited mainly by the availability of CO_2 at a reasonable cost. If CO_2 were more widely available in the future at a reasonable distance from existing oil fields as a result of limits on CO_2 emissions, more widespread use of CO_2 EOR could be anticipated. (See the section titled "Geologic Storage of CO_2," later in this chapter and the section titled "Oil and Gas Reservoirs" in Annex 7.A for additional discussion of the potential for CO_2 EOR to contribute to geologic storage of CO_2.)

Heavy oils are difficult to displace; hence, typical primary recovery of oil from such reservoirs is much lower than that of lighter oils. Heavy oil is typically recovered by injecting steam, which warms the oil and reduces its viscosity so that it can flow more easily into production wells. Steam for injection is typically generated by burning a portion of the oil produced or by burning natural gas in areas where air-quality restrictions limit use of the crude oil as a fuel. This technology is now relatively mature and has been applied widely in heavy-oil fields in California, for example. Dissolving CO_2 in heavy oil also reduces its viscosity, but the use of CO_2 to recover heavy oil has not been tested in field projects.

Residual zone EOR refers to the possibility that some of the oil that is found in the transition zone between water and oil at the base of a reservoir can also be recovered by CO_2 injection. This process is less well proven and likely more expensive than CO_2 injection in zones that have less water and more oil present.

The estimates of undiscovered conventional and EOR resources in Table 7.6 are based on assessments by the U.S. Geological Survey (USGS) and the U.S. Minerals Management Service (MMS). The estimates shown for technically recoverable resources are 33 percent of those amounts for conventional recovery and an additional 17 percent for EOR. Reserve growth refers to the observation that the amount of oil listed as proved reserves often increases over time; information obtained through development drilling in the field is used to refine initial estimates of oil in place.

There is currently no significant production of oil from tar sands in the United States, as the U.S. tar sand resource is modest. There is a much larger resource of tar sands in Canada, however, and it has shown significant growth in production. The technically recoverable Canadian resource is estimated at 173 billion barrels (RAND, 2008), and the EIA projects production rates of 2.1–3.6 million barrels per day in 2020 and 4 million barrels per day in 2030, depending on oil price.

The largest oil resource listed in Table 7.6 is from oil shales, but it is among the most uncertain. The estimated overall resource is very large (1.5–1.8 trillion barrels); one source has estimated that as much as a third of it could eventually be recovered by some combination of mining followed by surface retorting or in situ retorting (Bartis et al., 2005). There is currently no production of oil from shale in the United States, though a new process for in situ retorting based on electric heating of the shale in the subsurface is being tested (Shell, 2006). Environmental impacts associated with mining, limitations on availability of water for processing, and potential demand for electricity to be used for in situ retorting must be assessed before better-constrained estimates of recoverable quantities of oil from shales can be assembled. Also, current cost estimates for shale oil recovery are not well defined.

In the absence of CO_2 capture and storage, production of oil either by enhanced oil recovery methods or by conversion from tar sands or oil shales emits more CO_2 than does conventional oil production. This is shown in Figure 7.1, which provides estimates of the potential emissions that result from production and use of fuels from various primary fossil-fuel resources (Farrell and Brandt, 2006).[6] The fuels all have about the same CO_2 emissions when they are burned, but the energy requirements to recover and upgrade the hydrocarbons vary significantly. As an example, fuels from tar sands may ultimately emit about 40 percent more CO_2 than do fuels from conventional oil,[7] though the ranges of estimated emissions indicate that there are significant uncertainties in the values reported. These emissions can in principle be mitigated by large-scale carbon capture and storage (CCS), as noted above, or by the use of low-carbon technologies for process heat and hydrogen production. In addition, both surface mining and in situ production of tar sands disrupt large land areas, as would surface mining of oil shales, and the amounts of water required to process the fuels will also be a constraint in some areas. Thus, there are significant environmental issues associated with the recovery and processing of some of the unconventional hydrocarbon resources.

Although the U.S. oil resource base is large, future domestic production will depend on two factors. One is the decline in production from existing fields. The decline rate varies from field to field, but it is everywhere significant. For example, the EIA assumes that currently producing fields decline at the rate of 20 percent per year. New fields are assumed to peak after 2 to 4 years, stabilize for a period, and then decline at the 20 percent rate (EIA, 2008b). While the National Petro-

---

[6]For a discussion of emissions associated with various fuel conversions, see Chapter 5.

[7]Emissions of CO_2 result from the use of significant quantities of natural gas to provide process heat for separating the hydrocarbons from the sand and for making the hydrogen needed to upgrade the oils. These emissions could be reduced significantly in the future if nonfossil sources of electricity and process heat, such as nuclear, were used in the recovery and conversion processes.

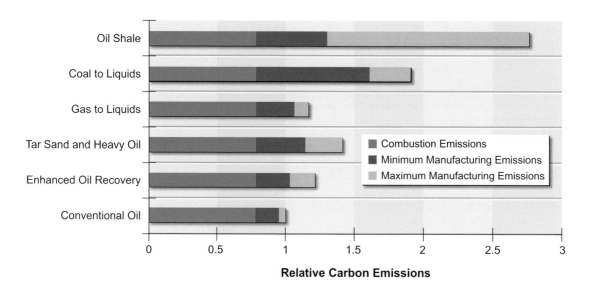

FIGURE 7.1 *Estimated relative CO₂ emissions of alternative sources of hydrocarbon fuels.*
Source: Farrell and Brandt, 2006.

leum Council (NPC) does not specify a decline rate, its report also notes the significance of declining production as fields mature (NPC, 2007).

The other factor that determines production is the ability to develop the resources listed in Table 7.6. This, in turn, depends on three key variables:

- *The pace at which technology can access increasingly challenging types of resources.* After 2020, the application of new methods will be required to offset the inevitable decline in production from existing large fields in the United States. NPC (2007) cites 11 significant technologies under development that should be available between 2015 and 2020 to meet this need (see Box 7.1). The expansion of CO_2 EOR is technically feasible, but it will depend on the availability of significant additional quantities of CO_2 (see the discussion on carbon capture from power plants, for example, elsewhere in this chapter) and on whether the infrastructure to deliver that CO_2 to the oil fields can be built.

BOX 7.1 **Continuing Advances in Oil and Gas Exploration and Production Technology**

Developing U.S. oil and natural gas resources depends critically on technology. The domestic resource base is lodged in geologic formations that make extraction more difficult, and they are often smaller (and therefore harder to find) than the more easily developed fields of the past. Substantial advances in technology have been made in the past few years, however. For example, deepwater offshore oil production has compensated for declines in shallow water offshore and in Alaska production. Natural gas production from unconventional resources now accounts for more than half of total domestic production. And the shift from two-dimensional to three-dimensional seismic technology has increased exploration-drilling success rates by 50 percent over a 10-year period (Bohi, 1998).

This trend toward more sophisticated technology must continue if domestic production rates of oil and gas are to be maintained, much less increased. Because essentially all of the technology that will be relevant before 2020 is being developed by the private sector, the AEF Committee has not conducted an independent assessment of the oil exploration and production technology. However, relying on the topic papers prepared for the National Petroleum Council report *Facing the Hard Truths About Energy* (NPC, 2007), it appears that appropriate development is under way.[1]

The critical technology need in oil production is the ability to manage fluids in complex underground reservoirs. These fluids involved are both the crude oil itself and materials such as CO_2 that are used in enhanced oil recovery. Table 7.5 summarizes the "highly significant" technologies that are currently being developed for conventional oil exploration and production (NPC, 2007, Topic Paper 19). In the view of committee members familiar with oil exploration and production, this summary table (and the more detailed discussion in the topic paper) is a reasonable reflection of the status of development. In general, it appears that these technologies, if developed successfully, will support the pace of resource development shown in Table 7.7.

In the case of natural gas, the chief technical challenge is to develop the resources contained in gas shale and other low-porosity formations. The necessary technologies involve the ability to drill horizontal wells and to fracture the shale formation to allow the natural gas to flow to the bore hole. These technologies advanced very significantly in the early years of this decade, which led to substantial increases in natural gas production from shale.

---

[1]See, especially, Topic Papers 19, 20, 21, and 26.

- *Economic feasibility.* The cost of exploiting alternative resources increases as they become more challenging (essentially from the top to the bottom of Table 7.6). Oil prices are set in a world market, even though the world price may be influenced by the actions of major producers, and historically, oil prices have been quite volatile. Such volatility can be a disincentive to the large and long-term investments needed to find and produce oil from technically challenging and increasingly costly resources.

- *Access to resources.* The resources listed in Table 7.6 include quantities of oil estimated to occur in the coastal plain of the Arctic National Wildlife Refuge (ANWR-1002 area), which is currently off limits to exploration and production, and parts of the outer continental shelf (OCS), for which policies on access for exploration and production are currently in flux (see "The Access Issue" subsection that follows for additional discussion).

Although predicting the level of domestic production that results from the confluence of these factors could be considered speculative, the EIA has estimated how oil production might be affected by changes in them. Table 7.7 summarizes the agency's most recent figures for several alternatives.[8]

Notwithstanding the considerable uncertainties involved in these estimates, it seems clear that the level of net domestic oil production is relatively insensitive to favorable developments in technology, higher world prices, and access to new resources. This is *not* to say that these factors are unimportant. Rather, it seems appropriate to conclude that because of the decline in currently (and future) producing oil fields, maintaining domestic production at something like current levels is a very challenging assignment. As a result, reducing consumption is likely to be the most important factor in decreasing domestic dependence on oil as an energy source.

---

[8]Considerable caution should be used in interpreting Table 7.7. For one thing, the cases are not additive. In some instances, they involve arbitrary changes to parameters in the reference case, and assumptions about physical properties are not explicit. The high-oil-price case is not built up from a cumulative supply curve in the EIA estimating procedure and thus should not be thought of as representing actual economics. Other sources offer different projections, but because the EIA reference case appears to lie near the middle of the range it is useful for comparison purposes. See the National Petroleum Council Data Warehouse (available on CD with the NPC report *Hard Truths* [NPC, 2007]) for a collection of forecasts from a variety of sources.

TABLE 7.7 Projected U.S. Crude Oil Production in Various Years

EIA Alternative Cases[a]	Projected Production (million bbl/d) (2007 annual U.S. production: 5.1 million bbl/d, not including natural gas liquids)		
	2010	2020	2030
Reference case	5.9	6.3	5.6
High oil price	5.9	6.4	6.4
Rapid technology	6.0	6.5	6
ANWR 1002 access	5.9	6.5	6.3
Access to all OCS	5.9	6.4	5.8

[a]The "Access to all OCS" case comes from EIA, 2007, while the other cases are from EIA, 2008a. See Appendix E of each document for a description of assumptions.

For the foreseeable future, U.S. reserves and production are likely to remain a modest fraction of world reserves and production.[9] Indeed, none of the changes in Table 7.7 would lift U.S. production above about 8 percent of current world totals.

Although this committee has not attempted to evaluate non-U.S. oil reserves and production, it should be noted that the tension between declining production from existing reserves and investment in new production exists worldwide. The 2008 *World Energy Outlook* published by the IEA (2008b) reviews the status of the world's largest existing oil fields and concludes that "field-by-field declines in oil production are accelerating . . . and barriers to upstream investment could constrain global oil supply." Referring to its scenario analysis, the report observes that "the projected increase in global oil output hinges on adequate and timely investment. Some 64 million barrels per day of additional gross capacity—the equivalent of almost six times that of Saudi Arabia today—needs to be brought on stream between 2007 and 2030" (IEA, 2008b). These uncertainties are reflected in the range of production estimates from various publicly available sources. According to an NPC review of estimates for 2030, world oil production could range from 90 to 120 million barrels per day, as compared with about 85 million barrels

---

[9]Other publicly available projections are consistent with these EIA estimates. See, for example, IEA (2008a). Data from private-sector sources (oil companies and consultants) available in the NPC data warehouse are, if anything, somewhat less optimistic.

per day today. The 2008 *World Energy Outlook* reference scenario projects 2030 oil production at 106 million barrels per day (IEA, 2008b).

In any case, countries that have much larger production potential than the United States does can more easily increase (or decrease) oil production by the amount potentially obtainable from U.S. areas, both restricted and unrestricted. It is for this reason that this country is more likely to be a price taker than a price setter.

Natural Gas

Unlike the situation with oil, the United States currently produces most of the natural gas it consumes (see Table 7.3). Moreover, its imports are almost entirely from Canada, with the result that North American production is able to meet North American demand. If increased U.S. production of natural gas were able to maintain this balance, the United Staes could limit imports of natural gas (in the form of liquefied natural gas, or LNG). If not, natural gas imports would increase and at some point could result in significant economic and security risks, much like those that presently exist in the oil market. As noted in the following discussion, whether the United States can or cannot increase its domestic production of natural gas is not yet clear.

Table 7.8 shows the various types of U.S. natural gas resources. Significant conventional gas resources are located both offshore and onshore, although much of the offshore resource is in deep water. Nonassociated conventional resources are not physically mingled with oil deposits. Unconventional gas resources are of three types. Tight gas sands and gas shales are formations with low porosity and thus require technology to fracture the structures for the gas to flow to producing wells. Coal-bed methane is natural gas trapped in coal deposits.

Natural gas hydrates (not included in Table 7.8) are a potentially large but poorly defined resource. Estimates of the total global resource range from 1 to 100 times the world resource of conventional natural gas (NPC, 2007, Topic Paper 24; Ruppel, 2007). Hydrates are materials in which water molecules form cages that can contain a guest molecule, in this case methane. Forming at temperatures above the freezing point of water and at high pressures, they are found in many ocean sediments around the world and in locations in the Arctic where land temperatures are low. Methods for recovery of hydrates are under investigation. Whether any recovery method can produce at rates large enough to allow commercial production over an extended period and with acceptable environmental

TABLE 7.8 U.S. Natural Gas Resources

	Trillion Cubic Feet (Tcf) (2007 annual U.S. production: 19.3 Tcf)
Proved reserves	204
Conventional gas resources	
Onshore (nonassociated)	286
Offshore (nonassociated)	214
Associated dissolved gas	130
Unconventional gas resources	
Tight gas sands	304
Coalbed methane	71
Gas shales	125

Source: EIA, 2008b, Table 50. Based on USGS and MMS data with adjustments for recent information. Does not include Alaska or off-limits OCS areas. While the Potential Gas Committee (PGC, 2006) uses somewhat different categories, the PGC aggregate estimate is consistent with the EIA estimate.

consequences has yet to be established (see Annex 7.A for additional discussion). Thus, while the resource is potentially large, it is unlikely to contribute significant production of natural gas by 2035 unless significant progress is made on developing economically feasible and environmentally acceptable recovery processes.

As is the case with oil, natural gas production levels are constrained by the tension between declining production from existing fields and the difficulty of bringing on new production. The EIA estimates that declines in natural gas fields are typically 30 percent per year, somewhat greater than the estimate for oil. And as with oil, the issues of technology, economics, and access determine the ability to bring on new production.

- Included in the proved reserves and estimates of technically recoverable resources are significant amounts of natural gas from unconventional geological formations (tight gas sands, gas shales, and coal-bed methane). Better than half of current natural gas onshore production comes from these resources, and they will remain the principal source of new production for the foreseeable future (see Figure 7.2).
- Producing from these formations does require advanced technology, though many of the methods being developed for oil production also are useful for natural gas production. Especially important for natural

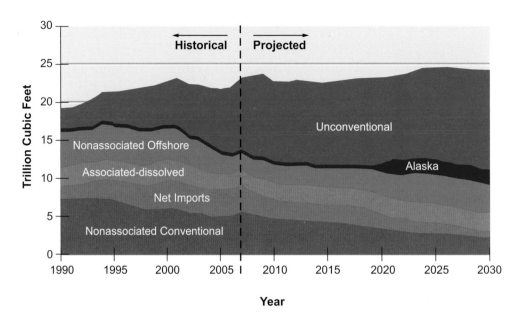

FIGURE 7.2 *U.S. Energy Information Administration reference case for U.S. natural gas production, showing the projected increase in the proportion of gas from unconventional sources along with the decline in gas from conventional sources. "Associated" refers to gas produced as a result of oil production.*
Source: EIA, 2009b.

gas are technologies for well drilling and completion in deep water and technologies for producing natural gas from low-porosity formations such as tight sands and shales.

- The price of natural gas has been volatile and will likely remain that way. This committee has not been able to develop a supply curve for natural gas production from publicly available data. However, it appears that at the lower end of the recent natural gas price range the production of gas shales and perhaps of some deepwater offshore resources is not economic. At the high end of the range, the private sector seems willing to invest in all of these types of gas resources.

- Potential natural gas reserves have until recently been off-limits along the Atlantic and Pacific coasts and in the eastern Gulf of Mexico. Their current status is in flux. Because limited data are available for evaluation of these areas, estimates of future production are necessarily uncertain, as with any estimate of undiscovered resources. The subsection titled "The Access Issue" addresses this issue and reports potential future-production estimates that do exist.

TABLE 7.9 Projected U.S. Natural Gas Production (trillion cuibic feet) in Various Years

EIA Alternative Cases[a]	2010	2020	2030
Reference case	19.8	20.2	20.0
High gas price	19.8	20.3	20.4
Rapid technology	19.8	21.0	21.3
Access to all OCS	19.8	21.5	21.6

Note: These estimates are subject to the same cautions as those regarding the earlier estimates for oil. Note also that private-sector estimates reported in the NPC database seem somewhat less optimistic. For example, the maximum estimate for 2020 among international oil companies is 18.9 trillion cubic feet.

[a]The "Access to all OCS" case comes from EIA (2007), while the other cases are from EIA (2008a). See Appendix E of each document for a description of the case assumptions.

Although the level of domestic production resulting from the confluence of these factors remains speculative, the EIA has estimated how natural gas production might be affected by changes in them. Table 7.9 summarizes EIA estimates regarding four alternatives.

According to these EIA estimates, maintaining domestic natural gas production, much less raising it above current levels, is challenging. However, resources in the OCS and new gas shale formations may have a significant upside production potential. Technology has recently made feasible the production of natural gas from shale formations in the Rockies, Mid-Continent, and Appalachian regions. Wood Mackenzie data (Snyder, 2008), for example, suggest a possible increase on the order of 3 Tcf per year by 2012, a level that can be maintained for several years. In the early release of the 2009 Annual Energy Outlook, the EIA reference case shows Lower 48 production of 21.6 Tcf in 2030. Thus, the upside potential for the deployment of new technology to exploit shale gas may be higher than the EIA's 2007 projections in Table 7.9.[10]

In any case, it is very important that domestic natural gas production keep

[10]Note that Figure 7.2 reflects EIA's 2009 early release projections (EIA, 2009b). Table 14 of EIA's updated reference-case forecast for the 2009 Annual Energy Outlook (AEO 2009) (April 2009) projects 23.03 Tcf of U.S. natural gas production in 2030. Shale gas is projected to contribute 3.66 Tcf of the total. This represents a doubling of shale gas production from 2007. Table A1 of the update shows declining natural gas imports between 2007 and 2030, suggesting that domestic supplies are robust over the period.

pace with domestic demand. Unlike oil, U.S. natural gas prices are not presently determined in a world market. But there is a growing world market in LNG, and if growth in domestic demand for natural gas exceeds growth in supply (even with expanded natural gas production from gas shales, for example), the United States may find itself beholden to that global market. In that case, increases in domestic demand would have to be satisfied, increasingly, by imports. Most of these imports would likely be in the form of LNG, which would require large capital investments in port facilities and regasification infrastructure. Moreover, global movements of LNG would increasingly result in a globally determined price for natural gas. At this writing, the delivered price of LNG in Japan (more than $17.10/GJ, or $18/million Btu),[11] roughly at parity with the price of oil based on energy content, and this is more than twice the U.S. price (~$7.60/GJ, or $8/million Btu).[12]

The Access Issue

Oil and gas exploration and production have been off-limits in some parts of the United States for a variety of policy reasons. Some 12 percent of U.S. petroleum resources and 20 percent of natural gas resources are believed to lie in these restricted areas. In late 2008, the president and Congress removed restrictions on access to previously restricted sections of the U.S. offshore resources, though a 2006 law banning drilling in the eastern Gulf of Mexico remains in effect (www.mms.gov/ooc/press/2008/FactSheet-MMSGOMSecurityActMARCH202008.htm). But how quickly offshore development will proceed, if it proceeds at all, is difficult to determine. For one thing, Congressional or Executive Branch action to reimpose the access ban remains a possibility. For another, individual states can intervene in development programs even without overriding a federal approval of a project—by preventing the oil or gas from coming on shore, for example. And the cost and technical difficulty of developing many of these resources can be significant (Durham, 2006). Thus the offshore access issue may remain an open policy question, at least for a while. Accordingly, this section provides background to help address that question.

---

[11]1 GJ = 0.948 million Btu.

[12]Recent prices in Japan have also been influenced by shutdowns of nuclear power plants pending review of earthquake safety. It is not clear how long these shutdowns will continue and what the natural gas price will be if demand for natural gas for electric power generation in Japan declines as a result of nuclear power plants going back on line.

TABLE 7.10 Estimated Undiscovered but Technically Recoverable Onshore Oil and Gas Resources on Federal Lands

	Oil (billion bbl)	Gas (Tcf)
Inaccessible[a]	19	94
Accessible with restrictions[a]	9.3	113
Accessible standard lease[a]	2.3	24
Total resources[a]	30.5	231
Northern Alaska total	17	67
National Petroleum Reserve-Alaska	9.3[b]	60[c]
Alaska National Wildlife Refuge (1002 Area)	7.7[c]	7[d]

[a]BLM, 2008.
[b]USGS, 2002.
[c]USGS, 1998.
[d]EIA, 2004.
Source: See www.blm.gov/wo/st/en/prog/energy/oil_and_gas/EPCA_III/EPCA_III_faq.html.

Table 7.10 reports estimates, compiled by the Bureau of Land Management from USGS and MMS sources, of the volumes of technically recoverable oil and gas for federal lands. The amounts shown are for 11 sedimentary basins, including the National Petroleum Reserve-Alaska (NPRA) and the Alaska National Wildlife Refuge 1002 (ANWR-1002) areas. The NPRA and ANWR-1002 estimates shown separately (but included in the 30.5 billion barrel estimate) in Table 7.10 are the largest components of the onshore, undiscovered, and technically recoverable resources. The NPRA estimate (9.3 billion barrels) is part of the estimate of undiscovered oil that is accessible with restrictions, and the ANWR-1002 estimate (7.7 billion barrels) is in the inaccessible category.

Comparison of these numbers with the scale of oil use is instructive: 2007 world oil consumption was about 85 million barrels per day (31 billion barrels per year); U.S. oil consumption was about 20.7 million barrels per day (7.6 billion barrels per year); and U.S. oil production was 6.9 million barrels per day (including natural gas liquids), which amounts to 2.5 billion barrels per year (British Petroleum, 2008). For natural gas, the corresponding 2007 numbers are world natural gas consumption at 104 Tcf, U.S. consumption at 23 Tcf, and U.S. production at 19.3 Tcf (British Petroleum, 2008).

The estimated undiscovered oil resources, which total 30.5 billion barrels, are included in the 76 billion barrels of undiscovered offshore resources listed in Table 7.6. The total gas resources listed, however, are not included in the natural gas resource estimates of Table 7.8. The resources listed as inaccessible are those that are estimated to lie within areas where exploration and production have been prohibited. These include lands that cannot be leased as a result of congressional or presidential action (including national parks, national monuments, and wilderness areas); lands that are not available for leasing based on decisions by the federal Bureau of Land Management (historical sites and endangered species habitats, for example); lands that are undergoing land-use planning or National Environmental Policy Act review; and areas that can be leased but with no surface occupancy (directional drilling might be able to access some resources, in which case they are included in the category of accessible with restrictions). Restrictions may include limits on drilling during a portion of the year or stipulations that require mitigation plans or exclude some areas within the lease from drilling. Operations in areas for which standard lease terms apply must observe pertinent environmental laws and regulations.

Table 7.11 gives related estimates for offshore resources that are located in areas that have not been open for leasing for exploration and production (NPC, 2007, Topic Paper 7, www.npchardtruthsreport.org/topic_papers.php). The largest undiscovered resources are estimated to be located in the restricted portions of the federal OCS. These estimated gas resources are in addition to those listed in Table 7.8. The estimated oil resources in Table 7.11 are included, however, in the estimates of Table 7.6.

The combined estimates of conventional onshore and offshore oil in areas that are now inaccessible or have been so until very recently comprise 32 percent (19 billion barrels onshore oil [Table 7.10], plus 19.3 billion barrels offshore oil [Table 7.11]) of the total estimated undiscovered conventional technically recov-

TABLE 7.11 Estimated Undiscovered but Technically Recoverable Offshore Oil and Gas in Areas Covered by Moratoriums

	Oil (billion bbl)	Gas (trillion cubic feet)
Eastern Gulf of Mexico OCS	3.7	22
Atlantic OCS	3.8	37
Pacific OCS	10.4	18
Great Lakes	0.4	5
State waters	1.0	2
Total resources	19.3	84

Source: NPC, 2007, Topic Paper 7, available at www.npchardtruthsreport.org/topic_papers.php.

erable oil shown in Table 7.6 (43 billion barrels onshore, plus 76 billion barrels offshore). As Table 7.6 indicates, however, the relatively high costs of developing some of the resources may limit the rate of development, though these costs are comparable to or lower than some of the costs of making liquid fuels from alternate sources (see Chapter 5). The estimates of undiscovered gas resources in the inaccessible areas (94 Tcf onshore [Table 7.10], plus 84 Tcf offshore [Table 7.11]) are about 28 percent in addition to the total conventional gas resources listed in Table 7.8 (630 Tcf) or about 16 percent in addition to the total of conventional and unconventional gas resources listed in Table 7.8 (1130 Tcf).

There is considerable uncertainty in these estimated volumes, as with any figures that purport to measure undiscovered resources. Geophysical data used to refine such estimates were last collected 25 or more years ago for the Pacific coast, the Atlantic coast, and portions of the eastern Gulf of Mexico. Since then, significant advances have been made in seismic technology, which could allow more accurate estimates of the size and location of potential accumulations. There is similar uncertainty in the rate of production that might be obtained from these areas if exploration and production were permitted. Offshore developments in deep water typically require extended time periods during which to begin production (5–7 years or more) if exploration is successful and more time to ramp up to full-scale production.

But even without considering new producing provinces, the substantial technology development for production in deep waters of the OCS—where leasing and drilling have been under way for some time—is projected to have a significant impact on U.S. oil production in the next decade. For example, in its 2008 reference case, the EIA projects that deepwater Gulf of Mexico conventional oil production will increase from about 1 million barrels per day in 2006 to a peak of 2 million barrels per day sometime between 2013 and 2019, declining thereafter to 1.6 million barrels per day in 2030 (EIA, 2008a, p. 79). (These quantities are similar to those being contemplated for production of liquid fuels from coal or biomass—see Chapter 5.) That increase in production, in turn, leads to a projected increase in total U.S. production from 5.1 million barrels per day in 2007 to a peak of 6.3 million barrels per day in 2018. Thus the increase in deepwater production more than offsets continuing declines in Alaska production and shallow offshore production, but only for a time. If leasing and development proceed in OCS areas that were previously off-limits, the technology improvements that have proved successful in deepwater Gulf of Mexico areas could be applied in those OCS areas as well.

EIA estimates (EIA, 2008a) of production rates from access to ANWR-1002 and the OCS showed increased U.S. production—more than the reference-case production of 6.3 and 5.6 million barrels per day—of about 200,000 and 100,000 barrels per day, respectively, in 2020 and about 700,000 and 200,000 barrels per day in 2030 (see Table 7.7). It is important to recognize that these estimated increments reflect both the increased production in the specified areas and the declines in production elsewhere. Mean production estimates cited in the NPC study (NPC, 2007, Topic Paper 7) for the ANWR-1002 area are 539,000 barrels per day in 2020 and 576,000 barrels per day in 2030. EIA estimates made in 2004 (EIA, 2004) showed somewhat larger estimated production for ANWR-1002, with assumed production starting in 2013 and peaking at 874,000 barrels per day in 2024.

While any additional oil production has some impact on oil price, as well as an obvious impact on the amount of oil imported into the United States, most observers have argued that the impact on oil price of net incremental U.S. production due to the opening of restricted areas will be small. Projected total production increases are modest compared to world demand (about 85 million barrels per day at present); they are projected by the EIA to grow to 96 million barrels per day in 2015 and 113 million barrels per day in 2030 (EIA, 2008d).

Oil prices are set in a global market, and both supply and demand depend on price, though supply responds slowly to high prices and demand usually responds faster. Short-term oil price volatility observed in recent months is a reflection of this dynamic, at least in part. But it is not known whether remote or offshore production will compete on costs with other sources of supply around the world, nor whether such resources will be developed in the first place, given the uncertainty as to future oil prices supporting development. As the EIA noted in its analysis of the impact of ANWR-1002 production, "Assuming that world oil markets continue to work as they do today, the Organization of Petroleum Exporting Countries could countermand any potential price impact of ANWR coastal plain production by reducing its exports by an equal amount" (EIA, 2004). Similar reasoning suggests that the impact of increased OCS production on world oil price in the long term would also be small.

It is possible that natural gas markets, which are becoming more global but still maintain regional differences, will respond differently to the potentially higher production quantities, although the magnitude of any response is uncertain (Baker Institute, 2008). The EIA estimates summarized in Table 7.9 suggest that access to restricted OCS, for example, might provide increased gas production of about 1.3 Tcf per year in 2020 and 1.6 Tcf in 2030 (U.S. consumption is about 23 Tcf per year at present), which could offset the need for some LNG imports.

A related discussion is under way concerning the potential environmental risks of developing oil or gas resources in locations such as the ANWR-1002 area, the National Petroleum Reserve, or the formerly restricted OCS. Technology improvements such as long-reach directional drilling have reduced the area required by surface facilities for drilling and production, but some surface impact is inevitable. Similarly, the use of subsea completions for deepwater oil and gas production, pipeline delivery of fluids to shore in place of tankers, and attention to modern MMS environmental regulations governing platforms have reduced the potential for adverse impacts in offshore production. However, there will always remain some risk, whether at the platform or at the land end of the undersea pipeline. In addition, close-in platforms have visual impacts.

In addition to the OCS and federal land resources discussed in this section, the increased interest in natural gas production from shale formations may create a need to balance energy and environmental values regarding this resource. Large shale formations in the mid-Continent and the Gulf Coast (e.g., Barnett and Haynesville) are located in areas where oil and gas are currently produced. Infrastructure exists in these areas, and public opinion is probably open to additional gas production. However, the Marcellus shale in Appalachia is spread over a wide area, where lack of infrastructure and fragmented land ownership make production from this area more challenging (Snyder, 2008).

Coal

Table 7.12 provides estimates of coal resources by coal rank. Anthracite and bituminous coals have the highest energy and carbon content, whereas subbituminous coals and lignites have lower energy content and larger moisture and ash content (NRC, 2007, Box 4.1). The table indicates that the United States has about 20 years of reserves in active mines, but a much larger resource would be available for production if new mines could be opened and if the rail infrastructure required to deliver coal—or, alternatively, if sufficient long-distance transmission lines for delivery of electricity generated at the mine mouth—could be put in place. Costs of coal production vary widely with geographic setting and the type of mining, but it is clear that costs are low enough that substantial quantities of coal can be produced at current coal prices.

TABLE 7.12 U.S. Coal Resources and Reserves

Coal (2005 Annual U.S. Production: 1 billion tons)[a]	Billion Tons
Recoverable reserves in active mines[a]	17.3
Recoverable reserves[a]	227
Demonstrated reserve base[a]	200
Identified resources[a]	1200
Proved Reserves[b]	
Anthracite and bituminous	123
Subbituminous and lignite	144

[a]NRC, 2007.
[b]British Petroleum, 2008.

The National Research Council has recently assessed the status of domestic coal resources (NRC, 2007). It concluded that:

> Federal policy makers require accurate and complete estimates of national coal reserves to formulate coherent national energy policies. Despite significant uncertainties in existing reserve estimates, it is clear that there is sufficient coal at current rates of production to meet anticipated needs through 2030. Further into the future, there is probably sufficient coal to meet the nation's needs for more than 100 years at current rates of consumption. . . . A combination of increased rates of production with more detailed reserve analyses that take into account location, quality, recoverability, and transportation issues may substantially reduce the number of years of supply. Future policy will continue to be developed in the absence of accurate estimates until more detailed reserve analyses—which take into account the full suite of geographical, geological, economic, legal, and environmental characteristics—are completed.

Even given the uncertainties in resource estimates, the United States likely has sufficient coal to meet projected needs. However, of all the fossil fuels, coal produces the largest amount of CO_2 per unit of energy released by combustion—about twice the emissions of natural gas, but can vary depending on coal rank—and mining has significant environmental impacts, which will limit its suitability for some locations. In any case, the estimates in Table 7.12 suggest that resource availability is not likely to be the constraint that sets the level of coal use.

Findings: Oil, Gas, and Coal Resources

Fossil-Fuel Resources and Production

The United States is not running out of oil anytime soon, but domestic oil production rates are unlikely to rise significantly. U.S. technically recoverable conventional oil resources are large relative to the country's liquid fuel demand. However, U.S. production capacity is limited by declining production in existing fields. Even with new technology, higher prices, and access to currently off-limits resources—none of which is certain—maintaining current production levels will be challenging.

The United States is not running out of natural gas anytime soon, and with favorable circumstances, domestic production could meet most of the domestic natural gas demand for many years. U.S. natural gas resources are large relative to demand, and current domestic production meets most of the domestic demand. Unconventional sources of natural gas are technically recoverable and appear to be large enough to meet domestic demand for several years. Doing so, however, would require both relatively high prices and moderate demand growth.

Unconventional oil from U.S. resources is not likely to result in significant new production volume before 2020. A large oil shale resource exists in some of the western states, but production from these reserves awaits technology demonstration and is highly unlikely before 2020. The U.S. tar sands resource is not large.

Crude oil production from Canadian tar sands is feasible now and likely to grow before 2020, but this resource has a larger carbon footprint than conventional resources have. Canadian tar sands production was 1.3 million barrels per day of crude oil in 2006, and it could grow to 4 million barrels per day by 2030. But with current technology, fuels derived from tar sands ultimately emit 15–40 percent more CO_2 than do fuels derived from conventional crude oil (see Farrell and Brandt, 2006).

Coal is abundant in the United States. Despite significant uncertainties in existing reserve sizes, there is sufficient coal at current rates of production to meet anticipated needs through 2030 and well beyond. More detailed analyses will be required, however, to derive accurate estimates of the impact of enhanced production on reserve life. There are also geographical, geological, economic, legal, and environmental constraints on the future use of coal.

Fossil-Fuel Supply and Demand

Changes in U.S. oil production—including from areas currently off-limits to drilling—are not likely to have a leading influence on world oil production.

Because U.S. crude oil reserves and production will remain a modest fraction of world reserves and production, the actions of other countries could have a greater impact. However, because U.S. oil demand is a large fraction of world oil demand, changes in U.S. demand are a significant factor in the market.

Greater domestic natural gas demand could boost U.S. reliance on LNG imports and cause a significant rise in domestic prices, though sustained significant increases in production of natural gas from shales could limit that reliance. If gains in production of natural gas from shales are not sufficient to meet heightened U.S. demand (such as for natural gas for electric power production), LNG could become the marginal supply to meet that demand. Eventually, LNG imports could grow to a point that linked the U.S. natural gas market to world LNG prices, which would be much higher than current U.S. prices.

Although domestic coal reserves are ample to 2030 and beyond, upward price pressures may exist. Growth in demand for electricity from coal-fired power plants, potential use of coal for producing liquid and gaseous fuels, the cost of opening new mines, and growth in export markets are examples of such pressures.

ELECTRIC POWER GENERATION WITH FOSSIL FUELS

Background on Electricity Generation and Carbon Dioxide Emissions

According to the EIA, U.S. electricity production in 2006 was 3727 terawatt-hours, with coal supplying 52 percent and natural gas 16 percent for a combined total of 68 percent, as shown in Table 7.13. The EIA has also made a projection of electricity generation in 2020 using its computer models, assuming a continuation of trends that were evident as of 2008.[13] Its reference scenario projects that total electricity generation in 2020 will be up about 16 percent from 2006, with a similar fuel breakdown.[14]

---

[13]EIA's scenarios do not include any changes from current policies; for example, none of the scenarios considered includes a price on carbon emissions.

[14]The EIA reference scenario is not a prediction. The agency has published data showing its performance over the years in projecting actual generation (EIA, 2008e), and the committee has reviewed this record for 10-year electricity projections. The committee found them to be reasonably accurate, with about a ±60 percent uncertainty range. But these projections were made dur-

TABLE 7.13 U.S. Electricity Generation by Fuel Type in 2006

Fuel Type	2006 Terawatt-Hours (percent)
Coal	1930 (52)
Petroleum	55 (01)
Natural gas	608 (16)
Nuclear power	787 (21)
Renewable sources	347 (09)
Conventional hydro	265 (07)
Other renewable	82 (02)
Total	3727

Source: EIA, 2008a, p. 131.

Table 7.14 shows data gathered by the U.S. Environmental Protection Agency (EPA) on 2006 U.S. greenhouse gas emissions, broken down by fuel type. Total emissions of all such gases were just over 7 gigatonnes of CO_2 equivalents, of which CO_2 itself accounted for about 85 percent; methane and nitrous oxide accounted for most of the rest. CO_2 emissions associated with energy consumption (as distinct from agricultural and other sources) accounted for 80 percent of all U.S. greenhouse gas emissions, of which 59 percent resulted from direct combustion as fuel (dominated by petroleum use in transportation) and 41 percent from fossil-fuel use in electricity generation. CO_2 emissions from electricity generation were dominated by coal (83 percent), but overall, the burning of coal for electricity accounted for only 27 percent of all U.S. greenhouse gas emissions in 2006.

Dividing the figures in Table 7.14 by those in Table 7.13, the committee finds that approximately 1.0 tonne of CO_2 was emitted per megawatt-hour of electricity produced from coal, and about 0.56 tonne of CO_2 was emitted per megawatt-hour of electricity produced from natural gas.

Looking Forward

Investor-owned utilities and independent power producers face difficult choices at present. They must invest in new power-generation assets to meet future demands for electricity and to replace some portion of the existing fleet of power plants as they are retired, but they must also consider what will happen if constraints

ing relatively stable energy conditions; given the recent turmoil in energy and financial markets, the uncertainty range is now presumably larger.

TABLE 7.14 U.S. CO_2 Emissions by Fuel Type in 2006

Source	Million Tonnes CO_2 Equivalent
Total U.S. greenhouse gas emissions	7054
Total CO_2 emissions	5983 (~85% of all greenhouse gas emissions)
CO_2 emissions from energy	5638
combustion	3310 (59%)
Electricity generation	2328 (41%)
Combustion CO_2 emissions by fuel	3310
Coal	133 (04%)
Natural gas	816 (25%)
Petroleum	2361 (71%)
CO_2 emissions from electricity generation, by fuel	2328
Coal	1932 (83%)
Natural gas	340 (15%)
Petroleum	56 (02%)

Source: EPA, 2008.

are placed on carbon emissions. Financial institutions are wary of lending for coal-fired power plants that do not include provisions for capturing CO_2, and some states have recently indicated that they will not approve the construction of coal-fired power plants without CCS. But some public utility commissions, which see their role as protecting consumers from unwarranted price increases, are reluctant to include the cost of such facilities in the rate base, absent a regulatory requirement.

Construction costs have risen rapidly in recent years, thereby increasing the capital cost of any power plant. The U.S. Department of Energy (DOE)-sponsored project FutureGen, which was to have demonstrated a coal gasification plant with carbon capture, was canceled at this writing because of high projected costs in favor of an alternative vision of supporting incremental carbon capture projects at several plants. At present there are no obvious choices as to the best designs for CO_2 capture. Meanwhile, although capital costs for natural gas plants are a fraction of those for coal or nuclear plants, the price of natural gas has increased substantially above historic levels and has shown some of the volatility of recent oil prices. Thus, the best choices among options for generating electricity are not at all clear at present.

The answers to the following three questions will determine the future of fossil-fuel power in the United States over the coming decades:

1. Will the United States undertake a large effort to reduce CO_2 emissions?
2. Will the technologies of CCS become commercially viable?
3. Will the domestic natural gas price be close to its highest recent value or its lowest recent value?

By 2020, decision makers will probably have sorted out the first question. It is inconceivable that CCS will prosper if there is not a large effort to reduce CO_2 emissions, because unless a significant cost is imposed on CO_2 emissions at a power plant it will nearly always be less expensive to vent the CO_2. The committee assumes here that government will formulate policies to reduce CO_2 emissions, thereby spurring already-existing technologies for generating electric power with reduced CO_2 emissions. The committee focuses here on pathways that deploy such technologies.

With a significant suite of demonstration plants, the country can also sort out the second question. Not enough is known yet to demand that all new plants be equipped with CCS, but much can be learned in the next decade.

The answer to the third question depends in part on the extent to which the U.S. market for natural gas links to the international market. That, in turn, depends, again in part, on the future role of natural gas in electric power generation. Thus the future mix of uses of natural gas and coal for electric power generation will depend sensitively on a combination of the constraints on carbon emissions, the costs of fuels, and the costs of conversion technologies. In particular, whether coal plays a larger or a smaller role in future electric power generation will depend strongly on whether CCS can be applied at the scale of many large power plants.

To examine these questions, the committee considers below three types of power plants: supercritical pulverized coal (PC), integrated gasification and combined cycle (IGCC) coal, and natural gas combined cycle (NGCC).

The PC/IGCC Competition

For large U.S. power generation projects, utilities and independent power producers are evaluating two ways of producing power from coal: PC and IGCC.[15]

---

[15]The committee focuses on coal here but notes that biomass can be substituted for limited quantities of coal in PC and IGCC plants without major changes in plant design. This approach can help alleviate limits on biomass conversion plant size arising from the need to collect biomass over a wide area and from seasonal availability. Biomass can also be used as a feedstock

PC plants use boilers to produce steam, which drives turbines to produce electricity. In its current form, this technology has been in use for over 50 years and continues to be improved. PC technology has progressed from subcritical to supercritical to the latest ultrasupercritical boilers; this is a designation that refers to the temperature and pressure of the steam, with higher values bringing higher efficiencies. As power plant conversion efficiencies increase, the amount of CO_2 emitted per unit of electricity generated declines.

Typical subcritical PC plants have thermal efficiencies of 33–37 percent (based on higher heating value of the fuel, 33–37 percent of the energy stored in the fuel is converted to electricity) and operate at temperatures up to 1025°F and typical steam pressures of 2400–2800 psi. Supercritical PC plants can achieve efficiencies of 37–42 percent at temperatures and pressures of 1050°F and 3530 psi, while ultrasupercritical PC plants are capable of 42–45 percent energy conversion at 1110–1140°F and 4650 psi (Katzer, 2008; MIT, 2007).

In the future, with advances in high-temperature materials and operating temperatures of 1400°F and above (Viswanathan et al., 2008), efficiencies could reach as high as 48 percent, though this would require major R&D breakthroughs.[16] In addition, operating plants often do not realize their full design efficiency, so a more realistic actual efficiency of a pulverized coal plant is likely closer to 40–44 percent without CCS, and perhaps 30 percent with CCS. Figure 7.3 shows that an ultrasupercritical boiler with an efficiency of 42 percent would reduce CO_2 emissions by about 12 percent compared with a standard subcritical boiler. Efficiency improvements of 1–3 percent are also possible through modernization at existing coal plants, but the required capital investment may not be attractive given other priorities. A succinct discussion of these and other variations

---

for gasification without coal. In either case, any CO_2 captured and stored leads to a reduction in atmospheric CO_2 concentration because the carbon present in the biomass was removed from the atmosphere by photosynthesis. A power plant that uses a mix of coal and biomass can therefore have zero net carbon emissions, or even negative net emissions. See the section "Future Biomass Power" below in this chapter, which addresses biomass-fueled power generation.

[16]This is a potential efficiency that might be achieved with steam pressures and temperatures of 5000 psi and 1400°F main steam, 1400°F reheat; however, the most robust current "ultrasupercritical" plants operate at pressures of around 4640 psi and temperatures of 1112–1130°F. The U.S. Department of Energy's National Energy Technology Laboratory, in collaboration with industrial consortia, is conducting research on advanced high-temperature materials (e.g., coatings and nickel-based alloys) for use in ultrasupercritical boilers and turbines (www.netl.doe.gov/technologies/coalpower/advresearch/ultrasupercritical.html).

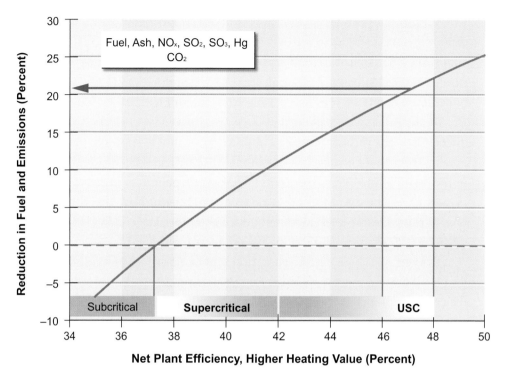

FIGURE 7.3 *An ultrasupercritical (USC) boiler with an efficiency of 46–48 percent would reduce CO_2 emissions by 20 percent compared with a standard subcritical boiler.*
Source: John Novak, Electric Power Research Institute, 2008.

on technologies currently in use for electric power generation from coal can be found in MIT (2007).

In principle, CO_2 can be captured from any of these PC power plants. Doing so requires use of some of the energy that would otherwise have been used to generate electricity; this fact is reflected in a reduction of the conversion efficiency of the plant. The diverted energy is used to separate CO_2 from the solvents used to capture it, to compress the CO_2, and for the power needed to move CO_2 and solvents through the plant. (See Annex 7.A for a description of some of the processes used to capture CO_2 from power plant combustion-product gases.)

The second approach, IGCC, is a technology for electricity generation that produces gas from coal to drive a high-efficiency gas turbine, whose hot exhaust then drives a smaller steam cycle similar to that of PC. The high-efficiency gas turbine process, which evolved from jet engine technology, can use either air or oxygen; the separation of oxygen from air at the front end creates a gas stream without nitrogen and leads to smaller and lower-cost plant components. Thus, for capture plants, there is no nitrogen to separate from the CO_2. Coal is converted in a reducing atmosphere to a gas known as synthesis gas, or syngas, which contains carbon monoxide (CO), CO_2, hydrogen (H_2), water vapor (H_2O), and traces of other components such as H_2S arising from the sulfur in coal.

In an IGCC power plant that does not capture CO_2, cleaned syngas goes directly to a gas turbine. To add a capture capability at a power plant, the syngas undergoes further chemical processing in a "shift" reactor, which converts most of the carbon into CO_2 rather than CO—thereby creating, in effect, a mixture of H_2 and CO_2. The two gases are then separated, power is obtained from a gas turbine burning the H_2 (with a diluent added to reduce the combustion temperature), while the CO_2 is pressurized and sent off-site for storage.[17]

Oxygen gas from an air separation unit can also be used to burn pulverized coal directly, a process that is known as oxyfuel combustion. In that case, the combustion products from electric power generation are CO_2 and water, plus small amounts of contaminants. In effect, the cost of air separation at the front end to produce O_2 for combustion is traded off against the cost of separation of CO_2 from N_2 at the back end. Both separations require additional capital and reduce net electricity generation. Removing the N_2 from the flow reduces the amount of flue gas, but some recycling of CO_2 is required to control combustion temperature. Another option, known as chemical looping, is also being investigated as a way to separate O_2 from N_2 and thus to avoid a subsequent separation of CO_2 from N_2 (see Annex 7.A for a description of this approach).

Chemical separations—CO_2 from N_2, CO_2 from H_2, or O_2 from N_2—lie at the heart of all these carbon capture schemes. At present, the separations are thermodynamically rather inefficient, and they represent the largest component of the incremental costs for CCS (Dooley et al., 2006; IPCC, 2005) (see Figure 7.A.6 in Annex 7.A for examples that compare capture costs with those of compression, CO_2 transportation, and injection). Because there is significant potential for improving the efficiency of capture and reducing both the costs and the energy penalties associated with capture, this area is an important component of research on CCS. For example, the current DOE program for research to improve separation technologies includes work on improved solvents, materials for mem-

---

[17]Gasification can also be the first step toward the production of synthetic transportation fuels (synfuels) or synthetic natural gas (SNG). In these cases, the shift reactor is used to tune the H_2:CO ratio for optimal synthesis. See Chapter 5.

brane separations, chemical looping, and use of ionic liquids and ion transport membranes (see Figure 7.A.3 in Annex 7.A). In the committee's opinion, the cost of CO_2 capture and the potential for reductions in cost are large enough that an aggressive R&D program on CO_2 capture could significantly affect the economics of deployment of CCS in the 2020–2035 timeframe.[18]

A variant on all capture schemes is "co-capture," in which the CO_2 sent to storage also contains other polluting gases that would otherwise have to be managed aboveground. Sulfur, for example, can be co-captured as SO_2 from PC plants or as H_2S from IGCC plants.

In an NGCC, natural gas is combusted in a high-efficiency gas turbine and the hot exhaust gases raise stream that is used to run a steam turbine. The options for CO_2 capture at an NGCC plant parallel those for coal plants.

The remainder of this section focuses on PC, IGCC, and NGCC plants to illustrate the range of costs of electricity with and without CCS. The committee chose those comparisons as a way to simplify a multifaceted discussion and because more cost data are available to support the analysis for PC and IGCC plants compared with, say, oxyfuel plants. It notes, however, that some reported estimates of costs of electricity for oxyfuel plants are similar to or even somewhat lower than those for PC plants (MIT, 2007), and hence the oxyfuel option should be considered as well. Demonstrations under way in Europe will provide useful additional information that will allow comparisons among the PC, IGCC, and oxyfuel plants with CCS.

R&D is supporting the continued development both of IGCC and PC technology, with the goal of lowering capital costs and improving efficiency, though how the two technologies will compete in a CO_2-constrained world is still uncertain. PC plants may be better suited to capture CO_2 from low-rank coals; the technology is supported by a large worldwide infrastructure for the design and construction of new units and the support of operating units. IGCC, on the other hand, promises lower CCS costs and higher efficiencies.

Complications of High-Price Oil and Natural Gas

High prices for oil and natural gas cause upward pressure on U.S. coal prices and production. For example, as gas prices rise, there is more room for higher coal

---

[18]A recent review of opportunities for improved carbon capture technologies may be found in Gibbins and Chalmers (2008).

prices even while maintaining the incremental advantage of coal-fired electricity over gas-fired electricity. The result may be an unchanged fraction of coal-fired power that is dispatched but at increased profits for coal producers. Increasing gas and oil prices also create new markets for coal, such as synthetic liquid fuels and synthetic natural gas, thereby potentially increasing demand for coal (see, for example, the discussion of alternative liquid transportation fuels in Chapter 5). In a carbon-constrained world, continued or expanded use of coal, whether for electric power generation or for manufacturing of liquid fuels, will require that large-scale CCS be a commercially viable approach to limiting CO_2 emissions.

In addition, total electric power generation could increase if electricity displaces oil in transportation. Plug-in hybrid or battery-electric vehicles are likely to be charged at night, when power is available at lower prices. At present, coal and nuclear power plants typically provide a larger fraction of the power generated at night, when peaking natural gas turbines are less likely to be in use. Nuclear plants are generally run at steady power output, while coal plants allow some downward adjustment of power levels at night. In areas of the country where coal is used to supply a large fraction of nighttime power and additional coal-fired capacity is available at lower cost than for other fuels, a shift to electric power for transportation could increase the fraction of electric power supplied by coal.

Fuel price volatility is also a major concern; it complicates investment decisions. Consider the fact that over the course of this study, U.S. natural gas prices rose above \$13/million Btu and fell to below \$4/million Btu. Such price swings have a dramatic effect on the competitiveness of natural-gas-fired power. For example, the committee calculated that at a price of \$6/million Btu, NGCC plants had the lowest levelized cost of electricity (LCOE) of any baseload generating option; at \$16/million Btu, however, they would have the highest.

Carbon Capture at Coal Plants in Operation Today

The current capacity of U.S. coal-fired electric power generation exceeds 300 GW, and there is a strong financial case for operating most of these PC units for the next 25 years or longer. The CO_2 emission rate for these plants is about 2 Gt CO_2 per year, which is about one-third of current U.S. CO_2 emissions and 7 percent of world emissions. If significant reductions in CO_2 emissions from the U.S. power sector are to be achieved on this timescale, consideration must be given to options for reducing emissions from existing subcritical PC plants. These options include: (1) making improvements to generating efficiency, thus reducing CO_2 emissions per unit electricity supplied; (2) retrofitting existing units with CCS capability; and (3) retiring subcritical units and replacing them with more efficient supercritical, oxyfuel, or IGCC units equipped with CCS. For the second and third options, of course, there must be suitable CO_2 storage sites within reasonable distances from the plants (on the order of 100 kilometers).

An analysis of these and other options can be found in MIT (2007, Appendix 3.E). Option 1 above is the least expensive, but the efficiency improvements are likely limited to just a few percentage points. Option 2 is a complex undertaking that involves much more than a simple in-line insertion of carbon-capture capability into the flue-gas stream. For example, the retrofit of an existing subcritical unit with amine capture can result in a derating (operation at below the rated maximum) of more than 40 percent (MIT, 2007). The steam required to regenerate the amine absorber severely unbalances the rest of the plant, causing the original steam turbine to operate far from its highest efficiency.

Option 3 involves either rebuilding the subcritical unit's core—replacing it with a more efficient supercritical, ultrasupercritical, or oxyfuel unit with CCS—or building an IGCC plant with CCS at the site. Although the total capital cost is higher for this latter option, the cost per kilowatt of generated electricity is expected to be about the same as a retrofit. Moreover, the power output can be maintained because the derating associated with CCS can be compensated for by the increased efficiency of the newer-generation technology. Rebuilds with new high-efficiency technology thus appear to be more attractive than retrofits are (MIT, 2007).

The introduction of CO_2 capture at existing PC units raises the same technology issues faced by new PC units, but the former has unique problems related to site constraints and steam management. A PC retrofit to scrub CO_2 from the flue gas by solvent absorption and desorption requires considerable space. Also, a significant quantity of energy must be used to remove the CO_2 from the solvent, which reduces the energy flow to the steam turbine. Hence, high levels of CO_2 capture affect turbine performance, requiring a rebalancing of the steam flow and possibly a new turbine. As for retrofit of any of the few currently operating IGCC units for CO_2 capture, this would require not only additional space for shift reactors but also the modification of the existing syngas-fueled turbine so it could burn hydrogen. A discussion of the complexities of how CCS might be implemented in the context of the existing fleet of coal power plants is beyond the scope of this report, but it is nevertheless a subject that merits further study.

Partial Capture of CO_2

The new fossil-fuel power plants just discussed would be designed to capture approximately 90 percent of the CO_2—in other words, to assure that 90 percent of the carbon atoms in the coal or natural gas fuel do not end up in the atmosphere. In the case of coal gasification, a small percentage of the captured carbon may end up in solid form, as char, but most of the captured carbon leaves the plant in pipes en route to disposal belowground. In carbon accounting, one may or may not include CO_2 emissions associated with "upstream" activity, such as the burning of fossil fuel associated with the energy to mine and transport coal. And one may or may not include emissions of "other greenhouse gases," notably unburned methane releases associated with coal mining.

Future fossil-fuel power plants may operate in policy environments either where a specific CO_2 capture percentage is specified or where a particular price is placed on each ton of CO_2 emitted. In the latter case, one can expect that the fraction of CO_2 captured will become a design variable. In determining this fraction, anticipated prices over the lifetime of the plant will be considered, as will the incremental investments and power-efficiency penalties associated with each extra percent of capture.

One can expect the shape of the marginal cost curve for CO_2 capture versus percent captured to rise sharply for capture percentages very close to 100 percent, as with any other pollution-control technology. At intermediate capture fractions, given the fixed costs in the capture and storage system (those incurred in permitting or CO_2 pipeline construction, for example), and given the economies of scale in capture components, the marginal cost curve may be relatively flat and descending. For some capture technologies, there are steps in such a curve that represent additional capital investments—e.g., the addition of a second shift reactor for precombustion capture. These are only general observations, however. Engineering analyses that accurately establish the shape of this marginal cost curve for specific capture technologies are now entering the public domain.[19]

CO_2 emissions reductions at existing power plants via improved energy efficiency may be realized through a wide variety of upgrades of equipment, especially if the plant is relatively old and inefficient to begin with. The cost curve for small emission reductions will show steps corresponding to opportunities

---

[19]A recent study by the DOE's National Energy Technology Laboratory suggests a linear relationship between cost and percent capture when current post-combustion amine-capture technology is used. The levelized cost of electricity increases by 2–7¢/kWh as capture increases from 30–90 percent (DOE, 2007).

to replace core components. As for CO_2 emissions reductions by means of CCS installed at existing coal plants, the degree of reduction achieved will depend highly on plant characteristics. Important variables are location of the plant relative to CO_2 storage sites, room at the plant site for additional equipment, and the age of the plant. In some instances, capture fractions far below 90 percent may well allow commercially viable CCS—if they allow boilers and steam turbines in existing PC units, and gas turbines in existing IGCC units, to continue to be used—whereas high capture fractions may require a total overhaul of the plant.

Major investments have been made in many of these existing plants to reduce pollution emissions (such as SO_x, NO_x, and Hg). Thus far, these investments have not been linked to strategies for dealing with CO_2 emissions.

Capture-Ready

The uncertainty about the scope and stringency of future U.S. policies aimed at reducing carbon emissions, as well as the high cost of retrofits, has led to a discussion of whether it makes sense for new coal plants in the interim to be built "capture-ready," in other words, capable of being economically retrofitted with CCS in the future. One recent analysis (MIT, 2007) suggests that this concept has a great deal of ambiguity and that design decisions—on equipment sizing, for example—made in anticipation of such policies are unlikely to be economically justified. However, the analysts suggest three guidelines for minimizing the costs associated with potential future constraints on CO_2 emissions:

- Building new plants with the most efficient technology that is economically viable
- Leaving space for future capture equipment, if possible
- Choosing plant sites while taking into account their proximity to carbon-storage repositories.

Cost Comparison of PC, IGCC, and NGCC Plants

Figures 7.4, 7.5, and 7.6 show estimated costs for new plants using several of the power generation options. These figures are based on an adaptation by researchers at the Princeton Environmental Institute (PEI)[20] of the analyses of power plants

---

[20]The PEI figures cited here correspond to a workbook that is available for download at cmi.princeton.edu/NRC_AEF_workbook.

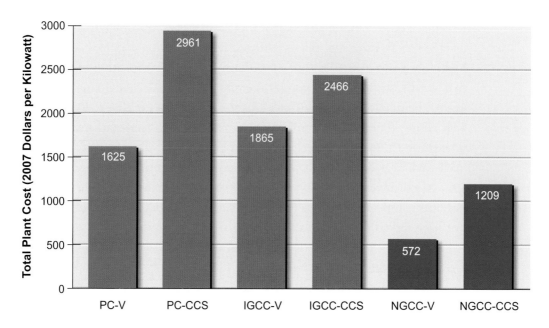

FIGURE 7.4 *Total estimated plant cost in 2007 dollars for three types of power plants—PC, IGCC, and NGCC—with and without CCS. These estimates, like those for other technologies, do not necessarily include all of the site-specific costs of building a plant nor all of the real-world contingencies that may be needed depending on economic conditions (see Box 7.2 for more discussion).*
Note: V refers to CO_2 vented.
Source: Princeton Environmental Institute.

done in the U.S. Department of Energy's National Energy Technology Laboratory (NETL) (NETL, 2007a).[21] The analysis describes the cost of building an "Nth plant," constructed after the large uncertainties in performance of plant components and systems have been resolved through experience. The cost of the Nth

---

[21]Capital costs for the components of natural gas and coal power plants in the NETL report were adopted by PEI without change, except for escalating capital costs from the end of 2006 to mid-2007. PEI then modified NETL results by using different financing assumptions (see Elecric Power Research Institute—Technical Assessment Guide [EPRI-TAG], 1993), using different operation and maintenance (O&M) assumptions, and including the CO_2 "overheads" (emissions upstream of power plants due to coal handling and transport, for example) reported in Argonne National Laboratory's GREET model (Argonne National Laboratory, 2008).

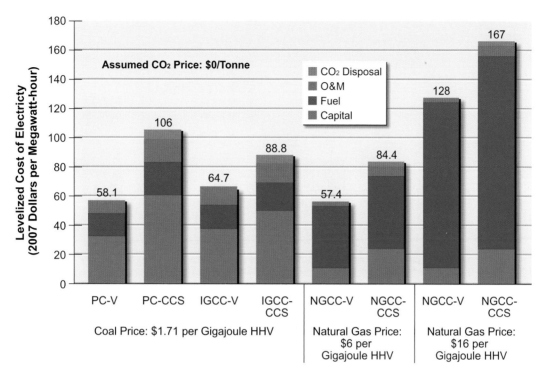

FIGURE 7.5 *Levelized cost of electricity (LCOE) estimated for various types of coal-fired and natural-gas-fired power plants at zero carbon price. These estimates, like those for other technologies, do not necessarily include all of the site-specific costs of building a plant nor all of the real-world contingencies that may be needed depending on economic conditions (see Box 7.2 for more discussion). The price of coal is fixed at $1.71/GJ, or $1.80/million Btu HHV (approximately equivalent to $50/tonne, depending on the energy content of the coal), but results for two natural gas prices are also shown ($6/GJ or $6.33/million Btu HHV, and $16/GJ, or $16.88/million Btu HHV) to illustrate how strongly the competitiveness of natural gas plants depends on fuel price. The cost shown for CO_2 disposal is estimated to be $6.30 per tonne CO_2 for PC-CCS and $6.80 per tonne CO_2 for IGCC-CCS and about $9 per tonne CO_2 for natural gas. See Annex 7.A for a discussion of variability and uncertainties in the cost of CO_2 disposal.*
Source: Princeton Environmental Institute.

plant is considerably less than the cost of any first-of-a-kind plant that could be built today. Typical values of N are between 5 and 10.[22]

---

[22]The estimation assumptions used here (exclusion of some contingencies that may be needed for first rather than Nth plants, contingencies that may be needed under some cost environments; various owner's costs; no additional escalation beyond 2007 dollars; and so on) may result in plant cost estimates in Figure 7.4 that are lower than other quoted cost estimates by 20 to 50 percent or more (see Box 7.2 for a more detailed discussion of possible reasons for differences

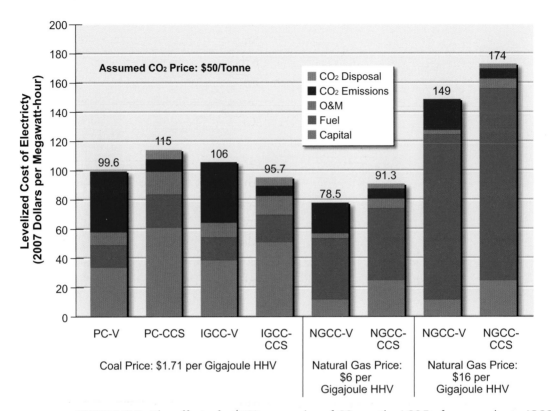

FIGURE 7.6 *The effect of a $50/tonne price of CO₂ on the LCOE of power plants. IGCC-CCS becomes the cheapest coal option, while the competitiveness of NGCC remains sensitive to fuel price. These estimates, like those for other technologies, do not necessarily include all of the site-specific costs of building a plant nor all of the real-world contingencies that may be needed depending on economic conditions (see Box 7.2 for more discussion).*
Source: Princeton Environmental Institute.

Although non-CO_2-capture PC and NGCC cases represent well-developed commercial technologies that have already reached the Nth plant level, the

---

among reported cost estimates). Midpoint LCOE estimates in Figure 7.5 may be low by comparison to costs based on the higher overall plant costs by 10 to 30 percent. (The percentages are different for the LCOE numbers, because (1) capital costs contribute only 80 percent or less to LCOE, and (2) a net 10 percent increase has already been added to LCOE as a result of the asymmetric range assigned to the uncertainty in the Nth plant and other assumptions.) It should be borne in mind when comparing the plant costs in Figure 7.4 to publicly quoted cost estimates for specific plants that the AEF figures do not assume any escalation above inflation during planning and construction. Other estimates often do so.

non-CO_2-capture IGCC plant costs are less mature. The post-combustion CO_2 removal technology of the PC and NGCC capture cases is immature as well. Capital and other cost estimates for these removal technologies have significant uncertainties associated with them and must be considered illustrative calculations rather than forecasts of commercial Nth-plant costs. The precombustion CO_2-removal technology for the IGCC capture case has a stronger commercial base.

Figure 7.4 shows capital costs (\$/kW) for three pairs of plants: supercritical PC, IGCC, and NGCC, each with and without CCS.[23] It is important to note that the capital cost figures presented here correspond to a specific set of assumptions used to obtain consistent comparisons across various technologies discussed in this report; the figures do not reflect all of the actual costs incurred in building a power plant today. Box 7.2, following, discusses many of the cost categories involved in estimating capital costs, including some that are not considered here.

Plant capacity in the reported estimates is taken to be 500–600 MW, all plants are assumed to have an 85 percent capacity factor, and the assumed percentage of capture is about 90 (see Annex 7.A for more discussion of the assumptions that underlie these estimates and those presented in Figures 7.4 and 7.5). Considering first only those plants without CCS, the capital cost of the NGCC plant (\$572/kW) is estimated to be about a third of the capital costs of the coal plants. The IGCC plant is estimated to be about 15 percent, or \$240/kW, more expensive than the supercritical PC plant. In this comparison, a single coal type is assumed—Illinois #6 bituminous coal. A relative increase in capital costs for the IGCC plant using lower-rank coals can be anticipated, thereby increasing the competitiveness of the PC. Overall, the uncertainty in cost estimates makes a 15 percent difference relatively insignificant.

The estimated incremental capital cost for CO_2 capture is about the same for natural gas and IGCC plants (\$600/kW) and less than half of the corresponding incremental cost for the PC plant.

Figure 7.5 shows the estimated levelized cost of electricity (LCOE, 2007\$/MWh) for the same three pairs as Figure 7.4, but for two assumed prices of natural gas. The LCOE takes into account the costs of capital, operation and

---

[23]Oxyfuel PC plants, which remain a viable option, were not modeled in NETL (2007a) and are also not modeled here.

BOX 7.2 *Comparison of Power Plant Capital Costs*

The final cost of a plant or other facility consists of several components. Costs that depend strongly on the specific site are typically not included in the estimates. Most of the others are related to the process or facility to be constructed on the site and thus are included.

A primary objective of the cost estimates developed by Princeton Environmental Institute (PEI) researchers and presented in this chapter, and in the chapter on alternative transportation fuels (Chapter 5), is to develop, with consistent assumptions, a set of cost estimates that can be compared for a range of different technologies for the conversion of coal and biomass to electric power and to liquid transportation fuels.

The first step is to estimate the total plant cost (TPC) for the Nth plant for each of the processes (technologies) under consideration. TPC, also referred to as the overnight cost, is the cost to construct the plant if it were put up "overnight." The TPC estimates are based on 2006 equipment quotes for all of the major pieces of equipment in the plant, as reported in NETL (2007a). These costs are first escalated to mid-2007 using the Chemical Engineering Plant Construction Cost Index. These base-cost estimates include cost of installation, materials, labor, some process and project contingencies, and balance of plant (BOP) costs. Where not included in the component equipment quotes, these costs are estimated from historical experience in the power industry. If not already included in the component quote numbers, a typical engineering contingency—ranging from 5 percent to 20 percent, depending on the component—was added. Summing these cost numbers provides a consistent set of TPC estimates.

PEI researchers calculated a levelized capital charge rate (LCCR) on installed capital using EPRI-TAG methodology (EPRI-TAG, 1993) assuming an owner's cost of 10 percent of TPC, a 55 percent:45 percent debt:equity split, real costs of debt and equity capital of 4.4 and 10.2 percent per year, and a 3-year (i.e., Nth plant) construction period. LCCR was 14.38 percent per year of the total plant investment (TPI), where TPI is the sum of TPC and the allowance for funds used during construction (AFDC), 7.16 percent of TPC. As a result, the cost of installed capital is 15.41 percent of TPC per year.

A number of cost components that could be significant under various economic environments and for early or first-of-a-kind (as opposed to "Nth") power plants are not included in the capital costs estimates, although some may be accounted for in computing levelized costs per kWh. These components may include the following:

- *Additional project contingency and risk.* Especially for relatively immature technologies, an additional total project cost contingency is often added to account for unforeseen or underestimated costs that could come up during construction of the facility. Some project contingency was included in the PEI estimates. However, the amount depends on the facility type, construction location, construction environment, and whether the contractor or utility bears the risks associated with cost increases. In the rising-construction-cost environment of the last several years, any quote to build would have a considerable project contingency included in order to protect from unforeseen construction-cost escalation. Twenty to 30 percent of TPC would not be unusual.

- *Profit for the engineering construction company.* This amount strongly depends on the construction-cost environment, the type of construction contract, and the construction firm. In a tight construction environment, the profit included in an estimate is minimized in order to secure work flow and maintain engineering and construction management capabilities in-house; in an expansive environment, beyond that assumed in the PEI estimates, profit margins can be considerable, some 5–10 percent of the job cost.

In addition to these project costs that may be appropriate in various cost environments, there are different accounting protocols for stating capital costs that may lead to apparent differences in plant cost estimates.

- *Interest during construction and minimal owner's costs as reflected in total capital requirement (TCR).* TCR is TPC, plus an AFDC, plus specific "owner's costs." Some observers report a TCR (see, for example, Booras, 2008), rather than TPC or TPI, because it is closer to what is reported to public utility commissions in project submissions. The experience is that for PC units the TCR/TPC multiplier is about 1.20, of which about 7 percent is additional owner's costs not included in TPC, and up to 16 percent is interest during construction (based on 4 years for construction and a high interest rate). This suggests that approximately 14 percent should be added to the PEI estimated value for TPI to be equivalent to the TCR number reported by Booras. Note that the PEI estimates account for moderate interest during 3 years of construction and some owner's costs (10 percent).
- *Owner's costs that are not included in TCR.* These costs can include, for example, those of dock or rail facilities, transmission lines, and transformer stations. They typically add another 15 percent to TCR.

A quote to a utility may have any of these estimated costs included in capital cost estimates, and they may not be separable in a simple form. Two examples illustrate the various estimated costs.

Example 1. Suppose a commercial quote for a PC-vent plant reports a TPC of $1630/kW.

- *Total capital required.* To obtain TCR from TPC we multiply by 1.20. TCR is $1630 × 1.20 = $1956/kW.
- *Contingency.* The HIS-CERA downstream construction index increased from 130 to 178, or 37 percent, from the beginning of 2007 to the beginning of 2008. By contrast, the Chemical Engineering Construction Cost Index, used in the

continued

BOX 7.2 *Continued*

estimates reported here, grew 8.6 percent over this time period. In light of such rapid change in the construction-cost environment, which is at least partly driven by construction company bids, any forward construction quote will have a significantly higher than usual contingency to protect the contractor from future cost escalation. Assume 25 percent: $1956 × 0.25 = $489/kW.

- *Profit.* Assume 10 percent of TCR: $1956 × 0.10 = $196/kW.
- *Other owner's costs.* Assume 15 percent: $1956 × 0.15 = $293/kW.

As a result, the total project bid to the utility could be $1956 + $489 + $196 + $293 = $2930/kW, almost 80 percent higher than the TPC for the same plant. This estimate of the bid to the utility is roughly consistent with recent quotes for PC plants with vented CO_2:

- AEP, Hempstead, Arizona, 600 MW SCPC/PRB, December 2006; $2800/kW
- Duke, Cliffside, North Carolina, 800 MW SCPC/PRB, March 2007; $3000/kW
- Sunflower, Holcomb, Kansas, 1400 kW, September 2007; $2575/kW
- American Municipal Power, Meigs Co., Ohio 1000 MW SCPC/Bit, October 2005; $2900/kW.

Example 2.[1] As discussed herein, it is important to note that TPC is invariably significantly lower than TCE, the total cash expended (in mixed-year dollars) at commissioning, which is the quantity commonly reported in the press. This is particularly true in periods of rapidly escalating costs. For example, NETL researchers have demonstrated that NETL's TPC of $1812/kW for a 640 MW GE radiant-quench coal IGCC with CO_2 venting (case IGCC-V here) is entirely consistent with the TCE of $3150/kW for the almost identical IGCC approved by the Indiana Utility Regulatory Commission on November 20, 2007, for Duke Energy's proposed Edwardsport facility. Starting with the base plant TPC of $1812/kW and simply adding equipment specific to Edwardsport (a selective catalytic reduction unit, an extra rail spur, and a transmission line, which raise the owner's cost from 10 to 15 percent of TPC); changing from merit shop to Midwest union labor rates (increasing labor costs by 40 percent); assuming a 4 percent annual escalation rate (specified in the project's front-end engineering and design study) but no additional capital cost escalation; and using a weighted nominal interest rate for calculating AFDC (before-tax weighted cost of capital, adding 3 percent per year on equity return as an owner's risk premium—essentially the same interest rate assumed by PEI) of 11.5 percent per year, with unequal annual spending over the 4-year construction period (10 percent, 40 percent, 30 percent, 20 percent)—the calculated project TCE is $3200/kW.

[1]Example 2 from J. Wimer, NETL, personal communication, June 2009.

maintenance, fuel, and CO_2 disposal.[24] Estimated CO_2 disposal costs are $6–7 per tonne CO_2 for the PC-CCS and the IGCC-CCS but about $9 per tonne CO_2 for NGCC-CCS, reflecting the smaller CO_2 output. In Figure 7.5, CO_2 emissions are cost free, but for the CCS plants, the carbon-capture technology is installed anyway. Given the uncertainties in the various estimates, small differences in LCOE should not be considered significant.

Concentrating first on the coal plants, the PC without CCS is about $7/MWh cheaper than the IGCC without CCS. By contrast, the PC with CCS is about $17/MWh more expensive than the IGCC with CCS. The order of LCOE for the four kinds of coal plants (from least to most expensive) in Figure 7.5 is also found in many other studies: PC without CCS, IGCC without CCS, IGCC with CCS, and PC with CCS. Under the committee's assumptions, when there is a high CO_2 price, IGCC with CCS has the lowest cost, but when there is a zero CO_2 price, as in Figure 7.5, PC without CCS has the lowest cost. Because all of these technologies use coal, albeit with different efficiencies, differences in their power costs are relatively insensitive to the cost of coal—here $1.71/GJ, or 1.80/million Btu—but they may be affected by the type of coal.

Where NGCC fits into these comparisons is a sensitive function of the price of natural gas. In Figure 7.4, the lower of the committee's two assumed prices is $6.00/GJ, or $6.33/million Btu. At this price—the one used in the NETL (2007a) study—the values of LCOE for the NGCC and PC plants are about the same as for non-CCS plants, whereas for CCS plants, the LCOE of the NGCC and IGCC plants are about the same. However, for natural gas at $16.00/GJ, or $16.88/million Btu,[25] the LCOE for every NGCC plant is much higher than that of any coal plant.

In Figure 7.5, the estimated increase in LCOE for CO_2 capture and storage

---

[24]The CO_2 disposal cost is the sum of the costs of a 100-km pipeline, wells of 2 km depth, and aboveground infrastructure at the injection site. A pipeline of this length dominates the total cost. For example, the IGCC-CCS plant emits at a peak rate of 11,200 tonnes of CO_2 per day. As a result, assuming an 85 percent capacity factor, a single dedicated pipeline carries 3.49 million tonnes CO_2 per year and costs about $780,000/km ($1.3 million per mile), contributing $4.29 per tonne CO_2 (or 65 percent) to the disposal cost (at a capital charge rate, including O&M, of 19 percent per year). An assumed maximum injection rate per well of 2,500 tonnes CO_2 per day leads to a requirement of five wells, which cost $5.8 million each and together account for $1.59 per tonne CO_2 (or 24 percent) of the disposal cost. The remainder of the $6.65 per tonne CO_2 disposal costs is associated with the disposal-site infrastructure. Monitoring costs are not included.

[25]$16/GJ was chosen to represent a reasonable upper bound on the price of natural gas.

(at the lower natural gas price) is $24/MWh for IGCC and $27/MWh for NGCC, but the increase is $48/MWh for the PC. However, because half as much CO_2 is produced per kilowatt-hour from a natural gas plant as from a coal plant, the incremental cost *per tonne of* CO_2 *captured and stored* is similar for the NGCC (at the lower gas price) and for the PC plant. A rising CO_2 price would therefore stimulate the introduction of CCS first at IGCC plants.

Figure 7.6 shows how a CO_2 price of $50 per tonne affects the estimated LCOE of the various power plants. A new segment (CO_2 emissions) is added to the bars in Figure 7.5 for this contribution to the cost, and relative costs are affected. IGCC with CCS is now the lowest-cost coal option. In other words, in this example a $50 per tonne CO_2 price is high enough to make a new IGCC plant with CCS more attractive than a new plant (either PC or IGCC) that vents its CO_2. The higher CO_2 price improves the competitiveness of NGCC relative to coal-fired plants, but the position of NGCC remains sensitive to fuel price. At $6.00/GJ, or $6.33/million Btu, for natural gas, a new NGCC plant with or without CCS is cheaper than any coal option is, but at $16/GJ, or $16.88/million Btu, for natural gas, NGCC becomes the most expensive option.

The costs presented in Figure 7.6 are in 2007 dollars and do not include further escalation.[26] The LCOEs for coal-based plants (PC and IGCC) are more sensitive to changes in construction costs than are NGCC plants; for example, a 10 percent change would alter PC/IGCC LCOE by 7–8 percent, but NGCC LCOE by only 2–3 percent. (The reverse is true for fuel prices, which are discussed elsewhere.) Although opinions differ about future trends of construction costs, these sharp differences in sensitivity seriously affect investment decisions.

A full model of competition among coal, natural gas, and other alternatives would need to allow for changes in the capital, operating, and fuel costs relative to those assumed in Figures 7.5 and 7.6. In particular, the capital and operating costs for the Nth plant with CCS shown in these figures would likely be higher than for the 10 × Nth plant (the 100th plant, if N is 10), to the extent that these costs fall as a result of experience and R&D.

Given the fact that there is no obvious choice of the best technology at this point, a portfolio of demonstrations will be required to investigate cost and performance variations as functions of coal types, separation technologies,

---

[26]Beyond 2007 (e.g., during construction), it is assumed that capital costs increase at a rate equal to that of economy-wide inflation. Thus, neither TPC nor TPI (see Box 7.2)—expressed in 2007 dollars—depend on the plant commissioning date.

approaches to combustion, and geologic storage settings (see the section titled "Geologic Storage of CO_2," below in this chapter). The cost range of these demonstrations will be significant, but it should be considered an investment in defining more accurately the alternative pathways for reducing CO_2 emissions from electric power generation with coal and natural gas.

Sensitivity of LCOE to Capital Cost and Coal Price

The sensitivity of LCOE to estimated capital cost is illustrated in Table 7.15 for carbon costs of $0, $50, and $100 per tonne of CO_2. The table compares the impacts on LCOE from three different scenarios: capital costs equal to, 20 percent below, and 30 percent above the estimates used to construct Figures 7.4–7.6. Because the fraction of LCOE attributable to capital cost varies with the fuel and plant types, the percentage change in LCOE also varies. Table 7.15 shows that the cost of electricity is more sensitive to capital cost variations in coal plants than in natural gas plants, for which fuel cost accounts for a greater fraction of LCOE. It should also be noted that the critical threshold on cost of CO_2 emissions, above which it is less expensive to install CCS than to emit the CO_2 and pay the associated cost, shifts as the capital cost changes (see Annex 7.A for additional discussion). Until more experience is gained in construction and operation of the various plant types with CCS, however, significant uncertainties in costs are likely to remain.

Table 7.16 shows the sensitivity of LCOE to doubling the price of coal, analogous to Table 7.15. LCOE is considerably less sensitive to coal price than to capital costs; the effect of doubling the price of coal on LCOE is comparable to that of increasing the capital cost by 30 percent.

The Competitiveness of Natural Gas

When capital costs rise and are uncertain, utilities historically choose to build natural gas plants (NGCC), which can be erected quickly (3–4 years versus the more typical 4–8 years for coal-fired power plants)[27] and require a much smaller commitment of capital. However, the competitiveness of natural gas power is vulnerable to a rising gas price, which can make this power too expensive to dispatch. Significant expansion of the use of natural gas for electric power generation in the

---

[27]Construction time depends on permitting and other site-specific issues.

TABLE 7.15 Effect of Changes in Plant Capital Cost on LCOE

| | At −20 Capital Cost | | | Estimate Original Capital Cost LCOE in \$/MWh | | | At +30 Capital Cost | | |
	LCOE in \$/MWh (% change)						LCOE in \$/MWh (% change)		
CO$_2$ Cost (2007\$/tonne)	0	50	100	0	50	100	0	50	100
Plant Type									
PC-V	50 (−15)	91 (−9)	133 (−6)	58	100	141	71 (+22)	112 (+13)	154 (+9.0)
PC-CCS	90 (−16)	98 (−15)	107 (−14)	106	115	123	131 (+24)	140 (+22)	148 (+20)
IGCC-V	55 (−15)	97 (−9)	138 (−7)	65	106	148	79 (+23)	121 (+14)	163 (+10)
IGCC-CCS	75 (−16)	82 (−15)	89 (−14)	89	96	103	110 (+24)	117 (+22)	124 (+20)
NGCC-V	54 (−5)	75 (−4)	97 (−3)	57	78	99	62 (+8)	83 (+6)	104 (+4)
NGCC-CCS	77 (−8)	84 (−8)	91 (−7)	84	91	98	95 (+12)	102 (+11)	109 (+11)

Note: Cost of natural gas = \$6/GJ, or \$6.33/million Btu.

United States may imply larger international trade in natural gas and the need for additional facilities for handling LNG in the quantities required.

Natural gas plants will be affected by a price on CO$_2$ emissions, even though only about half as much CO$_2$ is produced per kilowatt-hour when the power is produced from natural gas instead of coal. The options for CO$_2$ capture at an NGCC plant parallel those for coal plants. CO$_2$ is available for capture in the stack after burning the natural gas in air, although at a much more dilute concentration (at today's power plants, typically 3–5 percent instead of 12–15 percent). Alternatively, CO$_2$ can be captured prior to combustion after processing the natural gas with steam to produce CO$_2$ and H$_2$.

TABLE 7.16 Effect on LCOE of Doubling the Price of Coal at Various Types of Coal Plants

	$1.80 per Million Btu Coal Price (~$45/tonne)			$3.60 per Million Btu Coal Price (~$90/tonne)		
	LCOE in $/MWh			LCOE in $/MWh (% change)		
CO$_2$ Cost (2007 $/tonne)	0	50	100	0	50	100
Plant Type						
PC-V	58	100	141	74 (+27)	115 (+16)	157 (+11)
PC-CCS	106	115	123	129 (+21)	137 (+20)	146 (+18)
IGCC-V	65	106	148	81 (+25)	122 (+15)	164 (+11)
IGCC-CCS	89	96	103	108 (+21)	115 (+20)	122 (+18)

Future Coal Power

The levelized costs of electricity (2007$/MWh) for coal power plants, as shown in Figures 7.5 and 7.6, provide a starting point for estimating the contribution of fossil-fuel power plants to future U.S. electric power generation. Also needed are assumptions about retirement rates of existing plants and maximum possible build rates of new plants. The discussion that follows is based on the assumptions that advanced coal technologies with CCS technologies are developed successfully and deployed at a rate that the committee judges to be "aggressive but achievable"—that is, in line with maximum historical deployment rates. Note, however, that these assumptions are made in order to identify some key issues; they should not be taken as predictions of future technology deployment.

The rate of retirement of the 300 GW of existing U.S. coal plants, today emitting 2 billion tonnes of CO$_2$ per year, is a key variable. Only 4 GW of these 300 are expected to be retired by 2030, according to a reference scenario of the EIA. This scenario assumes both that the United States has not adopted any carbon policy and, effectively, that the price of CO$_2$ emissions is zero (EIA, 2008a). Little work has been done on how the retirement of coal plants will depend on the price of future CO$_2$ emissions. But at some price level for any given plant, it becomes economical to modify the plant to capture CO$_2$, to close the plant, or in some cases, to operate it only during periods of very high demand. As discussed above, the first alternative comes in two versions: retrofit (modest modification, retaining much of the original steam-cycle equipment) and repowering (replacing the existing power plant with either a new PC or an IGCC, retaining only site amenities such as coal yards, rail lines, and power lines).

A separate matter is the rate at which new coal-fired power plants that capture CO_2 can come on line, along with the associated rates at which CO_2 pipelines and storage facilities can be brought into operation. The retirement rate, after all, could exceed or be less than the rate of commissioning of new coal power with CCS. An economic model of U.S. power developed by the Electric Power Research Institute (EPRI, 2007) shows a response to U.S. CO_2 policy in which the existing coal fleet is retired between 2010 and 2040, while new coal power generation with CCS is brought on line at a nearly comparable rate. Between 2020 and 2040, about 15 GW per year of baseload coal power with CCS is introduced. This is about one-third faster than the rate at which baseload coal power was introduced in the United States in the 1970s, a peak period of construction of U.S. coal power.[28]

The committee's view of the maximum pace of introduction of new power plants with CCS, assuming a strong policy driver, is presented here. No distinction is made between PC and IGCC, as their projected cost differences are small compared to the uncertainties. The assumption is that the cost of natural gas is comparable to its higher value in Figures 7.4 and 7.5 ($16/GJ, or $16.88/million Btu), so that natural gas is not a competitor. The committee found that:

1. A demonstration period lasting until 2020 will be required to instill confidence in various capture and storage technologies and to develop state and federal policy governing CO_2 storage belowground. Such a learning process could be postponed to a future decade, but postponement would not significantly reduce its cost—the required learning requires full-scale demonstration plants. By 2020, about 10 GW of coal power with CCS would be operating, much of it still in the form of demonstration plants. In that year, operating at 75 percent capacity factor these plants would be producing about 60 TWh of electricity and,

[28]The EPRI analysis also explores a world in which CO_2 emissions are expensive but nonetheless one in which CCS never becomes commercialized. Natural gas is expensive, but nuclear and renewable energy remain more expensive than natural gas. In that world, total coal power nearly disappears by 2040 and natural gas, in spite of its high cost, takes up most of the shortfall in coal-fired production.

at a 90 percent capture rate, about 50 million tonnes CO_2 would be stored belowground.[29]

2. During the period between 2020 and 2025, about 5 GW of new capacity could be commissioned per year. At 85 percent capacity factor, the additional 25 GW operating in 2025 would be producing 190 TWh per year and sending 150 million tonnes CO_2 per year belowground.

3. During the years from 2025 to 2035, bracketing the maximum installation rate between 10 and 20 GW of new capacity brought on line per year seems aggressive but achievable. (A typical coal plant with CCS will have a capacity of about 500 MW, so the bracket corresponds to 20–40 new plants per year.) In 2035, the amount of coal power with CCS would reach either 135 GW or nearly twice that value—235 GW. Assuming an 85 percent capacity factor for all of these plants, output power in 2035 would be between 1000 and 1750 TWh; the higher value is nearly as large as the total electric power from coal in the United States today. Also for the higher value, about 1.5 billion tonnes of CO_2 per year would be captured and stored in 2035. This rate of construction of new coal-CCS plants might continue from 2035 to 2050.

4. In parallel with the construction of new coal-CCS plants, there will be coal plant retirements and there may also be retrofit and repowering of older coal plants. New coal plants with CCS and retrofit or repowering for CCS will compete for the same specialized labor, equipment, and belowground storage space. The committee's view of the upper limit of the retrofit plus repowering rate for coal plants is 10 GW per year from 2020 to 2035, so that by 2035, approximately half of today's coal plants (150 GW out of 300 GW) could be retrofitted or repowered with CCS. However, the committee believes that the maximum combined rate of construction of new and retrofitted/repowered plants would still be 20 GW per year.

At what CO_2 price will it be economical to replace an existing coal plant by repowering, for instance, replacing an old coal plant with an IGCC plant with CCS? Some perspective on this question can be gained from Figure 7.6. At $50 per tonne CO_2, the cost of power from a new IGCC with CCS is slightly

---

[29]Additional inputs producing these values are 35 percent power plant efficiency (HHV), 29 GJ (HHV) per tonne of coal, and 70 percent carbon content of coal by weight.

less than the cost of power from a new PC plant without CCS: $96/MWh versus $100/MWh. The cost of power from an old PC plant is likely to be less than from a new PC plant, primarily because capital costs have risen and because most of the capital costs of an old plant will have been written off. (Some cost of capital for the old plant will remain, however, if there have been recent retrofits—for example, to address conventional air emissions). In addition, depending on the degree to which CCS and IGCC technology have proven reliable and on what fraction of the design and development costs have been written off or amortized by the vendors, early users will pay an additional premium.

Acting in the opposite direction, making old-plant power more expensive, is the plant-efficiency comparison: the old coal plant will almost surely be less efficient. But assuming that capital issues dominate fuel issues, the break-even CO_2 price for the replacement of an old PC plant with a new IGCC-CCS plant is likely to be greater than $50 per tonne CO_2. But in some cases, CCS retrofit strategies well short of complete plant replacement may be more competitive than *either* keeping the old PC plant running without CCS investment or repowering it entirely. Complicating the economics of some of these decisions may be the triggering of provisions of the Clean Air Act (New Source Performance Standards), which require the addition of pollutant controls on non-CO_2 pollutants once a significant upgrade of a grandfathered facility takes place.

The main message here is that if CCS is commercialized by 2020, its role in the U.S. power mix could be expanded over the succeeding two or three decades, with the installed capacity of coal plants with CCS becoming comparable to that of current U.S. coal power, if not considerably larger. CO_2 storage capacity is probably adequate (see below the section titled "Geologic Storage of CO_2") for such a large deployment. As a result, the U.S. coal mining and electric power industries could remain at their current sizes or even grow throughout the next half century.

It must be noted that the prior estimates are upper limits. Coal demand associated with dedicated power plants with CCS would not reach these values if end-use efficiency were incorporated aggressively into the U.S. electric system (as explored in Chapter 4) and if the competitors for low-CO_2 power—including renewable and nuclear power—prosper as well. On the other hand, additional coal production would result from demands for synthetic fuels and synthetic natural gas (which might be produced with associated electric power production). Also, demand for both electricity and coal + CCS could increase as the result of a shift to electric-power-based transportation.

Finally, if domestic natural gas (e.g., from shale gas deposits) proves plentiful, and confidence grows that prices will remain in the range of $7–9/million Btu or lower for decades, as some commentators think may happen (CERA, 2009), then NGCC plants with CCS could compete economically with PC and IGCC plants with CCS. In such a world, the cheapest way to gain large CO_2 reductions would be to use NGCC + CCS to replace over time existing and future coal units.

Although a large shift in this direction would increase natural gas demand significantly and put upward pressure on prices, the committee still considers it wise to plan for a broad range of future natural gas prices and domestic availabilities. Consequently, the committee envisions some CCS projects involving NGCC technology being part of the recommended 10 GW of CCS demonstrations. The committee has not made a judgment about the mix of PC, IGCC, and NGCC plants with CCS that would be appropriate.

Future Biomass Power

Biomass (plant matter) carries stored energy content, retrievable for use by means of oxidation (burning), just as with fossil fuels. Photosynthesis is the source of the stored energy and carbon in both cases, but for fossil fuels the storage occurred millions of years ago. The committee considers here the prospect of large biomass power plants and, as for fossil-fuel power plants discussed earlier, we consider plants with and without CCS. Note that biomass plants with CCS scrub the atmosphere of CO_2, as the CO_2 sent underground was removed from the atmosphere during photosynthesis only a short while before (years in the case of trees, days to months in the cases of grasses, crops, or crop wastes).

Here the PEI group's analysis is used to extend the committe's investigations to biomass power plants, applying the same energy accounting principles that were used for the discussions above of coal and natural gas plants together with the same biomass assumptions made in Chapter 5. The committee considers stand-alone biopower plants as well as power plants in which biomass and coal jointly contribute to producing electric power. The notation is as follows: a BTP-V plant is a biomass-to-power plant with venting of CO_2; a CBTP-CCS plant is a coal-plus-biomass-to-power plant with CO_2 capture and storage, and so on. The committee considers BTP plants first and then, briefly, CBTP plants.

The feedstock in these examples is switchgrass, which when consumed at the power plant is assumed to have 15 percent moisture content. Biomass-fueled power plants are assumed to burn at most 1 million dry tons of biomass per year (the same assumption used in Chapter 5 for biomass-to-liquid plants), on the grounds that the collection logistics for a larger plant would be too complex and therefore too costly. The power plants have a capacity factor of 85 percent, as with the fossil-fuel plants discussed above. When the plants are working at full capacity, 3800 tons of "as received" biomass arrive each day. These plants are large—the switchgrass input power (higher heating value) is 700 thermal megawatts—but not as large as the fossil-fuel plants considered above, whose thermal input power ranges from 1000 to 1900 thermal megawatts.

The cost estimates are based on a model of an oxygen-blown-gasification-based combined-cycle power plant, both in its BTP-V and BTP-CCS configurations, with switchgrass as its fuel. The BTP-V plant has a thermal efficiency (higher heating value) of 42.6 percent and a peak power output of 298 MW. The corresponding BTP-CCS plant pays for its CO_2-capture feature with a reduced efficiency of 36.2 percent and a peak output of 253 MW. These plants have roughly half the power output of the coal and natural gas plants considered in the earlier discussions. Their efficiencies are about 4 percentage points higher than the efficiencies for the corresponding coal IGCC plants.

The results are displayed in Figures 7.7, 7.8, and 7.9, which are modifications of the Figures 7.4, 7.5, and 7.6 presented earlier, with the sole change being the addition of BTP data. Figure 7.7 shows the capital costs of the BTP-V and the BTP-CCS power plants. On a dollars-per-kilowatt basis (total plant costs divided by peak output), the BTP-CCS plant is 43 percent more expensive than the BTP-V plant is, $2529/kW versus $1768/kW. The power plants with biomass feedstock have approximately the same capital costs as the corresponding coal plants. A BTP plant would be less costly than a coal gasification plant of the same size would be because less oxygen is needed for biomass gasification (biomass, already containing oxygen, is much more reactive than is coal), because the biomass gasifier operates at a lower pressure and therefore requires less power for O_2 compression, and because sulfur removal is assumed not to be required for biomass gasification. However, the biopower plant here is smaller, thus losing out on economies of scale.

Carbon Issues

The carbon balance for these biomass power plants is interesting. Because switchgrass "as received" (i.e., with 15 percent by weight moisture content) is 40 percent

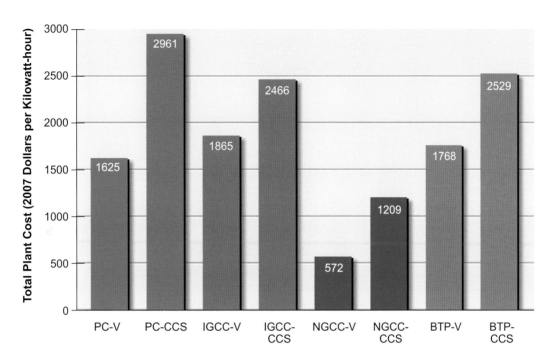

FIGURE 7.7 *Augmented version of Figure 7.4, with the addition of biomass power plants. These estimates, like those for other technologies, do not necessarily include all of the site-specific costs of building a plant nor all of the real-world contingencies that may be needed depending on economic conditions (see Box 7.2 for more discussion). Source: Princeton Environmental Institute.*

carbon by weight, 470,000 tonnes of carbon are in the annual input of switchgrass, as photosynthesis and respiration in the switchgrass has removed a net of 1.72 million tonnes CO_2 per year from the atmosphere.[30] As part of the gasification process, roughly 10 percent of the carbon in the switchgrass is assumed to be trapped in char—a solid waste product—assumed to remain unoxidized forever in a landfill. The carbon not in the char is oxidized to CO_2 and vented to the atmosphere whence it came, an annual CO_2 emission of 1.55 million tonnes CO_2 per year.

A full carbon accounting considers two "upstream" issues: fossil-carbon inputs and induced carbon storage in soil and roots. As discussed in Chapter 5, upstream fossil-carbon inputs resulting in additions of CO_2 to the atmosphere

---

[30]The net CO_2 removed from the atmosphere is 44/12 times the carbon fixed in the biomass because the ratio of the molecular weight of CO_2 to the atomic weight of carbon is 44:12.

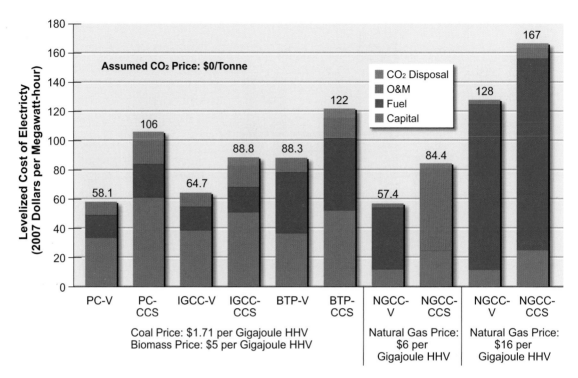

FIGURE 7.8 *Augmented version of Figure 7.5, with the addition of biomass power plants. These estimates, like those for other technologies, do not necessarily include all of the site-specific costs of building a plant nor all of the real-world contingencies that may be needed depending on economic conditions (see Box 7.2 for more discussion). Source: Princeton Environmental Institute.*

include direct inputs associated with tractor fuel and fertilizer, for example, as well as indirect inputs linked, say, to compensatory land clearing. To obtain *net* carbon inputs, these direct and indirect inputs are balanced against any buildup of carbon in soil and roots, which remove CO_2 from the atmosphere.

Values for switchgrass were taken from Argonne National Laboratory's GREET model, Version 1.8 (Argonne National Laboratory, 2008), where fossil-carbon inputs equal 10 percent of the carbon in the switchgrass and are thereby exactly balanced by char storage. GREET also assumes that carbon buildup in soil and roots is one-thirtieth of carbon acquisition from the atmosphere by plants. Thus, the net carbon flow for the BTP-V is, by a small amount, *out of* the atmosphere, corresponding to the buildup of carbon in soil and roots. CO_2 emissions associated with land clearing at the site and associated land clearing elsewhere are not considered.

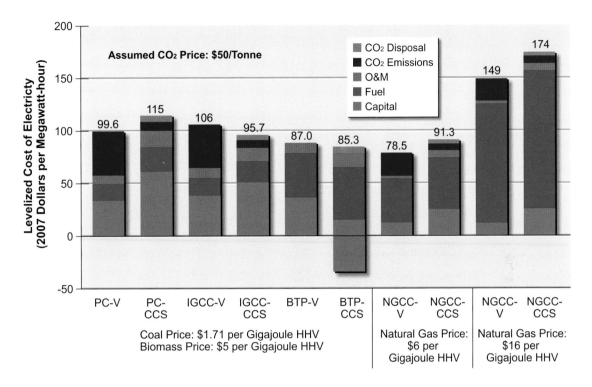

FIGURE 7.9 *Augmented version of Figure 7.6, with the addition of biomass power plants. In the case of BTP-CCS, in which CO_2 emissions costs are* negative *(–$37/MWh), the entire bar of positive costs has been accordingly lowered below zero. In that way, the top of the bar represents its* net *LCOE of $85/MWh and is thus directly comparable to the other cases. These estimates, like those for other technologies, do not necessarily include all of the site-specific costs of building a plant nor all of the real-world contingencies that may be needed depending on economic conditions (see Box 7.2 for more discussion).*
Source: Princeton Environmental Institute.

The BTP-CCS plant is designed to capture 86 percent of the 1.55 million tonnes CO_2 per year that the BTP-V plant vents and to then send it to storage. In other words, 1.33 million tonnes of CO_2 is permanently removed from the atmosphere by switchgrass photosynthesis. Taking into account the storage of captured CO_2, char storage, storage in soil and roots, and fossil-carbon inputs, the BTP-CCS plant actually removes 1.39 million tonnes CO_2 per year from the atmosphere. A 253 MW plant operating with an 85 percent power capacity produces 1.88 TWh per year, which can be restated as an intensity of CO_2 *removal* from the atmosphere: 740 g CO_2 per kWh. By contrast, the coal IGCC-V plant discussed in the earlier section *emits* about 830 g CO_2 per kWh. The BTP-CCS plant, installed kilowatt for installed kilowatt, nearly offsets the CO_2 emissions of an unvented new coal plant.

Levelized Costs

Figures 7.8 and 7.9 display the levelized cost of energy for these BTP plants as new vertical bars added to Figures 7.5 and 7.6, respectively, presented earlier. The assumed switchgrass cost is $5/GJ, or $5.27/million Btu (higher heating value), which is the same as $80 per ton of switchgrass "as received." Because the switchgrass energy cost is only slightly less than the lower of the two energy costs of natural gas ($6/GJ, or $6.33/million Btu), while the thermal efficiencies of the BTP plants are somewhat lower than the corresponding thermal efficiencies of the natural gas plants, the fuel contributions to the BTP plants, as shown both in Figures 7.8 and 7.9, are almost exactly the same as the fuel contributions to the $6/GJ NGCC plants. The higher capital costs for BTP relative to NGCC make biopower more expensive than $6/GJ natural gas power. Indeed, biopower is the most expensive of the alternatives—except for power with the higher price of natural gas in the vent cases and in the CCS cases of Figure 7.8, where the CO_2 emissions price is zero.

By contrast, in Figure 7.9, where the CO_2 emissions price is $50 per tonne CO_2, the two BTP plants have become competitive with the NGCC-V plant with low-priced natural gas, and they are less costly than the coal plants. The absence of a visible CO_2 emissions component to the BTP-V bar reflects its nearly carbon-neutral character, discussed previously. Comparing the BTP-CCS and the BTP-V plants, one sees that the extra capital cost of the BTP-CCS plant and its lower efficiency are approximately canceled by the $36.9/MWh credit for its removal of CO_2 from the atmosphere (shown as a bar segment extending below zero).

Total Contribution with 100 Megatons per Year of Biomass

Chapter 5 suggests that the total U.S. production rate for biomass could grow to 550 million dry tons (500 million dry tonnes) per year. Suppose that 100 million dry tons per year is used for power. Then 100 of our BTP power plants could be deployed (or a larger number of smaller plants, though on the principle of economies of scale, they would have higher costs). With 100 1-million-ton-per-year BTP-V plants, the total contribution to U.S. baseload power capacity would then be 30 GW, which is 10 percent of current coal baseload power capacity. Alternatively, assuming that these BTP power plants are BTP-CCS plants, the contribution would be 25 GW, but they would remove 140 million tonnes CO_2 per year from the atmosphere, which is about 7 percent of the CO_2 emission rate from today's coal power plants. These relatively small additions to U.S. electricity and reductions in U.S. CO_2 emissions could be multiplied by a factor of 5.5 if *all* of the 550 million dry tons of biomass per year were devoted to electric power, but then no biomass would be available for liquid fuel production.

Co-firing Biomass and Coal

In Chapter 5, synthetic fuels plants using a feedstock that is a combination of coal and biomass are considered at length (CBTL plants, where "L" stands for "liquids"). Here the committee considers the corresponding power plants powered by coal and biomass, which are denoted as either CBTP-V or CBTP-CCS. A CBTP-V plant will have lower net CO_2 emissions as its biomass fraction increases, and a CBTP-CCS plant with sufficient biomass fraction can have net negative emissions.

Some of the facilities considered in Chapter 5 produce substantial amounts both of synthetic liquids and electricity, with electricity regarded as a by-product. One result reported there is that coal-plus-biomass plants that capture CO_2 can provide competitive power and competitive fuel in a world where the oil price and the CO_2 emissions price are both high. In such a world, these CBT(P+L)-CCS plants may turn out to be strong competitors as providers of new U.S. power capacity.

From the societal perspective, the committee has identified four basic options for biomass utilization: BTP, CBTP, BTL, and CBTL, each with and without CCS. Sorting out their respective roles in a socioeconomic environment of high oil prices and strong constraints on CO_2 emissions is complicated. For example, transport may be powered by biomass fuels or by batteries charged from biomass-based electricity. A host of technological innovations still to come will determine relative roles in such competitions.

Supply Curves and Power Plant Mixes in 2020 and 2035[31]

"Supply curves," such as the ones shown in Figures 7.10 and 7.11, are ways of ordering different technologies by estimated cost while simultaneously showing

---

[31]This section does not consider the biomass-to-power plants discussed in the previous section.

estimates of the contribution to supply that each technology might make at various levels of demand. Because both costs and supply contributions are uncertain, these types of supply curves should not be viewed as forecasts but rather as illustrations of possible futures. It is then possible to examine the sensitivity of results of interest (such as carbon emissions) to parameters of interest (such as carbon and fuel price).

Note that the consumption of fossil electricity is determined by the intersection of the supply curve with an upper-left to lower-right demand curve (not shown). The demand curve is influenced by many factors, including the rate of improvement of energy efficiency.

Figure 7.10 shows baseload power supply curves for 2020 for two CO_2 emissions prices—$0 per tonne CO_2 and $50 per tonne CO_2—to illustrate the effect of the CO_2 price on the coal/natural gas competition among existing plants. The effect of the CO_2 price is to raise the cost of power from coal twice as much as the cost of power from natural gas. LCOEs in the absence of a CO_2 price are illustrative: 4¢/kWh for an existing coal plant and 8¢/kWh for an existing natural gas plant (a cost of natural gas power consistent only with a high natural gas price).[32] With these assumptions, a $50 per tonne CO_2 price is not sufficient to make natural gas power the less costly alternative. In fact, the two alternatives equalize at about $100 per tonne CO_2.

Figure 7.11 shows baseload power supply curves for 2035 for three CO_2 emissions prices: $0 per tonne CO_2, $50 per tonne CO_2, and $100 per tonne CO_2. This figure shows an expanded competition relative to Figure 7.9. In addition to existing coal and natural gas plants, there are new coal plants and new natural gas plants, with and without CCS—in all, a six-way competition. (Retrofits of existing coal plants, which would raise the cost of their power, are not included.) Maximum penetration rates for new plants are assumed, as discussed in the section below titled "Future Coal Power"; the cost of power from new plants is approximately that shown in Figures 7.5 and 7.6; and the same high natural gas price as in Figure 7.10 is assumed. The important result shown in Figure 7.11 is that the ordering of alternatives is affected by the CO_2 price. At $0 per tonne CO_2 and $50 per tonne CO_2, existing coal plants produce the least costly power, and the competition for second place is between new coal plants without CCS and existing natural gas plants. At $100 per tonne CO_2, however, new coal plants with CCS are

---

[32]Further discussion of supply-curve assumptions is found in Annex 7.A.

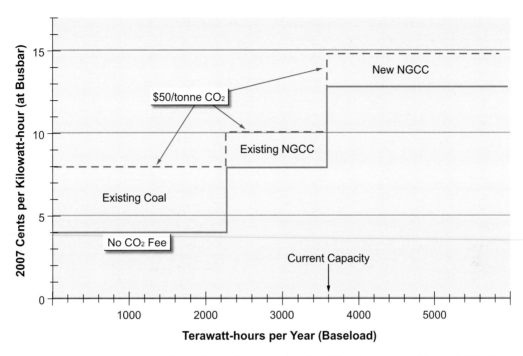

FIGURE 7.10 *Hypothetical supply curves for fossil baseload electricity in 2020, illustrating the coal/natural gas competition as a function of CO_2 price. The busbar cost for natural gas power in the absence of a CO_2 price is consistent only with a high natural gas price. Not shown are the perhaps 60 TWh of coal-CCS power produced from plants installed during the CCS evaluation period by 2020.*

the least-cost alternative, cheaper even than existing coal plants. New coal plants without CCS do not appear in Figure 7.11 at all for the $100 per tonne CO_2 case.

Figures 7.10 and 7.11 present *national* supply curves and *national* emissions, without taking into account state laws, regulations, and initiatives. The committee recognizes that many states, in the absence of a federal policy on CO_2 emissions, have already begun to take action. Three regional "cap and trade" initiatives, involving 23 states, have been formed to begin addressing CO_2 emissions (Cowart, 2008). In addition, some states have essentially put an infinite price on CO_2 emissions from coal plants. Florida, for instance, has ruled out a new uncontrolled coal plant, and California has issued a rule to ban the import of electricity generated with CO_2 emissions that are greater per kilowatt-hour than those of natural gas plants. (Little coal is consumed directly in California.) Although there may be constitutional challenges to such a rule based on the Commerce Clause of the

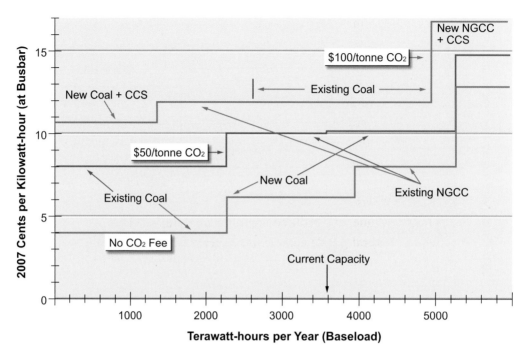

FIGURE 7.11 *Three hypothetical but illustrative supply curves for fossil baseload (85 percent capacity factor) electricity in 2035, based on CO_2 emissions prices of $0, $50, and $100 per tonne CO_2. The same high natural gas prices as in Figure 7.10 are assumed, successful CCS is assumed, and penetration rates of new coal plants are those discussed in the section titled "Future Coal Power." The committee assigns a ±10–25 percent uncertainty range to the vertical axis and a ±25 percent uncertainty to the horizontal axis.*

Constitution, it is probable that state actions will influence national supply curves to some extent. More generally, state policy will most likely have a major influence on federal policy, as it has in the past with non-CO_2 air-pollution control.

Findings: Electric Power Generation

The rate at which new fossil fuel power plants will be constructed strongly depends on the rate of penetration of energy efficiency and the rate of retirement of existing plants. Flat loads and near-zero retirements mean virtually no new construction. The benefits of new power plants—lower pollutant emissions and (if all goes well) CO_2 capture and storage—will accrue much more rapidly if retirements can be accelerated. Existing coal power plants emit 2 billion tonnes of CO_2 per year—one-third of total U.S. emissions—and natural gas power plants emit an additional one-third of a billion tonnes of CO_2 per year. It is difficult to see how

U.S. leadership in carbon mitigation can be established without aggressive retirements or retrofits. Investments in old plants to address CO_2 emissions and criteria air pollutants will be more cost-effective if they are coordinated.

Looking toward 2020 and perhaps beyond, the mix of coal and natural gas in new power generation is highly uncertain. Significant increases in capital costs, the form of potential regulation of CO_2 emissions, other licensing and regulatory issues, and the low availability of financing have made the construction of new coal-fired power plants less attractive. By contrast, the situation for new natural gas power plants appears favorable; they have lower construction cost, shorter construction time, and reduced environmental impact. However, the resulting growth in natural gas demand could increase LNG imports or bring on high-cost domestic gas, resulting in higher electricity costs.

Too little is known at present for determining which coal-based technology can best generate electricity after 2020 if CO_2 emissions are constrained. For equivalent levels of CO_2 capture, current estimates suggest that in many situations the IGCC plant is somewhat less costly than the PC plant is. However, falling costs for post-combustion capture at PC plants, the successful commercialization of burning coal in an oxygen environment (oxyfuel), and the reduced efficiency of IGCC for lower-rank coals all complicate the analysis. PC plants could thus be more competitive in some cases.

Carbon constraints can reorder the supply curve. Power-production technologies run the gamut on CO_2 emissions, venting from all to essentially none of the carbon contained in coal and natural gas fuels. Biomass incorporated into the fuel mix can, in principle, have *negative* carbon emissions. The cost of electricity across the range of these technologies therefore varies with the price of carbon. The variation is large enough to significantly move the cost break-even points among technologies.

Development of reliable cost and performance data needed for commercial deployment of new power-generation technologies requires the construction and operation of first commercial plants and the funding of innovative R&D. Commercial demonstration is needed in order to give vendors, investors, and other private-industry players the confidence that power plants incorporating advanced technologies can achieve the performance levels required under commercial terms.

Because of the variety of coal types and technology options, a diverse portfolio of demonstrations will be necessary.

An aggressive research, development, and demonstration (RD&D) program on carbon capture technology could have a major impact on the economics of coal-based electricity. Because existing technologies both for precombustion and post-combustion carbon capture impose large parasitic loads, reducing this inefficiency is a crucial research goal. Technical road maps for addressing this challenge exist, and they indicate that substantial progress could be made by 2020.

While these RD&D projects will be expensive, failure to initiate them in the near future will jeopardize the ability to widely deploy CCS technology in the 2020–2035 period. If the investments are not made now, they will have to be made later. Determination of which technologies offer the lowest-cost alternatives for reductions in CO_2 emissions from electric power generation will thus be delayed.

GEOLOGIC STORAGE OF CO_2

Potential Storage Sites

If significant quantities of captured CO_2 are to be stored, many subsurface locations will have to be found. Three principal storage settings are being considered: oil and gas reservoirs, deep formations that contain salt water, and coal beds too deep to be mined. Oil and gas reservoirs always have seal rocks that prevent the oil or gas from escaping; where available, they will be obvious first choices for storage locations.

Considerable practical experience for guiding future geologic storage projects has accumulated over the past three decades, as CO_2 has been used in enhanced oil recovery (EOR) operations to produce oil that would be left in the ground by conventional oil recovery methods (primary production and water injection). EOR projects in many oil fields (see Box 7.3) have demonstrated that CO_2 can be safely transported by pipeline and that it can be an effective agent for oil recovery. Technology required for CO_2 injection in gas fields is likely to be similar to that of oil recovery, and testing of injection of CO_2 into depleted gas fields is under way.

In many parts of the United States, however, oil and gas fields are not present, and use of other subsurface formations will be required if geologic storage

---

BOX 7.3 *CO₂ Capture and Storage Projects*

Figure 7.12 shows some of the CO_2 injection projects that are active around the world (IPCC, 2005). Commercial-scale projects that inject ~1 million tonnes per year are currently under way at Sleipner in the North Sea, Weyburn in Saskatchewan, In Salah in Algeria, Salt Creek in Wyoming, and Snøhvit in the Barent Sea. All but one of these projects (Weyburn) inject CO_2 that is separated from natural gas and would otherwise be vented to the atmosphere. The separation of CO_2 from natural gas is typically done with solvents such as amines—this step is required if the natural gas is to be purified sufficiently for sales (see Annex 7.A for details). At Sleipner, the gas is processed at an offshore platform, and the separated CO_2 is injected into a porous sandstone that contains salt water at a depth of 800–1000 m. At In Salah, the separated CO_2 is injected into the same formation that contains the natural gas, but below the gas zone so that salt water will be present. The Snøhvit project takes the gas ashore, makes liquified natural gas (LNG), and, as at In Salah, reinjects the separated CO_2 into a formation below the zone from which gas is produced. These projects indicate that additional early testing of geologic storage is possible where the CO_2 must be separated anyway.

The Weyburn project differs in that the CO_2 being injected comes from a coal gasification plant in North Dakota, and the CO_2 is used for a combination of enhanced oil recovery and CO_2 storage. That CO_2, along with about 1 percent H_2S separated with the CO_2, is transported by a 205-mile pipeline to the oil field; thus, the Weyburn project is an example of co-storage of CO_2 and H_2S. Salt Creek is also an enhanced oil recovery (EOR) project. It uses CO_2 separated from natural gas, as does a similar project at Rangely, Colorado. Numerous other EOR projects are currently under way in west Texas; they inject about 28 million tonnes per year, though mainly they use naturally occurring CO_2 that is brought by pipelines from Colorado and New Mexico. Considerable experience has been gained there in CO_2 transportation by pipeline and subsurface injection in the three decades of operation.

Many other pilot and demonstration projects, some of them quite large, are being planned. Descriptions for many of them can be found in IPCC (2005), at the IEA Greenhouse Gas Program Website (www.co2captureandstorage.info/co2db.php), and at Websites at MIT (sequestration.mit.edu/) and Stanford (pesd.stanford.edu/publications/pesd_carbon_storage_project_database).

---

is to be undertaken in any of them. Storage in deep formations that contain salt water is likely to be used in such locations. CO_2 injected into saline formations will eventually (in decades to centuries, although low-permeability settings could require thousands of years) dissolve in the brine, but rocks above the formation are required to contain the CO_2 during that time. Brine containing dissolved CO_2

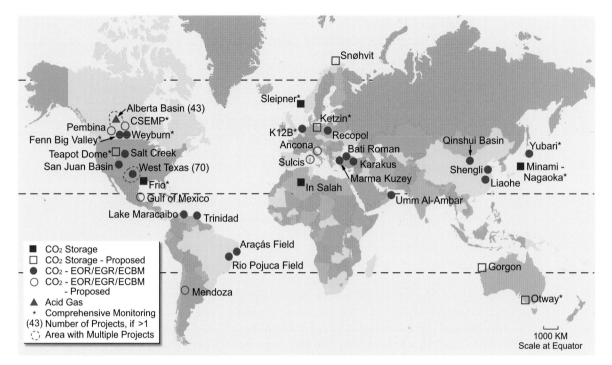

FIGURE 7.12 *CO_2 injection projects worldwide.*
Note: ECBM = enhanced coal bed methane; EGR = enhanced gas recovery;
EOR = enhanced oil recovery.
Source: IPCC, 2005.

is slightly denser than salt water alone, so once the CO_2 is dissolved the driving force for upward migration of CO_2 disappears. Several large-scale projects to test CO_2 injection into such formations are under way (IPCC, 2005; Figure 7.12), including a project at Sleipner in the North Sea and at In Salah in Algeria, where, as noted in Box 7.3, CO_2 separated from natural gas is injected into a sandstone formation above the natural gas reservoir and into the aquifer below the gas-bearing zone, respectively (IPCC, 2005). These tests both involve injection of about 1 million tonnes of CO_2 per year.

Storage in deep unminable coal beds has also been proposed, but it has been tested only in a very limited way (see Annex 7.A for a more detailed description of the attributes of the various potential storage settings).

Potential storage locations are widely distributed in the United States. For example, Figure 7.13 shows the locations of oil and gas provinces, areas with deep formations that contain salt water, and areas where coal is found; the locations of coal-fired power plants are shown as well. Dooley et al. (2006) concluded that about 95 percent of this country's 500 largest sources of CO_2 are located within 50 miles of a potential storage formation.

Estimates of the capacity for CO_2 storage in each of the settings have been made by regional DOE-supported research teams (NETL, 2007b): specifically, oil and gas reservoirs for 80 Gt CO_2; saline formations for 920–3430 Gt CO_2; and coal beds for 160–180 Gt CO_2. While the sedimentary rocks that might be suitable for CO_2 injection are widespread, not all locations will be appropriate. Sites will have to be carefully selected and evaluated. Suitable sites will have seal rocks that prevent vertical flow; sufficient pore space available that can be accessed without exceeding the maximum pressure (which could cause fracturing

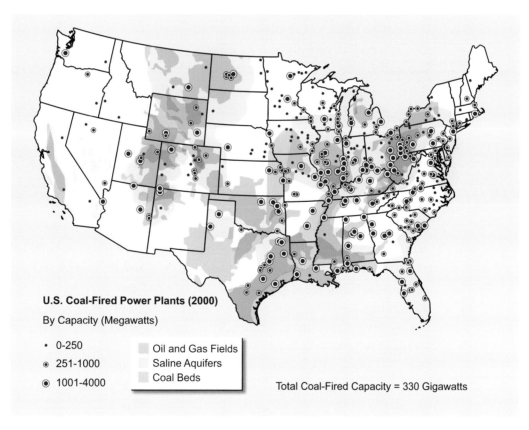

FIGURE 7.13 *Locations of coal-fired power plants and potential subsurface formations that could be used for geologic storage of CO_2.*
Source: MIT, 2007.

or leakage through the seal); rock properties that allow the CO_2 to be injected at a reasonable rate (more wells will be required for formations in which flow resistance is higher, which could increase costs over those used for the cost estimates reported in the section titled "Carbon Capture at Coal Plants in Operation Today"); and no potential leak paths. Appropriate sites will have rock layers above the storage zone that retain the injected CO_2 in the deep subsurface for times sufficient for physical mechanisms such as dissolution of CO_2 in brine and trapping of CO_2 as isolated bubbles to immobilize a large fraction of the CO_2, a period that is likely to be decades to centuries as noted preivously. Even given the constraints on specific storage sites, the estimated capacities just listed are large compared to U.S. emissions of 6 Gt CO_2 from energy use in 2007 (EIA, 2008c), an indication that sufficient capacity is likely to be available for geologic storage of CO_2.

Large-scale storage of CO_2 requires the integration of technologies, most of which have already been proven at commercial scale. Still, critical experience with CO_2 storage can be learned only by conducting many capture and storage projects in parallel. They must span the numerous types of coal, capture strategies, and storage sites, apply both to power and to synfuel plants, and entail storage both in deep saline aquifers and in hydrocarbon-bearing formations. Challenges to the management of a storage site for very large storage rates and quantities include (1) selecting enough well sites for high-volume injection and (2) monitoring movement of the CO_2 in the subsurface. Therefore, important information will be gained by conducting a number of projects in which CO_2 is injected over several years at a rate of at least 1 million tonnes CO_2 per year.

An indication of the very large scale of operation that might be required is provided by the emissions of a large coal-fired power plant. A 1000 MW coal plant will emit about 6 million tonnes of CO_2 per year. At typical densities of CO_2 in the subsurface, the volume of the CO_2 injected belowground is then about 300 million standard cubic feet per day, or 160,000 barrels per day in oil units. This volume is similar to the oil flow from a large oil field. A suite of projects can be designed to clarify the costs, risks, and environmental impacts of CO_2 capture and storage associated with coal and natural gas plants. This will enable informed judgment on whether such plants can become a significant contributor to the U.S. power system in a carbon-constrained world.

Work along these lines is being done with support from the DOE through a series of regional CO_2 storage partnerships (NETL, 2007b). Additional analysis will also be needed that links specific storage locations with potential sources, evaluates opportunities for capture, and considers how the CO_2 might be transported at the scale required.

CO_2 Storage Supply Curve

Figure 7.14 shows estimated cost as a function of the total quantity of CO_2 that could be captured and stored with current technologies at existing U.S. CO_2 sources. It indicates that if all the CO_2 emitted from stationary sources that could be stored at costs below about $50 per tonne were captured, the total emissions reduction would be about 1.5 billion tonnes CO_2, or about 20 percent of current emissions (Dooley et al., 2006). The storage cost shown as negative in Figure 7.14 stems from the use of CO_2 for EOR, which may make use of sources of CO_2 other than power plants (ammonia plants or natural gas purification, for example) in which CO_2 is already separated. About 28 million tonnes CO_2 per year is presently injected for EOR in oil fields in west Texas, though most of that CO_2 comes from natural underground sources.

Significant growth in the amount of CO_2 captured from anthropogenic sources would likely be put to use for EOR in regions where oil fields are within reasonable distance from a source. The existence of a CO_2 price would favor expansion of EOR in such locations as the magnitude of present EOR operations is constrained by the availability of CO_2 for injection. Thus combined EOR and CO_2 storage has the potential to lead to a significant expansion of EOR production from the 200,000 barrels per day produced now.

Findings: Geologic Storage

Long-term geologic storage of carbon dioxide appears to be technologically feasible, but it has yet to be demonstrated at the scale of a large power plant in a variety of geologic repositories.

A suite of projects can be designed to clarify the costs, risks, and environmental impacts of carbon storage. This would enable a determination of whether such plants can become significant contributors to the U.S. power system in a carbon-constrained world. Successful demonstration will require projects spanning the many types of coal, using several capture strategies, at a variety of storage sites, at both power and synfuel plants, and with storage both in deep saline aquifers and in hydrocarbon-bearing seams.

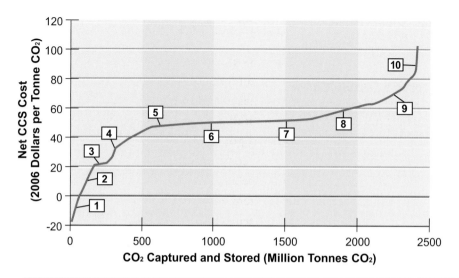

FIGURE 7.14 *Supply curve for geologic storage of CO$_2$. Net CCS costs shown include the costs of capture, transportation, and subsurface injection. See Figure 7.A.6 in Annex 7.A for a breakdown of these costs into four components: capture, compression, transport, and injection.*
Note: ECBM = enhanced coal bed methane; EOR = enhanced oil recovery.
Source: Dooley et al., 2006.

Significant expansion of domestic oil production via enhanced oil recovery could result from a price on CO$_2$. At least for light oils, CO$_2$ is the fluid of choice for EOR, but until now the level of EOR activity has been constrained by the availability of low-cost CO$_2$. If CO$_2$ emissions are constrained, EOR will be an attractive market for CO$_2$, as most of the CO$_2$ it uses will remain underground. Widespread realization of CO$_2$ capture opportunities will be required, along with an infrastructure of CO$_2$ pipelines.

ENVIRONMENTAL QUALITY AND SAFETY ISSUES

The use of coal, oil, and natural gas—involving exploration, extraction, transportation, processing (e.g., cleaning and refining), combustion, and recycling or disposal of petrochemical materials—has always produced environmental impacts.[33] They include air and water pollution that escape pollution controls in place today on vehicles, refineries, factories, and power plants, as well as discharges from sources without pervasive pollution controls, such as residences. The impacts also include changes to the landscape and ecosystem that remain after surface-mine reclamation, as well as ecosystem changes resulting from oil tanker spills. Because all of these impacts have been extensively reviewed, because Chapter 6 on renewable energy includes life-cycle analyses that have been carried out on fossil-generated electricity (particularly with respect to emissions), and because the National Research Council has released a report on energy externalities,[34] the AEF Committee does not repeat this work here.

From time to time, the well-studied impacts and the regulations under which they fall are summarized—e.g., in the environmental sections of the *Encyclopedia of Energy* (2004). But regulations, as well as results from environmental science, change every year. Up-to-date reports and statistics are more likely to be found on the websites of regulatory agencies,[35] government research laboratories and

---

[33]At each stage, the air and water pollution produced has the potential to impair human health. And in addition to effects such as stresses on wildlife, wildlife habitats, vegetation, and biota in rivers and streams, fossil-fuel-generated haze can also reduce visibility.

[34]*Hidden Cost of Energy: Unpriced Consequences of Energy Production and Use.* Available at http://www.nap.edu/catalog.php?record_id=12794.

[35]Examples include the Environmental Protection Agency (EPA) websites on air (www.epa.gov/oar/), water (www.epa.gov/OW/), and wastes (www.epa.gov/swerrims/); the Minerals Management Service (MMS) website on environmental assessment and regulation of offshore facilities (www.mms.gov/eppd/index.htm); the Department of the Interior's website on surface mine reclamation (www.osmre.gov/osm.htm); the website of the Council on Environmental Quality, which coordinates policy on environmental impact statements and assessments (www.nepa.gov/nepa/nepanet.htm); the website of the Office of Ocean and Coastal Resource Management (OCRM), a division of the National Oceanic and Atmospheric Administration (NOAA) (coastalmanagement.noaa.gov/); and the Fish and Wildlife Service's website on the endangered species program (www.fws.gov/endangered/).

advisory committees,[36] environmental and industry stakeholders,[37] and private research institutes.[38]

Much progress has been made in mitigating or eliminating impacts of fossil-fuel development and use, but environmental damage remains, including that caused by emissions from older power plants that have yet to be fully controlled. Airborne particulates are due, in part, to emissions from coal-powered power plants of gaseous SO_2, which forms into particulates downwind of an SO_2-emitting stack. As a result of such emissions and those from vehicles, industry, and other sources, there are still 208 counties in the United States that by December 2008 had not attained EPA limits on PM2.5 particulate air pollution (www.epa.gov/air/data/nonat.html?us~USA~United%20States). The number may fall to 52 by 2015, according to EPA modeling, with most of the residual regions in California.[39] However, even should the number of non-attainment counties shrink to zero, it would not necessarily mean that all health effects and annoyances will be eliminated because there is always debate about the proper standard. The EPA chose a PM2.5 standard in 2006 that was higher than the level recommended by its Clean Air Scientific Advisory Committee, although still within the advisory committee's range (Stokstad, 2006). The American Lung Association cites analysts who estimate that many thousands of premature deaths can be statistically related to pollution from particulates (ALA, 2009; Stokstad, 2006).

---

[36]Examples include the National Laboratories' website (www.energy.gov/organization/labs-techcenters.htm) and the reports of the National Petroleum Council, which advises the Secretary of Energy (www.npc.org/).

[37]Examples include the environment briefs of EPRI (mydocs.epri.com/docs/public/000000000001016774.pdf); the reports of the Union of Concerned Scientists (UCS) on coal power (www.ucsusa.org/clean_energy/); the reports and overviews of the National Resources Defense Council (NRDC) (www.nrdc.org/energy/); and the issue briefs of the National Mining Association (www.nma.org/issues/environment/default.asp), as well as other industry trade groups and environmental organizations.

[38]Examples include Resources for the Future (RFF) (www.rff.org/focus_areas/Pages/Energy_and_Climate.aspx); World Resources Institute (WRI) (www.wri.org/publications/climate); Worldwatch Institute (www.worldwatch.org/taxonomy/term/40); the Heritage Foundation (www.heritage.org/Research/EnergyandEnvironment/index.cfm); and the National Academies (www.national-academies.org).

[39]By 2015, the EPA projected the number of non-attainment counties to fall to 52 for PM2.5 (epa.gov/pm/pdfs/20061025_graphsmaps.pdf) under its Clean Air Interstate Rule (CAIR) program, which has now been overturned for application in Minnesota by the judiciary. The rule was considered too lenient (*Science* 321(5890; August 8):756-757, 2008). The EPA is reconsidering the application of the rule elsewhere (www.epa.gov/cair/pdfs/20090114fs.pdf).

Although difficult to estimate, the monetized estimates of environmental damage are not trivial, with most commentators finding coal to have the highest damage costs and natural gas the least. Thus there are regular attempts to either strengthen or change environmental regulations, which could alter the trajectory of energy development, extraction, and use of fossil fuels in the future. For instance, defining appropriate limits on the amount of very fine particulates that should be allowed to leave vehicle exhausts and combustion stacks is an ongoing area of health research. Similarly, increased production of natural gas from shale formations raises a variety of water quality issues.[40]

The AEF Committee is aware that government agencies, industry, and other stakeholders are currently working on many of these issues and that many of the potential problems will be resolved in the normal course of doing business. Still, the question arises as to whether or not existing laws, regulations, and enforcement capabilities will be sufficient to handle, both from a substantive and a public-perception viewpoint, the changes that may be coming over the next few decades within the fossil-fuels system.

In principle, a complex set of regulations that cover the use of fossil fuels is in place or can be changed to guide future fossil-fuel development. Examples of landmark federal legislation include the Clean Air Act, the Clean Water Act, the National Environmental Policy Act, and the Endangered Species Act. In addition, the Toxic Substances Control Act (TSCA), the Resource Conservation and Recovery Act (RCRA), the Comprehensive Environmental Response, Compensation and Liability Act (CERCLA), the Surface Mine Control and Reclamation Act of 1977 (SMCRA), the Outer Continental Shelf Lands Act, and the Coastal Zone Management Act (CZMA) are in place. Moreover, many states have passed legislation that affects use and development of energy facilities. The regulatory systems represented by all these legislative actions provide the opportunity for society to address emerging environmental concerns; the challenge is to make sure that the legislation is kept up-to-date and that funding for state and federal regulatory and enforcement programs keeps pace.

In addition, deploying many of the technologies discussed in this chapter will present environmental issues that are unfamiliar to the public or will lack appropriate regulatory frameworks. If not addressed properly and early enough, public

---

[40]See, for example, naturalgas.extension.psu.edu/Environmental.htm.

resistance or regulatory delays may put off or even curtail potential fossil-fuel developments.

The key issue areas to address are

1.   Capture and storage of CO_2
2.   Environmental and safety management necessitated by the increased use of coal, should the scale of coal-to-synthetic-natural-gas or coal-to-liquid programs grow significantly
3.   Environmental management of oil shale and tar sands
4.   Safety management of LNG terminals, increased tanker traffic, and extended networks of natural gas pipelines
5.   Water use.

These five areas are discussed in turn below.

Capture and Storage of CO_2

The major environmental issue facing fossil fuels today is the emission of greenhouse gases, particularly CO_2. Technologies discussed in the section above titled "The PC/IGCC Competition," such as pulverized coal combustion and coal gasification, offer a mechanism for capturing CO_2 in new coal-fired power plants. While natural gas–based electricity generation is already attractive for reasons spelled out in the section above titled "The Competitiveness of Natural Gas," the lack of regulations that would provide a greater incentive to mitigate CO_2 emissions, as well as great uncertainties about the cost of CCS, further discourages investment in coal plants.

In the CO_2 context, there may be lessons to be learned from reviewing the strengths and limitations of regulations of other air pollutants. For example, cap and trade programs are one of a number of methods for establishing a price on greenhouse gas emissions. Trading systems for atmospheric pollutants have a long history in the United States, and examples of their successes are described in Annex 7.A. Another method of establishing a price on carbon is levying a tax or a fee on CO_2 emissions.

The section above titled "The PC/IGCC Competition" explains why most PC and IGCC plants with CCS will be built after 2020, but there can easily be opportunities to capture CO_2 from other sources. As such, one new environmental challenge may be pipeline transport of CO_2 from its source to where it can be stored underground. Because safety issues associated with transportation of CO_2 would be novel to the public (see Annex 7.A), opposition to it could exceed that of natural gas pipelines, which have had difficulty being sited in crowded areas such as, the Northeast. Pipeline transport of CO_2 requires attention to route selection, overpressure protection, and leak detection (IPCC, 2005). Outside of high-population-density regions, opposition would likely be modest.

Storage of CO_2 is a way to mitigate CO_2 emissions for the time being, as discussed in the prior section titled "Geologic Storage of CO_2." Although there appears to be a large amount of storage capacity in the United States and elsewhere with suitably low initial leakage rates, long-term monitoring of leakage will be necessary. Further, there is a need for regulations concerning land-use compensation, underground storage rights, and long-term liability associated with CO_2 injection (Wilson et al., 2007a,b).

Public acceptance cannot be taken for granted; it must be won by performance. Regarding industry and major environmental groups in the United States, thus far they are supportive of CCS playing a major role in transitioning to energy systems with lower CO_2 emissions, assuming that safety questions related to possible releases are satisfactorily resolved. However, major environmental groups in Europe are concerned that CCS is not sustainable and that it may delay development of renewable-energy solutions.

The creation of a regulatory framework for geologic CO_2 storage is currently beginning, and there is also considerable experience with injecting CO_2 into oil formations for enhanced oil recovery, which has been regulated under rules for oil and gas production. Testing of geologic storage of CO_2 can be done under existing regulatory frameworks, but large-scale implementation will require significant further development. Work on such a regulatory framework is under way at the EPA.

In July 2008, the EPA issued proposed rules for regulation of underground injection of CO_2 under the Safe Drinking Water Control Act. These rules would create a new class of injection well for CO_2 within the Underground Injection Control program, and they also include requirements for storage-site characterization, injection-well design and testing, monitoring of project performance, and demonstration of financial responsibility; finalization of the rules is expected by late 2010 or early 2011. Such a regulatory structure is likely to continue to evolve in the decade ahead as the science and technology used to describe the behavior of CO_2 in the subsurface improves as a result of testing large-scale CO_2 injection (see Annex 7.A for additional discussion of the development of a regulatory framework).

Environmental and Safety Management of Increased Use of Coal

Should a dramatic increase occur in the use of coal, including more use for electricity generation, there could be more ash to be disposed of and greater total emissions of mercury and SO_2 (a precursor to airborne particulates); these effects could probably be mitigated by a tightening of limits on individual power plants, albeit with increased costs. There would also be new scrutiny of mine safety, mountaintop removal, and other forms of surface mining. Surface mining is subject to state and federal reclamation requirements, but adequacy of the requirements and enforcement is a constant source of contention.[41] For instance, the EPA recently announced a review of permit requests for mountaintop coal mining, citing serious concerns about potential harm to water quality.[42] Increased risk to endangered species from increased used of coal could be an issue as well. Increased use of coal also would mean greater risk of spills from coal ash impoundments, an issue now receiving active EPA attention.[43] In any case, the opening of new coal mines is likely to be much more expensive than it used to be, and new environmental/safety regulations to deal with the growth in coal-extraction rates will add to the cost.

As more U.S. coal makes its way throughout the world, it may be used in facilities less strictly regulated than those in the United States. Similarly, imported coal may not have the same upstream regulations likely to be in place in the United States.

Environmental Management of Oil Shale and Tar Sands

In addition to greater CO_2 emissions per unit of oil output from oil shale, there are environmental issues related to surface mining or in situ processing, including water management. Although Canada has a much larger tar sand resource than

---

[41]Environmental groups (e.g., www.sierraclub.org/MTR/downloads/brochure.pdf)) are particularly concerned about mountaintop removal and associated valley fills, which are largely confined to the Appalachian Mountains. Proponents of the practice point to its efficiency, the environmental benefit of the low-sulfur coal found there, and the resulting increase of flat land in areas where there is often little.

[42]See www.nytimes.com/2009/03/25/science/earth/25mining.html.

[43]"The EPA plans to gather coal ash impoundment information from electric utilities nationwide, conduct on-site assessments to determine structural integrity and vulnerabilities, order cleanup and repairs where needed, and develop new regulations for future safety." See www.ens-newswire.com/ens/mar2009/2009-03-09-093.asp.

does the United States, it is possible that tar sands from eastern Utah will be used one day for significant oil production. Producing oil from tar sands, like producing oil from shale, also produces a large amount of CO_2 before the oil is ever turned into transportation fuel and consumed, and there are impacts similar to those from extracting oil shale. Land and energy requirements are very high.

Government lands that will be made available for oil shale and tar sands leasing in the United States have not yet been selected (BLM, 2007a). Issues raised by exploration and extraction include pollution, impacts on scenic values, and impacts on fish and wildlife, including threatened and endangered species (BLM, 2007b).

This is an issue likely to be important in the 2020–2035 period, but planning for it during 2010–2020 will be essential.

Safety Management of Liquid Natural Gas Terminals, Increased Tanker Traffic, and Larger Natural Gas Pipeline Networks

Imports of liquid natural gas accounted for about 3 percent of U.S. natural gas supply in 2007 (British Petroleum, 2008). While LNG has been used safely around the world for many years, LNG storage facilities or tankers could now be vulnerable to terrorist attacks—perhaps the newest obstacle to siting LNG facilities.

Congress has passed legislation (the U.S. Maritime Transportation Security Act of 2002) requiring all ports to have federally approved security plans. Security assessments of LNG facilities and vessels are also required. Opposition to LNG is likely to be a regional, not a national, phenomenon. In fact, new LNG facilities are being constructed along the Gulf Coast, and more are being planned.

If LNG terminals were built far from load centers, natural gas from these terminals could be carried by pipeline. However, siting a pipeline along populated corridors can also be difficult; upgrading the capacity of existing lines may be easier.

Water Use

Population growth brings the need for more electricity as well as for more water, but these requirements often conflict, particularly on a regional basis. In the United States, those regions with the highest population growth are also those with the more severe water shortages. As a result, water use has become one of the most contentious issues in the siting of new electric power plants.

Power generation accounts for an estimated 40 percent of U.S. freshwater withdrawals (almost entirely at the once-through cooled plants) but only about 3 percent of total U.S. freshwater consumption. Power plants of nearly all types require water for a variety of uses within the plant. In addition, for fossil-fuel plants, including coal-fired IGCC plants, water is required in the extraction and processing of the fuels.

The withdrawal, consumption, and discharge of water from power plants all have an effect on the ecosystem. Aquatic life can be adversely affected by impingement on intake screens, by entrainment in the cooling water, or by thermal pollution from the discharge water. The primary effects of thermal pollution are direct thermal shock, changes in dissolved oxygen concentrations, and the mortality and redistribution of organisms in the local community. Additionally, chlorine and other chemicals that are added to condition the water or to prevent fouling of the cooling system can also impact ecosystems.

Most power plants in operation today use either once-through cooling or closed-cycle wet cooling.[44] Once-through cooling is used on about 40 percent of existing plants; closed-cycle wet cooling is used on nearly all the remaining 60 percent. Options for significantly reducing the use of freshwater by power plant cooling systems include the use of nonfreshwater sources as makeup water for closed-cycle wet cooling and the use of dry cooling or hybrid cooling systems.

Finding: Environmental Issues

A regulatory structure for carbon sequestration is needed and must be tested in the 2010–2020 timeframe if this technology is to be successfully implemented after 2020. Of course, there may be other important environmental issues associated with transforming the energy system, so agencies, stakeholders, and funders must be vigilant and strengthen their preparation for the changes that may be coming in fossil-fuel systems. Because these are difficult issues the participants may find the use of negotiated conflict resolution techniques to be helpful.

---

[44]In closed-cycle systems, cooling water is circulated through cooling towers to transfer heat to the atmosphere and then recirculated through the plant. However, losses occur due to evaporation and discharge, requiring the addition of makeup water.

REFERENCES

ALA (American Lung Association). 2009. State of the air. Washington, D.C. Available at www.lungusa.org.

Argonne National Laboratory. 2008. The Greenhouse Gases, Regulated Emissions, and Energy Use in Transportation (GREET) Model. Transportation Technology R&D Center. Available at www.transportation.anl.gov/modeling_simulation/GREET/index.html.

Baker Institute. 2008. Natural Gas in North America: Markets and Security. Baker Institute Policy Report 36. January. Available at www.bakerinstitute.org/publications/BI_PolicyReport_36.pdf.

Bartis, J.T., T. LaTourrette, L. Dixon, D.J. Peterson, and G. Cecchine. 2005. Oil Shale Development in the United States. U.S Department of Energy Report by RAND Corp. Available at www.rand.org/pubs/monographs/2005/RAND_MG414.pdf.

BLM (U.S. Bureau of Land Management). 2007a. Draft Oil Shale and Tar Sands Resource Management Plan (RMP) Amendments to Address Land Use Allocations in Colorado, Utah, and Wyoming and Programmatic Environmental Impact Statement (PEIS). Vol. 1. Washington, D.C.

BLM. 2007b. Draft Oil Shale and Tar Sands Resource Management Plan (RMP) Amendments to Address Land Use Allocations in Colorado, Utah, and Wyoming and Programmatic Environmental Impact Statement (PEIS). Vol. 2. Washington, D.C.

BLM. 2008. Inventory of Federal Oil and Gas Resources and Restrictions to Their Development. Available at www.blm.gov/wo/st/en/prog/energy/oil_and_gas/EPCA_III/EPCA_III_faq.html.

Bohi, D.R. 1998. Changing productivity in U.S. petroleum exploration and development. Resources for the Future Discussion Paper, Table 3.1. Available at www.rff.org/documents/RFF-DP-98-38.pdf.

Booras, G. 2008. Economic Assessment of Advanced Coal-Based Power Plants with CO_2 Capture. MIT Carbon Sequestration Forum, September 16.

British Petroleum. 2007. Statistical Review of World Energy 2007. London: BP. June.

British Petroleum. 2008. Statistical Review of World Energy 2008. London: BP. June.

CERA (Cambridge Energy Research Associates). 2009. Rising to the Challenge: A Study of North American Gas Supply to 2018. March. Available at www.cera.com/aspx/cda/public1/news/pressReleases/pressReleaseDetails.aspx?CID=10179.

Cowart. 2008. Carbon caps and efficiency resources. Testimony before the Select Committee on Energy Independence and Global Warming, U.S. House of Representatives Hearing on Energy Efficiency and Climate Policy, May 8.

DOE (U.S. Department of Energy). 2006a. Undeveloped Domestic Oil. Report by Advanced Resources International. Washington, D.C. February.

DOE. 2006b. Update to Undeveloped Domestic Oil. Available at www.fossil.energy.gov/programs/oilgas/publications/eor_co2/F_-_UndevDom_Oil_Res_FACT_SHEET.

DOE. 2007. Carbon Capture from Existing Coal-Fired Power Plants. DOE/NETL-401/110907. Washington, D.C. November.

Dooley, J.J., R.T. Dahowksi, C.L. Davidson, M.A. Wise, N. Gupta, S.H. Kim, and E.L. Malone. 2006. Carbon Dioxide Capture and Geologic Storage. Technical Report. Global Energy Technology Strategy Program, Battelle, Joint Global Change Research Institute.

Durham, L.S. 2006. GOM Lower Tertiary Play. American Association of Petroleum Geologists Explorer. November 11. Available at www.aapg.org/explorer/2006/11nov/jack_play.cfm.

EIA (U.S. Energy Information Administration). 1999. U.S. Coal Reserves: 1997 Update. Available at www.eia.doe.gov/cneaf/coal/reserves/chapter1.html#chapter1b.html.

EIA. 2004. Analysis of Oil and Gas Production in the Alaska National Wildlife Refuge. Available at www.tonto.eia.doe.gov/ftproot/service/sroiaf(2004)04.pdf.

EIA. 2007. Annual Energy Outlook 2007. DOE/EIA-0383(2007). Washington, D.C.: U.S. Department of Energy, Energy Information Administration.

EIA. 2008a. Annual Energy Outlook 2008. DOE/EIA-0383(2008). Washington, D.C.: U.S. Department of Energy, Energy Information Administration.

EIA. 2008b. Assumptions to the Annual Energy Outlook 2008. DOE/EIA-0554(2008). Washington, D.C.: U.S. Department of Energy, Energy Information Administration.

EIA. 2008c. Monthly Energy Review. DOE/EIA-0035(2008/04). Washington, D.C.: U.S. Department of Energy, Energy Information Administration.

EIA. 2008d. International Energy Outlook 2008. DOE/EIA-0484(2008). Washington, D.C.: U.S. Department of Energy, Energy Information Administration.

EIA. 2008e. Annual Energy Outlook Retrospective Review: Evaluation of Projections in Past Editions (1982-2008). DOE/EIA-06403. Washington, D.C.: U.S. Department of Energy, Energy Information Administration.

EIA. 2009a. Annual Energy Review 2008. DOE/EIA-0384(2008). Washington, D.C.: U.S. Department of Energy, Energy Information Administration.

EIA. 2009b. Annual Energy Outlook 2009. DOE/EIA-0383(2009). Washington, D.C.: U.S. Department of Energy, Energy Information Administration.

Encyclopedia of Energy. 2004. St. Louis, Mo.: Elsevier.

EPA (U.S. Environmental Protection Agency). 2008. U.S. Greenhouse Gas Inventory Reports. USEPA No. 430-R-08-005. April.

EPRI (Electric Power Research Institute)–TAG. 1993. Technical Assessment Guide (TAG) Electricity Supply—1993. TR-102276-V1R7. Palo Alto, Calif.: Electric Power Research Institute.

EPRI. 2007. The Power to Reduce CO_2 Emissions: The Full Portfolio. Fig. 4-3, p. 4-5, right panel. Palo Alto, Calif.: Electric Power Research Institute.

Farrell, A.E., and A.R. Brandt. 2006. Risks of the oil transition. Environmental Research Letters 1:014004.

Gibbins, J., and H. Chalmers. 2008. Carbon capture and storage. Energy Policy 36:4317-4322.

Hermann, W. 2006. Quantifying global energy resources. Energy 31(12; Sept.):1685-1702.

IEA (International Energy Agency). 2008a. Energy Technology Perspectives. Paris: IEA Publications.

IEA. 2008b. World Energy Outlook. Paris: IEA Publications.

IPCC (Intergovernmental Panel on Climate Change). 2005. Special Report on Carbon Capture and Storage. Prepared by Working Group III of the Intergovernmental Panel on Climate Change. New York and London: Cambridge University Press.

Katzer, J. 2008. Coal-based power generation with CCS. Presentation to the workshop of the Fossil Energy Subgroup of the AEF Committee, National Research Council, Washington, D.C., January 29–30.

Klusman, R.W. 2003. A geochemical perspective and assessment of leakage potential for a mature carbon dioxide–enhanced oil recovery project and as a prototype for carbon dioxide sequestration; Rangely Field, Colorado. American Association of Petroleum Geologists Bulletin 87(9):1485-1507. DOI: 10.1306/04220302032.

MIT (Massachusetts Institute of Technology). 2007. The Future of Coal: Options for a Carbon-Constrained World. Cambridge, Mass.: MIT.

NETL (National Energy Technology Laboratory). 2007a. Cost and Performance Baseline for Fossil Energy Plants. DOE/NETL-2007/1281, Revision 1. August.

NETL. 2007b. Carbon Sequestration Atlas of the United States and Canada. U.S. Department of Energy, Office of Fossil Energy. Available at www.netl.doe.gov/technologies/carbon_seq/refshelf/atlas/ATLAS.pdf.

New York Times. 2008. Natural gas prices fall as shale yields bounty. August 25.

NPC (National Petroleum Council). 2007. Facing the Hard Truths about Energy. Topic Papers Nos. 7, 19, 21, 24, and 26. Washington, D.C.: NPC.

NRC (National Research Council). 2007. Coal Research and Development to Support National Energy Policy. Washington, D.C.: The National Academies Press.

PGC (Potential Gas Committee). 2006. Potential Supply of Natural Gas in the United States. Potential Gas Agency, Colorado School of Mines. December 31. Available at www.mines.edu/research/pga/. Powerpoint slides available at www.aga.org/NR/rdonlyres/6CC4915E-D584-4B03-978A-7493E2FF2CF5/0/0709PGCSLIDES.PPT.

RAND Corp. 2008. Unconventional Fossil-Based Fuels: Economic and Environmental Trade-Offs. October. Available at www.rand.org/pubs/technical_reports/TR580/.

Ruppel, C. 2007. Tapping methane hydrates for unconventional natural gas. Elements 3: 193-199.

Shell (Shell Frontier Oil and Gas, Inc.). 2006. Oil Shale Test Project Plan of Operations. Prepared for the Bureau of Land Management. February 15. Available at www.blm.gov/pgdata/etc/medialib/blm/co/field_offices/white_river_field/oil_shale.Par.79837.File.dat/OSTPlanofOperations.pdf.

Snyder, J. 2008. Natural gas supply and demand: A widening gap: Presentation at the Electric Power Research Institute Summer Seminar. Beyond PRISM: Analysis to Action. August 4.

Stokstad, E. 2006. New particulate rules are anything but fine, say scientists. Science 311(5757):27.

USGS (U.S. Geological Survey). 1998. Arctic National Wildlife Refuge, 1002 Area. Petroleum Assessment, 1998, Including Economic Analysis. Available at www.pubs.usgs.gov/fs/fs-0028-01/fs-0028-01.htm.

USGS. 2002. Petroleum Resource Assessment of the National Petroleum Reserve in Alaska (NPRA). Available at www.pubs.usgs.gov/fs/2002/fs045-02/fs045-02.pdf.

Viswanathan, V., R. Purgert, and P. Rawls. 2008. Coal-fired power materials. Advanced Materials and Processes 166(8):47-49.

Wilson, E.J., S.J. Friedmann, and M.F. Pollak. 2007a. Research for deployment: Incorporating risk, regulation, and liability for carbon capture and sequestration. Environmental Science and Technology 41(17):5945-5952.

Wilson, E.J., M.F. Pollak, and G. Morgan. 2007b. Policy Brief: Regulation of Carbon Capture and Storage. International Risk Governance Council. Available at www.irgc.org/Expert-contributions-and-workshop.html.

ANNEX 7.A: FOSSIL FUELS

This annex expands on selected topics presented in Chapter 7. It is not intended as a stand-alone discussion.

Fossil Fuel Supply

World Oil and Gas Reserves and Resources

The United States has some 13 percent of the world's petroleum resource base, but only about 2 percent of global reserves. Converting resources to reserves requires new technology, with associated increases in production costs. Thus one factor that will determine whether this country can expect to exploit its resources in a significant way is whether it can convert them to reserves less expensively than can be done elsewhere in the world.

Data on the costs and technologies involved are largely unobtainable, in no small part because almost all of the world's petroleum resources and reserves are in the hands of national oil companies. However, the limited amount of available data suggests that the U.S. resource base is relatively high cost. For example, the ratio of proved reserves to annual production in the United States is 11.9, while for the world as a whole the ratio is 40.5 (British Petroleum, 2006). In other words, the rest of the world can maintain its current production of conventional crude oil from known reserves some four times longer than the United States can. This suggests that developing the U.S. resource base is a less economically competitive proposition than is continuing to produce from large reserves elsewhere.

Table 7.A.1 breaks down the resource base and reserve-to-production ratios by region. It shows that better than 65 percent of the world's conventional crude oil resources are concentrated in the Middle East, in non-OECD Europe (mostly in the Russian Federation), and in Central/South America (primarily Venezuela). It is in these three regions that the reserve-to-production ratio is largest. Although the estimates in Table 7.A.1 are very approximate both in magnitude and in regional distribution, it seems reasonable to draw the conclusion that the U.S. resource base is harder to develop than most of the crude oil elsewhere in the world.[1]

---

[1]Moreover, in these three regions the recoverable reserves are 25–50 percent of the resource base. This may suggest that applying new technology to convert resources to reserves in these regions is less challenging than in the United States. However, national oil companies, along with their national governments to varying degrees, determine the rates at which new reserves can

TABLE 7.A.1 Crude Oil Resources in Various Regions of the World

	Percent of World Conventional Crude Oil Resources	Reserve-to-Production Ratio in Region
Africa	5.4	32.1
Asia/Oceana	6.6	14.0
Canada	6.7	14.9
Central/South America	10.3	30.2
Middle East	39.0	79.5
OECD Europe	3.5	8.0
Non-OECD Europe	16.0	28.0
United States	12.5	11.9

Source: Calculated from the data warehouse in NPC (2007).

The situation with natural gas is somewhat similar, though regional data are not available. Globally, the reserve-to-production ratio for natural gas is around 60, while for the United States it is 11 (British Petroleum, 2006). Fifty-five percent of the world reserves are in Iran, Qatar, and the Russian Federation. The reserve-to-production ratio of the Russian Federation is more than 75, while the ratios both for Iran and for Qatar exceed 100.

Natural Gas Hydrates

Gas hydrates (in which a large amount of methane is trapped within the crystal structure of water ice) occur widely in marine sediments and on land in areas where temperatures are low enough to allow permafrost to exist. The presence of a methane molecule can stabilize a cage of water molecules at temperatures above the freezing temperature of water when the pressure is sufficiently high. This methane can come from biogenic sources or from thermogenic sources similar to those that generate the methane present in natural gas reservoirs. In either case, low temperatures (such as those that occur deep in the ocean or in the arctic) and high pressures (which occur at sufficient depths in either setting) are required in order for the hydrates to be stable. Because temperatures increase as the depth below the surface of sediment or land increases, hydrates are stable only for a lim-

be created by exploration and development, and they determine to what extent investments are made to deploy new technologies. There is considerable variability among the national oil companies in capacity for and willingness to make such investments.

ited range of depths. If sufficient methane is present in such settings, it will form a zone of free gas below the hydrate.

Estimates of the total hydrate resource vary widely (Ruppel, 2007). Some of the reasons for the variation include assumptions that hydrates exist throughout the stability zone when they may not, as well as assumptions that the fractions of pore space occupied by hydrate are larger than is often the case. Significant quantities of hydrate are likely to occur at relatively low concentration, which will make such resources difficult to recover at reasonable cost (Moridis et al., 2008).

Despite the uncertainties, the hydrate resource estimates are very large compared to natural gas reserves and production, though as with all estimates of a total resource, the amounts recovered at economically viable costs could be much lower. Ruppel reports that estimates of methane contained in hydrates in the Exclusive Economic Zone and the North Slope permafrost region of Alaska are about 150 times U.S. natural gas reserves and 900 times annual U.S. production of natural gas (primary sources of numerous resource estimates are also provided in Ruppel [2007]). A recently released summary of a Canadian study calls for additional research to establish whether Canadian hydrate resources are sufficient to warrant development efforts (Council of Canadian Academies, 2008).

The presence of hydrates in marine sediments can often be detected through seismic methods similar to those used to explore for oil or gas, though it can often be difficult to accurately establish how much of the hydrate is present. Seismic methods are less effective in permafrost regions on land because hydrate properties are similar to those of ice. In either case, drilling is usually required to quantify the resources.

Methane can be released from hydrates in several ways: by mining and then moving the hydrates to a zone of lower pressure or higher temperature; by recovering the methane that is released when heating the hydrates in place; by reducing the pressure; or by injecting an inhibitor (a chemical that causes the hydrates to become unstable) (see Ruppel [2007] for a more detailed description of these processes). Also, research is under way to exploit the fact that there is a thermodynamic driving force for CO_2 to displace CH_4 from its hydrate cage, but this process is far from field demonstration (Council of Canadian Academies, 2008).

Direct mining of hydrates is likely to be limited because of difficulties in handling the mined material and because large-scale mining would have significant environmental consequences: habitats would be disturbed both in marine sediments and on land for example. For subsurface settings, heating hydrates is energetically unfavorable because energy must be expended, not only to heat large quantities of sediment but also to break the bonds between the hydrates' water molecules. Inhibitor injection is likely to require significant quantities of relatively expensive chemicals, and managing subsurface flows so that the inhibitor reaches desired locations may be difficult.

As a result, pressure reduction appears to be the preferred method. One version of this process is to drill a well to release free gas that lies beneath the hydrate accumulation, thereby reducing the pressure so that additional hydrate above that zone dissociates and flows into the well. Several tests of such an approach have been performed at the Mallik site in Canada's Mackenzie Delta (Council of Canadian Academies, 2008; Moridis et al., 2008). That well was drilled into coarse-grained sediment with relatively high hydrate concentrations. While production tests have been of limited duration, initial gas flows were favorable.

Additional environmental and safety considerations will also arise. For hydrates that occur in unconsolidated sediments, the changes in pore pressure and gas volume as the hydrate dissociates could lead to slope failure (Jayasinghe and Grozic, 2007), and a large slope failure event could result in tsunami formation. The U.S. Department of Energy (DOE) is supporting research to delineate the physical mechanisms so that safe production methods could be devised (Allison, 2000). Drilling through hydrate zones in any setting requires precautions, as recognized in current arctic drilling activities.

Whether natural gas hydrates are produced in significant quantities will depend on three issues: development of exploration methods that can establish not only the location of hydrates but also the quality of the resource; economically viable recovery methods with acceptable environmental consequences; and the availability of infrastructure for transporting recovered gas to markets. Although research is under way on these issues, it is too early to tell how successful the efforts will be. Hence any significant recovery of hydrate resources is likely to occur after (perhaps well after) 2020.

Electric Power Generation

Because extensive analyses of technologies for generating electricity from coal have been published (see, for example, MIT, 2007; IPCC, 2005; and NETL, 2007a), no attempt is made to repeat them here except to emphasize a few points related to the findings of Chapter 7. Models presented for power generation in this report are based on those cited by the National Energy Technology Laboratory (NETL, 2007a).

Coal provides more than half of the electricity generated in the United States. The U.S. coal-based generating capacity is about 330 GW, and the average age of the coal power-plant fleet is 35-plus years (MIT, 2007). Even so, with current life-extension capabilities, the remaining service lives of many of the plants could be more than 30 years. It is currently expected that only about 4 GW of existing coal-fired generating capacity will be retired by 2030 (EIA, 2008).

Air-Blown Pulverized Coal Power Plants

Current coal plants burn air-blown pulverized coal (PC) to raise steam that drives a steam turbine. Of the more than 1000 boilers in the United States, about 100 are classed as supercritical (steam cycle up to ~3530 psi, 1050°F), with the remainder being subcritical units (steam cycle up to ~3200 psi, 1025°F). New U.S. plants built by 2015 are expected to be mostly supercritical, and advances in high-temperature materials could make ultrasupercritical PC units (perhaps exceeding temperatures of 1400°F) the norm after 2020.

Oxygen-Blown Coal Plants

One of the fundamental challenges in capturing the CO_2 produced by air-blown coal plants is the large amount of nitrogen in the flue gas, which reduces the concentration and partial pressure of the CO_2. If oxygen (~95 mole percent) is substituted for air in pulverized coal combustion, the nitrogen is largely eliminated, thereby raising the partial pressure of CO_2 in the flue gas and allowing it to be compressed after required cleanup for pipeline transport to an injection facility. This is called the oxyfuel process.

Alternatively, oxygen can be used to partially oxidize the coal in a gasifier to produce synthesis gas ($CO+H_2$) in an integrated gasification combined-cycle (IGCC) plant. After a further shift reaction with water to produce CO_2 and additional H_2, the CO_2 can be economically separated without combustion in a high-pressure stream and then readily compressed for transport through a pipeline and, finally, geological injection. This process is called oxygen-blown gasification.

Without carbon capture, both oxyfuel PC and oxygen-blown IGCC are more expensive than is a supercritical PC plant of a comparable size; but with carbon capture (assumed to be 90 percent), their cost is predicted to be somewhat lower than that of the supercritical plant (MIT, 2007).

Cost Estimates and Underlying Assumptions

Figures 7.4 to 7.6 are based on plants with a nominal capacity of 550 MW and a capacity factor of 0.85. The capacity factor times the rated power (output at full power, or capacity) equals the average power. A capacity factor of 0.85 means that the plant's annual output (kWh) is equivalent to the output of the plant running at its full capacity for 7446 instead of 8760 hours per year. Even a baseload power plant has less than 100 percent capacity factor because it is shut down occasionally for maintenance and because at some times (e.g., when power is very inexpensive) it simply does not run or runs at less than full power.

Here, the committee considers the differences between "dispatch cost" and "levelized cost" of electricity. Dispatch cost is the sum of variable operation and maintenance (O&M) cost, fuel cost, cost for CO_2 disposal, and cost for CO_2 emissions. Levelized cost is dispatch cost plus cost of installed capital plus cost of fixed O&M. In the Princeton Environmental Institute (PEI) work discussed earlier, total O&M (fixed plus variable) is assumed to be 4 percent per year of the total plant costs (TPC). The cost of installed capital (calculated using the Electric Power Research Institute [EPRI]-TAG methodology (EPRI-TAG, 1993), which assumes an owner's cost of 10 percent of TPC, a 55 percent:45 percent debt:equity split, and real costs of debt and equity capital of 4.4 and 10.2 percent per year, respectively) is 14.38 percent of the total plant investment (TPI), where TPI is the sum of TPC and the allowance for funds during a 3-year construction period (allowance for funds used during construction, or AFDC). AFDC is assumed to be 7.16 percent of TPC. As a result, cost of installed capital is 15.41 percent of TPC per year, or nearly four times total O&M.

Consider the costs of a new pulverized coal, CO_2 vented (PC-V) plant and a new pulverized coal with carbon capture and storage (PC-CCS) plant, as estimated in the PEI model used here and assuming that all O&M is variable O&M. At $0 per tonne CO_2, the dispatch costs are $24.5/MWh and $45.0/MWh, respectively. The three components of the dispatch cost are as follows: The O&M costs are $8.7/MWh and $15.9/MWh, respectively, reflecting the estimate that adding CCS will nearly double the total overnight plant cost—from $890 million ($1625/kW) to $1.62 billion ($2960/kW). The contribution of the fuel cost to the dispatch cost is $15.7/MWh for the venting plant and $22.6/MWh for the CCS plant. The fuel cost for the CCS is about 50 percent higher, reflecting the large energy consumption with today's post-combustion capture technology. Finally, the CCS plant is assumed to have paid for the pipelines and disposal wells required for disposal, even when the actual price for CO_2 emissions is zero; the CCS plant pays $6.3 per tonne CO_2, or $6.5/MWh, for CO_2 disposal.

The LCOE includes the capital costs, are noted above as being nearly in the ratio of 2:1. They are $33.6/MWh for the venting plant and $61.3/MWh for the CCS plant. As a result, the LCOE in the absence of a price on carbon is $58.1/MWh for the venting plant and $106.2/MWh for the CCS plant. These data are plotted in Figure 7.5.

With a $50/t$CO_2$ price added, the venting plant pays an extra $41.5/MWh for its emissions. The CCS plant, because it still emits some CO_2, incurs a cost of $8.5/MWh for its uncaptured emissions. Adding in these emissions costs, the dispatch cost for the venting plant becomes higher than that of the CCS plant, $66.0/MWh versus $53.5/MWh. However, the LCOE for the venting plant is less than that of the CCS plant, $99.6/MWh versus $114.8/MWh. The LCOE values are plotted in Figure 7.6. The crossover price of CO_2 for the LCOE is about $70 per tonne CO_2, somewhat above the $50 per tonne CO_2 price whose associated cost estimates are reported in Figure 7.6. For IGCC plants, however, both the dispatch cost and the LCOE for the venting plant are higher than for the CCS plant at $35 per tonne CO_2.

Carbon Capture Strategies

There are three broad classes of CO_2 capture strategies.

1. Post-combustion (end-of-pipe) capture from flue gas, after combustion in air. The concentration of CO_2 in the exhaust-gas stream ranges from 3–5 percent for gas turbines to 12–15 percent for coal-fired boilers. Nitrogen makes up most of the remainder of the flue gas (small amounts of other contaminants are also present).
2. Post-combustion with hardly any nitrogen present, either because combustion has occurred in oxygen or because of "chemical looping," in which the oxygen is provided by a regenerated metal oxide.
3. Precombustion capture, built on oxygen-blown gasification. For gasification plants in which CO_2 capture is the objective, air-blown gasification is typically not used because capturing CO_2 from a CO_2-N_2 gas mixture after gasification but before combustion would add considerable cost. Even when there is no CO_2-capture objective, oxygen-blown gasification is usually chosen at IGCC plants; even though there are additional costs for oxygen production, they are outweighed by the savings imparted by smaller gasifiers and downstream components.

An overview of these capture strategies is shown in Figure 7.A.1. All three are available, in principle, for all hydrocarbon sources, but relative costs differ, as do the energy-conversion routes that are favored. Gasification is probably the most competitive for high-rank coals, but this may reflect the lack of investment thus far in the development of gasifiers for low-rank coals. Petcoke is similar to a high-rank coal. Biomass may be co-fired with coal or petcoke, but to gasify a combination in which biomass represents a significant fraction of the total thermal input, there may need to be a separate biomass gasifier with its own feed-handling strategies.

CO_2 capture from natural-gas-fired power plants can be accomplished using any of the three strategies. One proposed project design features autothermal reforming of natural gas to make hydrogen, with combustion of the hydrogen

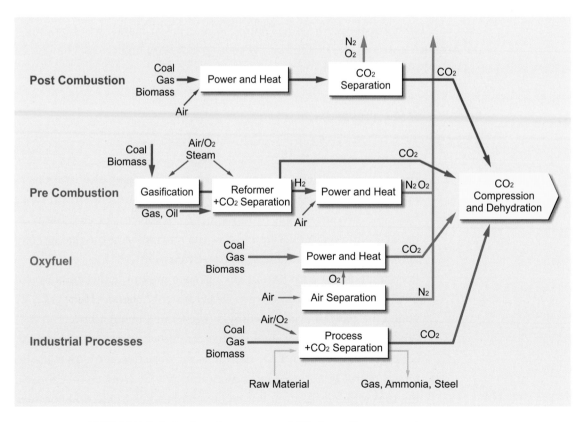

FIGURE 7.A.1 *Options for capture of CO_2 from flue-gas and process streams. Source: IPCC, 2005.*

to produce electric power and use of the CO_2 separated from the hydrogen to increase oil production.

The committee elaborates below on CO_2 separation, oxygen production, co-capture of CO_2 and other pollutants, and CO_2 compression.

CO_2 Separation

The first step in CCS is CO_2 capture from a gas mixture. CO_2 is currently separated at commercial scale where H_2 is generated from natural gas for petroleum-refining processes or for the manufacture of ammonia. However, the capture of CO_2 from large power plants has not yet been demonstrated. Commercial precombustion separation is typically based on physical adsorption in solvents (Selexol, Rectisol), while post-combustion separation involves chemical absorption (amines). Adsorption on solids such as activated carbon has also been demonstrated, but typically at small scale. Cryogenic and membrane separations are development frontiers. Selectivity and throughput rate are critical, though CO_2 purity can be much lower than in relatively demanding applications such as food production.

Typical separations have significant energy requirements, which are reflected in their costs. In chemisorption in an amine, for example, the gas mixture contacts the liquid-amine solution and CO_2 transfers to the liquid and reacts with the amine molecules, which releases significant amounts of heat. Nitrogen and other components remain in the gas phase, which is separated physically from the liquid. The amine solution containing CO_2 is then heated to release the CO_2. The combination of heat-transfer requirements and energy required to remove the CO_2 from the amine is a significant component of the cost of the separation.

In some settings (ammonia manufacturing, hydrogen production, and natural gas processing, for example), the CO_2 separation must be performed in order to make the product; thus the incremental cost of separation is zero. In others (electric power generation or cement, iron, or steel production), incremental separation costs are significant, and the requirements for energy to accomplish the separation lead to significant reductions in the overall thermal efficiency of the power plant.

Oxygen Production

Two of the three strategies described above (gasification and oxyfuels) involve an oxygen input, and the cost of oxygen is a significant component of the total cost. The oxygen demand with oxyfuels is about three times higher than with gasification, as complete combustion is required for oxyfuels, but only partial combustion is needed for gasification.

Extensive R&D programs are under way to find lower-cost sources of commercial oxygen than the conventional route of cryogenic separation. In particular, membrane separation at high temperature (using an ion-transport membrane) is an area of intense development. With high-temperature separation in place of cryogenic separation, optimizing thermal management would lead to reconfiguration of the capture plant.

Co-capture of CO_2 and Other Pollutants

Co-capture and co-storage of other gases with the CO_2 is a strategy that could potentially lower the costs of CCS by reducing the costs of pollution control aboveground. Sulfur co-capture as H_2S in precombustion systems and as SO_2 in post-combustion systems is straightforward technically because solvents used to capture CO_2 have similar affinities for H_2S and SO_2. However, depending on the location of the storage site, when H_2S is present, the licensing of transportation and storage of gas mixtures will inevitably be more complicated than the licensing of CO_2 alone. In Alberta, Canada, at about 30 locations, H_2S and CO_2 are removed together during the preparation of wellhead natural gas for insertion into the grid; they are co-stored belowground with extreme attention to safety because of the toxicity of H_2S. The CO_2 and H_2S are separated simultaneously at the Dakota Gasification Plant, and that mixture (containing about 97 percent CO_2, 1 percent H_2S, and small amounts of hydrocarbons) is transported by pipeline for injection at the Weyburn Field in Saskatchewan in a project that combines enhanced oil recovery (EOR) and CO_2 storage.

CO_2 Compression

By convention, the capture cost includes all incremental costs required to produce CO_2 at high pressure at the plant gate. The high pressure is needed for transportation (volumes of CO_2 are impractically large at low pressure) and for subsequent injection into porous geological formations 1 km or more below the surface. The costs of CO_2 compression include both the capital cost and the operating cost of the compressor, which is an internal load that reduces marketable output. CO_2 compression is a significant component of the combined capture and compression system cost, though compression costs are often less than separation costs.

Most applications at present in which CO_2 is separated commercially do not require CO_2 compression. For example, CO_2 for the food industry is useful at atmospheric pressure, and CO_2 separated from natural gas is vented at atmospheric pressure as well. Typical gasifiers, however, operate at high pressure (on the order of 60 bar). Therefore, a system optimization for CCS may exist that is based on separation of CO_2 at elevated pressure.

Effect of Carbon Capture on the Cost of Electricity

Design studies indicate that the addition of carbon capture and compression can have a significant effect both on the efficiency and on the cost of power plants. The largest source of efficiency reduction for air-blown PC is the energy required to recover the CO_2 from the amine solution (binding to the amine must be strong to capture CO_2 efficiently from a low-partial-pressure flue-gas stream). To compensate for this efficiency loss, corresponding increases in unit size and fuel-feed rate would be needed for the same power output. All in all, adding carbon capture to a new plant raises the total cost of electricity substantially, as seen in Figure 7.5. Further research, however, should lower many of these incremental costs. For example, it has been estimated that improvements in the technology of post-combustion carbon capture could reduce the cost of the PC-based alternative by 20–30 percent (MIT, 2007).

Other factors affecting the relative costs of electricity from PC and IGCC plants are coal type and quality. In general, coal type and quality will have greater effects on an IGCC system than on a PC system (MIT, 2007). IGCC functions best with dry high-carbon fuels such as bituminous coals and coke. Coals with high moisture content and low carbon content are most efficiently combusted in a PC unit. On the other hand, IGCC has inherent advantages for controlling emissions of criteria pollutants; cleanup can be accomplished at high pressures in the synthesis gas rather than at low pressures in the flue gas (MIT, 2007). Thus the future tightening of regulations on emissions of criteria pollutants such as mercury could favor IGCC.

Uncertainties in the Mix of New Coal Versus Natural Gas Generation

The uncertainties associated with building new coal plants (outlined in the section titled "Future Coal Power") and those for natural gas electric power generation (outlined in the section titled "The Competiveness of Natural Gas") have already led to significant increases in new natural-gas-fired plant capacity in the

United States, notably in the 2000–2004 period. For example, 60 GW of new natural gas capacity (peaking plus baseload) was added in 2002. However, over that same period, the real price of natural gas also grew by a factor of four; this price increase was enough to make electricity from gas-fired plants cost more than that from coal. Moreover, in many regions there was surplus electricity capacity, and electricity prices were not high enough to make natural gas generation economic. As a result, natural gas combined-cycle (NGCC) units averaged only about a 40 percent capacity factor in 2007 compared to well over 70 percent in 2002 (C. Bauer, NETL, personal communication). Natural gas prices experienced an increase in 2008 followed by a rapid decrease as the decline of economic activity reduced demand for natural gas. That variability in natural gas prices is the reason for the use of a wide range of natural gas prices in our analysis of the cost of producing electricity with natural gas.

As noted in the section "Natural Gas," there are significant interacting uncertainties regarding the future availability of natural gas, including unconventional sources such as gas shales; potential reductions in demand for electricity due to improvements in energy efficiency or increases in demand due to use of more electricity for transportation; the availability and cost of LNG on the world market; and the details of future regulation and price of CO_2 emissions. All of these factors and their inherent uncertainties will influence the future mix of new coal and natural gas electric power generation, and the uncertainties make it unlikely that precise forecasts of the energy mix made now will be accurate.

Assumptions Used in Developing Electricity Supply Curves

Here the committee provides more detail on the assumptions behind the supply curves in Figures 7.10 and 7.11 in particular and behind supply curves in general. Levelized costs of electricity ($/MWh) are needed; for new plants the data from Figures 7.5 and 7.6 and related calculations are used. Also needed are assumptions about the retirement rates of existing U.S. plants powered by coal as well as by other resources. One should expect the retirement rate to increase with the CO_2 emission price. Six competing baseload fossil-fuel power technologies are considered here: PCs, IGCCs, and NGCCs, with and without CCS. The time periods are 2010–2020, 2020–2035, and 2035–2050.

To develop supply curves for fossil-fuel power, one could explore at least the following five hypothetical choices:

1. *Either* the United States successfully implements a CO_2 mitigation strategy that results in a price for CO_2 emissions from power plants by 2020, and the price remains through 2050, *or* the United States has no effective CO_2 price. If there is a price, the committee alternately considers $50 per tonne CO_2 and $100 per tonne CO_2.
2. *Either* CO_2 storage is successfully launched at coal- and natural gas-fired power plants by 2020, with the resulting cost increments for power shown in Figures 7.5 and 7.6, *or* CCS does not succeed, resulting in cost increments for CCS storage several times higher than those shown in these two figures.
3. *Either* the natural gas price remains near $6.00/GJ ($6.33/million Btu), *or* it rises to and remains near $16.00/GJ ($16.88/million Btu).
4. *Either* the retirement rate at existing coal plants is negligible, *or* it is 3 percent per year after 2020.
5. *Either* the coal price remains at the low price assumed in Figures 7.5 and 7.6, *or* it becomes significantly more expensive.

Among the additional modeling choices, perhaps the most critical are the rates of reduction in capital costs resulting from experience and R&D. (For example, will the DOE's research program succeed in greatly reducing CCS incremental costs relative to those assumed in this chapter?) Build rates for CCS plants also need to be estimated, taking into account the availability of suitable storage locations. And regional analysis can underpin national analysis when it comes to retirement rates, fuel costs, and storage capacity.

Ultimately, to develop a view of actual deployment, supply curves must be joined with demand curves. In so doing, one must decide how demand for power, both at the regional and at the national level, will develop for each of the many alternative futures just described.

As an illustration of how build rates might affect supply curves for 2020 and 2035, consider Figure 7.10 and Figure 7.11, respectively. The committee assumes that 10 GW of coal+CCS facilities are built between now and 2020 as part of a CCS evaluation period; it also assumes that during this period no traditional coal units without carbon capture are built, given that the industry is awaiting certainty in climate policy. NGCC plants without CCS are assumed to be unaffected by such uncertainty. After 2020, coal plants either with or without CCS are assumed to be capable of being built at rates of 15 GW/yr if they are economically competitive and there is sufficient demand. The corresponding assumption for NGCC plants, with and without CCS, is 36 GW/yr. These figures are 33 percent higher than are historic rates during the peak building periods of 1969–1984 for coal plants and 2001–2004 for NGCC plants. Potential build rates for use in supply curves should be higher than actual build rates, which are affected by competition from other energy sources.

The committee takes capital costs of coal and CCS facilities to remain unchanged over time, using the costs in Figures 7.5 and 7.6 as a base, which means that the illustrative curves can certainly not be considered forecasts. Given the closeness of the cost estimates for PC+CCS and IGCC+CCS in Figures 7.5 and 7.6, the committee considers one generic coal+CCS facility rather than treating PC and IGCC separately. Only the case when coal is inexpensive and natural gas is costly (new plants face $1.71/GJ coal and $16/GJ natural gas) is considered, and there are no retirements.

The results for 2020, presented in Figure 7.10, show a typical staircase supply curve that highlights the large disparity between the costs of existing coal plants and those of other fossil-fueled electricity technologies, should coal prices remain low. Note that the consumption of electricity is determined by the intersection of the supply curve with a downward-sloping demand curve (not shown), which can be influenced by many factors, including policies related to energy efficiency. Actual consumption of fossil-fuel electricity in 2020 could be greater than or less than current consumption of electricity generated from fossil fuels. The X-axis of Figure 7.10 is arbitrarily cut off at a value higher than any plausible 2020 demand for baseload fossil-fuel power.

The results shown in Figure 7.11 continue the hypothetical analysis out to 2035. Three supply curves are shown, corresponding to three prices on CO_2 emissions: $0 per tonne CO_2, $50 per tonne CO_2, and $100 per tonne CO_2. At $100 per tonne CO_2, but not at $50 per tonne CO_2, new coal+CCS plants and—at very high electrical demand—even considerable numbers of NGCC+CCS are built. The cost of electricity across the range of power-production technologies therefore varies with the price of carbon. The variation is large enough to significantly move the cost break-even points among technologies.

Hypothetical supply curves such as those illustrated in Figures 7.10 and 7.11 can be used to explore the effect of different modeling assumptions and policy choices on total CO_2 emissions, as seen in Figure 7.A.2. The amount of CO_2 that would be emitted in 2035 is plotted for four cases. The base case has the coal price used in Figures 7.5 and 7.6, a CO_2 emissions price of $100 per tonne CO_2,

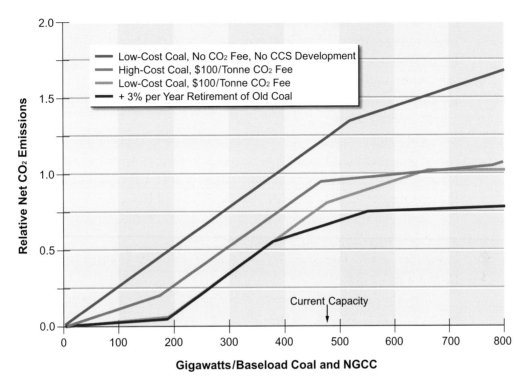

FIGURE 7.A.2 *CO_2 emissions from baseload fossil plants in 2035 relative to 2008 emissions as a function of baseload fossil electricity consumed in 2035. (High natural gas price and CCS successful, unless otherwise stated.)*

and no retirements of existing coal plants. In each of the other cases, just one of these assumptions is changed: (1) has a higher coal price—$100 per tonne, or about $3.70/million Btu; (2) has a CO_2 emissions price of zero; and (3) has a 3 percent per year retirement rate beginning in 2020.

The highest emissions curve in Figure 7.A.2, as expected, is for the case corresponding to the lowest supply curve of Figure 7.11, which has no price on CO_2 emissions. In the other three cases, where the CO_2 price is $100 per tonne CO_2, the total emissions never rise above current levels.

Note that this high-emission curve passes above 1.0 at the point on the power axis below that corresponding to today's current capacity. Indeed, if total demand for fossil fuel in 2035 were the same as today, the mix of coal and natural gas in power production would shift toward coal because new coal plants outcompete existing natural gas plants, given the assumed high price of natural gas and the absence of a price on CO_2 emissions.

The curve labeled "High-Cost Coal, $100/Tonne CO_2 Fee" in Figure 7.A.2

lies mostly above the curve labeled "Low-Cost Coal, $100/Tonne CO_2 Fee." One might have expected low-cost coal to bring about greater use of coal, and indeed it does, but the coal power that is encouraged by a low coal price and a $100 per tonne CO_2 price is coal power with CCS, and it is displacing natural gas power without CCS (which has higher emissions). As a result, greater competitiveness of coal in this instance means lower CO_2 emissions.

Once again, Figure 7.A.2 says nothing about how much fossil-produced electricity will actually be consumed in the target year 2035; that amount depends on the intersection point between the supply curve and an efficiency-dependent demand curve.

Clean Coal Research Plan and Deployment Schedule

Largely through the DOE, this country annually spends about $744 million on research, development, and demonstration (RD&D) for advanced coal technologies related to power generation. The research is wide ranging and dynamic. For example, on July 31, 2008, the DOE announced that $36 million would be awarded to 15 projects aimed at developing advanced carbon-capture technologies for the existing fleet of coal-fired power plants. Technologies involved included membranes, solvents, solid sorbents, oxycombustion, and chemical looping (www. netl.doe.gov/publications/press/2008/08030-CO2_Capture_Projects_Selected.html). Figure 7.A.3 shows that these technologies are expected to enable continued reductions in the cost of carbon capture over the next 20 years.

A recent review of the DOE's coal RD&D program (MIT, 2007) stressed the importance not only of research on innovative emerging technologies but also of government funding for first-of-a-kind commercial-scale demonstration projects. The AEF Committee notes that at this writing the DOE's program is being redefined. The committee judges that demonstrations of CCS integrated at the scale of a large power plant are important, as is continued R&D to improve separation technologies such as those listed in Figure 7.A.3.

Figure 7.A.4 shows one example of a timetable for RD&D for advanced coal technologies (involving both improved efficiency and carbon capture and storage) proposed by EPRI, which judges the timetable to be aggressive but achievable. For *pulverized-coal plants*, steady improvements in materials are projected to enable higher boiler and turbine temperatures and pressures; improvements in oxygen separation and post-combustion gas-separation membranes could enable ultrasupercritical designs with post-combustion CCS to be demonstrated at scale

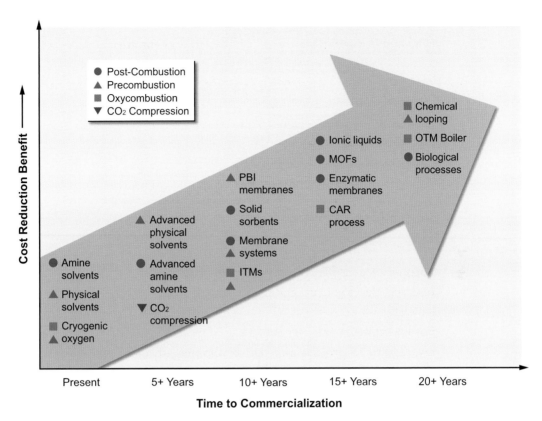

FIGURE 7.A.3 *Assessment of the estimated time to deployment of various carbon-capture technologies.*
Note: CAR = ceramic autothermal recovery; ITM = ion transport membrane; MOF = metal organic framework; OTM = oxygen transport membrane; PBI = poly[2,2′-(m-phenylene)-5,5′-bibenzimidazole].
Source: Bauer, 2008.

by 2025. For *integrated gasification combined-cycle* plants, improved gasifiers, precombustion gas-separation technologies, hydrogen turbine developments, and chilled ammonia methods of carbon capture could enable IGCC plants with CCS to be demonstrated by 2025. Integrated gasification fuel-cell plants, which could improve efficiency over gas turbines, could be demonstrated by about 2030. Finally, *CCS* could be fully demonstrated by about 2020, but three to five large-scale demonstration plants would be necessary to give vendors, investors, and private industry the confidence that the advanced technologies can be built and operated under normal commercial terms and conditions. While these specific

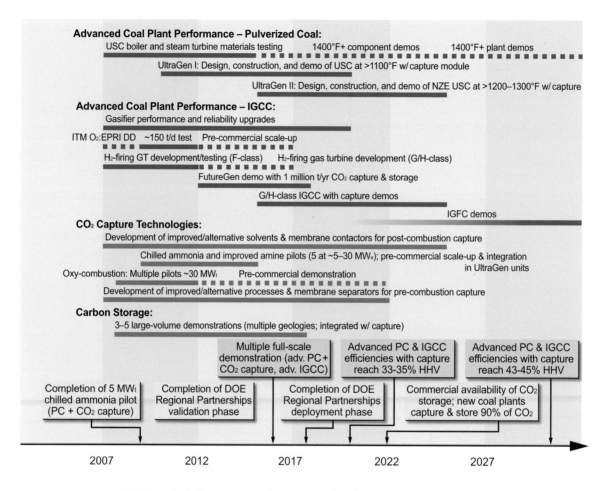

FIGURE 7.A.4 *One proposed research, development, and demonstration timetable for clean coal technologies.*
Source: EPRI presentation at the AEF Committee's fossil fuels workshop, 2008.

recommendations differ somewhat from the committee's estimates presented in Chapter 7 that about 10 GW of coal-fired electric power generation with CCS could be installed by 2020, they reflect a common view that there is a need to move to demonstration of CCS at large scale.

EPRI has estimated that the cumulative cost of the identified RD&D program would be $8 billion by 2017 and $17 billion by 2025, which is consistent with an earlier estimated need of $800–850 million per year (MIT, 2007). While many research projects are involved, the costs are dominated by the need to acquire years of experience with large-scale demonstration projects, both regarding advanced generation technologies and CCS. Further, it is necessary to initiate these projects immediately in order to meet the timetable in Figure 7.A.4. Major nontechnical challenges must also be addressed before carbon storage can become a reality, including development of appropriate regulations, resolution of legal issues—largely having to do with ownership of reservoirs and liability in case of leakage—and incorporation of appropriate monitoring regimes.

Geologic Storage of CO_2

A brief review of the case for rapid commercialization of CCS may be found in Sheppard and Socolow (2007). Three principal storage settings are being considered: oil and gas reservoirs; deep formations that contain salt water (saline aquifers); and coal beds too deep to be mined (see Figure 7.A.5). Varying amounts of

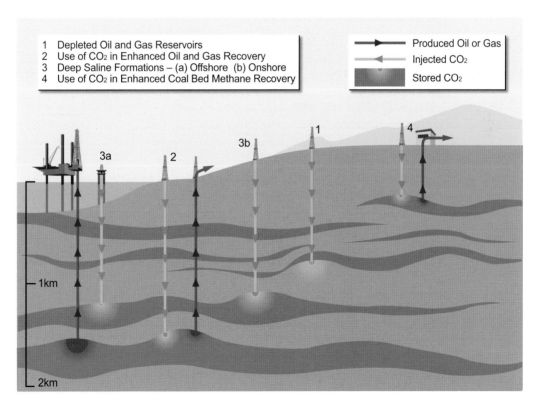

FIGURE 7.A.5 *Overview of carbon dioxide capture and storage. Source: IPCC, 2005, Fig. SPM.4.*

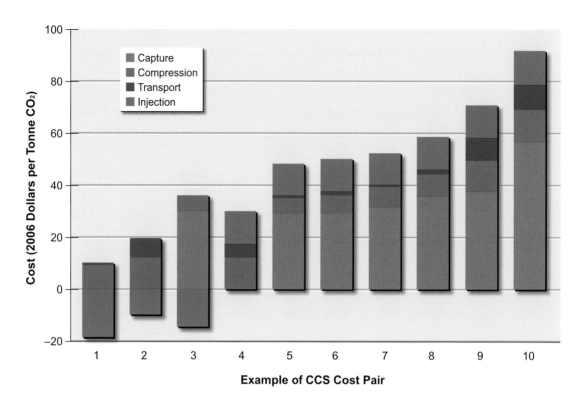

FIGURE 7.A.6 *Component costs for the storage systems listed in Figure 7.14. Source: Dooley et al., 2006.*

testing of all three settings have been done. Use of basalt formations and organic-rich shales has been proposed, but neither has been tested in the field (Dooley et al., 2006; NETL, 2007b). And while sedimentary rocks that might be suitable for CO_2 injection are widespread, not all locations would be appropriate. Storage in saline formations and coal beds will also require seal formations above the storage formation that prevent vertical migration of the CO_2 to the surface. Appropriate sites will have to be selected that have sufficient pore space available and that have rock properties that allow the CO_2 to be injected at a reasonable rate.

Figure 7.A.6 shows estimates of the cost components of CCS for various sources, sinks, and geographic distances between them. Note that there is no single homogeneous "CCS technology" or situation; economic viability will depend on specific source and sink characteristics. For situations in which the CO_2 is already separated (natural gas processing, H_2 production, or ammonia production, for example), the incremental separation cost is zero. About 6 percent of U.S.

emissions of CO_2 come from high-purity sources of this sort (Dooley et al., 2006), which provides opportunities for testing geologic storage at significant scale without requiring additional separations or constructing new power plants. In the limited instances where those sources are relatively close to locations at which EOR might be undertaken, the net cost of storage could be negative, as the revenue from oil sales would likely exceed the cost of storage. At coal-fired power plants, the largest sources in numbers and total emissions of CO_2, the cost of CO_2 capture typically exceeds the estimated costs of compression, transportation, and injection into the subsurface.

Dooley et al. (2006) estimated CO_2 capture costs ranging from zero (for plants that already separate a high-purity CO_2 stream) to $57 per tonne CO_2 (for a low-purity natural gas-fired combined-cycle power plant), and compression costs of $6–12 per tonne CO_2. Transportation costs were estimated to range from $0.2 to $10 per tonne CO_2, with the low-cost end of the range being for large-volume pipelines. Geologic storage costs are also likely to vary with the specific application. Dooley et al. estimated costs of minus $18 to plus $12 per tonne CO_2 for saline aquifer, EOR, and enhanced coal bed methane-injection projects, with the negative- and low-cost estimates applicable when cost recovery through sale of hydrocarbons is possible. Costs that are roughly consistent with these numbers are reported in the IPCC *Special Report on Carbon Capture and Storage* (IPCC, 2002), when corrections to translate 2002 costs to 2006 are made.

The forgoing estimates of potential costs of storage are "bottom-up," based largely on engineering estimates of expenses for transport, land purchase, drilling and sequestering, and capping wells. However, quantified factors based on engineering analysis may represent a lower bound on future costs. Uncertainty in the regulatory environment created by public resistance to CCS could result in costly delays in implementation at the project level, both during the demonstration phase over the next decade and even when CCS has attained full commercial-scale operation (Palmgren et al., 2004; Wilson et al., 2007; IRGC, 2008). Extra costs could be incurred at a given project site because of interruption of operations even at a *different* site, given that the technologies, monitoring, and regulation of storage are likely to be closely related across sites. Costs usually not taken into account also result from the likely need to secure storage rights a very large amount of belowground space for the lifetime of a facility (Socolow, 2005).

One feature of CCS that improves the odds of deployment evolving without major disruption is that many of the early CCS projects will be EOR projects. They would likely be located where the general population is already familiar with and tends to be positively disposed toward the oil and gas industry, and where there will be revenue streams of benefit to all royalty holders, including local and state governments (Anderson and Newell, 2004).

Risks associated with storage are often handled uneventfully in the normal course of events, with smooth and reliable licensing, operation, and monitoring that make for minimal regulatory delays. Carbon dioxide is routinely transported long distances, injected underground, and stored at present without much attention from either the public or policy makers. Similarly, natural gas and chemical storage are longtime facts of life (Reiner and Herzog, 2004), and serious accidents and leaks do not threaten operations, at least not on an industry-wide basis. But counterexamples, from Bhopal to Three-Mile Island to Yucca Mountain, are easily cited as well. In short, public reaction is unpredictable.

Oil and Gas Reservoirs

Most of the experience in CO_2 injection into the subsurface comes from oil fields. High-pressure CO_2 has been used for more than three decades for enhanced oil recovery (EOR), with the largest operations being in west Texas. Most of the CO_2 injected for EOR has come from natural underground CO_2 sources rather than anthropogenic sources, but some has been obtained from natural gas processing operations that remove CO_2 from the gas prior to sale.

Oil and gas reservoirs trap buoyant fluids that would otherwise escape to the surface, and hence the formations above the porous zones that contain the oil and gas should prevent vertical migration of CO_2 as well. While similar principles apply to injection of CO_2 into gas reservoirs, experience there is much more limited because the combination of gas prices and CO_2 costs has not favored enhanced gas recovery using CO_2. A test is currently under way, however, in the K12B gas reservoir in the Netherlands (IPCC, 2002).

In an oil- or gas-production operation, two key measures are critical: the amount of recoverable hydrocarbons, and the production rate per well. The analogy for CO_2 storage or disposal is to determine the total mass of CO_2 that can be injected into a target formation and the injection rate per well.

As an example, the Weyburn Field in Saskatchewan is injecting 95 million cubic feet per day of anthropogenic CO_2 (from the Great Plains Synfuels Plant in North Dakota) into 37 wells (IPCC, 2005). This field has a total hydrocarbon volume of 1.4 billion barrels, of which 330 million had been produced—about 23 percent of the original oil in place—at the time CO_2 injection commenced. It is anticipated that about 20 million tonnes of CO_2 will be injected and become permanently stored some 1400 m (4600 ft) underground over the 25-year lifetime of the EOR project (expected to produce an additional 130 million barrels of oil).

The Great Plains Synfuels Plant, constructed in 1984, produces a variety of feedstocks from coal for products that include fertilizer, pesticides, gasoline, resins, krypton and xenon gases, and liquid hydrogen, in addition to the carbon dioxide that is sold to the Weyburn Field for EOR. Over its lifetime, the Weyburn Field project will inject about one-third of the concurrent CO_2 output of a 1000 MW coal plant.

Other currently active EOR projects using CO_2 from natural underground sources have original-oil-in-place volumes ranging from about 40 million to 2 billion barrels and CO_2 injection rates of 50–100 million cubic feet per day (3000–5000 tonnes CO_2 per day). Typically, existing production wells can be transformed into injectors at less than the cost of drilling the new wells that would be required for sequestration in a saline aquifer. Also, EOR projects offset the injection costs with revenue from produced oil.

Using CO_2 for EOR has obvious benefits, but project locations, injection rates, and service lives may not be sufficient for EOR, by itself, to accommodate the lion's share of CO_2 emissions from power plants. Although there is far more capacity for storing CO_2 in saline aquifers, wherever storage through EOR is possible it should prove very attractive, given the potential for cost recovery and the use of at least a portion of an existing infrastructure within the oil fields.

Saline Aquifers

Saline-aquifer storage is expected to be the workhorse storage option in the United States (Dooley, 2006). Saline-aquifer storage has also been tested in the Sleipner Field of offshore Norway at a scale similar to that of the Great Plains example (IPCC, 2005). The Sleipner Field produces natural gas that contains CO_2, which is separated from the natural gas and reinjected into the very large Utsira Formation, which is sandstone. Because that formation has high permeability (fluids flow relatively easily through the rock), only one injection well is required to handle about 1 million tonnes per year of CO_2 (2700 tonnes per day). Seismic evidence collected periodically indicates that the CO_2 has been contained in the Utsira Formation. While there is enough experience to date to indicate that CO_2 injection into formations that contain salt water can be undertaken, the combination of technologies required to store CO_2 from a large coal-fired power plant has not yet been demonstrated at sufficient scale.

Coal Beds

Coal-bed storage is the least well understood of the three main storage options. The mechanism depends on the fact that CO_2 adsorbs onto coal surfaces, and it does so more strongly than does methane. In 2005 coal-bed methane production was 1.7 Tcf, about 9 percent of U.S. natural gas production (http://tonto.eia.doe.gov/dnav/ng/ng_enr_cbm_a_EPG0_r52_Bcf_a.htm). In a typical coal-bed methane project, water is removed to reduce pressure, and the methane released from the coal at the lower pressure flows through fractures in the coal to producing wells. CO_2 injected into a fractured coal bed replaces adsorbed methane, which creates the possibility of enhanced coal-bed methane production using CO_2. While some coals can take up significant quantities of CO_2, flow through the coal becomes more difficult as the CO_2 adsorbs. Injection of CO_2 into a coal bed was tested at the Allison Unit in New Mexico, where significant permeability reductions were observed (IPCC, 2005). More testing will be required at various scales before significant storage in coal beds is likely to occur.

Retention of CO_2 in the Subsurface

Subsurface formations that are appropriate for CO_2 storage will have rock layers above the storage zone that do not permit vertical flow. Those seal rocks, often shales or evaporites, will be needed to isolate the injected CO_2 from the near-surface region for an extended period during which several physical mechanisms act to immobilize the CO_2. When CO_2 dissolves in brine, for example, the resulting mixture is slightly denser than brine alone, and hence the driving force for upward migration of the CO_2 disappears, and the flow of the CO_2-laden brine away from the CO_2 zone helps dissolve the CO_2 more quickly than it would by diffusion alone. When brine invades areas formerly occupied by CO_2 as it dissolves, trapping of the CO_2 as isolated bubbles occurs. These bubbles cannot move under the small pressure gradients present. Dissolution and trapping happen on timescales that range from centuries to a few thousand years, depending on the permeability of the formation (Riaz et al., 2006; Ide et al., 2007). On longer timescales (multiple thousands of years) chemical reactions can convert some of the CO_2 to solid materials, depending on the composition of the brine and the minerals present in the rock.

Safe operations of storage sites will require that the amount of CO_2 allowed to escape from the deep storage zone to the near-surface environment be very small. Oil and gas reservoirs provide an example of the kind of storage settings that retain buoyant fluids for geologic time periods. Nevertheless, it is possible that some storage sites might leak, and hence the quantitative impact of leakage has been assessed. Based on IPCC emission scenarios, Hepple and Benson (2005) argue that overall rates of leakage less than 0.01 to 0.1 percent annually of the total amount stored would be sufficient to allow CCS to contribute effectively to the stabilization of CO_2 concentration in the atmosphere, depending on the target. If leakage occurs, it is more likely to happen relatively early in the life of a storage site, when pressures are highest around an injection well. Wells are the most likely leak path, but well leakage is readily detected and can be repaired.

Careful attention to leakage hazards will be required in any CCS project. At low concentrations in air, CO_2 is not dangerous. It is a normal component of air, and large power plants currently emit millions of tonnes per year directly to the atmosphere. At high concentrations, however, it is an asphyxiant and is toxic. A concentration of 4 percent CO_2 is immediately dangerous to health, and the NIOSH and OSHA exposure limits (NIOSH, 1996) are 5000 ppm (0.5 percent). Because CO_2 is denser than air is, designing and monitoring CO_2 pipelines and wells to make sure that leaking CO_2 does not collect in low-lying areas is essential. Storage security generally increases with time after injection ceases (IPCC, 2005), as the highest subsurface pressures relax, as CO_2 dissolves in brine, and as trapping of CO_2 occurs. Monitoring schemes such as those used at Sleipner and other field tests (Chadwick et al., 2008; Daley et al., 2008) can be used to determine whether the CO_2 is remaining isolated from the surface over time.

Nontechnical Issues with CCS

Whichever of the three main options are used, significant regulatory issues will have to be addressed if geologic storage is to be undertaken on a large scale. These issues include long-term ownership of the CO_2, liability exposures over time, requirements for the monitoring of storage sites, and regulations for safe operation. Figure 7.A.7 outlines the decision points associated with the life cycle of a storage facility. For a detailed discussion of the many issues that arise in site selection and project design and implementation, see Chapter 5 of IPCC (2005).

Site screening will include matching of potential CO_2 sources and sinks, with appropriate attention to the feasibility of separating the CO_2 and transporting it to the storage location. Similarly, attention must be given to understanding the subsurface characteristics: in particular, the potential storage capacity, the permeability of the formation (which will control injection rates and pressures), the

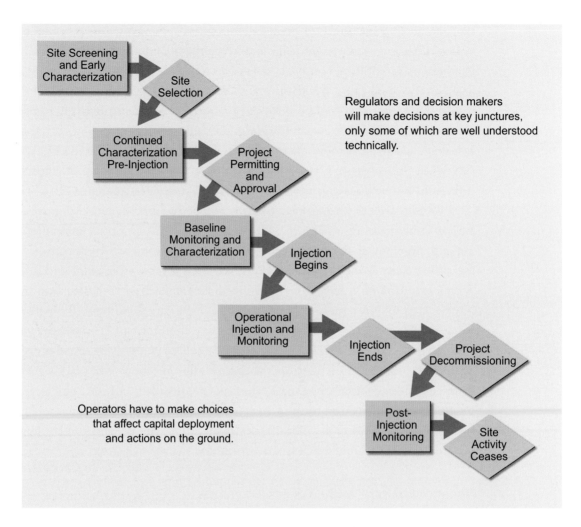

FIGURE 7.A.7 *Key steps in the implementation of a large-scale CO$_2$ storage project. Source: J. Friedmann, presentation at the AEF Committee's fossil fuels workshop, 2008.*

existence of appropriate barriers to vertical flow, and the absence of likely leak paths. Once potentially appropriate source/sink combinations have been identified, additional effort—including more detailed study of the properties of the geologic formation, the drilling of one or more test wells, and analysis of rock samples—will be required to refine the characterization of the subsurface. In that way, predictions of flow behavior and the long-term fate of injected CO$_2$ can be made. These predictions will be part of a permitting process involving some combination of local, state, and federal regulatory agencies, depending on the specific location (Wilson et al., 2007).

As part of this process, an appropriate level of project-performance monitoring will need to take place. A number of geophysical techniques are available for monitoring the movement of the CO_2 in the subsurface and the vicinity of the injection project (see Table 5.4 of IPCC [2005]). Which techniques are appropriate will depend on the geologic setting and on the project's stage. During the injection phase, monitoring activities will likely be more extensive in the initial years to ensure that the injected CO_2 is entering the intended formation and that surface leaks in the injection area do not occur (wells and the associated pipes and fittings are the most likely sources of leaks, but they can also be repaired most easily). Time-lapse seismic methods have been demonstrated for detection of the subsurface movement of the CO_2, and electromagnetic and gravity surveys may also be used in some settings. After injection of CO_2 ceases, there will still be a period in which gravitational forces cause the buoyant CO_2 to move in the subsurface, but the rate of movement will decline with time; hence the need for frequent monitoring activities will also decline.

Many issues associated with the development of appropriate regulatory processes remain to be resolved. In particular, what entity would bear long-term liability after injection has ceased? Testing of geologic storage of CO_2 is allowed under existing regulatory structures, but these regulations must be further developed in order to embrace large-scale implementation of CO_2 injection for the purpose of avoiding emissions of CO_2 to the atmosphere (Wilson et al., 2007). That development process is now in the beginning stages. In July 2008, the U.S. Environmental Protection Agency issued proposed rules for regulation of underground injection of CO_2 under the Safe Drinking Water Control Act (www.epa.gov/safewater/uic/wells_sequestration.html#regdevelopment). The rules would create a new class of injection well for CO_2 as part of the Underground Injection Control program. They would also include requirements for storage-site characterization, injection-well design and testing, monitoring of project performance, and demonstration of financial responsibility.

References for Annex 7.A

Allison, E. 2000. Department of Energy Methane Hydrate Research and Development Program: An update. Annals of the New York Academy of Sciences 912:437-440.

Anderson, S.R., and R. Newell. 2004. Prospects for carbon capture and storage technologies. Annual Review of Environment and Resources 29:109-142.

Bauer, C. 2008. Presentation to the workshop of the Fossil Energy Subgroup of the AEF Committee, National Research Council, Washington, D.C., January 29–30.

British Petroleum. 2006. BP Statistical Reviews of World Energy 2006. Available at www.bp.com/liveassets/bp_internet/Switzerland/Corporate_Switzerland/STAGING/local_assets/downloads_pdf/pq/pm_statistical_review_of_world_energy_full_report_2006.pdf.

Chadwick, R.A., D. Noy, R. Arts, and O. Eiken. 2008. Latest time-lapse seismic data from Sleipner yield new insights into CO_2 plume development. Presented at the Ninth Greenhouse Gas Technology Conference, Washington, D.C., November 16–20, 2008.

Council of Canadian Academies. 2008. Energy from Gas Hydrates: Assessing the Opportunities & Challenges for Canada.

Daley, T., L. Myer, J. Peterson, E. Majer, and G. Hoversten. 2008. Time-lapse cross-well seismic and VSP monitoring of injected CO_2 in a brine aquifer. Environmental Geology 54:1657-1665.

Dooley, J.J., R.T. Dahowksi, C.L. Davidson, M.A. Wise, N. Gupta, S.H. Kim, and E.L. Malone. 2006. Carbon Dioxide Capture and Geologic Storage. Technical Report. Global Energy Technology Strategy Program, Battelle, Joint Global Change Research Institute.

EIA (Energy Information Administration). 2008. Monthly Energy Review. DOE/EIA-0035(2008/04). Washington, D.C.: U.S. Department of Energy, Energy Information Administration.

EPRI (Electric Power Research Institute)–TAG. 1993. Technical Assessment Guide (TAG) Electricity Supply—1993. TR-102276-V1R7. Palo Alto, Calif.: Electric Power Research Institute.

Hepple, R.P., and S.M. Benson. 2005. Geologic storage of carbon dioxide as a climate change mitigation strategy: Performance requirements and the implications of surface seepage. Environmental Geology 47:576-585. DOI 10.1007/s00254-004-1181-2.

Ide, S.T., K. Jessen, and F.M. Orr, Jr. 2007. Storage of CO_2 in saline aquifers: Effects of gravity, viscous, and capillary forces on amount and timing of trapping. Journal of Greenhouse Gas Control 1:481-491.

IPCC (Intergovernmental Panel on Climate Change). 2005. Special Report on Carbon Capture and Storage. Prepared by Working Group III of the Intergovernmental Panel on Climate Change. New York and London: Cambridge University Press.

IRGC (International Risk Governance Council). 2008. Regulation of Carbon Capture and Storage. Available at www.irgc.org/IMG/pdf/Policy_Brief_CCS.pdf. Accessed May 4, 2009.

Jayasinghe, A.G., and J.L.H. Grozic. 2007. Gas hydrate dissociation under undrained unloading conditions (abstract). P. 61 in Submarine Mass Movements and Their Consequences. Vol. IGCP-511. UNESCO.

MIT (Massachusetts Institute of Technology). 2007. The Future of Coal: Options for a Carbon-Constrained World. Cambridge, Mass.: MIT.

Moridis, G.J., T. Collett, R. Boswell, M. Kurihara, M.T. Reagan, C. Koh, and E.D. Sloan. 2008. Toward production from gas hydrates: Current status, assessment of resources, and simulation-based evaluation of technology and potential. Paper SPE 114163. Presented at the SPE Unconventional Reservoirs Conference, Keystone, Colo., February 10–12, 2008.

NETL (National Energy Technology Laboratory). 2007a. Cost and Performance Baseline for Fossil Energy Plants. DOE/NETL-2007/1281, Revision 1. U.S. Department of Energy.

NETL. 2007b. Carbon Sequestration Atlas of the United States and Canada. U.S. Department of Energy, Office of Fossil Energy. Available at www.netl.doe.gov/technologies/carbon_seq/refshelf/atlas/ATLAS.pdf.

NIOSH (National Institute of Occupational Safety and Health). 1996. Documentation for Immediately Dangerous to Life or Health Concentrations/Carbon Dioxide. Available at www.cdc.gov/niosh/idlh/124389.html.

NPC (National Petroleum Council). 2007. Facing the Hard Truths About Energy: Topic Paper No. 19. Washington, D.C.: NPC.

Palmgren, C., M.G. Morgan, W. Bruine de Bruin, and D.W. Keith. 2004. Initial public perceptions of deep geological and oceanic disposal of carbon dioxide. Environmental Science and Technology 38:6441-6450.

Reiner, D.M., and H.J. Herzog. 2004. Developing a set of regulatory analogs for carbon sequestration. Energy 29:1561-1570.

Riaz, A., M. Hesse, H. Tchelepi, and F.M. Orr, Jr. 2006. Onset of convection in a gravitationally unstable, diffusive boundary layer in porous media. Journal of Fluid Mechanics 548:87-111.

Ruppel, C. 2007. Tapping methane hydrates for unconventional natural gas. Elements 3:193-199.

Sheppard, M.C., and R.H. Socolow. 2007. Sustaining fossil fuel use in a carbon-constrained world by rapid commercialization of carbon capture and sequestration. AIChE Journal 53:3022-3028.

Socolow, R.H. 2005. Can we bury global warming? Scientific American (July):49-55.

Wilson, E.J., S.J. Friedmann, and M.F. Pollak. 2007. Risk, regulation, and liability for carbon capture and sequestration. Environmental Science and Technology 41:5945-5952.

8

Nuclear Energy

Utilities in the United States have recently expressed renewed interest in adding new nuclear power plants to their mix of electricity generation sources. As of July 2009, the U.S. Nuclear Regulatory Commission (USNRC) had received 17 applications for combined construction and operating licenses[1] for 26 units, and it expects to receive a total of 22 applications for 33 units by the end of 2010.[2] The 104 currently operating nuclear plants (largely constructed in the 1970s and 1980s) contribute substantially to the U.S. electricity supply: nuclear power provides 19 percent of U.S. electricity as a whole and about 70 percent of electricity produced without greenhouse gas emissions from operations. These plants provide electricity safely and reliably, and they have operated with capacity factors greater than 90 percent over the last few years.[3] Still, hurdles remain, and no new nuclear plants have been ordered in the United States in more than 30 years.

This chapter discusses the prospects for the future use of nuclear power in the United States, including an assessment of future technologies, deployment

---

[1]Previously, the licensing process had two steps, construction and operation, each of which required a different license to be issued. The Combined Construction and Operating License is a part of the USNRC's new "streamlined" application process.

[2]The USNRC's lists of received and expected applications are available at www.nrc.gov/reactors/new-reactors/col.html and at www.nrc.gov/reactors/new-reactors/new-licensing-files/expected-new-rx-applications.pdf, respectively; accessed July 2009.

[3]The net capacity factor of a power plant is the ratio of the actual output of a power plant over a period of time and its projected output if it had operated at full nameplate capacity the entire time.

costs, and the barriers to and impacts of increased nuclear power plant deployments by 2020, by 2035, and by 2050.

Interest in new nuclear construction has also been growing around the globe, and with a new element: interest among countries that do not currently have nuclear plants. According to the International Atomic Energy Agency (IAEA), in excess of 40 new entrant countries have expressed interest, of which 20 are actively considering construction (IAEA, 2008a).

In addition, the IAEA has recently estimated that 24 of the 30 countries with existing nuclear plants intend to build new reactors—a departure from policies of the past few decades in many countries (IAEA, 2008a). Following the Chernobyl accident in 1986, Italy banned construction of new nuclear reactors; the governments of Sweden and Germany pledged to phase out their own nuclear plants; resistance to new construction in the United Kingdom was strong; and Spain put in place a moratorium on new construction. These attitudes are now changing, likely as a result of subsequent uneventful nuclear operations and growing concerns about climate change and future energy needs.

Thus, Italy has announced plans to build nuclear plants; Sweden, after shutting down two plants, intends to reverse the planned phase-out and construct new nuclear plants; and the Labor government in the United Kingdom has recently announced plans to replace 18 nuclear plants retiring by 2023 with new ones.[4] But this new outlook is not universal. The current head of the Spanish government remains opposed to nuclear power, and the current government in Germany still intends to shut down its 17 remaining nuclear plants. Meanwhile, new construction is planned or under way in Finland, France, and Japan, countries that never wavered in their support of nuclear power.

Overall, the IAEA projects that by 2030, world nuclear capacity could

---

[4]Press articles discussing these developments in more detail include "Recalled to half-life," *The Economist*, Feb. 12, 2009 (www.economist.com/world/europe/displaystory.cfm?story_id=13110000); "What Sweden's nuclear about-face means for Berlin," *Der Spiegel*, Feb. 6, 2009 (www.spiegel.de/international/world/0,1518,605957,00.html); "Italy seeks nuclear power revival with French help," *Reuters*, Feb. 24, 2009 (uk.reuters.com/article/oilRpt/idUK-LO72469220090224); "Spain must reconsider nuclear energy," *La Vanguardia*, Feb. 25, 2009 (www.eurotopics.net/en/search/results/archiv_article/ARTICLE458-0); "Governments across Europe embrace nuclear energy," ABC, Mar. 4, 2009 (www.abc.net.au/pm/content/2008/s2507565.htm); and "Europe looking set for a Nuclear Revival," *Your Industry News*, Mar. 6, 2009 (www.yourindustrynews.com/europe+looking+set+for+a+nuclear+revival_26046.html). These articles were accessed in July 2009.

increase by 27 percent under business-as–usual conditions, or in the agency's "high case,"[5] to nearly double, after accounting for retirements (IAEA, 2008b). Nonetheless, even in the high-case projections, nuclear power would rise only slightly as a percentage of total electricity generated worldwide—from 14.2 percent in 2007 to 14.4 percent in 2030—assuming business as usual for construction of fossil-fueled plants.

The handful of plants that could be built in the United States before 2020, given the long time needed for licensing and construction, would need to overcome several hurdles, including high construction costs, which have been rising rapidly across the energy sector in the last few years, and public concern about the long-term issues of storage and disposal of highly radioactive waste.[6] If these hurdles are overcome, if the first new plants are constructed on budget and on schedule, and if the generated electricity is competitive in the marketplace, the committee judges that it is likely that many more plants could follow these first plants. Otherwise, few new plants are likely to follow.

Existing federal incentives[7] for the first few nuclear plants may hasten initial construction. Even if this occurs, nuclear power's share of U.S. electricity generation is likely to drop over the next few decades. In fact, for nuclear power to maintain its current share—19 percent of U.S. electricity—the equivalent of 21

---

[5]The IAEA's high estimates (IAEA, 2008b) "reflect a moderate revival of nuclear power development that could result in particular from a more comprehensive comparative assessment of the different options for electricity generation, integrating economic, social, health and environmental aspects. They are based upon a review of national nuclear power programmes, assessing their technical and economic feasibility. They assume that some policy measures would be taken to facilitate the implementation of these programmes, such as strengthening of international cooperation, enhanced technology adaptation and transfer, and establishment of innovative funding mechanisms. These estimates also take into account the global concern over climate change caused by the increasing concentration of greenhouse gases in the atmosphere, and the signing of the Kyoto Protocol."

[6]Both nuclear plants and coal plants with carbon capture and storage (CCS) present intergenerational issues: nuclear plants because of the very long-lived radioactive waste, and coal with CCS because of the need for stored CO_2 to remain underground for long periods. However, the timescales differ by orders of magnitude. For radioactive waste, this timescale is on the order of a million years; for CO_2 it is likely significantly less because of the availability of natural mechanisms for removing CO_2 from the atmosphere (see Ha-Duong and Keith, 2003; Hepple and Benson, 2005).

[7]In addition to federal incentives for construction, the first few nuclear plants benefit from incentives for operation, such as the production tax credit. This is discussed in more detail in Box 8.5 in this chapter.

new 1.4 GW plants would need to be built by 2030 (not including new plants built to replace any that may be retired during this period), according to the reference-case projections of the U.S. Energy Information Administration (EIA, 2008).[8]

The amount of new U.S. nuclear generating capacity that could reasonably be added before 2020 is limited; however, if the first handful of new evolutionary plants (about 5 plants) are constructed and are successful, the potential for nuclear power after 2020 will have much increased. Thus, deployment of the first few nuclear plants would be an important first step toward ensuring a diversity of sources for future electric supply. It may prove to be important to keep the option of an expanded nuclear deployment open, particularly if carbon constraints are applied in the United States in the future.

TECHNOLOGIES

The existing nuclear plants in the United States were built with technology developed in the 1960s and 1970s. In the intervening decades, ways to make better use of the existing plants have been developed, as well as new technologies that are intended to improve safety and security, reduce cost, and decrease the amount of high-level nuclear waste generated, among other objectives. These technologies and their potential for deployment in the United States are explored in the following sections.

Improvements to Existing Nuclear Plants

Over the last few decades, there have been significant technical and operational improvements in existing nuclear power plants. These improvements have allowed nuclear power to maintain an approximately constant share of U.S. electrical capacity, even as demand has grown and no new plants have been constructed. This trend of increasing output from current plants is likely to continue over the coming decades and, before 2020, could result in additional nuclear capacity comparable to what could be produced by new plants. The potentials for improvements are focused in the following three areas:

---

[8]According to the EIA, U.S. electricity demand could rise by as much as 29 percent between 2008 and 2030. The reader is referred to footnote 14 of Chapter 7 of this report ("Fossil-Fuel Energy") for a discussion of uncertainty in EIA projections.

- Existing plants can be modified to increase their power output;
- Existing plants' operating lives can be extended; and
- Downtimes (periods when the plant is not producing power) can be further reduced.

Such improvements, which are far less expensive than constructing new nuclear plants and can be implemented comparatively rapidly, are discussed below.

Power Uprates

A plant's power output can be significantly increased (uprated) by replacing the fuel with higher-power-density/longer-lived fuel and by modifying major plant components. The latter includes, for example, replacing turbines and major heat exchangers with more efficient versions. Uprates are a cost-effective way to increase energy production: they typically cost hundreds of dollars per added kilowatt (kW) of capacity, compared to as much as $3000–6000 (overnight cost[9]) per kilowatt of electricity for new nuclear plants (see section on "Costs"). To date, 7.5 gigawatts-electric (GWe)[10]—amounting to about 7.5 percent of the current U.S. nuclear generating capacity—have been added through uprates.[11]

Many plants have already planned capacity additions. In 2008 alone, the USNRC approved 10 upgrades to existing plants, adding a total generating capacity equivalent to about half of one new nuclear plant. Eleven applications are pending, and the USNRC expects 40 more applications through 2013.[12] If

---

[9]Overnight cost is the cost of a construction project if no interest was incurred during construction, as if the project was completed "overnight." All costs are expressed in 2007 dollars.

[10]The electric power output of a nuclear power plant is often described in gigawatts-electric (or simply gigawatts [GW]). Similarly, the thermal power output of a nuclear plant is stated in gigawatts-thermal (GWt). The thermal power output is typically about three times the electric power output. This is because the thermal efficiency of nuclear plants (the efficiency of converting heat to electricity via a steam turbine generator) is typically around 33 percent.

[11]The USNRC's list of approved uprate applications is available at www.nrc.gov/reactors/operating/licensing/power-uprates/approved-applications.html; accessed July 2009.

[12]In 2008, applications were approved for capacity additions of about 2178 MWt. This would result in about 720 MWe of new electric generating capacity. New plants are assumed to have a capacity of 1.35 GWe. Pending applications represented a total of 973 MWe of capacity additions as of July 2009, and applications expected at that time represented 2075 MWe of capacity additions. The USNRC's lists of pending and expected applications are available at www.nrc.gov/reactors/operating/licensing/power-uprates/pending-applications.html (pending) and www.nrc.gov/reactors/operating/licensing/power-uprates/expected-applications.html (expected); accessed July 2009.

approved and undertaken, these uprates would add about 3 GWe—the equivalent of about 2 new nuclear plants—in the near term.

Operating License Extensions

More power can be also be generated over the lifetimes of existing plants by extending their operating licenses. In the United States, the initial license term for a nuclear power plant—40 years—is subject to extensions in increments of up to 20 years.[13] In the 1990s, the USNRC established a regulatory system to assess applications for such extended licenses.

In the majority of cases, the owners of the currently operating U.S. plants will seek to extend plant licenses for an additional 20 years, to 60 years' service in total. As of July 2009, 56 plants had received 20-year extensions, 16 plants were in the queue for approval, and 21 more had announced their intent to seek license extensions.[14] The original 40-year limit was not technically based, but some technical challenges are involved in extending operating licenses because some structures and components may have been engineered assuming a 40-year operating life. This limitation will be avoided in new plants, which are being designed to ensure that components with expected lifetimes of less than a projected plant life of 60 years can be replaced readily.

The industry has begun to assess whether it would be technically feasible and economic to extend current plant operating licenses for an *additional* 20-year period beyond 60 years (to 80 years). The plant modifications that might be required for another 20-year extension are potentially more difficult and expensive than those for the first 20-year extension. Degradation phenomena that affect the performance of plants operating for as long as 80 years are not well understood at a fundamental level, and further research is needed prior to decisions about further license extensions. At this point, it is not clear whether the option will be practical, although there will be strong economic incentives to pursue it.

The USNRC, the U.S. Department of Energy (DOE), and industry are considering what research and development (R&D) will need to be done to prepare for the possibility of extending plant operating licenses beyond 60 years. Although participants in an USNRC/DOE workshop held in February of 2008 "did not

---

[13]This was provided for in the Atomic Energy Act of 1954.

[14]The USNRC's list of current and expected operating life extensions is available at www.nrc. gov/reactors/operating/licensing/renewal/applications.html; accessed July 2009.

believe there is any compelling policy, regulatory, technical or industry issue pre-cluding future extended plant operation" (USNRC/DOE, 2008), many areas were identified where R&D should begin soon. They included irradiation effects on primary structures and components (such as the reactor vessel, reactor coolant system piping, steam generators, pressurizer, and coolant pumps), aging effects on safety-related concrete structures, aging effects on safety-related cable insulation, and inspection capabilities for aging mechanisms.

Much of the equipment that is of concern is embedded in the structure of the plant and would be expensive and time-consuming to replace. Thus many of the issues imposed by plant lifetime extensions and materials aging require ways of nondestructively assessing the status of operating plants. New scanning systems are being developed, but further research is needed, particularly in light of the regulatory decisions that could rely on these inspections.

Decreasing Downtimes

Finally, more power can be generated over the course of a year by reducing the periods when the plants are not producing electricity. Existing plants have been operated with increasing efficiency over time, and average plant capacity factors (averaged across all operating nuclear plants) have increased markedly, from 66 percent in 1990 to 91.8 percent in 2007 (NEI, 2008). Nuclear plant operators in the United States have succeeded in reducing downtimes primarily through increased on-line maintenance as well as through efforts to plan outage times so as to ensure that necessary work is done quickly and efficiently.

As a result of such improvements, refueling outages—which are also used to perform necessary maintenance on the reactor—were reduced to an average of 40 days in 2007 (averaged across all currently operating U.S. plants) from 104 days in 1990. Based on the accomplishments of the best-performing plants to date, in the future these downtimes may be reducible to an average of 25–30 days while maintaining currently high levels of safety and reliability.

Nuclear Reactor Technologies[15]

A nuclear reactor generates heat by sustaining and controlling nuclear fission, and that heat is converted to electricity. The dominant use of nuclear reactor technol-

---

[15]For a more thorough treatment of many of the issues reviewed briefly in this section, see Annex 8.A.

ogy is in commercial nuclear power plants, which contribute baseload[16] electric power generation.[17] Nuclear plants can each include one or more nuclear reactors.

The waste heat[18] from nuclear reactors can be utilized as well. For example, several countries, including Russia and the Ukraine, use nuclear reactors for cogeneration (or combined heat and power [CHP]). Particularly effective in cold regions, CHP uses waste heat from nuclear reactors to create steam, which is piped to heat surrounding areas. Such systems in nuclear plants have been discussed in the United States, but they are not currently deployed. In other countries (for example, Japan, India, and Pakistan), waste heat from nuclear plants is used for desalinization of seawater.

The majority of reactors used for electricity generation around the world are pressurized water reactors (PWRs) and boiling-water reactors (BWRs), reactors that are collectively referred to as light-water reactors (LWRs)—that is, they are thermal reactors (see Box 8.1) that use ordinary water both as the coolant and as the neutron moderator. These are the only reactor technologies currently used in the United States for commercial power production, where 69 PWRs and 35 BWRs are currently in service.

New nuclear reactor designs have been developed in the decades since these plants were deployed. In the sections that follow, the committee discusses these new designs, which are grouped into two categories:

- *Evolutionary reactor designs*, which are modifications that have evolved from LWR designs currently operating in the United States
- *Alternative reactor designs*, which range from more significant modifications of currently deployed designs to entirely different concepts

---

[16]Baseload power is the minimum power that must be supplied by electric generation or utility companies to satisfy the expected continuous requirements of their customers. Baseload power plants generally run at steady rates, although they might cycle somewhat to meet some variation in customer demand. Typically, large-scale nuclear, coal, or hydroelectric power plants supply baseload power.

[17]Nuclear reactors are also used for propulsion (particularly for naval vessels), for materials testing, and for the production of radioisotopes for medical, industrial, test, research, and teaching purposes. In the past, nuclear reactors have also been used in space missions (primarily by Russia, but also by the United States) and for nuclear weapons materials production. Nuclear reactors dedicated to the production of nuclear materials have been shut down in the United States. This report focuses on nuclear reactors used for commercial electricity generation.

[18]A significant amount of the heat generated in a thermal power plant is not used to generate electricity; rather, it is vented through a cooling system to the outside environment.

BOX 8.1 ***Fast Reactors and Thermal Reactors***

Nuclear reactors are often classified as "fast" or "thermal" reactors. This nomenclature refers to the energy of the neutrons that sustain the fission reaction. In fast nuclear reactors, the fission reaction is sustained by neutrons at higher energies ("fast" neutrons); in commonly deployed thermal reactors, such as light-water reactors, the fission reaction is sustained by lower-energy ("thermal" neutrons). Fast and thermal reactors are distinguished by the presence or absence of a material known as a "neutron moderator," or simply "moderator." This material is present in thermal reactors but not in fast reactors. Collisions with the moderator slow the neutrons emitted by fissioning nuclei to thermal energies.

In the next few decades, the majority of the new nuclear plants constructed in the United States will be based on evolutionary reactor designs. In most cases, alternative reactor designs will require significant development efforts before they can be ready for deployment.

Evolutionary Reactor Designs

Any new nuclear plants constructed before 2020 will be evolutionary designs that are modifications (often significant) of existing U.S. reactors. These designs are intended to improve plant safety, security, reliability, efficiency, and cost-effectiveness. Some evolutionary designs include passive safety features that rely on natural forces, such as gravity and natural circulation, to provide cooling in the case of an accident. These features are intended to reduce capital cost while further enhancing safety margins.

Several evolutionary reactor designs will be ready for deployment in the United States after the USNRC completes design certification.[19] In some cases, this could occur as soon as 2010 or 2011. Evolutionary reactors have already been built in Japan and South Korea, and they are under construction in India, France, and Finland. U.S. utilities have expressed potential interest in building plants with the following designs in the United States: the U.S. evolutionary power reactor (USEPR), the economic simplified boiling-water reactor (ESBWR), the advanced boiling-water reactor (ABWR), the AP-1000, and the advanced pressurized water

---

[19]Before a nuclear plant of a new design can be constructed in the United States, the design must first be certified by the USNRC.

reactor (APWR). These designs are all modifications of current-generation LWR designs.[20]

Because construction of new nuclear plants is likely to require a long lead time, the first deployment of evolutionary nuclear reactors in the United States is unlikely to be until after 2015. Typical construction times for foreign plants have ranged from 4 to 7 years for plants that began construction in the last decade (IAEA, 2008c). Lead times for licensing and large component fabrication can also run to years. Current plans (as of July 2009) suggest that about 5–9 new nuclear plants could be on line in the United States by 2020, and a more substantial deployment of these plants may occur after 2020 if these first plants built in the United States meet cost, schedule, and performance targets. Moreover, actual construction will also depend on many other factors, including comparative economics and electrical demand.

Further R&D over the next decade could lead to efficiency improvements both in existing reactors and in evolutionary LWRs. Some of the key areas for continuing research include the following:

- *Improved heat transfer materials,* such as high-temperature metal alloys, are being developed to improve efficiency by allowing for higher operating temperatures. Some of these materials may be available after 2025. Widespread application is likely between 2035 and 2050.
- *Coolant additives,* such as very dilute additions of nanoparticles, can improve the heat transfer capabilities of the coolant in current and evolutionary LWRs. Twenty years or more are likely needed to develop the additives and redesign current reactors for their use.
- *Annular fuel rods* could allow plants to produce significantly more power than traditional cylindrical fuel rods do. At least 10 years of work will be needed for regulatory approval and commercial-scale deployment in existing LWRs.

---

[20]The ABWR and AP-1000 designs are currently certified by the USNRC, but applications for amendment have been received for the AP-1000 and are expected for the ABWR. The USNRC is currently reviewing design certification applications for the ESBWR, the USEPR, and the US-APWR designs. The review of the amended AP-1000 design and the ESBWR is targeted for completion in 2010, and for the USEPR and US-APWR in 2011. Available at www.nrc.gov/reactors/new-reactors/design-cert.html; accessed July 2009.

- *Higher burn-up fuel* would allow a larger percentage of the fissionable content of the fuel to be used. Thus operating cycles could be prolonged, and the heat load[21] and total amount of used nuclear fuel[22] to be stored or disposed of could be reduced.[23] This is a program of continuous improvement, but for significant breakthroughs, basic research will be required, particularly on fuel-rod swelling due to buildup of fission products and the resulting risk of cladding breach.
- *Digital instrumentation and control (DI&C)* research offers opportunities to improve control systems and to enhance control-room designs so as to facilitate appropriate operator action when needed. New LWRs will have fully integrated DI&C, and more research will be needed on the safety implications of an increased reliance on digital systems. Understanding the full implications of DI&C is likely to prove to be a long-term effort, despite the reliance on DI&C in the near term.

These types of R&D could improve both current and evolutionary reactors. However, evolutionary reactor technology is technically ready for deployment, and no major additional R&D is needed for an expansion of nuclear power through 2020, and likely through 2035.

Alternative Reactor Designs

In addition to the evolutionary reactor designs just discussed, alternative nuclear reactor designs are being developed (and, in some countries, have been used).[24]

---

[21]When nuclear fuel is removed from the reactor after use, it not only is highly radioactive, but also emits heat. This amount of heat emitted is known as the "heat load" of the fuel.

[22]"Used nuclear fuel" (also referred to elsewhere as spent nuclear fuel, or SNF) refers to fuel that is removed from a nuclear reactor after use. As discussed later in this chapter, only a small fraction of the energy potentially available in the fuel is used.

[23]The total amount of used fuel to be disposed of would be reduced with higher burn-ups because fewer fuel assemblies would need to be used to produce the same power output. Although high burn-up decreases the amount of nuclear fuel remaining in the fuel assemblies after use, for the first century or so, heat and radioactivity are the major challenges for used fuel disposal. This initial heat and radioactivity are dominated by fission products, isotopes produced as a result of the fission of a massive atom such as U-235.

[24]For example, as mentioned previously, sodium-cooled and gas-cooled reactors have been in operation around the world for decades. These designs are significantly different from the light-water reactor (LWR) designs currently in use in the United States, and new U.S. deployments of these reactors are considered here as "alternative" designs.

These reactors range from more significant modifications of currently operating U.S. reactors to completely different concepts. Many new alternative reactor designs are intended to increase safety and efficiency and improve economic competitiveness, as well as to perform missions beyond electricity production.

The alternative reactor technologies that could be deployed in the United States include fast and thermal reactor designs. Both include modular designs as well as designs modified for high-temperature heat output (potentially for applications such as hydrogen production). Alternative fast reactor designs also include "burner" reactors—reactors intended to reduce the long-lived high-level radioactive waste burden by destroying transuranic[25] elements—and "breeder" reactors—reactors intended to create more fissile material than is consumed.

Some alternative thermal reactor designs, including small modular LWRs, could be deployed in the United States shortly before or after 2020. For example, NuScale, Inc., has expressed interest in deploying a 45-MWe design before 2020. In addition, under the Next Generation Nuclear Plant (NGNP) program, the DOE is continuing to develop a commercial-scale prototype very-high-temperature reactor (VHTR)[26] that would produce not only electricity but also process heat for industry. Hydrogen production is a possibility as well if materials—particularly for the heat exchangers and hydrogen process equipment—able to withstand the necessary high temperatures can be developed.[27] The DOE requested expressions of interest in April 2008 for a demonstration high-temperature nuclear plant that could produce hydrogen and electricity;[28] current plans are for start-up in 2018–2020.

---

[25] "Transuranic elements" (also known as transuranics or TRU) are elements with an atomic number greater than uranium—that is, having nuclei containing more than 92 protons. Examples of transuranics are neptunium (atomic number 93), plutonium (94), and americium (95).

[26] The NGNP will have somewhat lower outlet temperatures than originally envisioned for a VHTR.

[27] At a briefing of the Nuclear Energy Advisory Committee (NEAC) in September 2008, DOE staff stated that they had reduced their high-temperature goal to 700–800°C because they did not have materials suitable for operation at higher temperatures. This situation makes hydrogen production problematic for the near term.

[28] The DOE's request for expressions of interest for high-temperature nuclear plants is available at nuclear.gov/pdfFiles/NGNP_EOI.pdf; accessed July 2009.

In contrast, many R&D issues must be successfully addressed before fast reactors—particularly fast burner reactors—can be expected to make a contribution to U.S. energy production. A great deal of engineering development work will be required to move these reactor designs from the drawing board through prototypes and pilot plants to full-scale facilities. In addition, further study will be needed to improve reliability and reduce costs (some experts have estimated fast reactors may cost between 10 and 30 percent more than LWRs, as discussed in the section titled "Cost of Alternative Plant Designs and Fuel Cycles" later in the chapter).

While other types of fast reactors are under investigation, fast burner reactor technology has been emphasized in the United States because of concerns about high-level radioactive waste management. In principle, by using alternative fuel-cycle technologies, transuranics from used fuel can be incorporated into burner reactor fuel and then fissioned, as discussed in more detail in the following section. This option has the potential to reduce the volume and heat load of residual high-level radioactive waste that needs to be managed for very long times.[29] Fast reactor technologies are not new, and historically, those that have been deployed have experienced problems.[30] But it is the committee's judgment that, although deployment should not be pursued at present, the long-term potential provides justification for a continued R&D program on fast burner reactors and associated fuel-cycle technologies.[31] If this R&D is undertaken, the committee judges that the first generation of fast burner reactors to transmute nuclear wastes has the technological potential to come on line after 2025, and they could be deployed commercially after 2035 if they prove economically competitive.

---

[29]The volume of long-lived radioactive waste is not the only important consideration for the disposal; see footnote 42 for further discussion.

[30]These reactors demonstrated significantly less reliability than did LWRs, and they suffered from sodium leaks and fires. MONJU, a sodium-fueled fast reactor in Japan, suffered a sodium leak a year after being brought on line in 1994. In addition, the SuperPhenix reactor in France had many problems with sodium leaks, and it was shut down in 1998, having operated at full capacity for only 174 days. At present, only one fast reactor in the world (the BN-600 in Russia) is operating for electricity production.

[31]The Obama administration (as exemplified in the president's fiscal year 2010 budget request) intends to continue funding R&D for fast reactor technology (including fast burner reactors) but to discontinue the previous administration's plans for near-term deployment of these technologies.

Alternative Fuel Cycles[32]

Nuclear fuel cycles are divided into two major categories: once-through, in which the fuel is removed from the reactor after use and disposed of, and closed, in which the used fuel is recycled to extract more energy or to destroy undesirable isotopes. Recycling used fuel requires several steps, including chemical or electrochemical processing to separate the fissionable parts of the used fuel and to enable the fabrication of new fuel,[33] which can then be utilized in a reactor. To achieve high efficiency for burning or breeding, multiple repetitions of this process are required.

The United States currently uses a once-through fuel cycle, though U.S. policy on closing the nuclear fuel cycle has varied over time.[34] As of the writing of this report, the Obama administration had announced plans to pursue "long-term, science-based R&D . . . focused on the technical challenges of the back end of the nuclear fuel cycle" but not to pursue near-term commercial demonstration projects for closed fuel cycle technologies at present.[35]

Closed fuel cycle technologies (for either burning or breeding) are not needed to enable the near-term expansion of nuclear power in the United States, at least until 2050. Uranium supplies are sufficient to support a worldwide expansion of nuclear power using a once-through fuel cycle for the next century. Moreover, used fuel from even a greatly expanded nuclear fleet can be safely stored for up to a century (APS, 2007; Bunn, 2001), with or without a licensed geologic repository. In addition, a closed fuel cycle raises proliferation issues that

---

[32]The term "fuel cycle" describes the life cycle of a nuclear reactor's fuel. For a more thorough treatment of many of the issues reviewed in this section, the reader is referred to Annex 8.B ("Alternative Fuel Cycle Technologies").

[33]In addition to new fuel, it is also technically possible to form transuranic targets, which are specialized assemblies designed for burning transuranics in thermal reactors. This possibility is discussed in Annex 8.B.

[34]The Nixon administration supported closing the fuel cycle. The Ford and Carter administrations opposed it. Under the Reagan administration, reprocessing again became a possibility, but industry concluded that it was not economic. The first Bush administration followed the lead of the Reagan administration, but the Clinton administration opposed the use of reprocessing. The policy of the second Bush administration was to establish a geologic repository at Yucca Mountain, Nevada, for the disposal of used fuel, and it also wished to implement a program that would explore closing the fuel cycle in the longer term while pursuing a limited recycle option in the near term. As part of this program, a specific closed fuel cycle was selected for investigation by the DOE. There is no legal bar to reprocessing in the United States today.

[35]Available at www.neimagazine.com/story.asp?sectionCode=132&storyCode=2052719; accessed July 2009.

have not been resolved.[36] The benefits and drawbacks of deploying closed fuel cycles in the United States are currently being debated, as discussed in Box 8.2.

To implement closed fuel cycles, separations technologies are needed to process the used fuel so that it can be formed into new fuel. The current-generation technology for such recycling is plutonium and uranium extraction (known as PUREX), which is well understood and could be deployed in the United States after 2020,[37] but it carries a significant proliferation risk.[38] Alternatives to PUREX currently under investigation include both evolutionary modifications of PUREX and entirely different separations technologies. A modified PUREX technology currently under development that allows some amount of uranium to remain in the plutonium stream would be somewhat more proliferation resistant than is PUREX,[39] but it would not likely be commercially deployable until well after 2020.[40] Other separations technologies, intended to further improve proliferation resistance as well as to reduce the volume and long-term radioactivity of the waste, are even farther from the commercial deployment stage. These technologies include UREX+ (a suite of aqueous processes best suited for oxide fuels such as those used in LWRs) as well as electrochemical separations.[41] Neither process is likely to be available for commercial-scale deployment before 2035.

---

[36]The reader is referred to the "Impacts" section of this chapter for a more detailed discussion of uranium supplies, impacts of used fuel storage, and proliferation concerns.

[37]Although the PUREX technology is well understood, reprocessing plants have not been built in the United States in decades. Designing, licensing, and building a reprocessing plant is likely to push potential commercial deployment past 2020.

[38]In this context, "proliferation" refers to the spread of nuclear-weapons-related technology and know-how. Because PUREX involves the production of a stream of separated plutonium, the opportunity arises for this material to be diverted for use in a nuclear weapon. The United States is a nuclear weapons state, and the primary proliferation risk applies to the use of such technologies outside the United States—in countries that are not weapons states. However, there is also concern about theft of weapons-usable materials from reprocessing wherever it takes place. In addition, no reprocessing technology is completely proliferation resistant, and none of the technologies currently under development would be deployabled in nonnuclear weapons states without causing significant proliferation concerns. For further discussion of this issue, the reader is referred to the "Impacts" section of this chapter.

[39]In a recent report, the National Research Council concluded that small adjustments to this process could convert it to PUREX (NRC, 2008).

[40]Modified PUREX could technically be deployed in the United States shortly after 2020. However, higher cost projections for closed fuel cycles (compared to once-through fuel cycles) as well as political resistance are likely to push potential commercial deployment well past 2020.

[41]Electrochemical processing, also known as pyroprocessing, becomes more attractive if metal fuels are used for the burner reactors or if a preprocessing step is added for oxide fuels.

BOX 8.2 *Recycling of Used Nuclear Fuel*

Concerns about proliferation could discourage the United States from pursuing the commercialization of used fuel recycling at present.[1] Current technology for separations, plutonium and uranium extraction (PUREX), poses a proliferation risk, and modifications of PUREX to increase proliferation resistance do not greatly improve the situation. Some suggest that if the United States were to deploy recycle technologies (or, some argue, even pursue further R&D on them), it would become more difficult to stop other countries from doing the same. But others argue that the United States could positively influence recycling elsewhere by developing and deploying technologies that are more proliferation resistant than PUREX. Although future R&D may develop more proliferation resistant options, these options are highly unlikely to be *entirely* proliferation resistant. True proliferation control will require strong international arrangements to supplement technical advances, and developing such arrangements will require considerable time and effort.

It is the judgment of the committee that, at present, used fuel recycling does not appear to be a promising option for commercialization in the United States before 2035. However, the committee believes that a continuing R&D program on alternative fuel cycles is justified, as there may be a need for such technologies in the future.

---

[1] There is no bar in current law to prevent a private-sector company from seeking and obtaining a license from the USNRC to pursue recycling. One company (Areva) has indicated that it intends to pursue such a license (Energy Daily, 2008; Nuclear Fuel, 2009).

Further R&D will help to clarify the trade-offs between the risks and benefits involved in the use of recycle technology. For example, if proven technically successful and economic, burning fuel cycles (intended to reduce the volume of long-lived high-level radioactive waste) could, over the long term, substantially change the discussion on storage and disposal of radioactive waste. If a major fraction of the transuranics in high-level waste could be transmuted into shorter-lived fission products with half-lives of 1000 years or fewer, the waste-disposal challenge would involve managing the waste for thousands of years rather than hundreds of thousands of years. In addition, the number of geologic repositories needed to isolate long-lived high-level radioactive waste has the potential to be significantly reduced.[42] However, significant technology challenges must be overcome before

---

[42] The number of repositories needed is determined in large part by how closely stored waste can be packed. For about the first century, the heat and radioactivity that are the major chal-

burning fuel cycles could be ready for commercial deployment. Overall, it is the judgment of the committee that the potential benefits of burning fuel cycles are sufficient to justify continuing long-term R&D, but that the technologies are not yet ready for near-term deployment. Two major categories of burning fuel cycles—full recycle and limited recycle—as well as associated technology challenges are briefly discussed in the paragraphs that follow.[43]

A full recycle program (such as that envisioned by the second Bush administration) would involve processing used fuel, making new fuel using some of the recovered material, and using that fuel in fast burner reactors (discussed in the section titled "Nuclear Reactor Technologies"). This sequence would be repeated multiple times to destroy transuranics.[44] A fully closed fuel cycle would be designed to significantly reduce the volume of long-lived waste produced per kilowatt-hour, but this transmutation would never burn 100 percent of the long-lived isotopes. Hence a repository, or repositories, capable of sequestering very long-lived high-level civilian waste might still be needed.[45] In addition, a larger quantity of low-level waste[46] would be produced, primarily during used fuel processing and new fuel fabrication. Further R&D is needed in order for any fully closed fuel cycle to be ready for deployment, with long-term goals of this effort being reduction of the cost and proliferation risk of fuel cycle processes and their associated facilities. If such R&D were initiated, the committee judges that a fully closed fuel cycle could be reasonably deployed sometime after 2035 if shown to be economically competitive.

---

lenges for managing the high-level waste would be dominated by short-lived fission products. To achieve a substantial reduction in the number of repositories required, these fission products (in particular, cesium and strontium) would need to be separated from the waste destined for the geologic repository. Alternatively, the cesium and strontium potentially could remain in the waste and, in principle, the repository could be actively cooled for the first 100 years in order to achieve closer packing.

[43]Many of these technologies are discussed in more detail in Annex 8.B as well as in a previous National Research Council report on the DOE's nuclear energy R&D program (NRC, 2008).

[44]A large number of burner reactors would be required to enable full recycle; however, such a system has not been planned in detail, and the exact ratio of fast reactors to LWRs required is not well known.

[45]A repository for managing waste over hundreds of thousands of years would almost certainly be required for high-level defense waste.

[46]Low-level waste is a general term for a wide range of wastes having generally lower levels of radioactivity. See www.nrc.gov/reading-rm/basic-ref/glossary; accessed in July 2009.

Alternatively, options for burning transuranics using limited recycle in thermal reactors—such as inert matrix fuel and transuranic targets[47]—are also currently being investigated. Under these options, the used fuel from LWRs would be processed to separate plutonium and uranium from transuranics and other elements. In principle, new fuel or targets would then be formed (incorporating the transuranics to be destroyed), and some fraction of the transuranics would be burned in thermal reactors.[48] If successfully demonstrated and shown to be cost-effective, limited recycle could reduce the long-lived high-level waste burden without introducing the complication of fast reactors. (However, with repeated passes, a state of diminishing returns would be reached, and ultimately, a fast neutron spectrum would be required to continue to destroy transuranics.) For these technologies, more R&D as well as subsequent regulatory approval will be required if they are to be deployable between 2020 and 2035. As is the case with many of the alternative concepts, the economic viability of the approach is very uncertain.

Based on the preceding discussion, it is clear that pursuing alternative fuel cycle options (including burning fuel cycles) will require a resource-intensive and time-consuming R&D program. This finding is consistent with the conclusions of a recent National Research Council study that examined the DOE's nuclear energy R&D programs (NRC, 2008). Initially, further research would need to be done in comparing the various architectures for closing the fuel cycle; this effort would enable judicious selection of any specific architecture for eventual deployment. Moreover, the architecture for the fuel cycle would have to be coupled with a waste-disposal regime, and R&D would be needed before any of these fuel cycles would be ready for deployment, with the exception of mixed oxide (MOX)/PUREX. But that fuel cycle has significant proliferation risks. Indeed, any closed fuel cycle based on current designs is likely to be more expensive and to result in more proliferation risk than a once-through fuel cycle. Closed fuel cycle R&D should be directed toward solving these problems, and any alternative fuel cycle that is ultimately deployed should be designed to minimize stockpiles of separated weapons-usable materials.

---

[47]These options are discussed in more detail in Annex 8.B.

[48]Limited recycle is currently being applied outside the United States, where mixed-oxide (MOX) fuel is formed from used LWR fuel and utilized in commercial reactors. However, MOX fuel as currently implemented is not effective for destroying long-lived transuranics such as americium and neptunium, which are included in the waste stream. Under the limited recycle option, used MOX fuel is disposed of as high-level waste.

Fusion Energy

In principle, nuclear fusion[49] could offer a virtually unlimited supply of energy with significantly reduced (and shorter-lived) quantities of radioactive waste. Over the last 50 years, many countries (the United States, Russia, Japan, the United Kingdom, and others) have investigated the concept of controlled fusion for electricity production (NRC, 2004). There is a multinational effort under way to develop a "burning plasma"[50] machine, the International Thermonuclear Experimental Reactor (ITER), by 2025.[51] ITER is intended to provide the information needed to assess the practicality and cost of a fusion reactor. If successful, fusion reactors would be unlikely to be ready for commercial deployment until after 2050, absent some major breakthrough.

COSTS

The cost of uprating an existing nuclear plant to increase its power output can be reasonably well estimated; however, the costs of new nuclear technologies are uncertain. There has been recent interest in building evolutionary nuclear plants, for example, but companies' estimates of costs for construction vary widely. And the costs of alternative plants and fuel cycles are even less clear at this point. These cost issues are discussed in the following sections.

Costs of Improvements to Current Plants

Improving current nuclear plants for the purpose of increasing power output or extending operating lifetimes is significantly less expensive per kilowatt of capacity than constructing a new plant. Depending on the type of uprate, plant uprates can cost from hundreds of dollars to about $2000 per added kilowatt of capacity, while new plants could cost as much as $6000/kW (overnight cost), as noted in the section to follow. For a plant license extension to 60 years, there is the expense of developing the associated documentation (approximately $50–60 million), and

---

[49]In a fusion reaction, two light atomic nuclei combine to form a heavier nucleus. In doing so, energy is released that can be used to produce electricity.

[50]"The plasma is said to be burning when alpha particles from the fusion reactions provide the dominant heating of the plasma" (NRC, 2004, p. 1).

[51]This date may slip as the program moves beyond concept to construction.

in many cases there also are costs associated with replacing or modifying structures or components for longer operating life. As in the case of uprates, these expenses are small in comparison to those of new plants.

Costs of Electricity from Evolutionary Plants[52]

While the costs of building a nuclear power plant are relatively high, the costs for fuel, operations, and maintenance are relatively low. Because most nuclear power plants in operation in the United States have been fully amortized, the average operating cost of electricity from the current fleet of plants is modest—1.76¢/kWh in 2007—less on average than all other sources, with the exception of hydropower.[53] Although operating costs are likely to be low for new plants as well, the levelized cost of electricity (LCOE)[54] is likely to be relatively high because of the substantial construction costs. (See Box 8.3 for a discussion of the distinction between electricity cost and price.)

Recent cost estimates[55] for new nuclear plant construction differ by over a factor of two, in part because of the recent dramatic escalation in construction and materials costs that have affected construction costs for all types of energy facilities. Thus there is considerable uncertainty regarding any estimates now in the literature, as present conceptions of future costs are in flux. Another part of the uncertainty reflects the absence of recent U.S. experience.

The AEF Committee has developed estimates of the LCOE for new evolutionary nuclear plants using these recent cost estimates as a starting point and

---

[52]For a more thorough discussion of the committee's cost estimates, reviewed briefly in this section, the reader is referred to Annex 8.C ("Projected Costs for Evolutionary Nuclear Plants"). The estimates discussed in this section are limited to evolutionary reactor designs and assume a once-through fuel cycle.

[53]This information is available at www.nei.org/resourcesandstats/documentlibrary/reliable andaffordableenergy/graphicsandcharts/uselectricityproductioncosts/.

[54]The levelized cost of electricity at the busbar encompasses the total cost to the utility—including interest costs on outstanding capital investments, fuel costs, ongoing operation and maintenance (O&M) costs, and other expenses—of producing the power on a per-kilowatt-hour basis over the lifetime of the facility. This is not the same as the price of electricity to the consumer, particularly in states that have restructured their electricity markets.

[55]The range of estimates for the levelized cost of electricity is discussed in more detail in Annex 8.C. Multiple primary cost estimates were relied on by the committee (including Scroggs, 2008; Moody's Investor's Service, 2008; NEI, 2008b; Keystone Center, 2007; Harding, 2007).

BOX 8.3 *Levelized Cost of Electricity Versus Electricity Price*

In restructured markets, the price of electricity to the consumer is related to the cost of electricity to the distribution utility—as opposed to the cost to the merchant owner of the plant to produce that electricity. Utilities can either negotiate long-term contracts with independent power producers (IPPs) or buy electricity in the spot market from the IPPs.[1] In that market, the electricity price the utility must pay reflects the price of the most expensive electricity in the dispatched mix (the clearing price), rather than the levelized cost of electricity (LCOE) for a given plant. Thus for lower-priced sources of electricity, the utility may have to pay significantly more than the LCOE to the IPP, if the IPP can provide the power from a low-cost source. In recent years, the use of nuclear power plants has generally been very profitable for merchant producers because the prices they have obtained have generally been the much higher prices for electricity produced by natural gas plants.

---

[1] Utilities can also generate electricity using their own plants, particularly in traditional markets.

---

assuming that the plants come on line in 2020.[56] Estimates were obtained for two distinct cases: plants built by investor-owned utilities (IOUs) and those built by independent power producers (IPPs).[57] The cost of nuclear power at the busbar[58] is sensitive to the return on investment because of the high capital costs associ-

---

[56] The committee gathered ranges for the key modeling parameters from a variety of sources, with the help of a workshop that was convened in March 2008. Stakeholders in attendance reflected diverse viewpoints, including those prevalent in industry, nonprofits, and academia. The committee used these parameter ranges (discussed in detail in Annex 8.C) in a spreadsheet calculation based on the economic model developed for the 2007 study by the Keystone Center (2007) and supplemented by a Monte Carlo analysis. Thus, these costs are not forecasts or predictions, but rather the result of an analytical exercise based on available but imperfect data.

[57] Vertically integrated (typically investor-owned, but also municipal and public) utilities own generating plants as well as the transmission and distribution system that delivers the power to their customers. In the past, this was the dominant model, but restructuring of the electricity market in some states has transformed the industry. In restructured markets, generation, transmission, and distribution may be handled by different entities. For example, independent power producers (IPPs) may sell power to distribution utilities or even directly to end users.

[58] The "cost at the busbar" refers to the cost to the electricity producer; it does not include transmission or distribution costs.

ated with nuclear power.[59] In addition, a risk premium is likely to be expected by investors in plants built by IPPs because of the absence of the financial protections afforded to a regulated entity. For baseload electricity, cost comparisons between different options can be helpful to decisionmakers. The committee used different but comparable methods to estimate the LCOE for future nuclear and fossil plants (see Box 8.4).

For new nuclear plants that may be constructed between 2008 and 2020 the committee estimates that the LCOE from plants built by IOUs will fall between 8¢/kWh and 13¢/kWh and that the LCOE for plants built by IPPs will also be 8¢/kWh to 13¢/kWh, in 2007 dollars.[60] These ranges assume an overnight construction cost of between $3000 and $6000 per kilowatt, and a 4–7 year construction period.[61] These cost estimates also rely on several financial parameter ranges listed in Annex 8.C of this report, including a central debt-to-equity ratio of 60:40 for IPPs and 50:50 for IOUs. These estimates do not account for any current or future federal incentives for new plant construction.

In some cases, companies interested in building nuclear power plants have stated that their financial assumptions include an 80:20 debt-to-equity ratio (Turnage, 2008). Such a financing structure is likely to require federal loan guarantees—for example, those included in the Energy Policy Act of 2005 (discussed in more detail in Box 8.5). The committee estimated the LCOE of new nuclear plants using an 80:20 debt-to-equity ratio, and assuming that federal loan guarantees for 80 percent of the eligible project costs are acquired. These incentives could result in a significant reduction in financing costs, and ultimately a lower LCOE at the busbar: the estimated range decreases to 6–8¢/kWh both for

[59]The financial parameter ranges used for the cost calculations are shown in Table 8.C.1 in Annex 8.C.

[60]Although the costs of equity capital are likely to be cheaper for investor-owned utilities (IOUs), they are likely to take on a larger equity share than IPPs. For this and other reasons, including differences in the duration of equity repayment, the levelized cost of electricity (LCOE) for IOUs and IPPs turn out to be in the same range. However, it should be borne in mind that the ability of an IPP to compete in a restructured market depends more on the early year costs of electricity than the LCOE. Because the cost in the early years is generally greater than the LCOE, the IPP numbers here are not definitive in assessing the market competitiveness of IPP nuclear plants.

[61]These ranges encompass most of the values found in the open literature. A factor of 0.8 (derived using the Keystone spreadsheet used by the committee) was used to convert some all-in cost estimates to overnight costs, where appropriate, for comparison.

BOX 8.4 *Comparing the Methodologies Used to Determine Costs of New Nuclear and Fossil-Fuel Power Plants*

Nuclear and fossil-fuel-fired power plants provide baseload electricity supply, and a comparison of their potential cost ranges is likely to be helpful in guiding decision making. However, when making these comparisons using the data shown in this report, it should be noted that slightly different (but comparable) methodologies and assumptions have been used to estimate the ranges of potential LCOE from new fossil-fuel-fired power plants (with and without carbon capture and storage [CCS]) and from new nuclear power plants. (A discussion of the LCOE for intermittent renewable electricity sources, as well as of other energy technology options, such as energy efficiency technologies, can be found in Chapter 2.)

The methodologies for estimating the LCOE for nuclear plants and fossil-fuel plants differ, at least in part because different consultants assisted the committee in developing the LCOE estimates.

Although both nuclear and fossil-fuel plants provide baseload electricity and both of them are capital intensive, several of the underlying assumptions needed to calculate the LCOE are not identical. For example, a 20-year financing period was used to estimate the LCOE for new coal plants with CCS, while a 40-year financing period was used for new nuclear plants. A 20-year financing life is appropriate for a new technology such as CCS (whereby the first few plants may not operate for as long as later versions), while evolutionary light-water reactors are a more mature technology and thus more likely to operate for 40 to 60 years or beyond. In addition, the LCOE for coal plants with CCS drops between 2020 and 2035 as more experience is gained in building plants in the United States. The same reasoning has not been applied to nuclear plants, although some vendors expect that construction costs will be reduced over time, as there is more experience in constructing them. The LCOE for new nuclear plants does not change in current-year dollars between 2020 and 2035. Overall, the LCOE ranges for new coal plants with CCS and new evolutionary nuclear power plants appear to be comparable, as shown in Chapter 2 of this report.

IOUs and for IPPs.[62] The IPP first-year cost in this case is estimated to be between 7¢/kWh and 9¢/kWh. When the full 80 percent is guaranteed by the federal gov-

_____

[62]With the exception of the debt-to-equity ratio (80:20), the value used for return on debt (4 percent), and the addition of the loan guarantee fee required by the DOE, the assumptions are the same for this calculation as for the previous ranges. The details of these calculations can be found in Annex 8.C.

BOX 8.5 *Federal Incentives for New Nuclear Construction*

There are many policies that influence the viability of nuclear power in the United States.[1] In particular, several federal incentives for nuclear power are in place that could affect the potential for nuclear power plant construction. The primary ones are the federal loan guarantees and production tax credit (PTC) included in the Energy Policy Act of 2005 (EPAct05).[2] This law allows for a 1.8¢/kWh PTC for new nuclear power facilities for an 8-year period after the plant is placed in service (and before 2021) for the first 6000 MWe of installed capacity brought on line before 2021. This PTC could help the first few nuclear plants compete, but it does not change the cost of generating electricity. The loan guarantees are likely to have a larger effect.

EPAct05 allows the Secretary of Energy, after consultation with the Secretary of the Treasury, to provide loan guarantees for up to 80 percent of eligible project costs for nuclear plant construction. It is not yet clear if the $18.5 billion loan guarantee allocation for nuclear projects contained in the 2008 Energy and Water Development Appropriations Act will be sufficient to guarantee four to five new plants, which is the number the committee judges would be needed to demonstrate that new nuclear plants can be built on schedule and on budget in the United States. The DOE issued a loan guarantee solicitation announcement in June 2008, and the Part 1 applications that were filed in response to this solicitation requested a total of $122 billion.

To obtain a loan guarantee the licensee must pay a fee that is designed to cover the default risk, given a licensee's credit rating. This "loan guarantee subsidy fee" covers the estimated long-term cost to the government of the loan guarantee,[3] calculated on a net present value basis. The exact value of these fees has not been released by the DOE, but according to the agency's website (www.lgprogram.energy. gov, accessed May 12, 2009), they will be in accordance with the methods for calculating loan guarantee subsidy costs outlined in OMB Circular A-11, part 185.[4] Using information from that circular, Standard and Poor's has attempted to estimate potential ranges for the subsidy fee (although the precise methods of calculation of the fee are not publicly available). It found that, "[f]or example, if a 1,000 MW nuclear unit built at $6,000 per kilowatt, with 80% financing from the FFB, is rated 'BB-' with

---

ernment, the standard government loan-guarantee rules require that the government itself allocate and provide the capital for the investment, which is repaid by the entity receiving the guarantee; presumably over a 30-year period in this case. A fee is also charged (loan guarantee fee) to cover the risk of failure to repay the loan. The magnitude of this fee is to be estimated by the DOE based on guidance from credit rating agencies.

a recovery of 70%, the subsidy cost would be a substantial $288 million while a 'BB' rated project at the same recovery may have to pay about $192 million"[5] (Standard and Poor's, 2008).

These guarantees will allow a high percentage of debt compared to equity (as much as 80 percent), which means lower average financing costs, for two reasons. First, the cost of debt is less than the cost of equity. Second, a loan guarantee means that the interest on the debt will be less than the interest that would otherwise be required. The total reduction in financing costs could result in a significantly lower levelized cost of electricity (LCOE) at the busbar, as discussed in more detail in the "Costs" section of this chapter.

---

[1] A broad range of policies influences the nuclear power industry; this is also true for other energy technologies, such as coal-fired plants and wind turbines. For nuclear power plants in particular, these policies include, for example, the Price-Anderson Act and federal responsibility for the disposal of used nuclear fuel.

[2] EPAct05 provides loan guarantees for other technologies in addition to nuclear—for example, renewable-energy technologies. The PTC for nuclear generating units is the same as the ones currently available for wind and solar.

[3] The standard government loan guarantee rules require that the government itself allocate and provide the capital for investments in which the government provides a guarantee for 100 percent of the debt instrument (through the U.S. Department of the Treasury's Federal Financing Bank [FFB]).

[4] The U.S. Government Accountability Office (GAO) judged that "DOE's metric to assess the effectiveness of financing decisions containing the loss rate to 5 percent may not be realistic; it is far lower than the estimated loss rate of more than 25 percent that we calculated using the assumptions included in the fiscal year 2009 president's budget" (GAO, 2008). The GAO's calculation was performed using assumptions contained in Table 6 of the Federal Credit Supplement, fiscal year 2009. However, the Federal Credit Supplement assumptions "reflect an illustrative example for informational purposes only. The assumptions will be determined at the time of execution, and will reflect the actual terms and conditions of the loan and guarantee contracts." The committee judges that the budget assumptions are not necessarily appropriate for assessing the accuracy of the DOE's estimates.

[5] The recovery rate is defined as the value of the borrower's debt after it defaults. This is distinct from the default probability, which is encompassed in the credit rating of the company. The DOE is likely to base its assumed default probability on the ratings produced by several credit-rating agencies.

Overall, these costs seem very high compared to average wholesale electricity prices (5.7¢/kWh in 2007),[63] but they may not seem so high in the future if price trends for primary fuels continue or if constraints or fees are placed on carbon

---

[63] Wholesale electricity prices are distinct from the cost of electricity to the utility (cited previously). The wholesale price is the price of electricity to the utility when purchased in restructured

emissions.[64] Finally, it should be reemphasized that such calculations, while useful, are not predictions of the future.

Cost of Alternative Plant Designs and Fuel Cycles

It is difficult to project the cost of electricity generated from plants using alternative advanced nuclear plant designs and fuel cycles. Some alternative plant designs may offer cost decreases resulting from reduced quantities of steel and concrete used in construction (Peterson, 2008), but in general, plants incorporating alternative reactor designs are likely to be significantly more expensive to construct than LWRs are. For example, a Russian expert estimated that construction costs for sodium-cooled fast reactors could be 10–15 percent more expensive than LWRs, although this range was based on limited analysis (Ivanov, 2008); the DOE estimated that fast reactors (intended to be deployed as part of the Advanced Fuel Cycle Initiative program under the Bush administration) could be as much as 30 percent more expensive than LWRs (Lisowski, 2008). As for any electricity source, in addition to construction costs, the LCOE will also depend on the capacity factor of the deployed fast reactors and other considerations. If the decision is made to pursue fast reactors, further R&D to reduce costs will be valuable.

The LCOE for plants using alternative fuel cycles is likely to be higher than for those using once-through fuel cycles, though how much higher remains uncertain. For example, the LCOE for plants using limited recycle with current-generation technology is likely to be higher than from plants using the once-through fuel cycle; however, different studies have come to different conclusions about limited recycle's economic feasibility. In general, limited recycle using MOX/PUREX is likely to be competitive with a once-through fuel cycle only if the price of uranium is high and if the cost of reprocessing is relatively low. A study by the Massachusetts Institute of Technology concluded that limited recycle could cost approximately four times more than the once-through fuel cycle (MIT, 2003), and another study by Bunn et al. (2003) noted that at current uranium prices, limited recycle could increase the costs attributable to used fuel management by more than 80 percent. In contrast, a study by the Boston Consulting Group

---

electricity markets. See DOE/Energy Information Administration, available at www.eia.doe.gov/cneaf/electricity/wholesale/wholesalet2.xls; accessed July 2009.

[64]A comparison of the LCOE from various generating technologies (including coal, gas, and renewable technologies) can be found in Chapter 2 of this report. See especially Figure 2.10.

(BCG, 2006), which was funded by Areva, found that the cost of limited recycle using MOX/ PUREX in the United States could be comparable to the cost of the once-through fuel cycle, and an earlier study by OECD/NEA (1994) found that reprocessing was about 14 percent more expensive per kilowatt-hour of generated electricity than was the once-through fuel cycle.

Another example discussed earlier in this report is the case of a system using a fully closed fuel cycle (including fast reactors as well as fuel cycle plants). The LCOE for such a system remains speculative, but it is likely to be more expensive than the once-through approach, as a large number of fast reactors and reprocessing plants will be required. On the other hand, as discussed in the section on "Alternative Fuel Cycles," such a closed cycle would produce a smaller volume of long-lived high-level waste than the once-through fuel cycle produces, and the long-term heat load could be reduced owing to the destruction of a large fraction of transuranics in the used fuel. If fission products were also removed from the fuel and handled separately, the short-term heat load could also be reduced, potentially allowing closer packing of waste in a repository. In this case—although a quantitative analysis has yet to be done—the increased expense for the reprocessing, the fuel fabrication, and the fast reactors might be counterbalanced by reduced cost for waste disposal if one or more future repositories become unnecessary.

POTENTIAL FOR FUTURE DEPLOYMENT

The AEF Committee's estimates of the potential supply from nuclear power in 2020, 2035, and 2050 are discussed in this section and tabulated in Table 8.1. The committee has estimated the maximum deployment of new nuclear plants that could be built under an accelerated deployment program, as described in Part 1 of this report; however, no attempt has been made to predict what will in fact be built. Any such prediction is intrinsically uncertain because it depends on many factors, including the economic conditions in the United States and around the world over the coming decade.

The contribution of new nuclear power plants to the U.S. electricity supply before 2020 is likely to be limited because new plant construction requires a long lead time: it can take some 4 years to obtain a construction and operating license and 4–7 years to build the plant (consistent with current world trends, as shown

TABLE 8.1 Potential Supply of Nuclear-Generated Electricity in 2020, 2035, and 2050

	Additional Electric Supply Compared to 2009 (TWh/yr)	LCOE (¢/kWh)	Maximum Net U.S. Electric Supply from Nuclear Power[a] (TWh)	
			Many New Plants After 2020 ("Nuclear Renaissance")	No New Plants After 2020 ("Nuclear Stall")
2020	Uprates of current plants: 39–63 (Capacity: 5–8 GWe) New Plants: 55–95 (Capacity: 7–12 GWe)	Uprates: negligible additional cost New Plants: IOU: 8–13 IPP: 8–13 With federal loan guarantees and 80/20 financing: IOU: 6–8 IPP: 6–8	810 (current) + 94–158 (new) – 0 (retirements) Total Supply: 904–968	810 (current) + 94–158 (new) – 0 (retirements) Total Supply: 904–968
2035	Uprates of current plants: 39–63 (Capacity: 5–8 GWe) New Plants: 741–788 (Capacity: 94–100 GWe)	New Plants:[b] IOU: 8–13 IPP: 8–13 With federal loan guarantees and 80/20 financing: IOU: 6–8 IPP: 6–8	810 (current) + 780–851 (new) – 204–209 (retirements) Total Supply: 1381–1452	810 (current) + 94–158 (new) – 204–209 (retirements) Total Supply: 695–759

in Figure 8.1).[65] Thus, if the prospective owner/operator of a nuclear plant applied for a combined construction and operating license (COL) in 2009, the plant would be unlikely to produce electricity before 2017.[66]

---

[65]The estimate of 4–7 years is a committee judgment based on discussions with various stakeholders. Some vendors (Westinghouse and Areva, for example) estimate shorter construction times of approximately 3 years. These estimates involve on-site construction, with additional time required to build the modules, and they also separate commissioning and testing from construction to some extent. Still, once experience with new construction is acquired, it might be expected that the duration of construction could shorten.

[66]This judgment is consistent with recent estimates by the USNRC. It projects that evolutionary LWR designs currently under review will complete design certification in 2011–2012, and that the first construction and operating licenses (COLs) will be issued in 2012. The agency also projects that the first fuel loads will be added around 2016 (Johnson, 2008).

TABLE 8.1 Continued

	Additional Electric Supply Compared to 2009 (TWh/yr)	LCOE (¢/kWh)	Maximum Net U.S. Electric Supply from Nuclear Power[a] (TWh)	
			Many New Plants After 2020 ("Nuclear Renaissance")	No New Plants After 2020 ("Nuclear Stall")
2050	Uprates of current plants: 39–63 (Capacity: 5–8 GWe) New Plants: 1545–2381 (Capacity: 196–302 GWe)	Unknown	810 (current) + 1545–2381 (new) – 798–814 (retirements) Total Supply: 1541–2393	810 (current) + 94–158 (new) – 798–814 (retirements) Total Supply: 90–170

Note: New plants are assumed to be evolutionary designs and to have an average capacity of 1.35 GWe, except for the completion of the 1180 MWe Watts Bar-2 reactor. New plants are assumed to operate with an average capacity factor of 90 percent, and currently operating plants are assumed to continue to operate at an average capacity factor of 91 percent. Five to 9 new plants are assumed completed between 2009 and 2020; 3 per year between 2021 and 2025; 5 per year between 2026 and 2035; and 5–10 per year between 2036 and 2050. Retirements reflect the assumption that all currently operating plants receive operating license extensions to 60 years. PWRs are uprated by 10–12 percent and BWRs by 20–25 percent before being retired. All costs are expressed in constant 2007 U.S. dollars. GWe = gigawatts-electric; IOU = investor-owned utility; IPP = independent power producer; LCOE = levelized cost of electricity; TWh = terawatt-hours.

[a]Extending the operating licenses for a fraction of currently operating plants to 80 years would decrease the number of plants retired between 2035 and 2050. Without license extensions allowing for 80-year operating lifetimes, the last currently operating plant will retire before 2056.

[b]After 2020, the uncertainties in the parameter ranges are so large that costs cannot be reliably estimated. For illustrative purposes, the committee assumes no net change in real costs per kilowatt-hour after 2020. However, after 2020, many factors could affect the actual LCOE: construction costs may be reduced as experience is gained with the new designs; high rates of escalation in construction costs are likely to stabilize; and the successful construction and operation of several plants may cause financing to become more favorable. On the other hand, delays and other difficulties during construction could significantly increase costs.

It is the judgment of the committee that as many as 5–9 additional nuclear plants could be constructed by 2020.[67] This projection is based on two factors: first, the Tennessee Valley Authority recently approved a project to complete the Watts Bar-2 nuclear reactor, expected to be online by 2013;[68] second, the committee estimates that as many as 4–8 new evolutionary nuclear plants could be constructed by 2020, with a number of follow-on projects that would lag the initial plants by 2 or 3 years.

[67]The estimate of 5–9 plants reflects the committee's judgment of the technical limits on new capacity—what the nuclear industry is likely to be able to construct by 2020. Because this is a technical estimate, recent economic conditions have not been factored in.

[68]In 1988, TVA suspended construction of Unit 2 of the Watts Bar nuclear plant because of a reduction in the predicted growth of power demand.

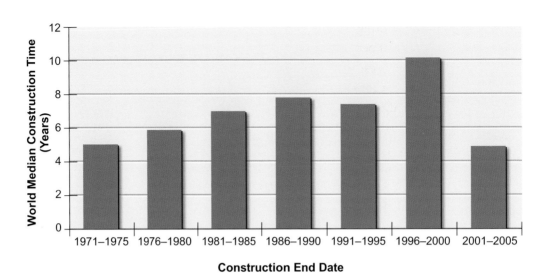

FIGURE 8.1 *Average construction times for nuclear plants around the world, 1971–2005. Construction times have been decreasing worldwide for plants started since the 1990s, with recent averages over the last 5 years between 4 and 6 years from first concrete to connection to the grid.*
Source: IAEA, 2007.

There are several reasonable scenarios that, in the short term, could result in 4–8 new plants. For example, each of the four major supplier teams could initiate one or two projects—one or two units of the same reactor design on a single site—once the licensing process was complete. Alternatively, several utilities could focus on one design to begin to build a fleet of standard plants; in this case, one vendor might build more than one pair of plants in the first wave. Given the currently announced plans, either scenario is plausible. These 4–8 new plants (with an average rating of 1350 MWe[69]) and the 1180 MWe Watts Bar completion project could increase the total U.S. nuclear generating capacity by 7–12 percent.[70]

Power uprates on existing plants could add another 5–8 percent to the U.S.

[69]Approximately half of the COL applications submitted to the NRC are for 1200 MWe reactor designs (the AP-1000), and most of the rest are for 1500 MWe reactors (the USEPR, ESBWR, and APWR). Two are for the 1300 MWe ABWR. The average capacity of the submitted applications as of July 2009 is around 1350 MWe. For the purposes of this report, this value is used as an estimate of the representative capacity for future reactors, although the value may change, particularly for the post–2020 period.

[70]Because none of the currently operating plants is likely to be retired by 2020, any new plants will add to the present nuclear capacity.

nuclear generating capacity by 2020, for a total increase (with new plant construction) of 12–20 percent by 2020. The maximum uprate likely with readily available technologies is estimated at 20–25 percent of the plant's power for BWRs and 10–12 percent for PWRs. If each currently operating plant in the United States is uprated by this maximum amount reasonable[71] (accounting for uprates that have already been performed), as much as 5–8 GW of additional power could be added in total. Nearly all of these capacity additions are likely to occur before 2020. As noted previously, the USNRC currently projects that it will receive applications for 3.0 GW of uprates by 2013.

It is likely, given the COL applications received by the USNRC as of July 2009, that most of the new units will be added at existing sites.[72] The advantages are significant, including reduced costs (because existing infrastructure is already available); the ability to connect to existing transmission lines (although the capacity of these lines may need to be expanded); and the existence of an operating organization. In addition, local populations and governments are more likely to be supportive.

However, existing sites may not be where the demand for power exists. Building at new sites will entail extra costs to purchase the site and prepare it for construction, as well as to build new transmission lines. Also, public concerns about safety and security may need to be addressed in the regions surrounding the new plant.

After 2020, there is significantly more uncertainty in the estimated supply from nuclear power plants. Assuming that all currently operating plants receive 20-year license renewals (for total service lives of 60 years), their operating licenses will begin to expire in 2028. Because the current nuclear plants are such low-cost power producers, there is a large economic incentive to extend their operating lives even further, by an additional 20 years, to 80 years. But, as noted earlier, because many technical challenges are still to be overcome it is not clear whether extending the lifetimes of current plants to 80 years will be possible.

If not, there will be a rapid drop in nuclear capacity between 2030 and 2050, as shown in Figure 8.2. By 2035, about 30 GW will be retired; by 2050, nearly all currently operating nuclear plants will be retired. Because of the long lead times

---

[71]Improvements or efficiency gains not yet identified have not been included in the calculation, and further improvements to existing plants may be possible.

[72]The USNRC's current list of received COL applications is available at www.nrc.gov/reactors/new-reactors/col.html; accessed July 2009.

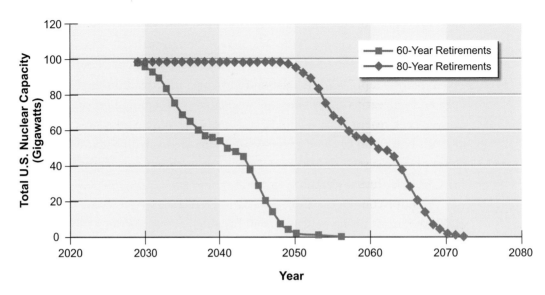

FIGURE 8.2 *Effect of operating life extensions on current U.S. nuclear generating capacity. Blue squares represent the generation capacity of currently operating nuclear power plants assuming license extensions to allow for 60-year operating lives. Green diamonds represent the capacity of the current fleet of plants assuming that all 104 plants receive license extensions to allow for 80-year operating lives.*
Source: USNRC, 2008.

involved in nuclear plant construction, companies will need to decide, by 2020 or shortly thereafter, whether to replace many of these plants with new nuclear reactors. However, companies are unlikely to be in a position to make such decisions with assurance at that time, unless several new U.S. nuclear plants will already have been added. It is likely that investors would want to observe whether they could be built in the United States on schedule and on budget while demonstrating safe and cost-effective operation. Thus, one purpose of providing federal loan guarantees is to acquire information needed by 2020 to make these decisions that will affect long-term U.S. electrical capacity.

If the first handful of new nuclear plants are constructed on schedule and on budget, and if they demonstrate safe and cost-effective operation, significantly more plant construction could follow between 2020 and 2050. However, if these first plants do not meet these requirements, few additional new plants are likely to be built.[73] To estimate the maximum number of nuclear power plants that could

---

[73]Although the text sets out the likely general trend, the future is not likely to divide clearly into just two alternative options. Generating companies will have to make decisions during the

be added after 2020, the AEF Committee relied on the historical build rates in the United States: about 5 reactors per year were constructed between 1965 and 1985 as nuclear power expanded its share of electric power generation and power demand grew rapidly. In the committee's judgment, a construction rate averaging 3 plants per year from 2021 to 2025 (to allow for learning) followed by a rate of 5 plants per year from 2026 to 2035 seems achievable. After 2035, assuming that electricity demand continues to expand, a construction rate of 5–10 plants per year could be sustained. Ultimately, however, the number of plants built will be influenced by future electricity demand, public attitudes about nuclear power, and the economic competitiveness of nuclear power compared to alternative sources of electricity.

POTENTIAL BARRIERS

Although there are several potential barriers to deployment of new nuclear power plants, the committee judges that these barriers can be reduced or eliminated if the first handful of plants are constructed on schedule and on budget, and they demonstrate initial safe and secure operation.

Economics

The large initial or upfront capital investment required for construction of new nuclear power plants could present a barrier to the expansion of nuclear power in the United States. Even for larger utilities, such a plant can represent a significant fraction of the company's net worth,[74] potentially putting the entire company at risk should the project be delayed substantially or costs escalate significantly. In addition, the substantial cost of constructing new plants is associated with a relatively high cost of electricity produced by these plants (in comparison to the cost

---

time that the early plants are being constructed as to how to meet emerging power demand. Because of this, they may commit to other new nuclear projects before the first few plants are completed. Thus, early favorable signals could lead to new orders even if the experience with new plants turns out to be negative. Alternatively, unfavorable signals, even coupled with later recovery, might delay new orders.

[74]In many cases, the market capitalization of the existing nuclear generating companies is $20 billion or less, whereas some recent cost estimates for individual new plants have exceeded $5 billion (for example, see Scroggs, 2008; Moody's Investor's Service, 2008).

associated with existing plants). The significance of this barrier will depend on the growth in demand for electricity, the cost of electricity from alternative sources, and what price (if any) is placed on carbon.

Financial markets are likely to be wary of investments in new nuclear plants until it is demonstrated that they can be constructed on budget and on schedule. Nuclear plants have not been built in the United States for decades, but there are unpleasant memories, because construction of some of the currently operating plants was associated with substantial cost overruns and delays. There is also a significant gap between when construction is initiated and when return on investment is realized.

Thus, it is likely that subsidies or financial guarantees that protect investors (as discussed in Box 8.5) will be required for the first few plants. But if these investments turn out to be financially favorable, the means to support construction of additional new plants is likely to be found. Innovative financial arrangements such as joint ventures, consolidation, and risk sharing among the participants may be required, however, and difficulties involved in working out these new financial structures (particularly in regulated utilities) could affect progress toward new construction.

Regulatory and Legislative Issues

All of the existing nuclear power plants in the United States were licensed using a two-step process: first, the USNRC issued a construction permit once it was satisfied with the preliminary plant design and the suitability of the site; second, USNRC staff undertook a detailed study of the plant after construction to determine whether to issue an operating license. This process has sometimes been blamed for extensive delays in operations and expensive retrofits of constructed plants.

In recent years, the licensing process for U.S. nuclear plants has been extensively revised. The new process allows reactor design certifications, early site permits, and COLs to be granted before the plant is constructed.

Utilities have shown interest in proceeding with applications for combined licenses, to the point that the processing of the current surge of applications could cause short-term delays in beginning new plant construction. In addition, the USNRC had initially expected that, in most cases, plant designs and sites (via early site permits) would be certified before a COL was sought for a plant.

However, many COL applications have been submitted for designs that are concurrently subject to design certification review, and nearly all of the proposed applications will not seek an early site permit. As a result, the orderly stepwise process that had originally been anticipated is not occurring. This means that the work associated with the issuance of a COL must be more extensive than anticipated and that design certifications in many cases are occurring in parallel with review of COL applications. This has complicated the licensing process for the USNRC.

Public Concerns

Public concerns about nuclear technologies, such as the safety and security of nuclear facilities and the disposal of nuclear waste, could pose a barrier to an expansion of nuclear power in the United States—at least in some areas of the country. Overall, however, recent U.S. polls have shown a majority of respondents[75] favoring the use of nuclear power to provide electricity, and a majority (51–67 percent) supporting the building of new nuclear plants.[76] This shift over the last two decades is likely the result of the improved performance and safety record of nuclear plant operations in the United States, as well as the growing concerns about the impacts of CO_2 emissions on climate change.[77] Public opinion may change (becoming either more or less supportive), depending on the performance and public perceptions of the new plants.

---

[75]In a 2009 Gallup poll, 59 percent of respondents favored the use of nuclear power for providing electricity. This number has been holding approximately steady since 2004 (Gallup, 2009), but there has been a recent rise. A 2008 Fox News poll found that 53 percent considered nuclear power to be a safe source of energy. See www.foxnews.com/story/0,2933,369827,00.html; accessed July 2009.

[76]A 2008 Zogby poll found that 67 percent of Americans surveyed support the building of new nuclear plants (23 percent were opposed, while 10 percent were unsure) (Zogby, 2008). A 2008 Fox News poll found that 51 percent supported the building of new plants. See www.foxnews.com/story/0,2933,369827,00.html; accessed July 2009.

[77]For example, in a recent poll, 71 percent of students who considered themselves environmentalists were in favor of the continued use of nuclear energy (Bisconti Research, Inc., 2006).

Safety and Security

Public concerns about the safety[78] and security[79] of nuclear power plants may serve as a barrier to new construction. This phenomenon has two aspects: some members of the public question whether it is possible to operate nuclear power plants safely and securely; others are concerned, even should safety and security be technically achievable, about the enforcement of regulations to adequately ensure them (UCS, 2008).

Public opinion about the safety of nuclear plants has become more positive over time: a 2006 poll showed that 60-plus percent of respondents believed nuclear power plants to be highly safe, in contrast to 35 percent with the same pollster in 1984 (Bisconti Research, Inc., 2006). However, any perceived slippage in safety or a reactor accident anywhere in the world could have an adverse impact on this currently favorable attitude. Similarly, whether security concerns will become a barrier to future development of nuclear power could depend on the level of terrorist activity in North America and possibly elsewhere, particularly if the activity specifically targets nuclear facilities.

Some members of the public are also concerned about the health effects on neighboring populations of small amounts of radiation released during routine nuclear plant operations. These emissions are typically several orders of magnitude below statutory limits and would not be expected to produce significantly increased health risks to people living near the plants compared to health risks if no plants were present. However, they can be of great concern to local citizens who may not have confidence in the regulatory limits.[80]

Disposition of Used Fuel

Public concerns about the federal government's failure to develop a final disposal pathway for commercial used fuel may also serve as a barrier to new construction. As discussed in more detail later in this chapter, from a technical perspective the absence of a geologic disposal facility does not present an impediment to contin-

---

[78]Safety is defined here as measures that would protect nuclear facilities against failure, damage, human error, or other accidents that would disperse radioactivity into the environment.

[79]Security is defined here as measures to protect nuclear facilities against sabotage, attacks, or theft.

[80]An example is the controversy over tritium leaks at the Braidwood plant in Will County, Illinois. News coverage of public concerns about these events can be found online at www.pbs.org/newshour/bb/environment/jan-june06/tritium_4-17.html; accessed July 2009.

ued operation of existing plants or to new construction. Technical analyses have shown that used fuel can be stored for up to a century in dry cask[81] storage at low risk of release of radioactive material (see the "Impacts" section following). Concerns remain, however, and 11 states have barred construction of new reactors until the nuclear waste problem is solved.[82] But such barriers may be softening, as some progress has recently been made on state bills introduced to overturn these bans.[83]

Domestic Technology and Skills Base

There could be shortages in certain parts and components (especially large forgings), as well as in trained craft and technical personnel, if nuclear power expands significantly worldwide. The population of suppliers of nuclear parts and components has become more limited over the last two decades, and the number of American Society of Mechanical Engineers (ASME) nuclear certificates[84] held around the world fell from nearly 600 in 1980 to about 200 in 2007 (ASME, 2008). There is also an insufficient supply of people with the requisite education or training at a time when vendors, contractors, architects, engineers, operators, and regulators will be seeking to build up their staffs. In addition, 35 percent of the personnel now working at U.S nuclear utilities will become eligible for retirement in the next 5–10 years (NEI, 2007).

---

[81]Used fuel is dry-stored in heavily shielded casks that use passive heat removal systems (conduction and convection) for cooling after it has been actively water cooled in a "used fuel pool" for at least 3 years. (See footnote 100 for a definition of used-fuel pools.) Storing and disposing of used fuel is discussed in the "Impacts" section of this chapter.

[82]Eleven states require that the USNRC make some finding regarding the potential for disposal of used nuclear fuel before an existing moratorium on new nuclear power plants within their borders is lifted. These states are Illinois, California, Wisconsin, Kentucky, Connecticut, Massachusetts, Maine, Oregon, West Virginia, Montana, and New Jersey. Minnesota prohibits new nuclear power plants altogether.

[83]Legislative hearings were held in Minnesota in March 2009 to consider the lifting of its ban. The lifting of the ban was approved by the state senate, and narrowly defeated by the state house; it will be introduced again next year. In Kentucky, a bill will be presented to the state house in 2009 that would strike down the moratorium; it was approved by the state senate and by a house committee. The governor of Wisconsin has expressed willingness to consider nuclear power as a possibility, although a bill to overturn the state ban was not taken up by the state senate (after passing the state house).

[84]The American Society of Mechanical Engineers maintains a nuclear certification program that provides a standard for quality assurance of construction materials, design, operation, inspection, and continuing maintenance of nuclear facilities.

These bottlenecks will be a particular problem for the construction of plants between now and 2020, though they should resolve themselves over time. Some expansion is already occurring to meet this demand: Japan Steel Works is working to double its capacity, and enrollment in universities' nuclear engineering departments is increasing. Economic incentives will eventually yield the resources in personnel and material to enable new construction to proceed, but there may be short-term dislocations as the worldwide economy adjusts.

IMPACTS

Given the small number of new nuclear plants likely to be built before 2020, their near-term impacts (compared to the currently operating fleet) are likely to be small.

Environmental[85]

Compared to other baseload electrical generation options, operating nuclear power plants have relatively few adverse environmental impacts, such as those derived from SO_x, NO_x, mercury, or CO_2 emissions. The magnitude of any environmental benefits of new nuclear plants will depend on the number of plants ultimately built, of course, as well as on the environmental profiles of the energy sources displaced.

Greenhouse Gas Emissions

The U.S. power sector overall is a significant source of greenhouse gas emissions, totaling roughly 2.4 billion tonnes of CO_2 in 2007 (see Figure 1.11 in Chapter 1). One of the environmental advantages of nuclear power is its small greenhouse gas footprint. In 2007, U.S. nuclear power plants were responsible for approximately 70 percent of the greenhouse-gas-free electricity production in the United States.[86] However, before 2020, new nuclear plants will contribute relatively little to reducing the total greenhouse gas emissions from the U.S. power sector because of the

[85]For a more thorough discussion of the topics briefly reviewed in this section, see Annex 8.D ("Environmental Impacts of Nuclear Technologies").

[86]This estimate was calculated by adding the nuclear and renewable contributions to U.S. electricity, and calculating the nuclear fraction of this total.

limited number of new plants that could be built (although the existing plants are likely to continue to contribute significantly). However, after 2035, if significant new construction has taken place during the preceding 15 years, the greenhouse gas emissions reduction could be substantial.

The AEF Committee uses a low case and a high case to estimate avoided CO_2 equivalent emissions. For the low case, the committee's estimate of potential new nuclear capacity (e.g., 5–9 plants by 2020) replaces an equal generating capacity of natural gas plants. For the high case, the committee's estimate replaces an equal generating capacity of traditional coal plants without CCS.[87] The CO_2-equivalent emissions of the current U.S. electric power supply (about 600 tonnes CO_2 equivalent per GWh) lie between the CO_2 emissions of natural gas plants (about 500 tonnes CO_2 equivalent per GWh) and coal plants (about 1000 tonnes CO_2 equivalent per GWh).

Thus, the deployment of 12–20 GWe of new nuclear capacity (through uprates of current plants and new plant construction) could avoid some 40–150 million tonnes of CO_2 equivalent per year by 2020 (0.04–6 percent of 2007 emissions).[88] In 2035, a deployment of 99–108 GWe of new nuclear capacity (including that deployed before 2020) could avoid 360–820 million tonnes of CO_2 equivalent per year (15–34 percent of 2007 emissions). As much as 730 million to 2.3 billion tonnes of CO_2 equivalent per year could be displaced in 2050 (30–96 percent of 2007 emissions). No assumptions about scheduled plant retirements are made in these estimates. The potential reductions are shown in Table 8.2, alongside projections for the total CO_2 emissions from the electric power sector in 2020, 2035, and 2050.

Nuclear power plants do emit a small quantity of CO_2 on a life-cycle basis—resulting largely from energy used for processes such as uranium enrichment and plant construction—but U.S. emissions from uranium enrichment are likely to decrease in the future because several energy-efficient gas centrifuge enrichment

---

[87]The source for the low and high supply estimates as well as capacity factors and other assumptions can be found in the section of this chapter on deployment of new nuclear plants.

[88]This calculation assumes that nuclear power plants emit 40 tonnes of CO_2 equivalent per GWh (including emissions from construction, mining, fuel fabrication, and other processes). The reader is referred to Annex 8.D of this report for an explanation of how the committee arrived at 40 tonnes of CO_2 equivalent per GWh. Nuclear plants are assumed to be operated at an average capacity factor of 90 percent. The maximum electricity supply from nuclear power (in TWh) in each of 3 years—2020, 2035, and 2050—is shown in Table 8.1.

TABLE 8.2 Avoided CO_2 Emissions for New Nuclear Capacity in 2020, 2035, and 2050

	Potential New Capacity Added (GWe)	Avoided CO_2 Emissions for This Deployment (million tonnes CO_2 equivalent per year)	EIA/DOE Total Projected CO_2 Emissions in Electric Power Sector (million tonnes CO_2 equivalent per year)
2020	12–20	40–150	2627
2035	99–108	360–820	3090
2050	201–310	730–2300	Not examined

Note: Projected CO_2-equivalent emissions for 2035 were extrapolated from DOE/Energy Information Administration (EIA) data for 2020–2030. Avoided CO_2-equivalent emissions assume that new nuclear capacity replaces, in the low case, an equivalent generating capacity of natural-gas-fired capacity without carbon capture and storage (CCS) and, in the high case, an equivalent generating capacity of traditional coal-fired capacity without CCS. The calculations assumed that natural gas plants emit about 500 g of CO_2 equivalent per kilowatt-hour, that coal plants emit about 1000 g of CO_2 equivalent per kilowatt-hour, and that nuclear power plants emit 40 g of CO_2 equivalent per kilowatt-hour on a life-cycle basis.
Source: EIA, 2008.

plants[89] are being constructed and planned in the United States.[90] By 2011, the two such plants expected to be on line may replace the current energy-intensive gaseous diffusion enrichment plant[91] at Paducah, Kentucky. In addition, if future sources of electric power used for fuel enrichment emit less CO_2, this will be reflected in the life-cycle emissions of operating nuclear plants.

Mining and Milling of Nuclear Fuel

Environmental impacts occur from the multiple processes involved in fabricating nuclear fuel. Several of these processes and their primary environmental impacts

---

[89]Gas centrifuge plants enrich uranium through a cascade of centrifuges, which utilize the very slight mass difference between U-235 and U-238 to separate the two isotopes. For a given output, this process requires significantly less energy than does enrichment by gaseous diffusion.

[90]The effect of these new enrichment facilities on the U.S. fuel supply is discussed in more detail in the section titled "Uranium Resources."

[91]Gaseous diffusion plants enrich uranium by forcing uranium hexafluoride through a cascade of many stages of semipermeable membranes. Like the gas centrifuge process, the gaseous diffusion process utilizes the slight mass difference between U-238 and U-235 to increase the percentage of U-235 in the final product.

are discussed in the paragraphs that follow. For more detail on the environmental impacts of these and other processes, the reader is referred to Annex 8.D.

The primary impact of *mining,* in which natural uranium is extracted from the earth, and *milling,* in which natural uranium is chemically converted to a dry and purified uranium concentrate, is the production of slightly radioactive byproducts known as mill tailings, which are disposed of in "tailings piles." Radon emissions from mill tailings were previously an issue of public concern in the United States. At present, uranium milled in the United States is subject to comprehensive regulation for the control of environmental impacts, including radon emissions, under the Uranium Mill Tailings Radiation Control Act of 1978. The majority of uranium is imported from nations such as Canada and Australia, which have regulations equivalent to those of the United States. However, 17 percent is imported from nations that may not have equivalent regulations, primarily Namibia and Kazakhstan (see www.eia.doe.gov/cneaf/nuclear/umar/table3.html; accessed July 2009).

In some locations, a process called in situ leach (ISL) mining has replaced hard-rock mining and milling of uranium. The use of ISL entails smaller amounts of mill tailings to be disposed of; however, there is potential for other environmental impacts, including groundwater contamination and increased water use.

An expanded deployment of nuclear power in the United States (particularly after 2020) may result in increased demand for uranium, with an associated increase in worldwide uranium mining and milling. If more mining is undertaken in the United States to meet increased domestic demand for uranium, domestic environmental impacts may rise.

Water Use

All thermal power plants use significant quantities of water during operation, primarily for cooling. Overall, the committee does not view water use to be a national barrier to an expansion of nuclear power plants in the United States. However, the water use and consumption of new plants may have significant local impacts, as would occur for any thermal power plant.

Nuclear power plants on average require more cooling water per kilowatt-hour of electricity produced than do fossil-fuel plants of comparable age, due to nuclear power plants' lower average thermal efficiency. Most U.S. power plants use one of two types of cooling processes: once-through cooling or closed-cycle wet cooling. In some instances, these wet cooling systems can use nonfreshwater sources such as seawater (if located on the coast), brackish water from wells or estuaries, agricultural runoff, "produced water" from oil and gas drilling operations, or treated municipal wastewater (Veil, 2007). The water consumption is often of concern with thermal power plants; however, the water use[92] can also be an issue, as cooling water is returned to the source at a higher temperature.

The water-use impacts of future nuclear plants will depend on where the new nuclear plants are sited (for example, along a coastline versus the arid Southwest) and what cooling technologies are employed. If deployed after 2020, alternative cooling technologies such as dry cooling[93] or hybrid cooling[94] could reduce water use compared to current technologies. Dry cooling has been used for some coal-fired plants,[95] but at present no commercial nuclear plants have been constructed using this technology; it is likely to have significant disadvantages, including higher costs, higher operating power requirements, and reductions in plant efficiency and capacity during hot-weather periods. Hybrid cooling may be used in several evolutionary nuclear plants proposed for construction in the United States in the near term, including the new reactor planned by UniStar for the Calvert Cliffs site in Maryland (Pelton, 2007).

Waste Management and Disposal

Electricity production by means of nuclear power results in several types of radioactive waste, all of which must ultimately be disposed of. They include waste from uranium mining and fuel production (just discussed); waste produced during operations (such as contaminated gloves, tools, water-purification filters and resins, and plant hardware); and used fuel. Additional waste will be generated when the plant itself is decommissioned.

The construction of new nuclear plants in the United States will cause the production of additional used fuel, other operational waste, and decommission-

---

[92]Water use refers to the amount of water used by the plant but returned to the source; water consumption refers to the amount of water used by the plant and not returned to the source.

[93]Dry cooling is usually accomplished with mechanical-draft air-cooled condensers, to which a turbine's exhaust steam is ducted through a series of large ducts, risers, and manifolds.

[94]Hybrid cooling systems typically consist of a dry cooling system operating in parallel with a conventional closed-cycle wet cooling system.

[95]For example, the Kogan Creek power station in Australia, a 735 MW coal-fired plant, uses dry cooling.

ing waste. The characteristics of these by-products will be similar to those being produced by current plants, with two possible exceptions. First, the burn-up of nuclear fuel will likely increase as new fuel designs are developed. This will reduce the mass of used fuel generated per unit of electricity generation, although it will increase the volume of transuranics and fission products (resulting in more radio-activity) per unit mass. Second, some new plants will have fewer cables, pipes, valves, and pumps than current-generation plants have, and in some cases less structural steel and concrete. This latter element could reduce decommissioning costs and time (as well as front-end construction costs and time), while also reducing the amount of material requiring disposal.

The majority of these wastes, including most of the radioactive decommissioning waste, can be disposed of in land disposal facilities.[96] However, higher-activity[97] wastes constitute the primary concern; in the case of used fuel in particular, the radioactivity is very long-lived and will need to be managed (though not necessarily actively) for hundreds of thousands of years. Thus concerns associated with managing this waste are intrinsically intergenerational. Facilities for disposal of higher-activity wastes (including but not limited to the used fuel) are not available, however, and plants are currently storing such wastes on-site until a disposal pathway is determined.[98]

The Nuclear Waste Policy Act provided that the disposal of used fuel from commercial nuclear power plants was a federal responsibility and that a deep geological repository would be built and operated by the federal government for this purpose.[99] The DOE filed an application in June 2008 to construct a repository at Yucca Mountain, Nevada. If that application (and a subsequent operating amendment) were approved by the USNRC and survived expected court

---

[96]A "land disposal facility" is a disposal facility located within a few tens of meters of the land surface for the disposal of radioactive wastes. A "geologic repository" is not considered a land disposal facility.

[97]Activity is the rate of decay of radioactive material per unit time.

[98]The DOE has developed an environmental impact statement (EIS) for the disposal of high-level waste and nuclear fuel at the Yucca Mountain site. Licensees have already prepared EISs for storage on their sites.

[99]Disposal is to be funded by a fee of $1/MWh, paid by the ratepayers of nuclear electricity-generation companies and collected in a federally administered fund. At the end of 2007, just over $27 billion had been credited to the fund from industry payments and interest, of which about $9 billion was spent to develop a repository. Current nuclear power generation adds about $800 million to the fund annually. (See www.ocrwm.doe.gov/about/budget/index.shtml; accessed July 2009.)

challenges, the DOE expected to open the repository after 2020. However, the prospects for the Yucca Mountain repository have been diminished and perhaps completely eliminated by the declared intent of the Obama administration not to pursue this disposal site. As stated by Energy Secretary Steven Chu before the Senate Budget Committee on March 11, 2009: "Yucca Mountain is not a workable option and . . . we will begin a thoughtful dialogue on a better solution for our nuclear waste storage needs" (www.congressional.energy.gov/documents/3-11-09_ Final_Testimony_(Chu).pdf; accessed July 2009). The fiscal year 2010 Presidential Budget Request begins the process of eliminating funding for the Yucca Mountain program.

The statutory limit for the amount of used fuel that was planned for disposal at Yucca Mountain is less than the amount of used fuel that will ultimately be produced by existing commercial reactors. Thus, even if Yucca Mountain were approved, a second geologic repository would be needed, or modification of the statutory limit for Yucca Mountain would be required, or the fuel cycle would have had to be altered. Political and technical issues make any of these options highly speculative.

As noted previously, technical analyses have shown that used fuel could be stored for up to a century in dry cask[100] storage at low risk (APS, 2007; Bunn, 2001).[101] It could be stored either at plant sites (the current practice in the United States) or at regional or national facilities. Interim storage has several advantages: the storage facilities can be monitored and maintained for indefinite periods of

---

[100]After removal from the reactor, used fuel is stored in water-filled pools (i.e., used-fuel pools) with active heat-removal systems. The water is an effective heat-transfer medium and also serves as an effective radiation shield. Used fuel can be moved into dry storage after at least 3 years of cooling in the pool, although most fuel being dry-stored is much older. Used fuel is dry-stored in heavily shielded casks that use passive heat-removal systems (conduction and convection) for cooling.

[101]The USNRC has previously made a generic determination that used fuel could be stored safely and without significant environmental impacts for at least 30 years beyond the licensed life of operation of a reactor at or away from the reactor site and that there was reasonable assurance that a disposal site would be available by 2025 (10 CFR 51.23). This generic determination meant that the environmental impact of such storage did not need to be considered in the environmental impact statement, environmental assessment, or other analysis prepared in connection with the issuance or amendment of a reactor license. The USNRC is now revisiting this generic determination and has sought comments on a proposed amendment (73 Fed. Reg. 59,547, Oct. 9, 2008). The commission proposed for public comment whether to modify its "waste confidence rule" to provide that used fuel can be stored safely and without significant environmental impacts until a disposal facility can reasonably be expected to be available.

time; and extended surface storage allows time for the radioactive decay of iso-topes with shorter half-lives, thereby reducing the heat load of the used fuel if it were later emplaced in a repository. Interim storage also could offer advantages if the United States decides to pursue reprocessing, but it has disadvantages as well: the siting and licensing of interim storage sites would likely present challenges, and eventual regional or national storage would require transport of used fuel from the plant sites.

Transportation of used fuel would result in additional expense, could engen-der political and public opposition, and would necessitate transporting the fuel a second time if a repository were eventually opened. In 2006, the National Research Council found that there are no fundamental technical barriers to the safe transport of used fuel and high-level waste in the United States. However, it also found that such transport faces a number of social and institutional chal-lenges (NRC, 2006).

Of course, interim storage only buys time. The United States must eventually find appropriate means for the long-term disposition of used fuel. However, dry cask storage gives us decades to define the path for such disposition.

Safety and Security

The safety and security of nuclear power involve not only the resistance to acci-dents and attacks on the plants themselves but also the safety and security of the associated fuel cycles—including the potential for proliferation of weapons-usable nuclear materials and technologies.

Resistance to Accidents and Attack[102]

It is possible that an accident at a nuclear power plant or an attack on it could result in off-site releases of radioactive material. There are two potential sources for such radioactive releases: the nuclear fuel in the reactor core, and the used fuel being actively cooled in water pools after having been removed from the reactor.[103] An accident or terrorist attack that disrupted the flow of coolant to the

---

[102]For a more thorough discussion of the topics briefly reviewed in this section, see Annex 8.E ("Safety and Security Impacts of Nuclear Technologies").

[103]After its removal from the reactor, used fuel continues to generate heat. Thus it must be stored in water-filled pools that have active cooling systems to remove this heat. The pools also have water-filtering systems to remove radioactive contamination.

reactor core could damage the fuel and release some of its radioactive materials to the environment. An accident or a terrorist attack on a used-fuel pool could have similar consequences.

The 5–9 new plants that could be built before 2020 would not significantly add to the safety impacts of the 104 plants already in place. As noted earlier in this chapter, a larger number of new plants could be deployed in the United States after 2020, as many as 65 between 2021 and 2035, and an additional 75–150 between 2036 and 2050. However, if the new plants meet design specifications, their safety and security impacts will likely be comparable to or lower than the impacts of the plants already in place.

The evolutionary and advanced plant designs described in this chapter have features that are designed to incrementally enhance safety and security over the existing fleet of plants, including better physical protection of the core and used-fuel pools. Some of the designs have cooling systems that rely more on natural forces such as gravity and convection—as opposed to the operation of pumps and valves—to maintain cooling. The designs also incorporate multiple independent safety systems to ensure reliability and improve survivability should there be an accident or an attack. The vendors of some evolutionary designs cite probabilistic risk assessment (PRA) evaluations to claim that their designs have a core damage frequency (CDF)[104] at least 10 times better than that of existing plants (Matzie, 2008; Parece, 2008).

Every U.S. nuclear plant has the capability to withstand an attack at the level of the Design Basis Threat (DBT),[105] which is approved by the USNRC and is subject to periodic force-on-force testing. Since the September 11, 2001, attacks, changes to access controls and enhancements of the DBT have resulted in the strengthening of security.

Attacks that are beyond the DBT (including aircraft attacks) are also a concern. The USNRC and the nuclear industry have undertaken analyses of existing plants to determine their vulnerability to aircraft attacks and have made modifications to the designs and operations to mitigate the consequences. Moreover, the

---

[104]Core damage frequency is an expression of the likelihood that an accident could cause the fuel in the reactor to be damaged.

[105]The DBT is a profile of the type, composition, and capabilities of an adversary. The USNRC and its licensees use the DBT as a basis for designing safeguards systems to protect against acts of radiological sabotage and to prevent the theft of potentially weapons-usable nuclear material. The DBT is described in Title 10, Section 73, of the Code of Federal Regulations [10 CFR 73].

USNRC recently promulgated a rule requiring applicants for new nuclear reactors to identify features and functional capabilities of their designs that would provide additional inherent protection from or avoid or mitigate the effects of an aircraft attack. The details of these analyses and modifications have not been released to the public (because of security concerns), and the committee has not reviewed this information.

Proliferation

Given that the United States is a nuclear weapons state, an expansion of nuclear power and associated fuel cycle technologies in this country does not directly affect the proliferation of nuclear weapons technology. The proliferation debate focuses primarily on this impact of nuclear power in other countries.

Nuclear power plants themselves are not a proliferation risk,[106] but nuclear fuel cycle technologies such as enrichment and reprocessing introduce the risk that weapons-usable material could be produced. This is possible because the same technologies used to enrich nuclear fuel for power plants (typically 4–5 percent U-235) can be applied to achieve higher enrichments for producing weapons-usable material. In addition, conventional fuel recycling technologies (i.e., the PUREX process) separate plutonium from uranium and other transuranics.

Some argue that any departure from a once-through fuel cycle—and the associated deployment of reprocessing technologies—in the United States could indirectly lead to an increased risk of nuclear weapons proliferation via a global expansion of such technologies; even the continued U.S. R&D on alternative fuel cycles, some suggest, could encourage such a global expansion. Others argue that other countries already seeking to pursue nuclear technologies are unlikely to be governed or constrained by U.S. approaches. They suggest instead that the United States actively involve itself in fuel cycle technology development in order to lead the way to more proliferation-resistant approaches.

Because proliferation prevention is driven primarily by international politics and understandings, the example set by the United States can play only a limited role in the proliferation arena. In addition, new technology can support, but not drive, strengthened international arrangements. Thus the course of development of commercial nuclear power in the United States is but one factor in the overall proliferation picture. Meanwhile, there are difficulties with current international

---

[106]A proliferation risk could arise from the theft of fresh fuel containing plutonium.

efforts, particularly as the International Atomic Energy Agency (IAEA)—the organization responsible for nuclear safeguards and inspections—has limited inspection
authority in some countries and, in any event, requires more resources to fulfill its
mission (IAEA, 2008d).

International plans have been suggested to mitigate proliferation risks associated with the fuel cycle, even while increasing the use of nuclear power worldwide. For example, the second Bush administration's Global Nuclear Energy
Partnership (GNEP) program sought to provide a mechanism guaranteeing fresh
nuclear fuel for civilian reactors and thus reducing the incentive for various countries to invest in enrichment plants. Another aspect of GNEP, in principle, was to
take back used fuel so that nations currently lacking reprocessing capability would
have less incentive to develop it. The position of the Obama administration on
the international aspects of the GNEP program has not yet been decided, and the
United States currently has no plans to take back any used fuel (with the exception of some classes of used research reactor fuel). The United States is unlikely
to develop such a plan until a solution has been found for handling its own used
fuel.

Uranium Resources

World uranium supplies will not be a barrier to the continuing operation of the
current fleet of plants or to the expansion of nuclear power in the time periods
considered in this report. The estimated supply of uranium is sufficient to supply the current and projected fleet of plants using a once-through fuel cycle for
more than a century (OECD/NEA, 2007); current world uranium reserves are
considered to be about 5.5 million tonnes, recoverable at a cost of up to $130/kg
(OECD/NEA, 2007). Undiscovered resources[107] are estimated at greater than
10.5 million tons.[108] However, an analysis of current and undiscovered uranium
resources indicates that exploitable resources are likely to be in the range of

---

[107]Undiscovered resources are estimates of resources that are ultimately expected to be found
based on geological characteristics of the discovered resources.

[108]These are not the only sources of uranium. For example, although currently not economical to extract, the amount of uranium found in the world's seawater is estimated at up to
4 billion tonnes (Garwin and Charpak, 2001). In Japan, R&D on "mining" uranium from seawater has found that the recovery cost could be some 5–10 times that of conventional mining of
uranium, though the researchers estimate that the cost could be reduced by half with improvements such as reduced equipment weight (Takanobu, 2001).

100–250 million tons at prices below $410/kg,[109] about 3.6 times the June 2009 monthly average uranium spot price of $114/kg.

In 2007, 14.2 million separative work units (SWU) of enriched uranium were purchased by the owners and operators of U.S. commercial nuclear power plants, only 11 percent of which was enriched domestically. The remaining 89 percent was enriched abroad, including a significant proportion from down-blended Russian highly enriched uranium (representing around 33 percent of the uranium used in U.S. commercial reactors [in U_3O_8-equivalent units]). The United States Enrichment Corporation, Inc. (USEC) operates the single commercial enrichment facility in the United States, a gaseous-diffusion plant in Paducah, Kentucky.

However, U.S. enrichment capacity is also unlikely to be a barrier to the continuing operation of the current fleet of plants or to the expansion of nuclear power in the time periods considered in this report. The Louisiana Energy Services Limited Parnership's National Enrichment Facility in New Mexico is scheduled to come on line in 2009, and the USEC America Centrifuge Plant is scheduled to come on line in 2010, achieving full production in 2012. Areva is planning a third facility at a site in Idaho, with construction beginning in 2011. GE is undertaking an engineering demonstration program to test and verify laser-enrichment technology for commercial-scale production.[110]

These new plants will produce a significant quantity of enriched uranium. The LES plant has a planned capacity of 3 million SWU per year, and the company recently (November 2008) announced that it will pursue an expansion to 5.9 million SWU; the USEC plant has a planned capacity of 3.8 million SWU per year; and the Areva facility has a planned initial capacity of 3 million SWU per year. The combined capacity of 10–13 SWU per year is nearly as much enriched uranium as was used by the 104 operating commercial U.S. nuclear plants in 2007.

Secondary sources of uranium (from government and commercial inventories, including dismantled nuclear warheads and re-enriched uranium tailings) are now depended on for 40 percent of the uranium used in world reactors.

---

[109]Uranium price cited was converted to 2007 dollars from 2001 dollars ($350/kg in 2001 dollars) using the Consumer Price Index conversion of 1.170 (www.uxc.com/review/uxc_Prices.aspx; accessed July 2009).

[110]The USNRC's description of this facility is available at www.nrc.gov/materials/fuel-cycle-fac/laser.html; accessed July 2009.

Because there is a long lead-time associated with bringing new primary resources into production, a short-term supply shortfall may develop as these secondary sources decline. However, the United States is presently down-blending 17 tonnes of highly enriched uranium to be available for a fuel bank,[111] and as more disarmament agreements are reached, it is likely that more Russian and U.S. highly enriched uranium will become available to be down-blended for use in power reactors.

FINDINGS

Companies in the United States are expressing renewed interest in building new nuclear power plants. Reasons cited include favorable recent experience with existing nuclear plants, particularly with regard to improved reliability and safety; concerns about natural gas prices; barriers to the construction of new coal-fired power plants; and concerns about the potential for future regulatory restrictions on CO_2 emissions. Like renewable sources, nuclear power plants produce no greenhouse gases during operations.

Thus there could be significant growth in this country's nuclear capacity in the years ahead, although substantial barriers—including high capital costs and lack of a means for the long-term disposition of used nuclear fuel—remain. The committee's major findings on the future deployment of nuclear technologies are given below:

> *Plant deployment:* **Until 2035, new U.S. nuclear power plants are likely to be based primarily on plant designs that are evolutionary modifications of currently operating U.S. plants.** Commercial deployment will depend largely on the economics of new plant construction.
> - *Evolutionary designs.* **Evolutionary nuclear plant designs are technically ready for commercial deployment now.** These designs incorporate features intended to improve operating efficiency, reliability, safety, and security.

---

[111]This was announced by Dennis Spurgeon, former Assistant Secretary for Nuclear Energy at the U.S. Department of Energy, at the IAEA 2 years ago as part of a discussion on assured fuel supply.

- *Alternative designs.* **Some alternative nuclear reactor designs may be ready for commercial deployment by 2035.** In the United States, high-temperature reactors for more efficient electricity generation or for industrial applications such as hydrogen production could be demonstrated by 2020 and deployed commercially 3–5 years later. Burner reactors intended for transmuting nuclear wastes could be demonstrated after 2025 and deployed commercially after 2035. Ultimately, commercial deployment of these technologies will depend in large part on whether they are proven to be economic.
- *Fusion.* **Fusion power technologies are unlikely to be ready for deployment during the time periods considered in this report, absent a significant technological breakthrough.**

Used fuel disposition: The disposition of used nuclear fuel remains unresolved. The committee has identified the major issues as

- *Storage.* **Used fuel can continue to be stored safely and securely in dry casks at operating U.S. nuclear plants, or at one or more centralized aboveground storage sites, for up to a century until a permanent disposal solution is available.** U.S. nuclear power plants produce enough spent fuel per year to fill about 400 dry casks. If all of the spent fuel currently in storage at U.S. commercial nuclear plants were to be stored together in dry casks 1.5 cask diameters apart, they would cover an area equivalent to about one-sixth of a square mile (see Annex 8.D).
- *Used fuel recycling.* **Reprocessing technologies could recycle fissile material in used fuel, thereby reducing the volume of long-lived high-level radioactive waste. Still, no technology completely eliminates the need for disposal facilities.** Two concepts for the recycling of used fuel have recently been under consideration in the United States:
  - *Partial recycle.* **A partial recycle program employing modifications to current-generation separation technology (PUREX) could be implemented after 2020.** The resulting mixed oxide (MOX) fuel would be recycled in light water reactors (LWRs). These modifications are intended to increase proliferation resistance (relative to PUREX) by preventing the separation of plutonium, but they do not substantially reduce the amount of long-lived waste requiring disposal.

—*Full recycle.* **A full recycle program employing alternative separation technologies and burner reactors is unlikely to be implemented in the United States until after 2035.** Such a program is aimed at extending existing uranium supplies, increasing proliferation resistance, and reducing the volume of long-lived high-level waste. Multiple recycles of used fuel and a large number of burner reactors will be needed to effectively transmute a significant fraction of the used fuel. Substantial R&D will be required before these technologies are ready for commercial-scale deployment. There is substantial uncertainty surrounding the economic viability of both approaches.

- *Geologic disposal.* **A permanent U.S. geologic disposal site for used fuel will not be available until after 2020.** The prospects for the previously proposed disposal site at Yucca Mountain, Nevada, are diminished by the declared intent of the Obama administration not to pursue this site. If ultimately pursued, the license application for Yucca Mountain would have to survive regulatory review by the USNRC and likely judicial challenges. As currently restricted by legislation, a repository at Yucca Mountain would not have sufficient capacity to handle all of the used fuel generated by currently operating nuclear plants; however, the site is estimated to be able to accommodate up to four times the legislated limit.

Cost: **The committee estimates that the levelized cost of electricity (LCOE) from new nuclear plants deployed by 2020 and built by investor-owned utilities (IOUs) or independent power producers (IPPs) could be 8–13¢/kWh (in 2007 dollars).** Federal loan guarantees and a financing structure incorporating 80 percent debt and 20 percent equity could result in a reduced LCOE of 6–8 cents for IOUs and IPPs. Under current legislation, loan guarantees are only sufficient for four to five plants. Calculated on the same basis, the cost of electricity from other baseload sources is also likely to increase by 2020—particularly if carbon constraints are imposed—and electricity from new nuclear plants may be cost competitive. The LCOE from nuclear plants and other power sources deployed *after* 2020 cannot be reliably estimated at present.

Supply: **The committee judges that five to nine new nuclear plants could be built in the United States by 2020.** Actual construction will depend on many factors, including economics and electricity demand. If these five to nine plants meet cost and performance requirements, more plants will probably follow; otherwise, few additional plants are likely to be built.

- **Up to an additional 20 GWe could be supplied by nuclear power plants by 2020, and as much as 88 GWe could be added to the U.S. nuclear capacity between 2021 and 2035.** These contributions, if achieved, would be significant: out of the 20 GW potentially supplied, new construction could potentially add up to 12 GWe of new capacity by 2020, while power uprates to existing plants could add up to 8 GWe. New construction alone could provide up to another 88 GWe by 2035. However, the actual supply added will depend on cost and other factors.

- **Unless many existing plants receive second 20-year license extensions—to allow for 80-year operating lifetimes—up to about 26 GWe of current U.S. nuclear capacity could be lost by the beginning of 2035 (assuming maximum uprates of all operating plants [see Table 8.1]) and nearly all of the remaining capacity could be lost by 2050 because of these plants' retirements.** Nearly all currently operating nuclear plants are likely to receive 20-year extensions to their current 40-year operating licenses, allowing for 60-year operating lifetimes. Work has begun to assess the technical feasibility and economic viability of extending licenses for an additional 20 years.

Potential barriers: **The potential barriers to expanding nuclear power in the United States are not technical. In fact, they are mainly associated with financial and societal concerns as well as current regulatory and infrastructural limitations.**

- *Financial.* **The high capital cost of new nuclear plants, the historically long construction times, and the lack of recent domestic experience with new construction create barriers to the deployments of these plants in the United States before 2020.** Financial incentives provided by the DOE could help to surmount such barriers for the first new plants. Market and policy forces (including potential carbon emissions regulations or fees) could help as well.

- *Societal.* **Public concerns about nuclear technologies, if widespread, could limit the expansion of nuclear power in the United States.**

—*Public support.* **Public support for construction of new nuclear plants has been increasing and in fact is at an all-time high, but that support could diminish if the new plants are not built on schedule and on budget and are not operated in a safe and secure manner.** Current U.S. polls show that a majority of Americans (ranging from 51 to 67 percent) support the building of new nuclear plants.

—*Waste disposal.* **The absence of a permanent disposal facility for used nuclear fuel does not present a technical barrier to new construction. However, there are political and societal barriers to selecting the location(s) for long-term used fuel storage.**

- *Regulatory.* **The licensing process for nuclear plants in the United States has been extensively revised. However, processing the surge of applications might cause delays in new plant construction.**

- *Infrastructure.* **Shortages of trained personnel (including nuclear engineers and skilled-crafts workers), as well as shortages of parts and components, could be a barrier to the construction of new nuclear plants through the early 2020s.** For example, the forging capacity for large components is limited. But these shortages are common to many parts of the power industry and should eventually be alleviated by market forces if new plant construction increases. At present, large forging capacity is expanding and nuclear engineering enrollments are rising.

CO_2 impacts: **Adding 12–20 GWe of nuclear capacity could avoid the emission of some 40–150 million tonnes of CO_2 equivalent per year in 2020, and adding 99–108 GWe could avoid 360–820 million tonnes of CO_2 equivalent per year in 2035.** This is a significant amount: the total emissions for the U.S. power sector were roughly 2.4 billion tonnes in 2007. The majority of a nuclear plant's relatively small life-cycle CO_2 emissions are generated during construction and fuel enrichment from electricity generated by fossil-fuel sources. These emissions should decrease in the future as more efficient enrichment technologies are deployed in the United States and as the sources of electric power used for fuel enrichment emit fewer greenhouse gases.

Uranium: **The world supply of uranium will not be a barrier to the continued operation of current plants or to an expansion of nuclear power over the time periods considered in this report.** Known uranium reserves will be able to supply the current and projected fleet of plants using a once-through fuel cycle for more than a century.

Proliferation: **Given the widespread international deployment of nuclear technologies, the proliferation impacts of the United States' choices for commercial nuclear power deployment are likely to be relatively minor if a once-through fuel cycle is used. However, the potential proliferation impacts of alternative U.S. fuel cycle choices remain a subject of debate among experts.** Proliferation prevention is driven primarily by strong international cooperation; technology to increase proliferation resistance, while important, can play only a limited role in reducing proliferation risk. Such technology can support but not replace strengthened international arrangements.

Research and development: **No major additional R&D is needed for an expansion of nuclear power through 2020 and likely through 2035.** However, there are still major R&D opportunities to improve nuclear technologies, including the following:

- *High burn-up fuel.* **Significantly increasing the maximum utilization of the reactor fuel's fissionable content requires a considerable R&D effort, the long-term irradiation of samples, and a sustained fuel qualification campaign.**
- *Reactor efficiency.* **R&D is needed on alternative coolants, coolant additives, and improved heat-transfer materials.**
- *High-temperature materials.* **R&D is needed on materials that can withstand the high temperatures likely to be required for hydrogen production.**
- *Alternative fuel cycles.* **Considerable R&D is needed before alternative fuel cycles will be ready for deployment. It is prudent to pursue such R&D, which is likely to be resource intensive and time-consuming, but to not initiate facility construction at present.** Increasing proliferation resistance as well as reducing the cost of fuel cycle processes and associated facilities will be a major goal of the R&D effort. Commercial-scale facilities are unlikely to be ready for deployment until after 2035.

REFERENCES

APS (American Physical Society). 2007. Consolidated Interim Storage of Commercial Spent Nuclear Fuel: A Technical and Programmatic Assessment. Washington, D.C.

ASME (American Society of Mechanical Engineers). 2008. Nuclear Component Certification Program. New York, N.Y.

BCG (Boston Consulting Group). 2006. Economic Assessment of Used Nuclear Fuel Management in the United States. Bethesda, Md.

Bisconti Research, Inc. 2006. Recent National Public Opinion Surveys About New Nuclear Power Plants. Washington, D.C.

Bunn, Matthew, John P. Holdren, Allison Macfarlane, Susan E. Pickett, Atsuyuki Suzuki, Tatsujiro Suzuki, and Jennifer Weeks. 2001. Interim Storage of Spent Nuclear Fuel: A Safe, Flexible, and Cost-Effective Approach to Spent Fuel Management. Cambridge, Mass.: Managing the Atom Project, Harvard University, and Project on Sociotechnics of Nuclear Energy, University of Tokyo. June. Available as of December 16, 2003, at http://bcsia.ksg.harvard.edu/BCSIA_content/documents/spentfuel.pdf.

Bunn, M., S. Fetter, J. Holdren, B. van der Zwaan. 2003. The Economics of Reprocessing vs. Direct Disposal of Spent Nuclear Fuel. Cambridge, Mass.: Project on Managing the Atom, Belfer Center for Science and International Affairs, John F. Kennedy School of Government, Harvard University.

DOE (U.S. Department of Energy). 2007. Notice of a Request for Expressions of Interest in an Advanced Burner Reactor to Support the Global Nuclear Energy Partnership. Federal Register 71 (August 7):44673.

EIA (U.S. Energy Information Administration). 2008. Annual Energy Outlook 2008. DOE/EIA-0383(2008). Washington, D.C.: U.S. Department of Energy, Energy Information Administration.

Energy Daily. 2008. Areva: States, communities want reprocessing plant. October 8.

Gallup. 2009. Support for nuclear energy inches up to new high. Available at www.gallup.com/poll/117025/Support-Nucler-Energy-Inches-New-High.aspx. Accessed July 2009.

GAO (U.S. Government Accountability Office). 2008. Department of Energy: New Loan Guarantee Program Should Complete Activities Necessary for Effective and Accountable Program Management. GAO-08-750. Washington, D.C.

Garwin, R.L., and G. Charpak. 2001. Megawatts and Megatons. New York: Alfred A. Knopf.

Harding, J. 2007. Economics of Nuclear Power and Proliferation Risks in a Carbon-Constrained World. Electricity Journal 20:65-76.

Ha-Duong, M., and D.W. Keith. 2003. Carbon storage: The economic efficiency of storing CO_2 in leaky reservoirs. Clean Technologies and Environmental Policy 5(3-4):181-189.

Hepple, R.P., and S.M. Benson. 2005. Geologic storage of carbon dioxide as a climate change mitigation strategy: Performance requirements and the implications of surface seepage. Environmental Geology 47:576-585.

IAEA (International Atomic Energy Agency). 2007. Milestones in the Development of a National Infrastructure for Nuclear Power. NG-G-3.1 2007. Vienna, Austria.

IAEA. 2008a. International Status and Prospects of Nuclear Power. Vienna, Austria. Available at www.iaea.org/Publications/Booklets/NuclearPower/np08.pdf. Accessed June 7, 2009.

IAEA. 2008b. Energy, Electricity, and Nuclear Power Estimates for the Period up to 2030. Vienna, Austria. Available at www.pub.iaea.org/MTCD/publications/pdf/RDSI-28_web.pdf.

IAEA. 2008c. Nuclear Power Reactors in the World. Vienna, Austria. Available at www. pub. iaea.org/MTCD/publications/pdf/RDS2-28_web.pdf.

IAEA. 2008d. Reinforcing the Global Order for Peace and Prosperity: The Role of the IAEA to 2020 and Beyond. Report prepared by an independent commission at the request of the director general of the IAEA.

Ivanov, V. 2008. Overview of the Russian sodium-cooled fast reactor program. Presentation to the Nuclear Subgroup of the AEF Committee, April.

Johnson, M. 2008. U.S. Nuclear Regulatory Commission activities on licensing of new reactors. Presentation to the National Academies' Board on Energy and Environmental Systems, Washington, D.C., December.

Keystone Center. 2007. Nuclear Power Joint Fact-Finding. Keystone, Colo.

Lisowski, P. 2008. Reactor design and fuel cycle choices. Presentation to the Nuclear Subgroup of the AEF Committee, Washington, D.C., April.

Matzie, Regis A. 2008. AP1000 nuclear power plant. Presentation at the MIT Nuclear Plant Safety Course, Cambridge, Mass.

MIT (Massachusetts Institute of Technology). 2003. The Future of Nuclear Power. Cambridge, Mass.

Moody's Investor's Service. 2008. New Nuclear Generating Capacity: Potential Credit Implications for U.S. Investor Owned Utilities. New York, N.Y.

NEI (Nuclear Energy Institute). 2007. Nuclear Industry's Comprehensive Approach Develops Skilled Work Force for the Future. Nuclear Energy Institute Fact Sheet. April. Washington, D.C.

NEI. 2008a. U.S. Nuclear Industry Capacity Factors. Available at www.nei.org/resources andstats/documentlibrary/reliableandaffordableenergy/graphicsandcharts/usnuclear industrycapacityfactors/. Accessed July 2009.

NEI. 2008b. New generating capacity costs in perspective. White paper. Washington, D.C.

NRC (National Research Council). 2004. Burning Plasma: Bringing a Star to Earth. Washington, D.C.: The National Academies Press.

NRC. 2006. Going the Distance? The Safe Transport of Spent Nuclear Fuel and High-Level Radioactive Waste in the United States. Washington, D.C.: The National Academies Press.

NRC. 2008. Review of DOE's Nuclear Research and Development Program. Washington, D.C.: The National Academies Press.

Nuclear Fuel. 2009. Much discussion, but no consensus, on U.S. reprocessing options. April 20.

OECD/NEA (Organisation for Economic Co-operation and Development/Nuclear Energy Agency). 1994. The Economics of the Nuclear Fuel Cycle. Paris, France.

OECD/NEA. 2007. Uranium 2007: Resources, Production, and Demand. Paris, France.

Parece, Martin. 2008. U.S. EPR overview. Presentation to the MIT Nuclear Plant Safety Course, Cambridge, Mass.

Pelton, T. 2007. Nuclear power has new shape. Baltimore Sun. December 25.

Peterson, P. 2008. Nuclear power's future: Advanced technologies and fuel cycles. Presentation to the Nuclear Subgroup of the AEF Committee, Washington, D.C., April.

Scroggs, S. 2008. COLA (Combined License Application Content) Engineering Evaluation of Current Technology Options for New Nuclear Power Generation. Testimony before the Florida Public Service Commission. Docket No. 070650.

Standard and Poor's. 2008. Update on the U.S. Department of Energy Loan Guarantee Program and Standard and Poor's Rating Considerations. October 7. New York, N.Y.

Takanobu, S., T. Masao, S. Tadao, S. Takao, U. Masaki, and K. Ryoichy. 2001. Recovery systems for uranium from seawater with Fibrons adsorbent and its preliminary cost estimation. Journal of the Atomic Energy Society of Japan 43(10):1010-1016.

Turnage, Joe. 2008. New nuclear development: Part of the strategy for a lower carbon energy future. Presentation to Center for Strategic and International Studies, July 31.

UCS (Union of Concerned Scientists). 2008. Nuclear Power in a Warming World: Assessing the Risks, Addressing the Challenges. Cambridge, Mass.

USNRC (U.S. Nuclear Regulatory Commission). 2008. Information Digest 2007–2008. NUREG-1350, Vol. 19. Washington, D.C. Available at www.nrc.gov/reading-rm/doc-collections/nuregs/staff/sr1350/v19/sr1350v19.pdf. Accessed June 7, 2009.

USNRC/DOE. 2008. Category 2 Public Workshop Co-sponsored by the USNRC and DOE on R&D Issues for Nuclear Power Plant Life Extension During Second and Subsequent License Renewal Periods, February 19–21.

Veil, J.A. 2007. Use of Reclaimed Water for Power Plant Cooling, Report No. ANL/EVS/R-07/3, Argonne National Laboratory; Use of Degraded Water Sources as Cooling Water in Power Plants, Report No. 1005359. Palo Alto, Calif.: EPRI.

Zogby International. 2008. Zogby Poll: 67% Favor Building New Nuclear Power Plants in U.S. Available at www.zogby.com/search/ReadNews.cfm?ID=1515. Accessed June 7, 2009.

ANNEX 8.A: NUCLEAR REACTOR TECHNOLOGIES

The design of nuclear reactors has changed in the decades since the currently operating U.S. plants were deployed. The committee divides these new designs into two categories:

- Evolutionary modifications of current U.S. designs, which are light water reactors (LWRs); and
- Alternative reactor designs, which range from more significant modifications of currently deployed designs to entirely different concepts.

In the next few decades, the majority of the new nuclear plants constructed will be based on evolutionary reactor designs. In most cases, alternative reactor designs require significant development efforts before they will be ready for deployment.

Evolutionary Nuclear Reactor Designs

Evolutionary nuclear reactor designs incorporate modifications to currently operating LWR designs intended to make the reactors simpler and safer. For example, to prevent accidents and mitigate their effects, current LWRs utilize active safety systems that require safety-grade AC power and cooling water. In place of active systems, some new LWR designs include passive safety features (relying on gravity, natural circulation, or pressurized water tanks) to avoid the need for safety-grade AC power and cooling water systems and thereby reduce the core damage frequency (CDF).[1] Other new designs provide modified active systems and claim similar reductions in CDF.

These modifications are intended to result in improved safety. The vendors of some evolutionary designs state that probabilistic risk assessment evaluations show that they have a CDF that is better than that of existing plants by a factor of 10 or more. For example, Areva has a design target for CDF that is less than 10^{-6} events per year for its U.S. evolutionary power reactor[2] design (Parece, 2008); Westinghouse claims a CDF of 5.1×10^{-7} events per reactor per year for its

---

[1] "Core damage frequency" is an expression of the likelihood that, given the way a reactor is designed and operated, an accident could cause the fuel in the reactor to be damaged.

[2] Areva's design is referred to as the "European pressurized water reactor" in Europe but as the "U.S. evolutionary power reactor" in the United States.

AP-1000 design (Matzie, 2008). For comparison, recent probabilistic risk assessments show that CDF for current plants is between 10^{-5} and 10^{-6} events per reactor per year (Sheron, 2008).[3]

Several evolutionary LWRs are operating or are being built around the world: an advanced boiling water reactor (ABWR) is operating and two large (1540 MWe) advanced pressurized water reactors (APWRs) are planned in Japan; the first Korean APR-1400 reactor (modeled after the ABB CE System 80+, a 1300 MWe APWR), was scheduled to begin construction in November of 2008;[4] a European pressurized water reactor (EPR) is being constructed in Finland, with a second in Brittany, France; and the AP-1000 design is being constructed in China. Some selected examples of evolutionary reactor designs are listed in Table 8.A.1.

Some new LWR designs have been certified for use in the United States, and construction of plants based on these designs could begin once the applications are approved. If new reactors are built (pending regulatory approval), some U.S. owner-operators have plans to use the AP-1000 design; others have identified the USEPR, the ABWR, the economic simplified boiling water reactor, or the APWR as their reactor of choice.

Alternative Nuclear Reactor Designs

In addition to the evolutionary reactor designs just discussed, alternative nuclear reactor designs are being developed (and in some cases, are already in use).[5] These reactors range from dramatic modifications of currently operating U.S. reactors to completely different concepts. Many new alternative reactor designs are intended to increase safety and efficiency. Some designs are intended for other purposes,

---

[3]These results are attributed to the USNRC State of the Art Reactor Consequence Analysis assessment. Final results from this study are planned for release in 2009 (Sheron, 2008).

[4]The Japanese APWR and the Korean APR-1400 exemplify a trend in Japan, South Korea, and China, where countries are designing and building their own reactors. As India is not a signatory to the nuclear nonproliferation treaty it was not eligible to receive imports of nuclear technology from other countries, and it had to design and build its own reactors. Under pressure from the United States, the Nuclear Supplier Group (a group of nuclear supplier countries that seeks to contribute to the nonproliferation of nuclear weapons through the implementation of guidelines for nuclear exports and nuclear related exports) recently voted to allow India access, and the United States recently rescinded its prohibition.

[5]For example, sodium-cooled and gas-cooled reactors have been in operation around the world for decades. New U.S. deployments of these reactors are considered here as "alternative" designs.

TABLE 8.A.1 Selected Examples of Evolutionary Reactor Designs

Design	Supplier	Features	Ready for Deployment in United States Before 2020?	Built Outside the United States?	Planned for Deployment in the United States?
ABWR[a]	Toshiba/GE	1371 MWe BWR	Yes	Yes; Japan	Yes
US APWR[b]	Mitsubishi Heavy Industries	1600 MWe PWR	Yes	Yes; Japan	Yes
VVER-1200	AtomEnergoProm	1200 MWe PWR	No	Proposed to be built in Russia	No
SWR 1000	Framatome	1254 MWe BWR	No	No	No
ESBWR[a]	GE	1550 MWe passive safety features BWR	Yes	No	Yes
AP-1000[a]	Westinghouse	1117 MWe passive safety features PWR	Yes	Yes; China	Yes
USEPR[b]	Areva	1600 MWe PWR	Yes	Yes; Finland and France	Yes

Note: Another example of an evolutionary reactor is Westinghouse's BWR 90+. However, this plant is not planned for deployment in the near future and is likely to require further development. ABWR = advanced boiling-water reactor; APWR = advanced pressurized water reactor; BWR = boiling-water reactor; ESBWR = economic simplified boiling-water reactor; PWR = pressurized water reactor; USEPR = U.S. evolutionary power reactor.

[a]Design certified by the U.S. Nuclear Regulatory Commission (USNRC). Amendments to the design certifications have been submitted to the USNRC for the ABWR and the AP-1000.

[b]The USNRC is currently reviewing design.

Sources: U.S. DOE Energy Information Administration (www.eia.doe.gov/cneaf/nuclear/page/analysis/nucenviss2.html; accessed May 12, 2009); and Areva, SWR 1000: The Boiling Water Reactor with a New Safety Concept (available at www.areva-np.com/common/liblocal/docs/Brochure/SWR1000_new_safety_concept.pdf; accessed July 2009).

such as reactors with significantly smaller generating capacities (potentially to supply power to countries with smaller grids);[6] reactors intended to reduce the long-lived high-level nuclear waste burden by destroying transuranic elements; and high-temperature designs intended to provide process heat to industry and/or to produce hydrogen. Some specific examples of new alternative reactor concepts are described in Table 8.A.2.

In 2002, the United States led the formation of the Generation IV Interna-

---

[6]"A generally accepted principle is that a single power plant should represent no more than 5–10 percent of the total installed capacity" (IAEA, 2007, p. 39).

TABLE 8.A.2 Examples of Alternative Reactor Concepts Being Studied or Developed

Reactor	Reactor Type	Capacity (MWe/MWt)	Originator	Notes
AHTR	Molten salt coolant	2400 MWt	ORNL	Core outlet temperature can be 1000°C; at conceptual design stage; coated-particle graphite matrix fuel
FTBR	SFR	400–600 MWe	India	Runs on thorium fuel cycle
SSTAR	LFR	10–100 MWe	LLNL	Contained completely within sealed container with fuel for 30 years; currently in development.
KLT-40C	PWR	30–35 MWe and up to 200 MWt	Russia	To be built on boats to reach locations on the remote northern coast of Russia; plans announced to build in July 2005
CAREM	PWR	27 MWe/100 MWt	Argentina/INVAP	
SMART	PWR	330 MWe	KAERI	
NP-300	PWR	100–300 MWe	Areva	To be used for electricity generation or desalination; based on submarine PWR
GT-MHR	GCR	288 MWe	General Atomics	No current plans for certification in the United States; modular reactor
HTR-PM	PBMR	200 MWt	Chinergy	Not of interest to the United States
BN-800	SFR	800 MWe	Russia	Being built in Russia based on BN-600; will be running in Russia; no plans to deploy in the United States
NuScale	PWR	45 MWe	NuScale Power, Inc.	Plans to file for design certification with USNRC in 2010; modular reactor with passive safety features

tional Forum (GIF), a 10-nation (plus the European Union) organization, to lay out a path for development of the next generation of nuclear plants.[7] Both thermal and fast reactor designs were considered. Six reactor design concepts were

[7]Reactor concepts for the next generation of nuclear plants are also being studied or developed independently by some nations who are not members of the GIF. For example, Russia is independently working on alternative reactor technologies. The BN-800, a commercial-scale sodium-cooled fast reactor design, is currently under construction. After 2025, four to five more are planned with capacities of up to 1 GW (Ivanov, 2008).

TABLE 8.A.2 Continued

Reactor	Reactor Type	Capacity (MWe/MWt)	Originator	Notes
Terrapower	Traveling-wave reactor		Terrapower, LLC	
Toshiba 4S	SFR	10 MWe/30 MWt 50 MWe/135 MWt	Toshiba	Under consideration in Alaska; plans to file for design certification with USNRC in 2010
Hyperion	Uranium hydride as fuel and moderator	25 MWe	Hyperion, Inc.	Has discussed design certification with USNRC
PBMR	Gas-cooled PBMR	180 MWe	PMBR Pty., Ltd.	Planned to be built in South Africa
ACR-700	CANDU PHWR	700 MWe	AECL	

Note: AECL = Atomic Energy Canada Ltd.; AHTR = advanced high temperature reactor; CANDU PHWR = Canada deuterium uranium pressurized heavy water reactor; CAREM = advanced small nuclear power plant; FTBR = fast thorium breeder reactor; GCR = gas-cooled reactor; INVAP = Investigaciones Aplicadas Sociedad del Estado; KAERI = Korean Atomic Energy Research Institute; LFR = lead-cooled fast reactor; LLNL = Lawrence Livermore National Laboratory; MWe = megawatts-electric; MWt = megawatts-thermal; ORNL = Oak Ridge National Laboratory; PBMR = pebble-bed modular reactor; PWR = pressurized water reactor; SFR = sodium-cooled fast reactor; SMART = system integrated modular advanced reactor; SSTAR = small, sealed, transportable autonomous reactor; USNRC = U.S. Nuclear Regulatory Commission.
Sources: Forsberg et al., 2004; Jagannathan and Pal, 2008; Smith et al., 2008; Pederson, 1998; Beliav and Polunichev, 1998; www.invap.net/nuclear/carem/desc_tec-e.html; IEA, 2002; www.hyperionpowergeneration.com/; //criepi.denken. or.jp/en/e_publication/a2004/04kiban18.pdf; www.nuscalepower.com/ri-Nuclear-Regulatory-Info-And-Process.php; www. intellectualventures.com/docs/terrapower/IV-Introducing%20Terrapower_3_6_09.pdf; Ivanov, 2008. All websites above last accessed on May 12, 2009.

selected for further examination by the participating countries: the very-high-temperature reactor, the supercritical water-cooled reactor, the lead-cooled fast reactor, the sodium-cooled fast reactor, the gas-cooled fast reactor, and the molten salt reactor (GIF/NERAC, 2005). The characteristics of these reactors are summarized in Table 8.A.3.

Around the world, several alternative reactor designs are in use or planned: for example, two new types of gas-cooled reactors are planned or operating, and two sodium-cooled fast reactors are planned to be operating in the near future (in addition to the one that is currently operating). In the following sections, a number of examples of alternative reactor designs are discussed.

TABLE 8.A.3 Reactors Selected for Examination by the Generation IV International Forum

Reactor	Description	Size (MWe/MWt)	GIF Nation Lead	Neutron Spectrum	Notes
VHTR	Helium-cooled with graphite moderator and ceramic fuel	400-600 MWt up to 300 MWe[a]	United States	Thermal	Core outlet temperature approaching 950–1000°C
SCWR	Cooled and moderated by supercritical water	1700 MWe	Japan	Thermal	Improved efficiency; core outlet temperature of 500°C
LFR	Molten lead coolant	50-1200 MWe	United States at lower priority than VHTR	Fast	The DOE phased out the U.S. R&D for this concept at the end of 2005
SFR	Liquid sodium coolant	150-1500 MWe	Japan leading effort with United States and France	Fast	
GFR	Cooled by helium or carbon dioxide	288 MWe	France	Fast	
MSR	Coolant is molten salt mixture: choice of salts (Na/Zr/F for burning; Li/Be/F for breeding) and fuels (U-238 or Th-232 fertile feed)	1000 MWe	France leading effort with United States and European Community	Fast	Limited programs are under way to evaluate concept outside the United States

Note: GIF = Generation IV International Forum; GFR = gas-cooled fast reactor; LFR = lead-cooled fast reactor; MSR = molten salt reactor; MWe = megawatts-electric; MWt = megawatts-thermal; SCWR = supercritical water reactor; SFR = sodium-cooled fast reactor; VHTR = very-high-temperature reactor.

[a]"The VHTR can also generate electricity with high efficiency, over 50 percent at 1000°C" (GIF/NERAC, 2005).

Sources: GIF/NERAC, 2005; Bennett, 2008.

Gas-cooled Reactors

New gas-cooled reactor designs include the pebble-bed modular reactor (PBMR) (which uses tennis-ball-sized fuel spheres that incorporate a carbon moderator), and gas-cooled reactors with hexagonal block fuel.[8] Several PBMRs have been constructed or are planned: two PBMRs were built and operated in Germany; a small (10 MW) PBMR is operating in China (the HTR-10); the Chinese have announced plans to build two new 200 MWt PBMRs; and PBMR Pty. Ltd. is planning to build a 165 MWe demonstration plant for Eskom, the South African utility, that is expected to come on line within the next 10 years. The Japanese operate a 60 MW graphite-moderated test reactor at the Oarai Site (the HTTR).

Small Modular Reactors

In the United States, some companies have expressed interest in submitting applications for design certification of alternative reactor designs within the next few years. Most of these designs are for small modular reactors. For example, NuScale, Inc. has plans to apply to the USNRC for design certification for their 45 MWe modular LWR design in 2010, and to apply for a COL in parallel with this process. NuScale projects that their first facility may be operational by 2015 or 2016 (www.nuscalepower.com/ri-Nuclear-Regulatory-Info-And-Process. php; accessed July 2009). Toshiba also plans to apply for design certification for the Toshiba 4S reactor (a sodium-cooled reactor) in 2010 (www.eia.doe.gov/ cneaf/nuclear/page/analysis/nucenviss2.html#_ftn12; http://criepi.denken.or.jp/en/ e_publication/a2004/04kiban18.pdf; accessed July 2009). This reactor is under consideration for use as a power source in remote areas of Alaska. Hyperion, Inc. is also in discussions with the USNRC about design certification for its 25 MWe sealed uranium hydride-fueled reactor,[9] and it has stated an intent to apply in 2012 (Johnson, 2008). However, the USNRC has stated that resources will first be applied to the operating license applications that have been submitted for evolutionary LWRs, and as time permits and resources are available, USNRC staff are conducting activities related to alternative designs (Johnson, 2008). The USNRC

---

[8]Several previous-generation block-graphite fueled reactors were built, such as Peach Bottom 1 in Pennsylvania and Fort St. Vrain in Colorado. Both have been shut down.

[9]The reactor is intended to be maintained underground. It would be unearthed every 5 years to be shipped to the factory for refueling. For more information, see www.hyperionpower generation.com/product.html.

has indicated that the design certification for these designs may be prolonged due to agency unfamiliarity with the reactor designs.

Fast Reactors

Several fast reactor designs, both "burning" and "breeding" types, are being researched. In the past, some fast reactors—notably sodium-cooled fast reactors (SFRs)—have been deployed around the world. In the past there have been problems with some of these plants, particularly with sodium leaks.[10] However, an SFR in Japan, MONJU, is planned to come back on line in 2009, and Russia is currently proceeding with a second SFR, the BN-800, the construction of which began in July 2006 (IAEA, 2008).

In particular, fast reactor designs intended to reduce the quantity of long-lived high-level waste by transmitting long-lived radioisotopes into shorter-lived isotopes as part of a closed fuel cycle ("burner reactors") are under development.[11] Much of the research on these designs is funded through the U.S. Department of Energy (DOE). Under the second Bush administration, support of alternative reactors was split between the GIF program and the Advanced Fuel Cycle Initiative and Global Nuclear Energy Partnership (AFCI/GNEP). As of the writing of this report, the Obama administration had not yet officially released detailed plans for fast burner reactor and nuclear fuel cycle programs. However, in April 2009, the DOE's Office of Nuclear Energy issued a statement that it plans to structure its nuclear fuel cycle program to concentrate on "long-term, science-based R&D . . . focused on the technical challenges of the back-end of the nuclear fuel cycle"[12] and not on near-term technology deployment.

Significant R&D will be required before burner reactors are ready for commercial deployment. For example, highly precise nuclear measurements are needed to reduce uncertainties and define relevant characteristics, such as the fission and capture cross sections for actinides, and substantial new data will be needed to

---

[10]The MONJU reactor in Japan suffered a sodium leak a year after being brought on line in 1994. In addition, the SuperPhenix reactor in France had many problems with sodium leaks and was shut down in 1998, having operated at full capacity for only 174 days.

[11]Under the second Bush administration, the DOE was investigating fast reactor designs to function as burner reactors as part of the Advanced Fuel Cycle Initiative (AFCI) (the technology development program associated with the Global Nuclear Energy Partnership [GNEP] program).

[12]Statement available at Nuclear Engineering International's website: http://www.neimagazine.com/story.asp?sectionCode=132&storyCode=2052719; accessed July 2009.

optimize system performance and economy. Improved safety, reliability, and economics will also be needed for long-term commercialization of burner reactor technologies. Thus it is the judgment of the committee that prototype burner reactors could come on line after 2025. However, developing the design, constructing prototypes and getting design certifications, testing fuel (if new type fuel is to be used), and licensing will likely push commercial operation until after 2035. Ultimately, commercial deployment of these technologies will depend upon their proving to be economic.

As mentioned previously, there is a decades-long experience with sodium-cooled fast reactors around the world, and one is currently producing electricity (the BN-600 in Russia). However, further research will be needed on fuel forms and fabrication in order to deploy these reactors as burner reactors. In August 2007, the DOE invited industry to provide concepts for burner reactors and selected four teams based on a competition (DOE, 2006). All the prototype concepts included sodium-cooled fast reactors, and one also included gas-cooled reactors (Lisowski, 2008). However, according to an April 2009 press release (mentioned previously), under the Obama administration, the DOE plans to "no longer pursu[e] near-term commercial demonstration projects."

Very-High-Temperature Reactors

Under the Next Generation Nuclear Plant (NGNP) program, the DOE is developing a commercial-scale prototype very-high-temperature reactor. The NGNP is planned to have somewhat lower outlet termperatures than were originally envisioned for this reactor. NGNP could produce electricity as well as high-temperature process heat for use by industry. Hydrogen production is also a possibility if economically acceptable materials that can withstand the necessary high temperatures (850–1000°C) can be developed, particularly for the heat exchangers and hydrogen process equipment. There is a significant potential demand for process heat in industry, over a wide range of temperatures. However, current LWRs cannot provide the needed temperature levels. Figure 8.A.1 provides a summary of U.S. process heat use, the typical temperatures at which industry utilizes process heat, and the temperatures available from various reactor technologies. The DOE requested expressions of interest in April 2008 for a demonstration plant able to produce both hydrogen and electricity (DOE, 2008a). Current plans are for start-up in 2018–2020.

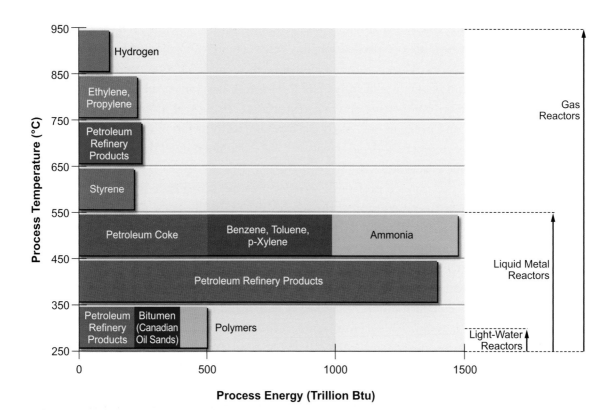

FIGURE 8.A.1 *Major energy-intensive industries and typical temperatures at which they use process heat.*
Source: Data from Alberta Department of Energy, 2007; Chenier, 2002; DOE, 2000; Gary et al., 2007; Moorhouse, 2007; NREL, 2001.

Research and Development Opportunities

Although R&D is not needed to deploy evolutionary nuclear plants in the near term, there are many R&D opportunities remaining for evolutionary LWR technologies (some of which could potentially be used in existing plants) and for alternative reactor technologies. Some of the major opportunities are discussed in this section.

Alternative Coolants

The thermodynamic efficiency of power plants is primarily constrained by the temperature of the coolant as it exits the reactor. Thermodynamic efficiency can be improved through the use of reactor coolants that allow higher coolant operating temperatures than does water. These include liquid metals such as sodium or lead, gases such as helium and carbon dioxide, molten salts, and operating coolants at supercritical conditions such as supercritical water or carbon dioxide. The major R&D challenges that need to be addressed include the associated chemical and metallurgical effects of these coolants on wetted materials, the fluid properties of these coolants, their radiation resistance, industrial scale handling, and safety. There is active international R&D in all of these areas. Test reactors and the first prototypes of new reactors using gas and liquid metal coolants are likely to be operable in some countries by 2020 or shortly thereafter.

Efficiency improvements in currently operating and evolutionary LWRs may be able to be gained by using coolant additives. The use of such additives could enhance heat transfer and potentially suppress phenomena that currently limit heat transfer and power density. Work has begun into the use of very dilute additions of nanoparticles to coolant water, and initial tests have been encouraging, suggesting that their use allows higher heat fluxes to be tolerated. Many R&D opportunities remain, including characterization of the enhanced heat transfer effect under realistic operating and transient conditions, metallurgical and chemical capability with other materials wetted by the coolant, radiation resistance, neutron absorption properties, and safety and environmental issues. Coolant additives (along with the associated redesign of reactors to adopt their use) are likely to be ready for commercial deployment after 2035.

Improved Heat Transfer Materials

As just noted, higher temperatures generally improve efficiency. At higher temperatures, improved materials are needed to contain the coolant and act as heat-transfer surfaces. High-temperature metal alloys developed for use in other applications such as combustion facilities and ceramics are being considered for improved heat transfer materials. Remaining R&D challenges for these materials include producing large quantities in the needed product form; improving fabricability, acceptance, and in-service inspection; understanding radiation effects; and fragility. Work is currently under way to use these materials with the alternative coolants previously described and to replace materials used in existing LWRs—for example, replacing the metallic tubes currently used for fuel cladding with ceramic tubes. After attractive advanced materials are identified, typically 15–20 years are required before the materials are commercially deployed. Since some materials of interest have been identified now, some new materials may be available after 2025, with a higher probability of successful widespread application after 2035.

Alternative Fuel Design for Light-Water Reactors

Improvements in reactor performance can be gained by improving the fuel—for example, by increasing the maximum utilization of the reactor fuel's fissionable content (its "burn-up") or by using fuel geometries with greater efficiencies.

The development of significantly higher burn-up fuel for LWRs could allow operating cycles to be prolonged; it could also allow the long-term heat load of the used nuclear fuel and the total amount to be stored or disposed of to be reduced.[13] R&D to increase fuel burn-up would focus on the materials issues associated with fuel integrity under long-term exposure to ionizing radiation as well as mechanical design issues which limit fuel lifetimes. For example, one issue requiring R&D is swelling of the higher burn-up fuel rods due to build-up of fission products, and the resulting risk of cladding breach. The development of higher burn-up fuel is a program of continuous improvement, but in order for significant breakthroughs to occur, considerable basic research, long-term irradiation of samples, and a sustained fuel qualification campaign are needed.

In current LWRs, the fuel rods have a cylindrical geometry in which coolant flows around the outside of the rods. An annular shape would increase the surface area of the rod in contact with the coolant by 60 percent, because coolant would flow through the center of the rod as well as along the outside surface. This would allow the coolant to be heated more efficiently, and fewer fuel rods could produce more power. Studies suggest that new plants may be able to achieve up to 50 percent more core power by using annular rather than cylindrical fuel rods (Kazimi et al., 2005). Annular fuel rods could be used in the current fleet of LWRs. Some modifications will be needed in these plants (for example, larger reactor coolant pumps, a larger pressurizer, and additional or greater capacity high pressure injection); however, the same containment, reactor vessel, and the majority of current equipment and piping could be used. There is interest in commercializing this technology (Westinghouse, 2006), but commercial-scale deployment in existing LWRs is unlikely to occur before 2020.

Degradation Phenomena

Many experts now believe that it is possible that both existing and advanced plants might be able to run for extended periods, perhaps as long as 80 years.

---

[13]The amount of used fuel would be reduced because fewer fuel assemblies would be needed to produce the same amount of power.

The phenomena that will affect performance over these time periods are not well understood at a fundamental level (e.g., stress corrosion and cracking).

Technical questions raised by lifetime extension are driven by material aging issues requiring techniques for nondestructively assessing the status of operating plants. For example, nondestructive examination (NDE) techniques and systems have been used to examine systems including outlet nozzle safe-end welds, BWR internals, and reactor vessel head nozzles. Recent NDE developments have focused on phased array, modeling and Lamb wave methods. New scanning systems are being developed for the efficient delivery of these techniques, but further research is needed, particularly in light of the heightened regulatory implications for these inspections (Westinghouse, 2006).

Digital Instrumentation and Control

The application of digital instrumentation and control offers great opportunities to improve control systems and control room designs. Nonetheless, there are challenges as well. There may be undetected bugs in software and the failure mechanisms as well as unanticipated interactions among various pieces of software and hardware. There is a need and an opportunity for research to more fully understand the inevitable increased reliance on digital systems.

Advanced Simulation Codes

Much of the existing reactor system technology relies on a detailed understanding of the performance of physical objects like fuel and pressure boundary materials, and many important environmental and in-service effects on the materials are empirically based. Because of this, the development of new designs is typically very expensive and time-consuming because extensive testing is needed. Advanced simulation codes (rooted in the large computers and sophisticated analysis approaches developed to simulate nuclear weapons) may provide increased understanding and more rapid application of new technologies; they may also better exploit existing materials and designs. R&D challenges include the development of the needed codes and their validation and verification. The validation and verification will involve ensuring that the codes meet design and regulatory requirements while substantially reducing the amount of testing and detailed post-testing examination. It is very difficult to estimate how rapidly and effectively the use of advanced simulation codes may progress. Many of the currently used codes for nuclear design and analysis originated decades ago, and the introduction of new families of codes in a given area (such as loss of coolant analysis) has taken over 10 years. This suggests that significantly more effective use of advanced simulation codes is unlikely to occur before 2020.

ANNEX 8.B: ALTERNATIVE FUEL CYCLE TECHNOLOGIES

The life cycle of the fuel that is used in a nuclear reactor (referred to as the "fuel cycle") can fall broadly into one of two categories:

- *A once-through fuel cycle*, in which the used fuel exiting the reactor is destined for permanent disposal. The used fuel is removed from the reactor after achieving design burn-up and only a small fraction of the energy potentially available in the fuel is obtained.
- *A closed fuel cycle*, in which more energy is extracted from the used fuel by processing it to separate the uranium and plutonium for reuse and to remove fission products.[14] The other transuranics[15] may also be reused or disposed of with the fission products.

The vast majority of nuclear-generated electricity in the world is produced using a once-through fuel cycle. The United States currently uses a once-through uranium fuel cycle; in this annex the committee focuses on alternatives to this fuel cycle.

Types of Closed Fuel Cycles

Closed fuel cycles fall into two major categories: (1) fuel cycles designed to produce at least as much new fissionable material as is destroyed in producing energy ("breeding fuel cycles"); and (2) fuel cycles designed to reduce the quantity of high-level nuclear waste ultimately requiring geologic disposal ("burning fuel cycles"). In either case, the used fuel is recycled, requiring chemical or electro-chemical processing to separate the fissionable parts of the used fuel and new fuel to be fabricated. The new fuel is then inserted into another reactor for additional power generation. These steps have to be repeated a number of times to achieve

---

[14]"Fission products" are isotopes produced as a result of the fission of a massive atom such as U-235.

[15]"Transuranic elements" (also known as "transuranics" or "TRU") are elements with an atomic number greater than uranium—that is, having nuclei containing more than 92 protons. Examples of transuranics are neptunium (atomic number 93), plutonium (94), and americium (95). The most important transuranic isotopes in used nuclear fuel are Np-237, Pu-239, Pu-240, Pu-241, Am-241, Am-243, Cm-242 through Cm-248, and Cf-249 through Cf-252.

the desired efficiency, for instance, to produce sufficient new fuel or to destroy sufficient undesirable isotopes.

There is continuing interest in closed fuel cycles due to several concerns: (1) extending available supplies of uranium; (2) the potential for reducing the amount of long-lived high-level radioactive waste that must be disposed of; and (3) impeding the potential expansion of proliferation risky reprocessing technologies by developing less risky technologies which can be used in their place. Breeding and burning fuel cycles are ways to address these concerns, and are discussed in the paragraphs that follow.

Available supplies of uranium can be extended significantly through the use of a breeding fuel cycle. New fissile material can be produced in a reactor when fertile isotopes (such as Th-232 and U-238) are bombarded by neutrons, converting them to fissile isotopes (such as U-233 and Pu-239) via neutron capture. Thus, if fertile isotopes are irradiated, new fissionable material can be created in the process of producing power in the reactor. A breeding fuel cycle is designed to create at least as much new fissionable material by neutron capture (for example, the fertile isotope U-238 is converted to the fissionable isotope Pu-239) as is destroyed by fissioning isotopes such as U-235 and Pu-239 to generate power. Ultimately, the fuel is removed from the reactor and reprocessed into new fuel incorporating this fissionable material.

Much larger supplies of fresh uranium are needed to maintain a once-through fuel cycle. If the use of nuclear power worldwide increases dramatically in the 21st century, some have expressed concern that this may put a strain on available resources of mined uranium. The use of closed fuel cycles could extend current supplies. However, as is discussed in the main text of Chapter 8, known uranium reserves will be able to supply an expanded fleet of plants using a once-through fuel cycle for the current century (OECD/NEA, 2007). This fact, combined with concerns about radioactive waste management has led to an emphasis on burning fuel cycles (as opposed to breeding fuel cycles) in the United States.

The amount of long-lived high-level radioactive waste could be reduced through the use of a burning fuel cycle, which can be designed to fission transuranic elements (or transuranics) contained in the used fuel, leaving behind shorter-lived elements, known as fission products, in their place. In principle, the transuranics contained in used fuel can then be incorporated into new fuel and fissioned in burner reactors. (Breeder reactors and LWRs can also destroy transuranics, but they are not specifically designed for this purpose.) Burning fuel cycles have the potential to significantly shorten the time for management of the resulting radioactive waste, as the resulting fission products typically have half-lives less than 1000 years, while some transuranics have much longer half-lives, as long as hundreds of thousands of years. The volume of long-lived high-level waste that is ultimately destined for deep geologic disposal could also be reduced; similarly, the long-term heat load could be reduced owing to the destruction of a large fraction of the spent-fuel transuranics. However, this does not reduce the short-term heat load in a repository, which for the first century is dominated by fission products. To achieve closer packing of the used fuel assemblies in a repository, these fission products (in particular, cesium and strontium) would need to be separated from the waste. Thus, in order to significantly reduce the number of repositories required, it is likely that strontium and cesium would need to be separated from the high-level waste and dealt with separately, or in principle, the repository could be actively cooled for approximately the first 100 years.

Burning fuel cycles can be further separated into limited recycle and full recycle, as illustrated in Figure 8.B.1. Under limited recycle, the used fuel from LWRs is chemically or electrochemically processed to separate fissionable material from transuranics and fission products. Fuel or a target could potentially be formed using the fissionable material and/or transuranics, and the fuel or target is used in thermal reactors. There are several possible fuel forms for limited recycle:

- Mixed-oxide (MOX) fuel consists of about 7–9 percent plutonium mixed with uranium oxide. This fuel type is currently in use outside the United States, for example, in France and the United Kingdom.[16] MOX fuel can be used to fuel commercial LWRs, but its production using currently available technologies poses a greater proliferation risk than is desired in the United States.
- Inert matrix fuel (IMF) is a proposed fuel form that consists of transuranics included with a neutron-transparent material (such as zirconium oxide). With the IMF option, spent LWR fuel would be reprocessed into such a fuel form and then inserted in the present fleet of LWRs, where the transuranics would be partially consumed. About 20 percent of the LWR reactor core would be IMF and 80 percent would be uranium

---

[16]MOX fuel is also planned to be used to dispose of surplus plutonium from the U.S. and Russian nuclear weapons programs. In the United States, the Mixed Oxide Fuel Fabrication Facility at the Savannah River Site in South Carolina began construction in August 2007 and is designed to turn 3.5 tonnes per year of weapons-grade plutonium into MOX fuel assemblies.

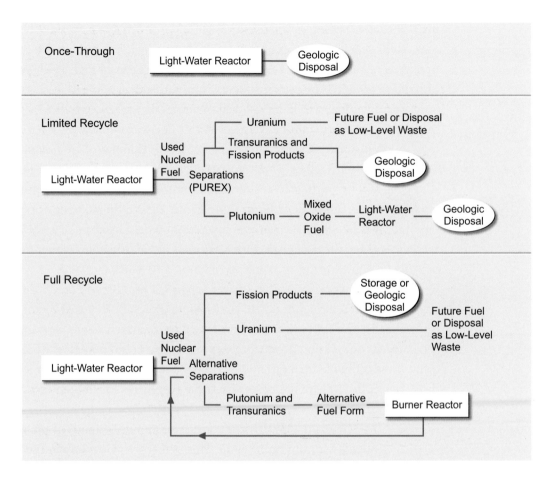

FIGURE 8.B.1 *Nuclear fuel cycles. (Top) In the once-through fuel cycle, used light-water reactor (LWR) fuel is sent directly to geologic disposal. (Middle) Under limited recycle using the plutonium and uranium extraction (PUREX) process, the used LWR fuel is chemically processed to separate uranium, transuranics, and fission products from plutonium. The uranium is used or disposed of; the transuranics and fission products are disposed of. The separated plutonium is formed into mixed oxide fuel (MOX), which is sent to geologic disposal after use. (Bottom) Under full recycle, the used LWR fuel is sent through alternative separations technologies (for example, UREX+ or electrochemical reprocessing) that separate the plutonium from uranium and fission products. In most separations processes for full recycle, the transuranics remain with the plutonium, which is formed into advanced fuel forms. This fuel is used in burner reactors. When this fuel is removed from the reactor, it is returned to advanced separations, and new fuel is fabricated from the remaining plutonium. This process is repeated for multiple recycles, until the transuranics are sufficiently consumed.*

oxide fuel.[17] The advantage of IMF fuel is that it has little or no U-238. The U-238 isotope, abundant in present LWR (uranium oxide) fuel, has the disadvantage of absorbing neutrons and being converted into transuranic isotopes with long-lived radioactivity.

- Transuranic targets, which are currently under development to burn transuranics in thermal reactors. One example may be the americium target option (Maldague et al., 1995). Spent LWR fuel would be reprocessed to extract the americium, which would be formed into americium oxide target rods and inserted into the present fleet of LWRs. About 10 to 20 percent of the LWR reactor core would be americium target rods; the remaining core would be made up of uranium oxide or MOX fuel.

Limited recycle using these technologies has the potential to reduce transuranics in the resulting high-level waste without introducing the complication of fast reactors; however, with repeated passes, a state of diminishing returns would be reached. At this point, a fast neutron spectrum would be required to continue the destruction of transuranics.

In contrast, under full recycle, used fuel would be processed to separate transuranics from fission products. Then, fuel fabrication facilities would be used to incorporate the transuranics into fuel for burner reactors. Finally, when the spent burner reactor fuel is removed from the reactor, it would be reprocessed and the recycled fuel used in burner reactors again. The last step would need to be repeated many times to significantly reduce transuranic content, and a number of burner reactors and reprocessing plants would need to be constructed for a fully closed fuel cycle to be effective. This type of full recycle and the associated reduction in transuranic content would be necessary to vastly reduce the amount of long-lived high-level waste as well as the number of repositories required to isolate high-level waste.

Alternative Separations Technologies

To implement either a burning fuel cycle or a breeding fuel cycle, separations technologies are needed to recycle (or reprocess) used nuclear fuel. These technologies

---

[17]The fuel assembly can be designed so that the IMF fuel rods are removable and can be shipped to a geologic repository while the remaining fuel rods are reprocessed.

are used to extract fissionable material and sometimes transuranics from the used fuel. Several technologies are discussed in the section that follows.

Current-generation technologies for recycling used fuel—in use in France, Japan, and the United Kingdom—are based on a process developed during the Manhattan Project called plutonium and uranium extraction (PUREX). This is an aqueous chemical process used to separate plutonium and uranium from used nuclear fuel. As a part of the process, PUREX produces a separated stream of plutonium, which can pose a proliferation and theft risk. Before commercial reprocessing was discontinued in the 1970s, a U.S. company operated a PUREX plant.

Modifications of PUREX are being developed that would allow some amount of uranium to remain in the plutonium stream to increase proliferation resistance. However, the National Research Council concluded that small modifications to the process could allow the generation of a separated plutonium stream (NRC, 2008). Modified PUREX could be commercially deployed in the United States well after 2020 (Lisowski, 2008).

The primary reprocessing technology that has been under investigation as part of the DOE's Advanced Fuel Cycle Initiative is UREX+ (DOE, 2007). UREX+ is most easily applied to oxide fuel—the fuel used in LWRs. UREX+, which is aqueous in nature, is actually a suite of processes in which uranium is extracted from transuranics along with other specifically targeted fission products. These processes (e.g., UREX+1, UREX+2) are described in Table 8.B.1. For example, in the UREX+1 process, cesium, strontium, and technetium in addition to transuranics, are extracted from the used fuel and separated from the remaining fission products. Flow sheets for UREX+1 have been developed and unit operations have occurred at engineering scale. However, the UREX separations and reprocessing approaches still must be proven out, and an integrated engineering-scale demonstration has not occurred. The technology is considered to be at the level of "proof of principle," and commercial-scale deployment is not likely before 2035.

Electrochemical separation[18] is a reprocessing technology that becomes more attractive if metal fuels are used for burner reactors or if a pre-processing step is added for oxide fuels. Electrochemical separation processes recover transuranic materials for recycle by electrochemical or selective oxidation and reduction processes using molten salts and liquid metals as the process solvent. Electrochemical

[18]This process is also known as pyroprocessing.

TABLE 8.B.1 UREX+ Processes

Process	Product 1	Product 2	Product 3	Product 4	Product 5	Product 6	Product 7
UREX+1	U	Tc	Cs and Sr	TRU and lanthanide fission products	Other fission products		
UREX+1a	U	Tc	Cs and Sr	TRU	Other fission products (including lanthanides)		
UREX+2	U	Tc	Cs and Sr	Pu and Np	Am, Cm, and lanthanide fission products	Other fission products	
UREX+3	U	Tc	Cs and Sr	Pu and Np	Am and Cm	Other fission products (including lanthanides)	
UREX+4	U	Tc	Cs and Sr	Pu and Np	Am	Cm	Other fission products

Note: Am = americium; Cm = curium; Cs = cesium; Np = neptunium; Pu = plutonium; Sr = strontium; Tc = technetium; TRU = transuranic elements; U = uranium.
Source: Idaho National Laboratory, August 10, 2006. Available at www-fp.mcs.anl.gov/nprcsafc/Presentations/NucPhysConf.pdf; accessed May 12, 2009.

separation is also considered to be in the proof-of-principle stage of development, and commercial-scale deployment of this technology is unlikely before 2035.[19]

Thorium Fuel Cycles

In addition to the fuel cycles described above, closed fuel cycles using thorium (an element approximately three times more abundant in nature than uranium) are possible. The use of thorium-based fuel cycles has been studied for about 30 years, but on a much smaller scale than for uranium or uranium/plutonium cycles (see

---

[19]Electrochemical separation has been used on a small scale for the Experimental Breeder Reactor-II (EBR-II) fuel and blanket at Argonne National Laboratory.

www.world-nuclear.org/info/inf62.html; accessed July 2009). With only around 0.8 percent of the world's uranium reserves, but about 32 percent of the world's thorium reserves, India has shown significant interest in developing the thorium fuel cycle. Russia and Norway have also shown some interest.

The thorium fuel cycle may offer some proliferation advantages over fuel cycles involving uranium. However, the thorium fuel cycle is technically more complicated than uranium or uranium/plutonium fuel cycles. By absorbing thermal neutrons, thorium-232 produces fissile uranium-233, which can be used as a nuclear fuel. Because the thorium itself is not fissile, a breeding and reprocessing phase must be introduced. In addition, fuel production and reprocessing are complicated by the higher melting point of thorium oxide (ThO_2) as compared to uranium oxide, and by the fact that the irradiated fuel is highly radioactive. Finally, a three-stream process for separating uranium, plutonium, and thorium from used fuel, though viable, is yet to be developed (IAEA, 2005). The committee judges that current experience with thorium fuel and the associated fuel cycle is very limited, and this fuel cycle is still under development. It would not be ready for deployment in the United States until after 2035.

In addition to the potential for closed fuel cycles using thorium, once-through fuel cycles using a mixture of thorium and uranium have been considered (Radkowsky, 1999; Kazimi, 2003). The fissioning of the uranium would provide the neutrons needed to initiate the conversion of thorium-232 to fissile uranium-233. It is possible that such a fuel cycle could be deployed using current LWRs. This fuel cycle could increase proliferation resistance over the once-through uranium fuel cycle because the resulting used fuel would display a significant reduction in plutonium content compared to the used fuel from a conventional reactor. Such technology is currently being developed by Thorium Power, Ltd. Their current plans are to deploy this fuel type in a lead-test assembly in a 1 GW reactor within the next 2–3 years (see www.thoriumpower.com/files/Thorium-Power-Ltd.-Information-Kit.pdf; accessed July 2009).

Research and Development Opportunities

To deploy alternative fuel cycles in the United States, further R&D is needed. Several areas in which this research may be needed are described below.

R&D for Fully Closed Fuel Cycles

If the choice is made to pursue the option of a fully closed fuel cycle, considerable R&D is needed, particularly on fuel design, separations processes, fuel fabrication (particularly of highly radioactive recycle fuel), alternative reactors, and fuel qualification.

In addition to fuel design, development and testing work still will need to be done on specific fuel forms and types before these fuel types will be ready for deployment. Further investigation will be required on the economic and process efficiency of fuel separation processes, as well as the relative resistance to diversion of weapons-capable materials. A major R&D effort will be needed for fuel qualification of recycle fuel, as the isotopic content of the recycle fuel changes on every pass through the reactor. This fuel must thus be qualified for a range of relevant parameters or re-qualified at each pass to avoid damaging the reactor. Waste streams, waste forms, and waste disposal will also require further research.

These R&D needs are very long-term and will not allow these technologies to be commercially deployed in the United States until after 2035. Similarly, a significant number of R&D challenges still need to be overcome. Many processes have been demonstrated at the laboratory scale, and some bench-scale demonstration projects have been successfully completed. However, before commercial-scale deployment is a viable option, integrated engineering-scale demonstration will also need to be completed. Studies are beginning to determine the optimum system configuration (the number of separations plants needed to support the LWRs and the burner reactors) and costs.

R&D for Limited Recycle

As noted previously, the development and use of alternative limited recycle options (such as inert matrix fuel or transuranics targets) could reduce the transuranic waste burden on a future geologic repository without the complication of introducing fast reactors. More R&D as well as subsequent regulatory approval will be required in order for these technologies to be deployable between 2020 and 2035.

Simulation and Modeling

Although significant systems analysis and comparison of once-through and closed fuel cycles has been done, further research in this regard will be essential. In addition, work on modeling and simulations will be needed, from high-level alternative system evaluations, through assessing combined nuclear and chemical processes, and detailed fuel design, to evaluation and qualification. The results of these R&D efforts are unlikely to be of any impact prior to 2020.

ANNEX 8.C: PROJECTED COSTS FOR EVOLUTIONARY NUCLEAR PLANTS

In view of companies' recent interest in building new plants, there has been a great deal of effort expended to estimate the levelized cost of electricity (LCOE)[20] and overnight construction costs[21] for new nuclear power plants in the United States. There is no recent domestic experience to draw on, however. Moreover, at the time of this writing, it is not yet clear what effect the financial crisis of 2008–2009 will have on investment decisions regarding nuclear power. The committee's analysis does not account for or explicitly address the impacts of the financial crisis.

Over the last few years, cost estimates in the open literature have varied by more than a factor of two. Recent estimates of the overnight cost for new construction have ranged from \$2400/kW to as much as \$6000/kW (NEI, 2008; Moody's Investor's Service, 2008). This range can be explained by several factors:

- Recent cost escalation of commodities, which affects all new construction;
- Uncertainty due to lack of recent builds; and
- Different assumptions made in estimating these costs.

Based on a range of overnight costs drawn from the open literature, the AEF Committee has produced an estimate[22] of the range for the LCOE for nuclear power plants deployed in the United States before 2020.[23] These estimates were produced using the financial model developed for the Keystone *Nuclear Power*

---

[20]The "levelized cost of electricity" at the busbar encompasses the cost to the utility of producing the power on a per-kilowatt-hour basis over the lifetime of the facility, including interest on outstanding capital investments, fuel, ongoing operating and maintenance (O&M) costs, and other expenses. See Box 2.3 in Chapter 2.

[21]"Overnight cost" is the cost of a construction project if no interest is incurred during construction. Several of the cost estimates discussed were originally expressed in terms of an all-in cost, which includes the interest incurred during construction and some owners' costs (e.g., costs of preparing the site). For purposes of comparison, all-in estimates were converted to overnight costs by multiplying by a factor of 0.8, derived from the Keystone financial model (Keystone Center, 2007).

[22]Note that as with all projections, the committee's are unavoidably based on assumptions that cannot be validated. In addition, the costs are rapidly changing and may go up and down; these numbers are not predictions or forecasts.

[23]Transmission costs are not included in the cost estimates discussed here. An average estimate of transmission costs for three deregulated utility districts is estimated at about 4¢/kwh (Newcomer et al., 2008).

Joint Fact-Finding report (Keystone Center, 2007)[24] and ranges for key modeling parameters gathered from a variety of sources.[25]

The primary modeling parameters and the ranges used are summarized in Table 8.C.1. These ranges are intended to bracket the views of a variety of analyses. For example, the range of overnight construction costs assumed by the committee falls between \$3100/kW and \$6000/kW, and it includes all published estimates of which the committee is aware, when standardized to current conditions. (Examples include NEI, 2008; Moody's Investor's Service, 2007; Harding, 2007; Keystone Center, 2007; MIT, 2003, 2009; University of Chicago, 2004; Scroggs, 2008; and TVA, 2005.) New nuclear plants are assumed to begin to deliver electricity in 2020, although a few might come on line earlier.

Due to the large up-front capital investment required, the LCOE for new nuclear plants is sensitive to the assumptions made for the financing of construction costs. All currently operating plants were built either by publicly owned[26] or by investor-owned regulated utilities (IOU). However, recent restructuring of the power market has enabled companies known as independent power producers (IPPs) to provide generation services independent of utilities in some areas. IPPs are considered to face higher risk for several reasons, which leads investors to expect a risk premium on their investment. Market competition makes IPPs more sensitive to operational problems; they face direct market competition in a way that IOUs do not. For example, they are unable to pass on unexpected costs to ratepayers as an IOU might (see Box 8.C.1). In the committee's analysis, financing parameters were treated separately for plants constructed by IPPs and IOUs. These financing parameters are summarized in Table 8.C.2.

The committee has used a range of 7–8 percent (in current dollars) for the rate of return on debt, reflecting the current trading range for debt securities on

[24]The effective capital charge rate for the committee's analysis, derived using the Keystone financial model, was verified to be consistent with that estimated by the Electric Power Research Institute in a recent publication (2008b), when assumptions are made equivalent.

[25]To aid in reviewing and evaluating the parameters affecting the cost estimates and determining their approximate uncertainties, the AEF Committee convened a workshop on the cost of electricity from new nuclear power plants in March 2008, including experts from industry, academia, and nonprofit institutions. A detailed discussion of the treatment of these parameters can be found at sites.nationalacademies.org/Energy/Energy_051536.

[26]Publicly owned utilities include both cooperative and municipal utilities.

TABLE 8.C.1 Parameter Ranges Used in Cost Calculations

Parameter	Low End	Distribution Average	High End
Overnight cost ($/kwe)[a]	3000	4500	6000
Escalation rate (%) during construction[b]			
Before 2013	–4	0	4
After 2013	–4	0	4
Life-of-plant capacity factor (%)[c]	75	90	95
Construction time (years)	4[d]	5.5	7
Decommissioning cost ($ million per unit)	250	625	1,000
Waste disposal cost[e] (¢/kwh)	0.05	0.1	0.2
Fuel costs (¢/kwh)	0.8	1.25	1.7
Uranium prices ($/lb)	20	85	150
Enrichment prices ($/kg SWU)	130	190	250
Life of plant (years)[f]	30	40	50
Cost of regulation and licensing ($ million)	50	100	150

Note: The parameters in the analysis were (in most cases) assumed to be independent and uniformly distributed between an upper and a lower range. The values in the table show the allowed range for each parameter. For example, the high case does not necessarily assume a $6000/kWe overnight cost and a 4 percent real escalation; 2.5 percent inflation is assumed. All costs are given in 2007 dollars. IOU = investor-owned utility; IPP = independent power producer; SWU = separative work unit.

[a]The capital cost range analyzed is a flat range from $3000/kWh to $6000/kWh. The use of a log-normal distribution made no significant difference in the final LCOE estimate.

[b]On average, no additional inflation was assumed in the nuclear sector above and beyond economy-wide inflation; however, an uncertainty of 4 percent was used in the calculations.

[c]The average capacity factor assumes that the lessons learned over the last few decades that have resulted in increasing capacity factors at existing nuclear plants will carry over to evolutionary designs. If plant life is extended, it may no longer be appropriate to continue to assume a 90 percent capacity factor. Also, extending the life may add costs not considered.

[d]In the 4-year-construction case, some costs have been accounted for in pre-construction. A flat distribution from 4–7 years was used in the analysis. This range is intended to account for the possibility of delays in construction. A log-normal distribution (which adds a low probability tail to higher and higher construction times) was examined, but ultimately not used; the use of this distribution made little difference in the final LCOE estimate.

[e]This value assumes isolation in a repository and does not account for the possibility that reprocessing becomes the method for waste isolation. Congress has the right to adjust this charge to utilities if needed, and this may occur; however, it is the judgment of the committee that this fee is unlikely to change before 2020.

[f]The physical life of the plant is typically distinct from the life of the debt. The median costs computed in this section assume a 40-year plant life. Extending the life to 50 years drops the levelized cost further, but by less than 0.2¢/kWh for an IPP and less than 0.26¢/kWh for an IOU.

BOX 8.C.1 *Effect on the Cost of Electricity of Rate-Basing Construction Work in Progress*

Unlike independent power producers (IPPs), investor-owned utilities (IOUs) are in some cases able to expense interest on construction work in progress (CWIP) for new-generation facilities as it is incurred and factor it into customer rates. Although CWIP may not affect the levelized cost of electricity (LCOE) from a new power plant, it can have a significant effect on a utility's decision process. It can substantially improve the utility's liquidity during construction, as interest costs are immediately recovered.

The initial increase in rates when a new capital-intensive plant comes on line can be very important to ratepayers, public utility commissions (PUCs), the media, utility executives, and their boards. CWIP can reduce this "rate shock" and is therefore an important factor in deciding whether to go ahead with a major capital investment. On the other hand, CWIP itself has immediate rate effects, as interest on construction work is included in rates before new power is generated. These issues can be challenging for state PUCs.

Finally, although the price of electricity to the consumer averaged over time may be lower if CWIP is used, the net cost to society may be the same. CWIP shifts some risks to ratepayers that, without CWIP, the owner-operators and their investors would bear directly.

TABLE 8.C.2 Financial Parameter Ranges Used in Wholesale Cost Calculations

Parameter	IPPs	IOUs	Plants with 80:20 Financing and Federal Loan Guarantees
Debt-to-equity ratio[a]	50:50–70:30	45:55–55:45	80:20
Return on debt (%)	8 ± 2	8 ± 2	4.5
Return on equity (%)	14.5 ± 5.5	12 ± 4	IPP: 14.5 ± 5.5 IOU: 12 ± 4

Note: These values show the central range of the estimates. These numbers are allowed to vary in the calculations to account for uncertainty. The values are nominal, and 2.5 percent inflation is assumed. IOU = investor-owned utility; IPP = independent power producer.

[a]Although 80:20 debt-to-equity ratios have been discussed by some IPPs, this financing structure is not explicitly included in the main analysis. Separate estimates for this structure (including loan guarantee assumptions) are found in the text.

utilities, as estimated prior to the recent economic slowdown.[27] The committee has used a range of 11–15 percent for the return on equity (also in current dollars),[28] with the lower value applying to IOUs and the upper value applying to IPPs, following discussions with Wall Street financial analysts (J. Asselstine, personal communication, 2008).

The committee estimates that the LCOE from new nuclear plants built by IPPs could be between 8¢/kWh and 13¢/kWh; for IOUs, the LCOE is also likely to be between 8¢/kWh and 13¢/kWh.[29] These ranges are 80 percent confidence ranges (from 10 percent to 90 percent.) These calculations do not take into account federal incentives for nuclear power, such as loan guarantees or production tax credits. Nearly all of the recent estimates of the range of LCOE from new nuclear power plants of which the committee is aware overlap with these ranges, as shown in Table 8.C.3.

Some IPPs and IOUs (for example, UniStar; see Turnage, 2008) have displayed an interest in a financing structure of 80 percent debt and 20 percent equity. The Energy Policy Act of 2005 (EPAct05) allows the Secretary of Energy to provide federal loan guarantees for up to 80 percent of eligible project costs after consultation with the Secretary of the Treasury. These loan guarantees are likely to be necessary to achieve such a financing structure, as it is unlikely that companies will be able to acquire loans for 80 percent of the project cost without them (J. Asselstine, personal communication, 2008). These incentives could result in a significant reduction in financing costs and, ultimately, a lower LCOE at the busbar: the estimated range decreases to 6¢/kWh to 8¢/kWh both for IOUs and for IPPs.[30] The first-year cost for IPPs in this case is estimated to be slightly higher, between 7¢/kWh and 9¢/kWh. The committee's assumptions for this financing structure are shown in Table 8.C.3.

---

[27]Whether or not nuclear plants can be considered typical utility investments can be debated, due to their history of schedule and budget overruns during construction and the lack of recent construction experience.

[28]The 4–7 years required for building a nuclear power plant requires that the financing reflect a long-term average. The return on equity is unlikely to vary by more than 1–2 percentage points over the next 20 years according to the financial analysts the committee consulted (J. Asselstine, personal communication, 2008); however, some economists expect the return rates to return eventually to their historical averages. The committee has accounted for this by assigning an uncertainty range to the return on equity used in the calculations.

[29]These estimates are national averages, and regional costs could be higher or lower.

[30]With the exception of the debt-to-equity ratio, the return on debt (4.5 percent), and the loan guarantee fee required by the DOE, the assumptions are the same for this calculation as for the previous ranges.

TABLE 8.C.3 Levelized Cost of Electricity for New Nuclear Construction

Source	Cost of Electricity in 2007 Dollars (¢/kWh)	Notes
NEI, 2008	6–8	First-year cost for IPP; includes loan guarantees and financing with 80 percent debt, 20 percent equity
	10–12	First-year cost for IOU; includes rate-basing CWIP
	7–9	IOU LCOE; includes rate-basing CWIP
Harding, 2007	9–12	LCOE
Keystone, 2007	8–11	LCOE
University of Chicago, 2004	5–8 (5–7¢/kWh in 2003 dollars)	LCOE
MIT, 2003	8–9 (7–8¢/kWh in 2002 dollars)	LCOE
MIT, 2009	8	LCOE
EPRI, 2008	7	LCOE
Energy and Environmental Economics, Inc. (for California PUC), 2008[a]	15	IPP LCOE
This report	8–13	IPP LCOE
	8–13	IOU LCOE
	6–8	IPP or IOU LCOE; includes loan guarantees and financing with 80 percent debt, 20 percent equity
	7–9	First year cost for IPP; includes loan guarantees and financing with 80 percent debt, 20 percent equity

Note: LCOE values not originally expressed in 2007 dollars were converted to 2007 dollars using the consumer price index. All costs have been rounded to the nearest cent. EPRI = Electric Power Research Institute; IOU = investor-owned utility; IPP = independent power producer; LCOE = levelized cost of electricity; MIT = Massachusetts Institute of Technology; NEI = Nuclear Energy Institute; PUC = public utility commission.

[a]In the 2008 study by Energy and Environmental Economics (E3) done for the California Public Utility Commission, their costs estimated how much a fixed-term purchase power agreement would cost for electricity from an IPP, whereas the AEF Committee is looking at a longer-term levelized cost. This accounts for about 10 percent of the difference in the cost estimates. In addition, E3 did not separately change the depreciation schedule for nuclear plants, which accounts for another 3 percent of the difference.

To obtain a loan guarantee, a fee must be paid by the licensee that is sufficient to cover the default risk, given a licensee's credit rating. This fee equals the "loan guarantee subsidy fee," which covers the estimated long-term cost to the government of the loan guarantee, calculated on a net present value basis. Using information from that circular, Standard and Poor's has attempted to estimate potential ranges for subsidy fees, (although the precise methods of calculation are not publicly available). They find that, "[f]or example, if a 1000 MW nuclear unit built at $6000 per kilowatt, with 80% financing from the FFB [Federal Financing Bank], is rated 'BB-' with a recovery of 70%, the subsidy cost would be a substantial $288 million while a 'BB' rated project at the same recovery may have to pay about $192 million" (S&P, 2008). For the purposes of the committee's estimates, the loan guarantee fee was estimated to be 5 percent of the principal. The committee has also assumed that in this case low interest rates will be available, and these calculations assume 4.5 percent return on debt.[31]

The committee's calculations do not explicitly take into account the possibility that vendors could offer significant cost reductions for the first few plants offered to induce a commitment for additional units in the future or potentially capture additional sales. Such incentives could include fixing the price for all or a major portion of the work; providing selected services or equipment at deep discounts; providing especially favorable financial terms; and providing or arranging for low-cost loans or loan guarantees from financial partners or from international sources of funds. These incentives could help to overcome the barriers to construction of the first few plants. However, the terms are likely to be specific to each project and will not be known until the deal is made and publicized. Thus, these effects cannot be built into the generic cost models discussed in this report.

---

[31]DOE is authorized to provide guarantees for loans covering up to 80 percent of the total project cost. When the government provides a guarantee for 100 percent of the debt instrument, the standard government loan-guarantee rules require that the government itself allocate and provide the capital for the investment (through the Department of the Treasury's Federal Financing Bank [FFB]), which is then repaid by the entity receiving the guarantee over the period of the loan. If an entity other than the FFB provides the loan, there is no federal money that changes hands at the outset. The program is intended to be revenue-neutral to the government; that is, the company benefiting from the guarantee is required to pay a fee to cover the risk of failure to repay the loan, as well as the administrative costs. DOE is authorized to provide $18.5 billion in loan guarantees for nuclear power facilities, but it is not yet clear whether this allocation will be sufficient for the four to five plants the committee judges will be needed to demonstrate whether new nuclear plants can be built on schedule and on budget. The DOE has found it difficult to implement the program, in part because of the challenge associated with estimating the appropriate fee.

ANNEX 8.D: ENVIRONMENTAL IMPACTS OF NUCLEAR TECHNOLOGIES

Electricity generated from nuclear power plants is associated with fewer negative environmental impacts (including fewer carbon dioxide, SO_x, NO_x, and mercury emissions) than is electricity generated from fossil-fuel plants. However, the environmental impacts from the nuclear fuel cycle are not negligible. This annex discusses the environmental impacts of nuclear power plants and associated fuel cycle technologies as well as the potential for additional impacts from an expanded nuclear deployment.

Greenhouse Gas Emissions

In operation, nuclear power plants emit essentially no greenhouse gases. However, CO_2 is emitted during nuclear fuel production (particularly enrichment, which accounts for most of the life-cycle CO_2 emissions) and during plant construction. Current estimates of life-cycle CO_2 emissions show wide variation, primarily due to three factors:

- *Method of enrichment assumed.* The gaseous diffusion enrichment process (currently in use in the United States and France) uses approximately 40 percent more electricity than gas centrifuge enrichment (currently in use in Russia and the United Kingdom, as well as other countries) per separative work unit (SWU).[32]
- *Source of electricity for enrichment.* Variations in the generation mix used to produce the electricity required for enrichment processes can produce significant variations in life-cycle CO_2 emissions.[33]
- *Life-cycle analysis (LCA) methods.* Different studies use different methods of life-cycle analysis. For example, Fthenakis and Kim (2007) note that economic input/output (EIO) analyses can produce significantly

---

[32]An SWU, or "separative work unit," is a unit which represents the amount of uranium processed and the degree to which it is enriched; as such it is the extent of increase in the concentration of the U-235 isotope relative to U-238.

[33]Uranium enriched in the United States, for example, has far higher associated carbon emissions than does uranium enriched in France or in the United Kingdom. The gaseous diffusion plant in Paducah, Kentucky, is electricity-intensive and draws from the heavily fossil-fuel-based electricity generation sources in the Ohio Valley. In contrast, the electricity used in the French gaseous diffusion plant is 94 percent nuclear, and the British gas centrifuge enrichment plant uses nothing but nuclear-generated electricity.

larger estimates for construction and operation than those produced using process-based analyses.[34]

The committee concurs with the conclusion reached by Fthenakis and Kim (2007) that life-cycle CO_2 emissions for nuclear plants, assuming that the current U.S. nuclear fuel cycle is maintained, could range from 16 to 55 g CO_2 equivalent per kilowatt-hour.[35] For comparison, coal plants without carbon capture and sequestration produce an average of 1000 g CO_2 equivalent per kilowatt-hour. This range includes many of the published life-cycle analyses the committee is aware of, with the notable exception of several European studies that estimate lower emissions (including the life-cycle estimate of 8 g CO_2 equivalent per kilowatt-hour used by the Organisation for Economic Co-operation and Development/Nuclear Energy Agency [OECD/NEA, 2008]) due primarily to the use of gas centrifuge enrichment.[36] The full range of life-cycle analyses reviewed is shown in Figure 8.D.1.

In the future, these life-cycle emissions associated with nuclear plants should decrease in the United States. If the sources of electric power used for fuel enrichment emit fewer greenhouse gases, emissions will be reduced for the nuclear fuel cycle. In addition, future nuclear power plants may require fewer materials and less labor to construct, which will also reduce life-cycle emissions. Finally, two gas centrifuge fuel enrichment plants are being constructed in the United States: one by the Louisiana Energy Services Limited Partnership (LES) in New Mexico, and one by the United States Enrichment Corporation, Inc. (USEC) in Ohio. These plants are planned to come on line in 2009 and 2010, with the latter to achieve full power by 2012. Areva is also planning to begin construction on a third gas

---

[34]"Economic input-output analysis" is a type of economic analysis in which the interdependence of an economy's various productive sectors is observed by viewing the product of each industry both as a commodity for consumption and as a factor in the production of itself and other goods. In this case, process-based analysis refers to the life-cycle assessment from manufacture to disposal. All inputs and outputs (within the boundaries of the analysis) are considered for all the phases of the life cycle.

[35]Their analysis assumes that a once-through fuel cycle is maintained. This range encompasses variation in the generation sources producing the electricity used for domestic enrichment (fossil-intensive versus less fossil-intensive); life-cycle analysis methodology used; and assumptions about plant operation.

[36]References include: Fthenakis and Kim, 2007; ACA, 2001; Vattenfall, 2005; Dones, 2003; Dones et al., 2005; Hondo, 2005; Tokimatsu et al., 2006; ExternE, 1998; British Energy, 2005; White, 1998; and Storm van Leeuwen and Smith, 2007.

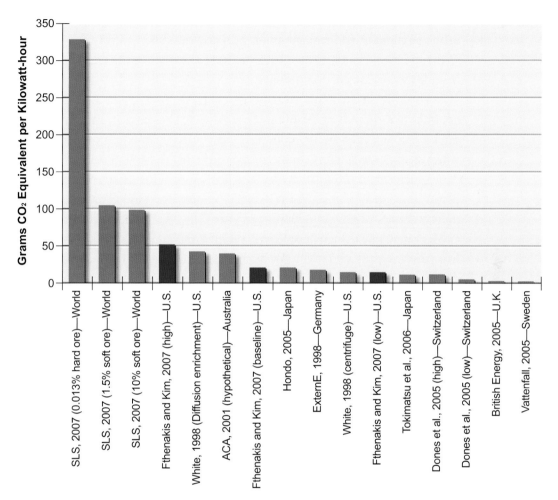

FIGURE 8.D.1 *Life-cycle CO_2 emissions for nuclear power plants. These estimates were gathered from the open literature. The red bars represent the estimates of Fthnakis and Kim for nuclear power plants built and operated in the United States. The estimates below this range include European and Japanese estimates that assume that nearly all fuel enrichment is done via gas centrifuge; this would not be the case in the United States in the near future. The estimates above this range were from a single source (Storm van Leeuwen and Smith, 2007). The highest estimate includes lower-quality uranium ore than the committee judges is likely to be needed in the near future. In addition, for these three estimates, a different type of life-cycle analysis was used, which may not be directly comparable with other estimates (both for nuclear and other generating options). For comparison, traditional coal plants emit approximately 1000 g CO_2 equivalent per kWh of electricity produced.*
Sources: Fthenakis and Kim, 2007; ACA, 2001; Vattenfall, 2005; Dones, 2003; Dones et al., 2005; Hondo, 2005; Tokimatsu et al., 2006; ExternE, 1998; British Energy, 2005; White, 1998; and Storm van Leeuwen and Smith, 2007 (SLS, 2007).

centrifuge plant in 2011. These plants may replace the energy-intensive gaseous diffusion plant at Paducah, Kentucky. In addition, General Electric (GE) (or an affiliate) has indicated an intention to build a facility in the United States deploying a laser enrichment technology.

As noted in the main text of this report, deploying new nuclear plants could have a significant effect on the total CO_2 emissions in the United States after 2020, but they are likely to have little effect before then. In 2007, the total emissions for the U.S. power sector were roughly 2.4 billion tonnes of CO_2. The deployment of 12–20 GWe of new nuclear capacity by 2020 (including both new plants and capacity increases at existing plants) could avoid as much as 40–150 million tonnes of CO_2 equivalent per year.[37] In 2035, a deployment of 100–108 GWe of new nuclear capacity could avoid 360–820 million tonnes of CO_2 equivalent per year, also assuming that new nuclear capacity is replacing an equivalent capacity of coal plants. However, by 2050, as much as 730–2300 million tonnes of CO_2 equivalent per year could be displaced.

Impacts on Waste from Production of Nuclear Fuel

There are environmental impacts from the multiple processes involved in producing nuclear fuel. These processes include:

- Mining, in which natural uranium is extracted from the ground;
- Milling, in which natural uranium is chemically converted to a dry, purified uranium concentrate: uranium octaoxide (U_3O_8), or "yellowcake";
- Conversion, in which the U_3O_8 is chemically converted to uranium hexafluoride (UF_6) gas for enrichment;

---

[37]These calculations assumed a high case and a low case. In the low case, the committee's low estimates for potential new nuclear supply by 2020, 2035, and 2050 replace an equal generating capacity of natural gas plants emitting 500 tonnes CO_2 equivalent per GWh. In the high case, the committee's high estimate for potential new nuclear supply in each of those 3 years replaces an equal generating capacity of traditional coal plants emitting about 1000 tonnes of CO_2 equivalent per GWh. The committee assumes that nuclear power plants emit 40 tonnes of CO_2 equivalent per GWh on a life-cycle basis (this includes emissions from construction, mining, fuel fabrication, and other processes). This is the average of the 24–55 tonnes CO_2 equivalent per GWh discussed in this annex. The nuclear plants are assumed to operate at an average capacity factor of 90 percent.

- Enrichment, in which the concentration by weight of the U-235 isotope is increased;
- Fabrication of fuel.

Aside from greenhouse gases produced, the primary environmental impact from these processes involves waste from the mining and milling processes—a slightly radioactive by-product known as mill tailings. The tailings contain about 85 percent of the natural radioactivity in unprocessed uranium ore, from radioactive thorium, radium, and radon. Mill tailings also contain low levels of nonradioactive toxic heavy metals (such as chromium, lead, molybdenum, and vanadium) that were present in the ore, and they can contain toxic chemicals used in the milling process. Approximately 200 pounds of mill tailings are typically produced for each pound of natural uranium (DOE, 1997).

Radon emissions from mill tailings due to radioactive decay of uranium were previously an issue of public concern in the United States. Mill tailings are subject to comprehensive regulation in the United States under the Uranium Mill Tailings Radiation Control Act of 1978, with the result that radon emissions from tailings piles are now strictly limited and other releases are tightly controlled.

In some locations, a process called in situ leach (ISL) mining has replaced hard-rock mining and milling of uranium. Conventional mining entails removing ore-bearing rock from the ground and processing it to retrieve the uranium ore. ISL involves recovering the minerals from the ground by injecting a leaching liquid (typically native groundwater mixed with a complexing agent) into the ground in one location and pumping this liquid (which contains dissolved uranium) out of the ground in another location. ISL is in use or planned for use in several locations in the United States (www.eia.doe.gov/cneaf/nuclear/dupr/qupd_tbl4.html; accessed July 2009). In order for ISL to be used, the uranium must occur in permeable rock. In many cases, the uranium occurs in permeable sandstone aquifers. The use of ISL results in smaller amounts of mill tailings to be disposed of. Nevertheless there is potential for the environmental impacts such as groundwater contamination if the leaching liquid spreads outside the uranium deposit; the removal of hundreds of millions of gallons of water from the aquifer (particularly in dry areas); and final contamination of the aquifer that is difficult to impossible to remediate once the mining operation is complete.

At present, very little uranium is mined in the United States, although there has recently been a significant upsurge in mining activity. In 2003, 997 tonnes of U_3O_8 were produced from U.S. mines; in 2007, 2057 tonnes of U_3O_8 were produced from U.S. mines (EIA, 2007). This remains a small fraction of the approximately 49,000 tonnes of U_3O_8 produced in the world in 2007 (42,000 tonnes in 2003 [www.world-nuclear.org/info/inf23.html; accessed July 2009]).

This foreign dependence does not appear to represent a security risk, as there are extensive uranium resources in Canada and Australia. In 2007, 33 percent of the uranium purchased by owners and operators of U.S. civilian nuclear power reactors was imported from Russia, and it was primarily produced from down-blended[38] Russian weapons-grade uranium; 88 percent of the remaining uranium was mined and milled outside the United States, primarily in Canada (21 percent of uranium purchased in 2007) and Australia (23 percent of the uranium purchased in 2007).[39]

The process of conversion results in less waste than from mining and milling, and these by-products are characterized by the presence of thorium, radium, and radon gas. Finally, enrichment separates natural uranium into enriched uranium for use in power plants and depleted uranium (DU). The DU must be disposed of. Because of its large density and relatively low radioactivity levels, some depleted uranium is used for commercial applications, such as ballast in commercial aircraft and ships, and in military applications, such as armor and armor-piercing munitions. Significant inventories of DU in the form of UF_6 remain for disposition at enrichment sites, and these materials present potential health and environmental risks because they are maintained in the form of UF_6.

The expanded deployment of nuclear power in the United States (particularly after 2020) may result in increased demand for uranium with an associated increase in worldwide uranium mining and milling. However, as noted previously, very little uranium is mined in the United States, and few nuclear plants are likely to be constructed in the United States before 2020. Thus, domestic environmental impacts related to the front end of the nuclear fuel cycle due to an increased

---

[38]"Down-blending" refers to a process in which low enriched uranium (reactor grade) is produced from highly enriched uranium (weapons grade).

[39]Environmental regulations for mill tailings are equivalent to those in the United States in both Canada and Australia. However, the majority of uranium purchased in 2007 by owners and operators of U.S. nuclear plants that was not domestically produced (8 percent) or imported from Russia, Canada, or Australia was imported from Namibia, Kazakhstan, and Uzbekistan (13 percent). These nations may not have equivalent regulations. (Less than 1 percent was imported from the Czech Republic in 2007. Data were not included on the origin of the remaining 2 percent of the uranium—around 889,000 lb of U_3O_8 equivalent—in the EIA's statistics.) See EIA (www.eia.doe.gov/cneaf/nuclear/umar/table3.html; accessed July 2009).

deployment will be small before 2020. These impacts may increase if these initial plants are successful and many more plants are constructed after 2020, and if more mining is undertaken in the United States to meet an increased demand for uranium.

Impacts During Operations

The environmental impacts of nuclear power plants during operations in many cases are similar to those of other large thermal power plants and relate largely to water use and consumption for heat management. These impacts arise from the cooling systems of large plants. Some routine radioactive emissions also occur during plant operations.

Water Use

Thermal power plants typically use significant quantities of water during operation, primarily for cooling. The amount of water required can create problems if the location does not have an adequate water supply, or if power output at some sites must be constrained to comply with permit limitations on the temperature increase that can be accepted in the receiving waters. The amount of cooling required is determined by the thermal efficiency of the plant; nuclear power plants on average require more cooling water per kilowatt-hour of electricity produced than do fossil-fuel plants of comparable age (due to nuclear power plants' lower average thermal efficiency.)

Most U.S. power plants use one of two types of cooling processes: once-through cooling or closed-cycle wet cooling.[40] Once-through cooling is rarely, if ever, used on plants built after the 1970s, as a result of environmental restrictions imposed by Section 316 of the U.S. Clean Water Act governing thermal discharges (Section 316[a]) and intake losses (Section 316[b]).

---

[40]Once-through cooling withdraws water from natural water bodies and uses it to absorb heat. It is then returned to natural receiving waters at a higher temperature (typically 8–17°C) than that at which it was withdrawn. Closed-cycle wet cooling circulates a similar amount of cooling water through the steam condenser but then cools the water in a mechanical- or natural-draft cooling tower by evaporating a small fraction of the flow (approximately 1 to 2 percent) and recirculates the cooled water back to the condenser. The water withdrawn from the source water body is only that required to make up the amount lost to evaporation in the cooling tower plus blowdown (water discharged from the cooling system to maintain acceptable circulating water quality).

There is a distinction between water use and water consumption; in many cases much of the water is returned to the source after it is used for cooling, albeit at a higher temperature. A nuclear plant using once-through cooling uses about 95,000–227,000 liters of water per megawatt-hour and consumes about 1,500 liters/MWh, whereas a nuclear plant using closed-cycle wet cooling uses about 3,000–4,200 liters of water per megawatt-hour of electricity produced and consumes about 2,700 liters/MWh. For comparison, a coal-fired power plant using once-through cooling uses about 76,000–189,000 liters of water per megawatt-hour and consumes about 1,100 liters/MWh, and a coal plant using cooling towers uses about 1,900–2,000 liters of water per megawatt-hour of electricity produced and consumes about 1,800 liters/MWh.

After 2020, alternatives such as dry cooling may be able to reduce water use further. Dry cooling is usually accomplished with mechanical-draft air-cooled condensers to which turbine exhaust steam is ducted through a series of large ducts, risers, and manifolds. Dry cooling still has significant disadvantages, including higher costs, higher operating power requirements, and reductions in plant efficiency and capacity during periods of hot weather. Dry cooling has been used for some coal-fired plants,[41] but at present, no nuclear plants have been constructed using this technology.

Hybrid cooling, which typically consists of a dry cooling system operating in parallel with a conventional closed-cycle wet cooling system, is an alternative that is finding increased use at some new coal-fired plants. A hybrid cooling system was built in 1988 at the Neckarwestheim Nuclear Plant in Germany. Hybrid cooling is also proposed for use in several evolutionary nuclear plants intended to be built in the United States in the near term, including the new reactor proposed by UniStar for the Calvert Cliffs site in Maryland (Pelton, 2007).

The water use impacts of future nuclear plants will depend on where the plants are sited and what cooling technologies are employed. Water use and consumption will be a consideration in siting new nuclear plants in areas such as the American southwest with growing populations but limited water supplies. In some instances, wet cooling systems can use nonfreshwater sources such as seawater (if located on the coast), brackish water from wells or estuaries, agricultural runoff, produced water from oil and gas drilling operations, or treated municipal wastewater (Veil, 2007).

---

[41]For example, the Kogan Creek power station in Australia, a 735 MW coal-fired plant, uses dry cooling.

Routine Radioactive Emissions

Some citizens are concerned by routine radioactive emissions that occur during plant operations.[42] These emissions of radiation originate from routine operations of nuclear power plants and largely consist of neutron activation products in the cooling water, fuel rod leaks, and radioactive contaminants from atmospheric emission of fission gases (particularly noble gases). Each U.S. nuclear plant is required to monitor and report these emissions to the USNRC in an annual report as a condition of maintaining its license.[43] Both gaseous and liquid releases are reported in units of the amount of radiation released and the resultant dose to the hypothetical maximum exposed individual. The reports take into account any interim used fuel storage on the site as well as the operation of the plant itself, and the releases are limited by the license of the plant.[44] These emissions are typically several orders of magnitude below statutory limits and would not be expected to produce meaningful health risks to people living near the plants. Nonetheless, these emissions can be of great concern to local citizens who may not have confidence in statutory limits, as seen in the controversy over tritium leaks at the Braidwood plant in Will County, Illinois.[45]

Disposal of Used Nuclear Fuel and Other Waste

The operation of a nuclear power plant generates several types of radioactive waste, which must be stored and eventually disposed of. These include:

---

[42]Coal plants also produce radioactive emissions, primarily from radioactive thorium and uranium that is naturally present in coal. When coal is burned into fly ash, the uranium and thorium are concentrated at up to 10 times their original levels. Some of these materials may escape with other particulates from an operating coal plant. As a result, the radioactive emissions from a coal plant may exceed those from a nuclear plant with an equivalent capacity.

[43]Typical examples of such reports are given in Public Service Enterprise Group (PSEG, 2007) and Entergy (2007).

[44]Detection of higher than usual releases, still well below statutory limits, can be helpful in identifying leaks needing attention. The data from the annual reports show that the absolute releases and resultant doses are typically several orders of magnitude below the statutory limits.

[45]News coverage of public concerns about the tritium leaks at the Braidwood nuclear power plant in Illinois are available at www.pbs.org/newshour/bb/environment/jan-june06/tritium_4-17.html; accessed July 2009. Further information about the actions taken can be found at the Illinois EPA website: www.epa.state.il.us/community-relations/fact-sheets/exelon-braidwood/exelon-braidwood-2.html; accessed July 2009.

- Used fuel from the nuclear reactor (also referred to as *spent fuel*);
- Other radioactive waste generated during plant operations; and
- Radioactive decommissioning waste resulting from the demolition of the plant after permanent shut down.

Because of their radioactivity, these wastes pose unique challenges. At present, there is inadequate disposal capacity for many types of radioactive wastes, including used nuclear fuel. The following sections and Table 8.D.1 provide further detail on the management and disposal of these wastes in the United States.

Used Nuclear Fuel

The 104 currently operating nuclear plants in the United States generate about 2200 metric tons of uranium (MTU)[46] per year of used nuclear fuel, an inert but highly radioactive solid. Because used fuel is highly radioactive, it must be handled using remote-handling equipment, stored in highly shielded facilities, and disposed of in a manner that is designed to sequester it from the environment such that predicted doses of certain potentially exposed people are below specified regulatory limits.

The major constituents of used fuel are uranium, transuranic elements produced by neutron capture, and fission products produced by neutron-induced or spontaneous fission of uranium and transuranic elements. The great majority of the radionuclides found in used fuel are relatively short-lived and decay to low levels over decades; for example, the radioactivity from Cs and Sr decreases rapidly a few decades after discharge from the reactor. The toxicity of used fuel as a function of time is shown in Figure 8.D.2. However, some long-lived actinides and fission products are potentially toxic for many thousands of years, and, with a once-through fuel cycle, used fuel will need to be managed (though not necessarily actively) for hundreds of thousands of years. Thus, concerns associated with managing used fuel are intrinsically intergenerational.

---

[46]Used fuel quantities are expressed in terms of "metric tons of uranium" (MTU) contained in the fuel before it is irradiated.

TABLE 8.D.1 Management and Disposal of Radioactive Waste from U.S. Nuclear Power Plants

	Waste Generation at U.S. Nuclear Power Plants		
	Used Fuel	Other Operating Waste	Decommissioning Waste
Annual waste generation	2200 MTU (2007)	3,834 m^3 of LLW generated by nuclear power industry (1998 data)	
Radioactivity	Long-lived, highly radioactive	Mostly short-lived, low-to-intermediate radioactivity; small volumes of long-lived highly radioactive waste	Mostly short-lived, low-to-intermediate radioactivity; small volumes of long-lived highly radioactive waste
Storage	Pools: about 58,000 MTU at 65 operating sites, 9 sites with no operating reactors, and one centralized storage site. Dry casks: about 10,500 MTU in about 900 dry casks at 40 sites	Dry storage (drums and casks) at plant sites; storage of Greater-Than-Class-C waste in pools and casks	No storage of Class A, B, C waste; storage of Greater-Than-Class-C waste in pools and casks
Disposal	Deep underground repositories	Land disposal facilities for Class A, B, C waste; no disposal pathway for Greater-Than-Class-C waste	Land disposal facilities for Class A, B, C waste; no disposal pathway for Greater-Than-Class-C waste[a]
Current availability of storage	Adequate wet and dry storage available on-site	Adequate storage available on-site	Waste can be stored on-site during decommissioning
Current availability of disposal	None	Adequate for Class A waste; limited for Class B, C waste; none for Greater-Than-Class-C waste	Adequate for Class A waste; limited for Class B, C waste; none for Greater-Than-Class-C waste

Note: MTU = metric tons of uranium.
    [a]In terms of radioactivity, low-level radioactive waste is classified as A, B, C, or Greater Than Class C (in order of ascending hazard) based on activities of specific radionuclides. See 10 CFR 61 for exact definitions.
Sources: Used fuel quantities: NEI written communication; NEI website (www.nei.org); U.S. Department of Energy Manifest Information Management Systems (mims.apps.em.doe.gov; accessed July 2009). Storage sites: APS, 2007.

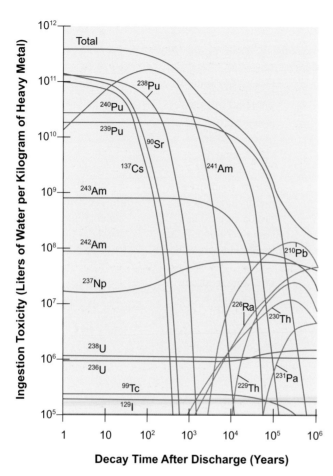

FIGURE 8.D.2 *Toxicity of nuclides in used fuel from a light-water reactor. Toxicity is defined here as the volume of water required to dilute the radionuclide to its maximum permissible concentration per unit mass of the radionuclide. High index numbers denote more toxic radionuclides—that is, more water is required to dilute these radionuclides to "safe" levels. The toxicity levels shown in this figure are for direct human ingestion of used fuel and therefore would not necessarily apply for other exposure pathways. For example, radionuclide toxicities for exposures from groundwater would be dominated by isotopes that are soluble and not sorbed completely by the host rock.*
Source: Oak Ridge National Laboratory.

Final Disposal

The final disposal of used fuel (as well as other high-level waste[47]) has been studied by the federal government since the 1950s. However, a decision as to how to permanently dispose of this material was not made until 1982. The Nuclear Waste Policy Act of 1982 provided that the disposal of used fuel from commercial nuclear power plants was a federal responsibility, as well as that the federal government would construct and operate a deep geological repository for this purpose.[48] In 1987, the Nuclear Waste Policy Act Amendments Act directed the federal government to investigate Yucca Mountain, Nevada, as the nation's first disposal site.

The 1987 Amendments Act required that the DOE, the agency responsible for siting, constructing, and operating a repository at Yucca Mountain, begin receiving commercial used fuel for disposal at Yucca Mountain no later than January 31, 1998; however, there have been delays. The DOE filed a license to construct the repository in June 2008. If that application (and a subsequent operating amendment) is ultimately approved by the USNRC (see Box 8.D.1) and survives expected court challenges, the DOE previously expected to open the repository sometime after 2020. However, the prospects for the Yucca Mountain repository are obviously diminished by the declared intent of the Obama administration not to pursue this disposal site. The FY 2010 Presidential Budget Request reduces the funding for the Yucca Mountain program to a level deemed necessary to respond to USNRC queries during the Yucca Mountain license review process.

To accommodate the used fuel from current plants, a second geologic repository may need to be constructed, or if Yucca Mountain is ultimately pursued,

---

[47]"High-level waste" (HLW) consists of radioactive materials at the end of a useful life cycle that should be properly disposed of, including (1) the highly radioactive material resulting from the reprocessing of used nuclear fuel, including liquid waste directly in reprocessing and any solid material derived from such liquid waste that contains fission products in concentrations; (2) irradiated reactor fuel; and (3) other highly radioactive material that the U.S. Nuclear Regulatory Commission, consistent with existing law, determines by rule requires permanent isolation (USNRC online glossary, available at www.nrc.gov/reading-rm/basic-ref/glossary.html; accessed July 2009).

[48]Disposal is to be funded by a waste management fee levied at one dollar per MWh, paid by the ratepayers of nuclear electricity generation companies, and collected in a federally administered waste management fund. At the end of 2007, just over $27 billion had been collected from ratepayers and credited to the fund from industry payments and interest. About $9 billion has been spent to develop a repository (www.ocrwm.doe.gov/about/budget/index.shtml; accessed July 2009).

BOX 8.D.1 *Radiation Exposure Limits at Yucca Mountain*

The U.S. Nuclear Regulatory Commission's (USNRC's) approval of the Department of Energy's (DOE's) application to construct a repository at Yucca Mountain is predicated on a demonstration by the DOE that the repository will satisfy regulatory requirements. In 2001, the Environmental Protection Agency (EPA) published standards for the disposal of used nuclear fuel and high-level radioactive waste at the geologic repository planned at Yucca Mountain, Nevada. The standard was remanded because it was not based on and consistent with the 1995 report *Technical Bases for Yucca Mountain Standards* (NRC, 1995). In 2005, the EPA proposed a revised standard and it was promulgated in September 2008. The revised standard provides for a separate dose limit (100 millirem/yr) to be applied beyond 10,000 years up to 1 million years. In February 2009 the USNRC published its final rule incorporating the EPA standards into the USNRC regulations for Yucca Mountain.

---

federal legislation may need to be modified to increase its capacity. (About 68,500 MTU of used fuel is currently in storage at U.S. plant sites, while storage was allocated for only 63,000 MTU of commercial used fuel at Yucca Mountain.[49]) If, as expected, nearly all of the 104 currently operating nuclear power plants in the United States receive 20-year license extensions (for a total of 60 years of operation), the eventual inventory of used fuel could reach 138,000 MTU, even without construction of new plants (USNRC, 2008).[50]

The Nuclear Waste Policy Act requires the Secretary of Energy to report to Congress no later than January 1, 2010, concerning the need for a second geologic repository. In December 2008, Energy Secretary Samuel Bodman transmitted this report to the President and the Congress; it stated that "unless Congress raises or eliminates the current statutory capacity limit . . . [on Yucca Mountain], a second repository will be needed" (available at www.energy.gov/news/6791.htm; accessed

---

[49]The statutory capacity limit at Yucca Mountain is 70,000 MTU; 7,000 MTU of this was allocated for DOE used fuel and HLW (DOE, 2008b).

[50]In making this estimate, it is assumed that each of the 104 currently operating reactors produces 21 MTU of used fuel per year from 2009 until the expiration of its 20-year license extension (for a total of 60 years of operation). About 69,300 MTU of used fuel would be generated in the future; along with the 68,500 MTU of used fuel currently in storage, this leads to a total of about 138,000 MTU of spent fuel over the lifetimes of these reactors.

July 2009). Given the experience with Yucca Mountain, even if a second repository is proposed, it is the judgment of the committee that a second repository almost certainly could not be operational until after 2035.

Interim Storage

Until a final repository becomes available, from a technical perspective, analyses have shown that used fuel can be stored for many decades in dry cask storage at low risk, or as long as society is willing to devote the attention and resources to managing it (NRC, 2001). There are two basic options for such storage:

- Continued aboveground storage at plant sites, initially in pools and ultimately in dry casks.[51]
- Centralized interim storage in dry casks at one or more regional sites or at a single national site.

Extended interim storage, whether at a plant or centralized facility, has several potential advantages. Extended surface storage would allow time for the radioactive decay of isotopes with shorter half-lives, which would reduce the heat loads if the used fuel were eventually emplaced in a repository. In addition, the facilities can be actively monitored and maintained for an indefinite period of time.

At present, used fuel is being stored at currently operating plant sites; this practice could be continued.[52] As of 2009, approximately 58,000 MTU of used fuel was in storage in pools at 75 sites, and about 10,500 MTU was in dry cask storage at 40 sites. Used-fuel pools at most plants are at or near their storage

---

[51]After removal from the reactor, used fuel is stored for several years in water-filled pools with active heat removal systems. The water is an effective heat transfer medium and also serves as an effective radiation shield. Used fuel can be moved into dry storage after at least 3 years of cooling in the pool, although most fuel being dry stored is much older. Used fuel is dry-stored in heavily shielded casks that use passive heat removal systems (conduction and convection) for cooling.

[52]As of November 2008, 40 operating generally licensed independent fuel storage installations (ISFSI) existed in the United States, and there were 15 specifically licensed ISFSIs at or away from reactor sites (see www.nr.c.gov/waste/spent-fuelstorage/locations.html; accessed July 2009). Thirty states have at least one ISFSI. The federal government is paying some plant operators' expenses for extended on-site storage of used fuel because of the delay in opening the repository.

capacities,[53] so the quantity of used fuel in dry cask storage is expected to increase in the future. However, there is enough room at existing sites to continue to store used fuel in dry casks, even if 60-year life extensions for all operating plants are granted (APS, 2007). All current U.S. commercial used fuel stored in vertical dry casks (spaced 16.5 feet, or about 1.5 cask diameters, apart) would cover an area of about 4.4 million feet, or about one-sixth of a square mile.[54] Once the plant shuts down and is decommissioned the used fuel at the site would become "stranded." The plant operator would remain responsible for managing this fuel and would not be able to terminate the plant license until the fuel was moved off-site. At present, used fuel is stranded at six plants in the United States. The U.S. government holds ultimate responsibility for the disposition of this stranded used fuel and at least some of the costs of its storage.

Alternatively, used fuel could be collected in several regional facilities or one national interim storage facility. Development of centralized storage could reduce pressure to working out the obstacles to geologic storage, but it would also have disadvantages. Regional storage would require transport of used fuel from the plant sites, which would require additional expense and could raise public concerns, and if a repository is eventually opened, the fuel may need to be transported a second time. In addition, it requires significant time to identify and license a site. For example, the licensing process for the centralized storage site proposed by Private Fuel Storage, LLC at Skull Valley, Utah, required almost 9 years from the filing of the license application with the USNRC until a draft license was issued.[55] Thus, it is unlikely that sufficient facilities could be identified and licensed before 2020.

Other Operating Wastes

In addition to used nuclear fuel, other radioactive wastes are generated during nuclear power plant operations. In the United States, these wastes are defined

---

[53]Plant operators leave enough open space in the pool so that all of the fuel in the reactor core can be transferred into the pool if necessary.

[54]This assumes casks with a diameter of 11.04 ft stacked vertically next to one another (open packing) and spaced 1.5 cask diameters apart. The cask diameter was taken from specification for the Holtec International Hi-Star/Hi-Storm storage overpack for its used fuel storage system (Holtec International March 2005 Shipping and Storage Cask Data for Used Nuclear Fuel).

[55]This facility is designed to be able to provide storage for up to 40,000 MTU of used fuel. However, the Interior Department has blocked moving forward with this facility by refusing to grant permits.

by exclusion and are referred to as low-level wastes (LLW).[56] These wastes have much lower radioactivity than does used fuel, and that radioactivity decays to background levels[57] in less than 500 years (about 95 percent decays to background levels within 100 years or less.) These wastes include items like contaminated gloves, personal protective clothing, tools, water purification filters and resins, and plant hardware.

LLW is typically characterized both by volume and by radioactivity (measured in curies for specific isotopes). The nuclear power industry produces about 3800 m³ of LLW per year. The median for each plant is about 21 m³ for PWRs and 79 m³ for BWRs. This volume of waste is equivalent to 33.6 average-sized refrigerators per year for each PWR and 126.4 average-sized refrigerators per year for each BWR.

In terms of radioactivity, low-level radioactive waste is classified as A, B, C, or Greater-Than-Class-C (in order of ascending hazard) based on activities of specific radionuclides.[58] About 90 percent of the LLW produced by U.S. nuclear plant operations is low-activity Class A waste; this waste can be disposed of in land disposal facilities. There are limited commercial disposal sites for these wastes in the United States. Higher activity wastes (Class B and C waste) can also be disposed of in land disposal facilities, but appropriate facilities are not available to generators in all states.[59] Due to these limitations and the small volume of these wastes, many plants are storing these wastes on-site. Wastes with even higher activities (Greater-Than-Class-C waste) currently have no approved disposal pathway.[60] This waste, which typically consists of neutron-activated metal plant hardware, is also being stored on-site at plants, usually in the used-fuel pools or in dry casks. This waste will remain at these sites until a disposal facility is available to accept it.

In recent years, the nuclear industry has made an effort to reduce the amount

---

[56]That is, low-level wastes are the wastes that do not fall into other regulatory categories, such as used nuclear fuel, HLW, or transuranic waste. These wastes are generated in many physical and chemical forms and levels of radioactive contamination.

[57]At sea level, typical natural background radiation levels are around 3 millisieverts (300 millirem) per year.

[58]Activity is the rate of decay of radioactive material per unit time. The activity levels for each class of low-level waste are specified in 10 CFR 61.

[59]Some disposal facilities are operated by state compacts and are open only to member states.

[60]The final disposal for Greater-Than-Class-C waste is a DOE responsibility. The DOE is developing an environmental impact statement for the disposal of this material.

of LLW produced. Cost for disposal is by volume and activity, and waste is processed to reduce volume by supercompaction.[61]

Decommissioning Wastes

The decommissioning of a nuclear plant after it has been permanently shut down produces large volumes of waste, some of which is radioactive.[62] Decommissioning can occur immediately after shutdown, or plants may be put into a safe storage condition for a number of years to allow time for radioactive decay. In the decommissioning process, fuel is removed from the reactor core, the reactor core internals and reactor vessel are removed, other radioactively contaminated parts of the plant (e.g., contaminated piping and equipment, contaminated concrete) are removed or decontaminated, and finally, the plant is demolished. This decommissioning waste is distinct from the stranded fuel, which as discussed previously is used fuel remaining at the site after final decommissioning of the power plant.

The waste produced during the decommissioning process consists of both nonradioactive and radioactive waste. Much of the waste is uncontaminated and can be recycled (e.g., steel or concrete) or disposed of in a landfill. Radioactively contaminated waste must be disposed of in a land disposal facility (for Class A, B, and C waste, as described previously). This waste includes most of the radioactively contaminated materials from the plant, including the reactor vessel. Some of the metal components of the reactor core are Greater-Than-Class-C wastes and currently have no approved disposal pathway. They must continue to be stored on-site until these materials can be removed to a disposal facility.

Adequate funds are assured to complete this process for all U.S. plants, as every licensee is required to contribute to a USNRC-supervised fund for this purpose during the period of operations. To date in the United States, 23 commercial

---

[61]Supercompaction involves compressing metallic drums, then placing the compressed drums into a larger overpack to reduce the volume disposed. Volume reduction efficiencies typically range from 4:1 to 10:1.

[62]The amount of waste to be disposed of and the fraction of radioactive to nonradioactive waste vary by site. For example, the Electric Power Research Institute (EPRI) estimated that to complete the Maine Yankee nuclear plant decommissioning, in total, 246 million pounds of radioactive waste would need to be shipped off-site, the majority being radioactively contaminated concrete versus 151 million pounds of nonradioactive waste. In contrast, the majority of the waste that was disposed of in the decommissioning of the Big Rock Point reactor was made up of "clean" concrete (Carraway and Wills, 2001; EPRI, 2005).

nuclear power reactors have been shut down, and 9 of these have completed the decommissioning process.

Impacts from New and Expanded Deployment

Two types of impacts might result from the deployment of future nuclear technologies:

- The construction of new nuclear plants in the United States will result in the generation of additional used fuel, other operational waste, and decommissioning waste.
- New waste forms requiring disposal may emerge from alternative fuel cycle technologies.

New nuclear power plants will produce waste that is similar to the waste produced by current plants, with two possible exceptions. First, the burn-up of nuclear fuel will likely increase as new fuel designs are developed. This will reduce the amount of used fuel generated per unit of electricity production.[63] However, the higher burn-up fuel will contain more heat-generating radioactive isotopes and may have to be actively cooled (in used-fuel pools) for longer periods of time. Second, advanced plant designs generally use fewer cables, pipes, valves, and pumps than current generation plants use and, in some cases, less structural steel and concrete. This could reduce decommissioning costs and time (as well as front-end construction costs and time) and also reduce the volume of material requiring disposal.

New nuclear plants will also generate additional used fuel. Assuming a once-through fuel cycle, if 5 to 9 new plants are constructed between 2009 and 2020, the quantity of used fuel produced each year in the United States could produce an additional 105 to 189 MTU of used fuel annually, an increase of at most 9 percent in 2020. However, if a larger number of plants are built after 2020, this amount could increase significantly. If 70 to 74 new plants are built by 2035, the amount of used fuel produced annually would increase by 71 percent between 2009 and 2035 (assuming that the operating licenses of all existing plants are extended to

---

[63]The first generation of commercial power reactors achieved fuel burn-ups of 20,000 to 25,000 megawatt-days per metric ton of uranium (MWd/MTU). At present, commercial power reactors can achieve up to about 60,000 MWd/MTU. Future goals are to achieve as much as 100,000 MWd/MTU, which would increase fuel efficiency by about 40 percent.

60 years.)[64] Given the current impasse on used fuel disposal, it should be anticipated that any new plant will be constructed with an eye toward the possibility that extended on-site storage of fuel may be required. This suggests that such storage could be incorporated into the design of new plants.

In contrast, technologies such as advanced fuel cycles may produce waste forms that are different from those produced by current U.S. plants. For advanced fuel cycles, various waste streams emerge from the separations processes. These can include separated strontium and cesium, technetium, claddings, and hulls, along with the remaining fission products. These waste streams will require specialized waste forms. For example, in the UREX+ process, the technetium isotope that is separated from used fuel would be relatively mobile if emplaced in the Yucca Mountain geologic setting unless placed in a specially designed waste form. Separation of technetium allows it to be separately handled in a specially designed waste form. The cesium and strontium isotopes in the used fuel have comparatively short half-lives and, if separated from the high-level waste for the repository, could potentially be stored in less costly aboveground or near-surface sites. These isotopes will have essentially decayed away in a few hundred years. In general, waste form certification is at the proof-of-principle stage (DOE, 2007).

---

[64]For new plants, operators have to sign a contract with the DOE to take title to used fuel.

ANNEX 8.E: SAFETY AND SECURITY IMPACTS OF NUCLEAR TECHNOLOGIES

The primary impact of concern in the event of an accident or an intentional attack on a nuclear power plant is the same: major off-site releases of radioactive material. This section examines potential safety[65] and security[66] impacts arising from the operation of nuclear power plants in the United States. The following discussion is drawn from a recent National Research Council report on the safety and security of commercial spent nuclear fuel storage (NRC, 2006).

There are two potential sources for off-site radioactive releases: the nuclear fuel in the reactor core and in used fuel storage. An accident or terrorist attack that disrupts cooling of the fuel could damage the fuel and release radioactive material to the environment. The fuel in the reactor core of a nuclear plant generates substantial quantities of heat and radioactivity. The plant's cooling system is designed to remove this heat from the core so it can be used for electricity generation. A loss of coolant would cause temperatures in the core to increase, even after the reactor is shut down.[67] At about 1000°C, the fuel cladding[68] would begin to oxidize rapidly in the presence of air or steam (if the core did not remain covered with water). This exothermic reaction releases large quantities of heat that would further raise temperatures. At about 1800°C, the cladding and fuel would begin to melt, releasing radioactive gases and aerosols into the core. These radioactive materials could be released to the surrounding environment if the reactor pressure vessel[69] and the containment[70] were to fail. Such releases could endanger local populations and contaminate the environment.

An accident or terrorist attack on a used-fuel pool could have similar con-

---

[65]"Safety" is defined here as measures that would protect nuclear facilities against failure, damage, human error, or other accidents that would disperse radioactivity into the environment.

[66]"Security" is defined here as measures to protect nuclear facilities against sabotage, attacks, or theft.

[67]This "heat" is the product of radioactive decay in the fuel.

[68]"Fuel cladding" is a thin-walled metal tube that forms the outer jacket of a nuclear fuel rod. It prevents corrosion of the nuclear fuel and the release of fission products into the coolant.

[69]The "reactor pressure vessel" is a thick-walled cylindrical steel vessel enclosing the reactor core in a nuclear power plant.

[70]A "containment building" is a steel or reinforced concrete structure enclosing a nuclear reactor. The containment building is typically an airtight steel structure enclosing the reactor, sealed off from the outside atmosphere and attached to a concrete shield. In the United States, the design and thickness of the containment and the shield are governed by federal regulations (10 CFR 50.55a).

sequences. After its removal from the reactor, used fuel continues to generate heat and must be actively cooled. Fuel is stored in water-filled pools that have active cooling systems to remove this heat and water filtering systems to remove radioactive contamination. An accident or terrorist attack that results in the loss of coolant from the pool could raise fuel temperatures, possibly resulting in cladding oxidation and fuel melting with a consequent release of radioactive gases and aerosols. These processes would likely unfold more slowly than would the events following a disruption of core coolant, because the used fuel stored in pools generally has lower rates of heat generation. Consequently, plant operators would have more time to implement backup cooling measures.

The pools themselves are constructed with thick reinforced concrete walls and stainless steel liners. A 2006 National Research Council report concluded that successful terrorist attacks on used-fuel pools would be difficult, and "an attack that damages a power plant or its spent fuel storage facilities would not necessarily result in the release of any radioactivity to the environment" (NRC, 2006, p. 6). The report also noted that used fuel in dry cask storage poses considerably less risk.

Nuclear plants have backup systems and procedures designed to prevent or mitigate the consequences from the accidental disruption of coolant flow to the reactor core or used-fuel pool. For example, the reactor containment is designed to limit the release of any radioactive material from the reactor core in the event of an accident. Plants have multiple backup supplies of cooling water as well as emergency cooling systems that can flood or spray the fuel in the core with water. They also have backup sources of water for the used-fuel pools, and water sprays could be deployed to cool the fuel even if the pool could not be refilled (NRC, 2006). In addition to these backup systems, plant operators are required to perform probabilistic analyses to understand and mitigate the consequences of accidental disruptions of core cooling, as well as to develop and implement plans to notify authorities and residents living near their plants in the event of emergencies.

Safety

Efforts to improve safety in U.S. plants have focused in part on reducing the probability of the most likely sequences of events or failures that could result in a radioactive release. Most early nuclear plants were designed to conform to particular design rules, such as an insistence that the design incorporate multiple barriers and provide the means to prevent an event from resulting in a radioactive release to the public, as well as conservative engineering assumptions as to the capabilities of materials and equipment. Over time, an analysis technique termed probabilistic risk assessment (PRA) was developed that allows the systematic evaluation of the various sequences of events or failures that could result in a release and the determination of the probability that any given sequence might arise. PRAs suggest that the likelihood of an accidental release in the United States from the currently operating reactors is small. According to the Reactor Safety Study undertaken by the U.S. Atomic Energy Commission in 1975, the probability of such an occurrence was estimated at one in 17,000 per reactor per year (USNRC, 1975). However, more recent studies have concluded that the core damage frequency is between 10^{-5} and 10^{-6} events per reactor per year (Sheron, 2008).[71] Extensive efforts have been undertaken by the USNRC and the licensees to consider the accident sequences presenting the greatest risk and to implement measures to thwart them.

Some critics contend that safety problems continue to arise associated with nuclear reactors in the United States due to inadequate enforcement of standards by the USNRC (UCS, 2008), pointing to 36 instances that have occurred since 1979 where individual reactors have been shut down for more than a year to restore safety standards (Lochbaum, 2006). However, it should be noted that these shutdowns to restore standards were initiated by the USNRC.

Security

In addition to reactor accidents, after the attacks of September 11, 2001, terrorist threats to nuclear power plants have become a concern. As noted above, the primary concern is that a terrorist attack on a nuclear reactor might result in a radioactive release to the surrounding area.

Every U.S. nuclear plant has a security plan that must be approved by the USNRC to respond to an attack at the level of the Design Basis Threat (DBT)[72] or below. The details of the DBT are not available to the public, but it is described as an attack carried out by a well-armed land force aided by a knowledgeable insider. The plants defend against this threat primarily through the use of a

---

[71]These results are attributed to the USNRC State of the Art Reactor Consequence Analysis assessment. Final results from this study are planned for release in 2009 (Sheron, 2008).

[72]The DBT is a profile of the type, composition, and capabilities of an adversary. The USNRC and its licensees use the DBT as a basis for designing safeguards systems to protect against acts of radiological sabotage and to prevent the theft of special nuclear material. The DBT is described in Title 10, Section 73, of the Code of Federal Regulations [10 CFR 73].

layered security system involving access controls and requirements, physical barriers (including standoff protection for bombs), armed guards, and armored firing positions.

Attacks that are beyond the DBT are also a concern, including, in particular, air attacks. The industry and its regulator (the USNRC) have stated that defending against these types of attacks is the federal government's responsibility,[73] not that of the plant operator.

Since the September 11, 2001, attacks, the USNRC and the nuclear industry have undertaken analyses of existing plants to determine their vulnerability to aircraft attacks and have made modifications to the designs and operations to mitigate the consequences of such attacks. In addition, the DBT has been increasd in severity, with the result that the capacity to withstand terrorist attacks of all types has been enhanced. U.S. plant operators report that they have spent in excess of a billion dollars on physical upgrades and security since September 11, 2001 (www.nei.org/keyissues/safetyandsecurity/factsheets/powerplantsecurity; accessed July 2009). These include changes to plant access controls, operating procedures, and other security measures. The details of these analyses and modifications have not been released to the public (due to security concerns), and the committee has not reviewed this information.

Impacts from Expanded or New Deployments

New evolutionary nuclear plant designs are intended to improve both safety and security over currently operating plant designs. Some modern designs for reactors of the types that are proposed for near-term construction in the United States (discussed in Annex 8.A) promise to reduce core damage frequency by a factor of 10 to 100 from the probability of such an event in an existing plant. In addition, these designs include enhanced physical protection of the core and used-fuel pools intended to reduce their vulnerabilities to beyond-DBT attacks such as air attacks; designs of core cooling systems that rely on passive systems (using gravity and natural circulation) to maintain cooling in the case of an accident or terrorist attack; and the design and placement of multiple independent safety systems to provide spatial redundancy intended to improve survivability in accidents or attacks (and also to allow some maintenance on these systems to occur while the

[73]Measures have been taken to defend against these kinds of attacks, including increased security at airports, locks on cockpit doors, and armed air marshals and pilots.

plant is operating). In addition, the USNRC recently promulgated a rule requiring applicants for new nuclear reactors to identify features and functional capabilities of their designs that would provide additional inherent protection from or mitigate the effects of aircraft attacks. Plants are either acceptable as designed or will be upgraded.

REFERENCES FOR ANNEXES 8.A–8.E

ACA (Australian Coal Association). 2001. ACARP-Coal in Sustainable Society. Available at ciss.com.au/ref/static/reports/public/acarp/acarp1.html.

Alberta Department of Energy. 2007. Alberta's Oil Sands 2006. December. Available at www.energy.alberta.ca/OilSands/pdfs/osgenbrf.pdf.

APS (American Physical Society). 2007. Consolidated Interim Storage of Commercial Spent Nuclear Fuel: A Technical and Programmatic Assessment. Washington, D.C.

Beliav, V., and V. Polunichev. 1998. Basic Safety Principles of KLT-40C Reactor Plants. IAEA-TECDOC-1172. International Atomic Energy Agency.

Bennett, R. 2008. Prospects for and barriers to emerging technologies: Advanced reactors. Presentation to the AEF Committee, January.

British Energy. 2005. Environmental Product Declaration for Torness Nuclear Power Station. Executive Summary available at www.british-energy.com/documents/EPD_Exec_Summary.pdf; last accessed July 2009.

Carraway, T., and B. Wills. 2001. Decontamination and decommissioning of Big Rock Point nuclear power plant. WM01 Conference, February 25–March 1, 2001, Tucson, Ariz.

Chenier, Philip J. 2002. Survey of Industrial Chemistry. Third Edition. New York: Kluwer Academic Plenum Publishers.

DOE (U.S. Department of Energy). 1997. Linking Legacies: Connecting the Cold War Nuclear Weapons Production Processes to Their Environmental Consequences. Washington, D.C.

DOE. 2000. Energy and Environmental Profile of the U.S. Chemical Industry. U.S. Department of Energy, Office of Industrial Technology. Washington, D.C. May.

DOE. 2006. Federal Register 71(151; Aug. 7):44673.

DOE. 2007. GNEP Technology Development Plan. Available at nuclear.inl.gov/gnep/d/gnep-tdp.pdf.

DOE. 2008a. Request for Expressions of Interest for the Next Generation Nuclear Plant Program. Available at nuclear.gov/pdfFiles/NGNP_EOI.pdf.

DOE. 2008b. USDOE Final Supplemental Environmental Impact Statement for a Geologic Repository for the Disposal of Spent Nuclear Fuel and High-Level Radioactive Waste at Yucca Mountain, Nye County, Nevada. DOE/EIS-0250F-S1. Washington, D.C.

Dones, R. 2003. Kernenergie. In: Sachbilanzen von Energiesystemen: Grundlagen fuer den oekologischen Vergleich von Energiesystemen und den Einbezug von Energiesystemen in Oekobilanzen fuer die Schweiz. Paul Scherrer Institut Villigen, Swiss Centre for Life Cycle Inventories, Duebendorf, CH.

Dones, R., H. Heck, M. Emmenegger, and N. Jungbluth. 2005. Life cycle inventories for the nuclear and natural gas energy systems and example of uncertainty analysis. International Journal of Life Cycle Analyses 10(1):10-23.

EIA (Energy Information Administration). 2007. Domestic Uranium Production Report (2003-2007). Form EIA-851A.

Energy and Environmental Economics, Inc. 2008. CPUC GHG Modeling Stage 1 Documentation. Attachment B. San Francisco, Calif. Available at http://www.ethree.com/GHG/R0604009_Attachment_B_v2.pdf.

Entergy. 2007. Entergy letter to USNRC of February 28, 2007, forwarding Annual Radioactive Effluent Report for 2006 Arkansas Nuclear One–Units 1 and 2, Docket Nos. 50-313, 50-368, and 72.13(IFSI). License Nos. DPR-51 and NPF-6.

EPRI (Electric Power Research Institute). 1992. Advanced Light Water Reactor Utility Requirements Document. EPRI NP-6780-L. Palo Alto, Calif.

EPRI. 2002. Water and Sustainability (Volume 3): U.S. Water Consumption for Power Production—the Next Half Century. Palo Alto, Calif.

EPRI. 2005. Maine Yankee Decommissioning Experience Report: Detailed Experiences 1997–2004. Palo Alto, Calif.

EPRI. 2008. Technology Assessment Guide for Central Station Power. Palo Alto, Calif.

ExternE 1998. ExternE: Externalities of Energy. Published by European Commission, Directorate-General XII, Science Research and Development. Office for Official Publications of the European Communities, L-2920 Luxembourg. Results are also available at http://www.externe.info/.

Forsberg, C., P. Peterson, and L. Ott. 2004. The Advanced High-Temperature Reactor (AHTR) for Producing Hydrogen to Manufacture Liquid Fuels. Available at www.ornl.gov/~webworks/cppr/y2001/pres/121482.pdf.

Fthenakis and Kim 2007. Greenhouse-gas emissions from solar electric- and nuclear power: A life-cycle study. Energy Policy 35:2549-2557.

Gary, James H., Glenn E. Handwerk, and Mark J. Kaiser. 2007. Petroleum Refining: Technology and Economics. Fifth Edition. New York: Taylor and Francis Group, LLC.

GIF/NERAC (Generation IV International Forum/U.S. Department of Energy Nuclear Energy Research Advisory Committee). 2005. A Technology Roadmap for Generation IV Nuclear Energy Systems. GIF-002-00. Washington, D.C. Available at gif.inel.gov/roadmap/pdfs/gen_iv_roadmap.pdf.

Harding, J. 2007. Economics of Nuclear Power and Proliferation Risks in a Carbon-Constrained World. Electricity Journal 20(10):65-76.

Hondo, H. 2005. Life cycle GHG emission analysis of power generation systems: Japanese case. Energy 30:2042-2056.

IAEA (International Atomic Energy Agency). 2005. Thorium Fuel Cycle—Potential Benefits and Challenges. IAEA-TECDOC-1450. Vienna, Austria. Available at //www-pub.iaea.org/MTCD/publications/PDF/TE_1450_web.pdf.

IAEA. 2007. Milestones in the Development of a National Infrastructure for Nuclear Power. NG-G-3.1 2007. Vienna, Austria.

IAEA. 2008. Nuclear Power Reactors in the World. Vienna, Austria.

IEA (International Energy Agency). 2002. Innovative Nuclear Reactor Development, Opportunities for International Co-operation. Paris, France.

Ivanov, V. 2008. Overview of the Russian sodium-cooled fast reactor program. Presentation to the AEF Committee, April.

Jagannathan, V., and U. Pal. 2008. Towards an intrinsically safe and economic thorium breeder reactor. Energy Conversion and Management 47(17):2781-2793.

Johnson, Michael. 2008. US Nuclear Regulatory Commission activities on licensing of new reactors. Presentation to the National Academies' Board on Energy and Environmental Systems, Washington, D.C., December.

Kazimi, M.S. 2003. Thorium fuel for nuclear energy—An unconventional tactic might one day ease concerns that spent fuel could be used to make a bomb. American Scientist 91:408-415.

Kazimi, M.S., and P. Hejzlar. 2005. High Performance Fuel Design for Next Generation PWRs: 4th Annual Report. Cambridge, Mass.

Keystone Center. 2007. Nuclear Power Joint Fact-Finding. Keystone, Colo.

Lisowski, P. 2008. US DOE, Reactor design and fuel cycle choices. Presentation to the AEF Committee, Washington, D.C., April.

Lochbaum, D. 2006. Walking a Nuclear Tightrope: Unlearned Lessons of Year-plus Reactor Outages. Union of Concerned Scientists, Cambridge, Mass. Available at www.ucsusa.org/assets/documents/nuclear-power/nuclear_tightrope_report_hires.pdf.

Maldague, T., S. Pilate, A. Renard, A. Harislur, H. Mouney, and M. Rome. 1998. Recycling Schemes of Americium Targets in PWR/MOX Cores. Proceedings of the Fifth OECD/NEA Meeting on Actinides and Fission Product Partitioning and Transmutation, Mol, Belgium. Available at www.nea.fr/html/trw/docs/mol98/session4/SIVpaper3.pdf.

Matzie, Regis A., 2008. AP1000 nuclear power plant. Presentation at the MIT Nuclear Plant Safety Course, Cambridge, Mass.

MIT (Massachusetts Institute of Technology). 2003. The Future of Nuclear Power. Cambridge, Mass. Available at web.mit.edu/nuclearpower/.

MIT. 2009. Update of the MIT 2003 Future of Nuclear Power. Cambridge, Mass. Available at web.mit.edu/nuclearpower/pdf/nuclearpower-update2009.pdf.

Moody's Investor's Service. 2007. New Nuclear Generation in the United States. New York, N.Y.

Moody's Investor's Service. 2008. New Nuclear Generating Capacity: Potential Credit Implications for U.S. Investor Owned Utilities. New York, N.Y.

Moorhouse, Jeremy, and Bruce Peachey. 2007. Cogeneration and the Alberta oil sands—cogeneration benefits are maximized with extraction and upgrading integration. Cogeneration and On-Site Power Production 8(4; July).

NEI (Nuclear Energy Institute) 2008. New generating capacity costs in perspective. White paper. Washington, D.C.

Newcomer, A., S. Blumsack, J. Apt, L. Lave, and M.G. Morgan. 2008. Short-run effects of a price on carbon dioxide emissions for U.S. electric generators. Environmental Science and Technology 42(9):3139-3144.

NRC (National Research Council). 1995. Technical Bases for Yucca Mountain Standards. Washington, D.C.: National Academy Press.

NRC. 2001. Disposition of High-Level Waste and Spent Nuclear Fuel: The Continuing Societal and Technical Challenges. Washington, D.C.: National Academy Press.

NRC. 2006. Safety and Security of Commercial Spent Nuclear Fuel Storage. Washington, D.C.: The National Academies Press.

NRC. 2008. Review of DOE's Nuclear Research and Development Program. Washington, D.C.: The National Academies Press.

NREL (National Renewable Energy Laboratory). 2001. Life Cycle Assessment of Hydrogen Production via Natural Gas Steam Reforming. NREL/TP-570-27637. Golden, Colo.

OECD/NEA (Organisation for Economic Co-operation and Development/Nuclear Energy Agency). 2007. Uranium 2007: Resources, Production, and Demand. Paris, France.

OECD/NEA. 2008. Nuclear Energy Outlook. Paris, France.

Parece, Martin. 2008. U.S. EPR overview. Presentation to the MIT Nuclear Plant Safety Course, Cambridge, Mass.

Pederson, T. 1998. BWR 90—the advanced BWR of the 1990s. Nuclear Engineering and Design 180(1; March):53-66.

Pelton, Tom. 2007. Nuclear power has new shape. Baltimore Sun, December 25.

PSEG (Public Service Enterprise Group). 2007. Letter to the USNRC dated April 26, 2007, transmitting 2006 radioactive effluent release report for Hope Creek Generating Station. Facility Operating License No. NPF-57, Docket No. 50-34.

Radkowsky, A. 1999. Using thorium in a commercial nuclear fuel cycle: How to do it. Nuclear Engineering International (January 8). Available at www.neimagazine.com/story.asp?storycode=1000169.

Scroggs, S. 2008. COLA (Combined License Application Content) Engineering Evaluation of Current Technology Options for New Nuclear Power Generation. Testimony before Florida Public Service Commission. Docket No. 070650.

Sheron, B. 2008. Technical issues in nuclear reactor regulation. Presentation at the MIT Nuclear Plant Safety Course, Cambridge, Mass.

Smith, Craig F., William G. Halsey, Neil W. Brown, James J. Sienicki, Anton Moisseytsev, and David C. Wade. 2008. SSTAR: The US lead-cooled fast reactor (LFR). Journal of Nuclear Materials 376(3):255-259.

S&P (Standard and Poor's) 2008. Update on the the U.S. Dept. of Energy Loan Guarantee Program and Standard and Poor's Rating Considerations. October 7. New York, N.Y.

Storm van Leeuwen, J.W., and P. Smith. 2007. Nuclear Power: The Energy Balance. Available at www.stormsmith.nl.

Tokimatsu, K., T. Kosugi, T. Asami, E. Williams, and Y. Kaya. 2006. Evaluation of lifecycle CO_2 emissions from Japanese electric power sector. Energy Policy 34:833-852.

Turnage, Joe. 2008. New nuclear development: Part of the strategy for a lower carbon energy future. Presentation to Center for Strategic and International Studies, July 31.

TVA (Tennessee Valley Authority). 2005. ABWR Cost/Schedule/COL Project at TVA's Bellefonte Site. DE-AI07-04ID14620. Available at ne.doe.gov/np2010/reports/mainreportA115.pdf..

UCS (Union of Concerned Scientists). 2008. Nuclear Power in a Warming World: Assessing the Risks, Addressing the Challenges. Cambridge, Mass.

University of Chicago. 2004. The Economic Future of Nuclear Power. Chicago, Ill.

USNRC (U.S. Nuclear Regulatory Commission). 1975. Wash-1400 NUREG-74/014 Reactor Safety Study. October. Washington, D.C.

USNRC. 2008. Information Digest 2007-2008. NUREG-1350, Vol. 19. Washington, D.C. Available at www.nrc.gov/reading-rm/doc-collections/nuregs/staff/sr1350/v19/sr1350v19.pdf. Accessed June 7, 2009.

Vattenfall, A.B., 2005. Vattenfall AB Generation Nordic Countries Certified Environmental Product Declaration of Electricity from Ringhals NPP. Available at www.vattenfall.com/www/vf_com/vf_com/Gemeinsame_Inhalte/DOCUMENT/360168vatt/386246envi/2005-EPD-Ringhals.pdf.

Veil, J.A. 2007. Use of Reclaimed Water for Power Plant Cooling, Report No. ANL/EVS/R-07/3, Argonne National Laboratory; Use of Degraded Water Sources as Cooling Water in Power Plants, Report No. 1005359. Palo Alto, Calif.: EPRI.

Westinghouse. 2006. Westinghouse World View. Issue 2. Pittsburgh, Pa.

White, S.W. 1998. Net Energy Payback and CO_2 Emissions from Helium-3 Fusion and Wind Electrical Power Plants. Ph.D. dissertation. University of Wisconsin-Madison.

9 Electricity Transmission and Distribution

Electric power transmission and distribution (T&D) in the United States, the vital link between generating stations and customers, is in urgent need of expansion and upgrading. Growing loads and aging equipment are stressing the system and increasing the risk of widespread blackouts. Modern society depends on reliable and economic delivery of electricity.

Recent concerns about T&D systems have stemmed from inadequate investment to meet growing demand, the limited ability of those systems to accommodate renewable-energy sources that generate electricity intermittently, and vulnerability to major blackouts involving cascading failures. More-over, effective and significant utilization of intermittent renewable generation located away from major load centers cannot be accomplished without significant additions to the transmission system. In addition, distribution systems often are incompatible with demand-side options that might otherwise be economical. Modernization of electric T&D systems could alleviate all of these concerns.

The U.S. T&D system has been called the world's largest machine and part of the greatest engineering achievement of the 20th century (NAE, 2003). This massive system delivers power from the nearly 3000 power plants in the United States to virtually every building and facility in the nation.

This chapter reviews the status of current T&D systems and discusses the potential for modernizing them (thus creating the "modern grid"). The focus is on the technologies involved—their potential performance, costs, and impacts—and potential barriers to such a deployment in the United States over the next several decades.

BACKGROUND

The Current Transmission and Distribution System

T&D involves two distinct but connected systems (as shown in Figure 9.1):

- *The high-voltage transmission system* (or grid) transmits electric power from generation plants through 163,000 miles of high-voltage (230 kilovolts [kV] up to 765 kV) electrical conductors and more than 15,000 transmission substations. The transmission system is configured as a network, meaning that power has multiple paths to follow from the generator to the distribution substation.[1]
- *The distribution system* contains millions of miles of lower-voltage electrical conductors that receive power from the grid at distribution substations. The power is then delivered to 131 million customers via the distribution system. In contrast to the transmission system, the distribution system usually is radial, meaning that there is only one path from the distribution substation to a given consumer.

The U.S. T&D system includes a wide variety of organizational structures, technologies, economic drivers, and forms of regulatory oversight. Federal, state, and municipal governments and customer-owned cooperatives all own parts of these systems, but approximately 80 percent of power transactions occur on lines owned by investor-owned regulated utilities (IOUs). These fully integrated utilities own generating plants as well as the T&D systems that deliver the power to their customers. In the past, this was the dominant model, but deregulation in some states has transformed the industry. In deregulated areas, generation, transmission, and distribution may be handled by different entities. For example, independent power producers (IPPs) may sell power to distribution utilities, or even directly to end users, using the transmission system as a common carrier (as shown in Figure 9.2).

The Federal Energy Regulatory Commission (FERC) has long had the authority to regulate financial aspects of the transmission of electricity in inter-

---

[1]"Distribution substations" connect the high-voltage transmission system to the lower-voltage distribution system via transformers. The system includes 60,000 distribution substations. "Transmission substations" connect two or more transmission lines.

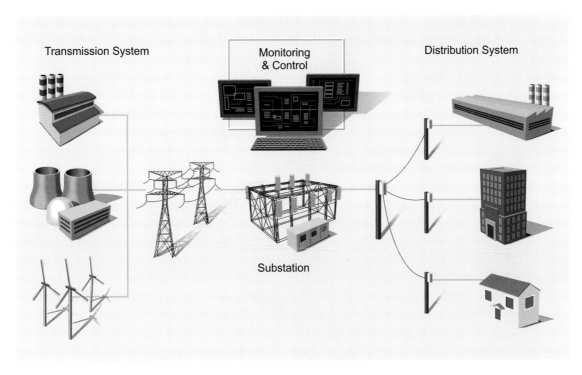

FIGURE 9.1 *The current T&D system comprises two distinct but connected systems: transmission and distribution.*
Source: Courtesy of NETL Modern Grid Team.

state commerce. The Energy Policy Act of 2005 expanded FERC's mandate, giving it the authority to impose mandatory reliability standards on the bulk transmission system and to impose penalties on entities that manipulate electricity markets. As part of its new authority, FERC has in turn granted the North American Electric Reliability Corporation (NERC)—a private organization created by the utility industry in 1968 to advise on reliability—the authority to develop and enforce reliability standards. The National Institute of Standards and Technology also is involved in developing standards for the grid.

In some areas, independent system operators/regional transmission operators (ISO/RTOs) are responsible for operating the transmission system reliably, including constantly dispatching power to balance demand with supply and monitoring the power flows over transmission lines owned by other public or private entities. The ISO/RTOs, with oversight by FERC and NERC, monitor their systems' capac-

Regulated Utility #1   **Regulated Utility #2**   **Independent Power Producer**

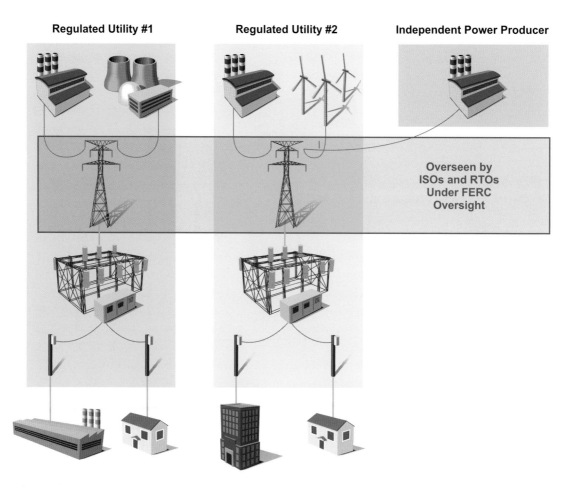

Overseen by
ISOs and RTOs
Under FERC
Oversight

FIGURE 9.2 *Key players in the T&D system. Power is produced by regulated investor-owned utilities (IOUs), which own the majority of the T&D systems, and in some areas by independent power producers (IPPs). IOUs typically provide electricity to end users through their own distribution systems, while IPPs sell to a utility or purchase transmission services to deliver electric power directly to an end user. There are also utilities that are federally or locally owned, such as municipal and rural co-ops. Most of these utilities own generating plants as well as T&D lines.*
Source: Courtesy of NETL Modern Grid Team.

ities and conduct the wholesale market to clear short-term transactions.[2] There are nine ISO/RTOs in North America, as shown in Figure 9.3. Seven of the nine come

---

[2]Market-clearing transactions match the available supply of electric power at a clearing price that matches the demand.

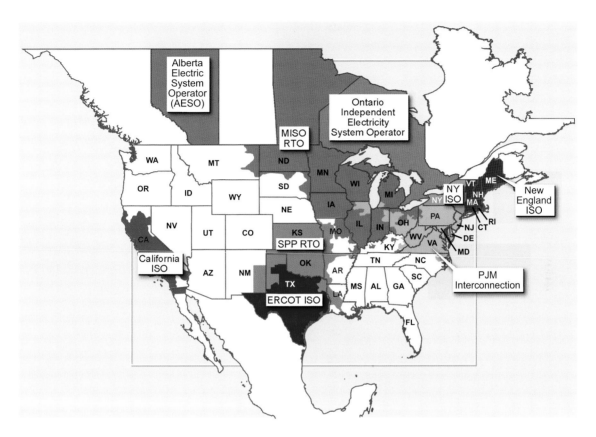

FIGURE 9.3 *Independent System Operators (ISO) and Regional Transmission Organizations (RTO) in North America. Regions in which the power industry has been restructured, such as Texas, the Northeast, the Upper Midwest, and much of California, are colored. In these areas, ISO/RTOs are responsible for operating the transmission system. In the white regions, where the industry has not been restructured, vertically integrated power utilities continue to operate the transmission system.*
Source: North American Electric Reliability Corporation.

under FERC's reliability oversight. The remaining two are subject to Canadian regulations.

Operationally, the electric transmission systems of the United States and Canada are divided into four large regions known as "interconnections," as shown in Figure 9.4:

- The Eastern Interconnection, which includes most of the United States and Canada from the Rocky Mountains to the Atlantic coast;

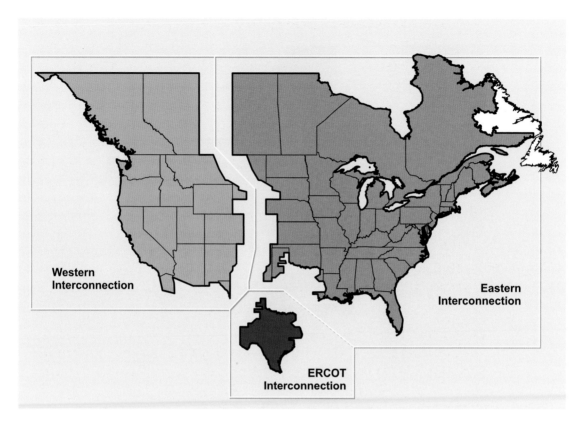

FIGURE 9.4 *North American power interconnections. The Quebec Interconnection is shown as part of the Eastern Interconnection because operations are coordinated. Source: North American Electric Reliability Corporation.*

- The Western Interconnection, which extends from the Pacific coast to the Rockies;
- The ERCOT Interconnection, which encompasses most of Texas;
- The Quebec Interconnection, which is shown in Figure 9.4 as part of the Eastern Interconnection because they are operated jointly.

Within each interconnection, all generators operate in synchronism with each other. That is, the 60-Hertz alternating current (AC) is exactly in phase across the entire interconnection. While all interconnections operate at 60 Hz, no attempt is made to synchronize them with each other. Electricity is transmitted between interconnections, but that is done by converting to direct current (DC) and then back to AC.

Controlling the dynamic behavior of this interconnected transmission system presents an engineering and operational challenge. Demand for electricity is constantly changing as millions of consumers turn on and off appliances and industrial equipment. The generation of and demand for electricity are balanced regionally by about 140 balancing authorities to ensure that voltage and frequency are maintained within narrow limits (typically 5 percent for voltage and 0.02 Hz for frequency). If more power is drawn from the grid than is being pumped into it, the frequency and voltage will decrease, and vice versa. If the voltage or frequency strays too far from its prescribed level, the resulting stresses can lead to system collapse and possibly damage to power system equipment.

Problems with the Current System

Most U.S. transmission lines and substations were constructed more than 40 years ago and are based on 1950s' technology, but demands on the electric power system have increased significantly over the years. Since 1990, electricity generation has risen from about 3 trillion kilowatt-hours (kWh) to about 4 trillion in 2007. Long-distance transmission has grown even faster for reliability and economic reasons, including new competitive wholesale markets for electricity, but few new transmission lines have been built to handle this growth.[3]

Figure 9.5 shows transmission investment from 1975 to 2007. From 1985 through 1995, transmission investment was fairly stable at the level of about $4.5 billion per year. Although this was about $2 billion per year lower than during the previous decade, reserve margins[4] were adequate because of prior over-building and slow growth in demand. However, in the late 1990s, the restructuring and re-regulation of the U.S. transmission system led to a decrease in invest-

---

[3]The stress on the U.S. transmission system that was brought about by wholesale electric competition was described by Linn Draper, chairman and CEO of American Electric Power, during his testimony before the House Energy and Water Committee shortly after the August 14, 2003, blackout: "In the five-year period during which wholesale competition first gained momentum, the number of wholesale transactions in the U.S. went from 25,000 to 2 million—an 80-fold increase." Another factor increasing demand for transmission is the difficulty of building generating facilities near load centers because of pubic opposition. Ironically, new transmission lines also are the object of considerable public opposition even while the need for them is increased by opposition to generating stations.

[4]Reserve margin is the amount of transmission capacity available above the maximum power expected to be delivered over the system. Some margin is necessary to allow for unexpected loads or outages on the system.

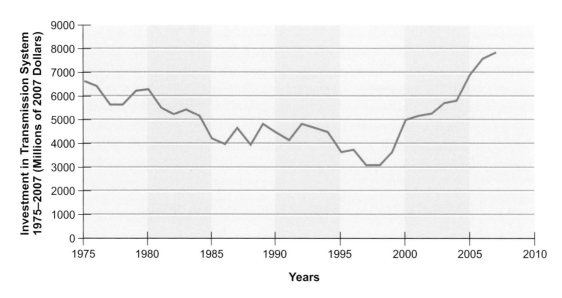

FIGURE 9.5 *Transmission investment by integrated and stand-alone transmission companies. The IOU data cover only 80 percent of the transmission system. All investment is shown in 2007 dollars. Data were adjusted as necessary using the Handy-Whitman index of Public Utility Construction Costs.*
Sources: 1975–2003 from EEI, 2005; 2000–2007 from Owens, 2008.

ment. This decrease "was principally due to uncertainty in the rate of return on investment (and whether it would be modified or disallowed in future years) offered to transmission owners/investors" (EPRI, 2004). Transmission investment averaged about $3 billion per year from 1995 to 2000.

The deficit of the late 1990s is still affecting reliability; it has contributed to transmission bottlenecks and other transmission deficiencies throughout North America, even with the more recent upward trend in transmission expenditures since 2000. According to NERC, the transmission system is being operated at or near its physical limits more of the time (Nevius, 2008). Stressed grids have less reserve margin for handling disturbances. Figure 9.6 shows the increase in transmission loading relief events. (TLR is a measure of when scheduled transmission requests could not be accommodated.[5])

Inadequate system maintenance and repair also have contributed to an

---

[5]Transmission loading relief (TLR) is a sequence of actions taken to avoid or remedy potential reliability concerns associated with the transmission system. Calls for TLRs involve problems

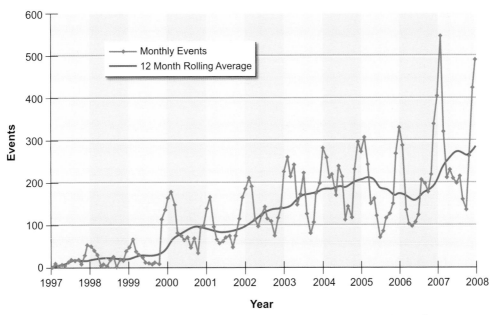

FIGURE 9.6 *Transmission loading relief (TLR) events. The number of TLR events is not an outage measure; it is the number of times a congestion limit is reached. Although this measure has been used to characterize transmission reliability, congestion limits can be reached purely for market reasons.*
Source: See www.nerc.com/docs/oc/scs/logs/trends.htm.

increase in the likelihood of major transmission system failures (EPRI, 2004), and the number of such disturbances has in fact been increasing in recent years, as shown in Figure 9.7. Of greatest concern is the risk of these disturbances cascading over large portions of the T&D systems. The 2003 blackouts in the world's two largest grids—the North American Eastern Interconnection and the West European Interconnection—resulted from such cascading failures (see Box 9.1). Each event affected 50 million people.

Another result of diminished investment in transmission is that the manufacturing of associated equipment has largely disappeared from the United States, along with commercial research and development (R&D) for transmission equipment (including transformers, switchgear, and high-voltage DC [HVDC] technology). Today, essentially all large power-transmission equipment is imported from Europe and Japan. This could become a potentially

---

that require intervention on the transmission system. These may or may not result in transmission outages or outages to customers.

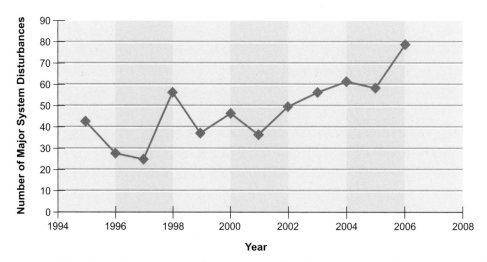

FIGURE 9.7 *Major transmission system disturbances reported to NERC. Disturbances include electric service interruptions, unusual occurrences, demand and voltage reductions, public appeals, fuel supply problems, and acts of sabotage that can affect the reliability of the bulk electric systems.*
Source: Compiled from data in NERC, 1994, 1995, 1996, 1997, 1998, 1999, 2000, 2001, 2002, 2003, 2004, 2005, and 2006.

serious problem, especially with long lead-time components, in case of major natural disaster or terrorist attack.

Modernization is progressing much more rapidly abroad. For example, China and India are building 800 kV HVDC and 1000 kV AC transmission lines, along with the underlying high-power infrastructure. About 30 high-power HVDC projects are under construction in Europe, including many submarine cable connections to increase utilization of offshore wind power. Two-way metering is common in Europe because it helps to maximize the potential of rooftop photovoltaics, which are being heavily promoted in Germany and other countries. Although the United States has vast potential for wind and solar generation, there is no consensus or plan for how this power could be transmitted to load centers.

While expenditures on the replacement and new construction of American T&D assets have increased recently (see Figure 9.5), grid assets are aging, and investments are still not keeping pace with the growing demand for electric power and power marketing. To meet these challenges, transmission

BOX 9.1 *The Northeast Blackout of August 14, 2003*

A modern T&D system could have helped to avoid the circumstances that initiated the August 2003 Northeast blackout. Two major issues contributed to this blackout: first, the operators did not know the system was in trouble; and second, there was poor communication between the utilities operating the transmission lines—First Energy and American Electric Power—and also between these utilities and the ISO responsible for the area (the Midwest Independent System Operator). The U.S.-Canada Power System Outage Task Force (2004) noted that four major factors contributed to the blackout:

1. Inadequate system understanding,
2. Inadequate situational awareness,
3. Inadequate tree trimming,
4. Inadequate reactive power control diagnostic support.

A modern T&D system could have provided better understanding of the state of the system, better communications, and, ultimately, better controls. Adequate monitoring, communication, and dynamic reactive power support during the initial voltage sag could have helped to prevent lines from overloading, heating up, and sagging excessively. Operators would have been better informed, and online real-time dynamic contingency analysis of potential system collapse would have helped operators stay aware of possible risks and actions to be taken in response. Finally, automatic actions could have been taken to island (isolate) portions of the system and prevent the ultimate cascading event (which spread the localized outage across much of the northeast United States and Canada). The system could also have been restored much more rapidly if a modern grid had been in place.

systems must be modernized—a complex but vital undertaking.[6] However, orders for modern transmission technologies remain low, largely because they are perceived to be risky and uneconomic,[7] as discussed in more detail later in this chapter. Thus if business continues as usual, investment will focus on new construction to meet peak load growth, which is projected to increase

---

[6]Modernization is defined here as the deployment of a suite of technologies (described in the coming sections) that will enable the T&D systems to meet a variety of challenges, particularly the seven characteristics (adapted from NETL, 2007d) discussed in more detail in the section titled "A Modern Electric T&D System."

[7]This view was presented repeatedly to the committee by industry representatives, including those representing Southern California Edison Co., Areva, ABB, and Siemens.

by 0.7 percent per year to 2030[8] according to the reference case of the DOE Energy Information Administration (EIA, 2008) and the replacement of aging components with equivalent technology.

Distribution systems are in better condition. Reliability, for example, has increased steadily over the last 7 years—in part because these systems have to be enlarged to handle new consumers. Public utility commissions usually provide revenue incentives based on indexes, shown in Annex 9.A, that directly measure customer service reliability. Consequently, the distribution companies have improved or at least held steady their customer outage statistics.[9]

Growth provides the opportunity for distribution companies to introduce new and smarter technologies on a limited basis before undertaking a wider application. For example, a utility can introduce modern, smart technologies on a substation-by-substation basis as it is determined that portions of the distribution network need upgrading. The nature of the distribution system allows upgrades to be done in such "modular" steps.

Addressing the Problems

T&D systems will require considerable investment just to maintain current capabilities and reliability, and the use of new technology could make the grid considerably more resilient. For example, the present system of local automatic controls overseen by human operators at regionally based control centers is not able to adequately foresee that disturbances in Cleveland can black out New York, Toronto, and Detroit, or that transmission outages in Switzerland can black out all of Italy. Modern communications and controls can move much faster to diagnose problems and bypass or isolate them. The same technology can provide cost benefits by maximizing power flows and integrating power from renewable energy sources.

New technology is an important part of the answer to the challenges facing the grid, but policy and regulatory changes will also be needed, particularly with

---

[8]Some lowering of this number may be possible with aggressive electricity end-use efficiency measures.

[9]The steady reliability of the distribution system does not contradict the increasing congestion and increasing number of system disturbances on the transmission system. Outages on the transmission system do not necessarily result in outages on the distribution system, as the transmission system is a network. This means that if one path is closed, there are alternative paths for power to flow to the consumer.

respect to the transmission system. Policies regarding T&D systems are varied and imposed by many entities; there is significant public resistance to siting new transmission lines; and the business cases for utilities to invest in modern grid processes and technologies are often incomplete, as societal costs and benefits are not typically internalized in companies' decision making. For example, the cost of not having power when it is needed is far greater to the user than the lost revenues to the utility that cannot provide it. Recognizing the value of a reliable, efficient, and flexible grid, and supporting the investments to make that possible, may require a national-level strategy.

As discussed below in this chapter, expanding and modernizing distribution systems will require considerably more investment than for transmission systems. Much of the expansion will be noncontroversial because it will be required to meet growing loads and can be done without much impact on people who do not directly benefit from it. In addition, modernization of distribution can be achieved on a more limited basis than for transmission, which will require coordination across many systems. Therefore, the emphasis in this chapter will be on transmission.

A MODERN ELECTRIC T&D SYSTEM

A modern T&D system should have capabilities beyond the reach of current systems through their incorporation of new technologies (hardware and software). They must also be expanded to meet future needs. New technologies such as power electronics, real-time thermal rating of transmission lines, and composite conductors can allow an increase in power flow on the existing T&D system, but new lines also will be needed.

Modern T&D systems are intended to provide effective operation, asset optimization, and systems planning capabilities under routine conditions and emergency response and fast restoration after a system failure. The characteristics required to achieve these performance standards are as follows:[10]

---

[10]Adapted from characteristics defined by the National Energy Technology Laboratory (NETL, 2007c,d) and discussed in more detail in the annex.

1. *Accommodates all generation and storage options.* A modern transmission infrastructure would include emerging technologies such as large-scale variable power sources and advanced energy storage devices. For example, it could smooth the variability of power from remotely located intermittent renewable resources[11] and maintain reactive power[12] on the system. The distribution system should be able to accommodate increasing amounts of distributed generation—often variable (such as rooftop photovoltaic devices)—and smaller-scale advanced energy storage devices.

2. *Enables wholesale power markets.* A modern T&D system should be enlarged to handle increased long-distance power flows and equipped with new communications and control capabilities to manage the vast amount of information required for wholesale power transactions (NETL, 2007a). In addition, the distribution system should enable the end user to participate in power markets by allowing self-generation opportunities.

3. *Is self-healing.* A modern T&D system would incorporate methods to automatically stop outages before they spread, thereby preventing major system collapses.[13] If a major system did collapse, the means would be available to isolate the problem, prevent it from spreading, and restore it rapidly and effectively. A modern T&D system would be able to monitor the state of the system, communicate key information to control centers, and take appropriate action automatically.

4. *Motivates and includes the customer.* The modern distribution system would empower customers to make end-use decisions that increase

---

[11]For variable renewable electricity sources to make up 10–20 percent or more of the total generating capacity of the interconnection, increased flexibility will be needed in the electric T&D systems.

[12]All equipment, lines, and loads have inductances and capacitance that in an AC system take power during half of each cycle and deliver it back during the other half cycle; hence, they load the lines and equipment but do not deliver net power. This is called "reactive power." It is not useful power that is measured by the electric meters in most homes, but it must be monitored and supplied by the utility as needed. Otherwise, the grid can become seriously unbalanced.

[13]A major system collapse can occur when a system becomes unbalanced—for example, when a major line is lost and other lines become overloaded as more power flows through them. As these lines are shut down by protective devices, the disturbance can propagate throughout the system, leaving large areas without power.

energy efficiency, help in load-leveling,[14] and enable residential and small-scale power generation. It would allow for self-generation and storage as well as for customers to participate in an interactive mode by responding to price signals.

5. *Provides high power quality where needed.* The modern distribution system would be capable of supplying higher "power quality"[15] where needed for a digital society that increasingly relies on sensitive microprocessor-based devices in homes, offices, commercial buildings, and industrial facilities. The highest power quality is not necessarily cost-effective for all users, so some may still need to provide additional sources of power, standby generation, or other devices that can ride through minor electrical disturbances on either the transmission or the distribution system.

6. *Is secure.* The modern T&D infrastructure would be minimally vulnerable to human error, natural disasters, and physical and cyber attacks. Resilience would be built into each element, and the overall system would be designed to deter, detect, respond to, and recover from any plausible disruption. The modern transmission system would also reduce the consequences of a successful attack through its self-healing and "islanding"[16] capabilities.

7. *Optimizes assets and operates efficiently.* A modern transmission system would utilize power lines as efficiently as possible, integrating and coordinating assets to maximize their overall function in an economical way.

These characteristics cannot be fully achieved by introducing individual modern technologies in isolation. Key technologies (such as high-speed measurements and communications and automated controls, discussed in the sections that follow) must be integrated using a systems approach designed to meet performance

---

[14]"Load-leveling" is a process for better matching generation with wide swings in demand during the day by storing energy when demand is low and using it later to meet peak demand.

[15]"Power quality" refers to the voltage, frequency, and harmonic content (frequencies that are integer multiples of the fundamental 60 Hz frequency) of the electricity supply. All these factors must be kept within tight bounds.

[16]When a large system collapses, some areas within its region may have a balance of generation and load. If those areas are able to disconnect from the collapsing system, they will remain powered—a process known as "islanding."

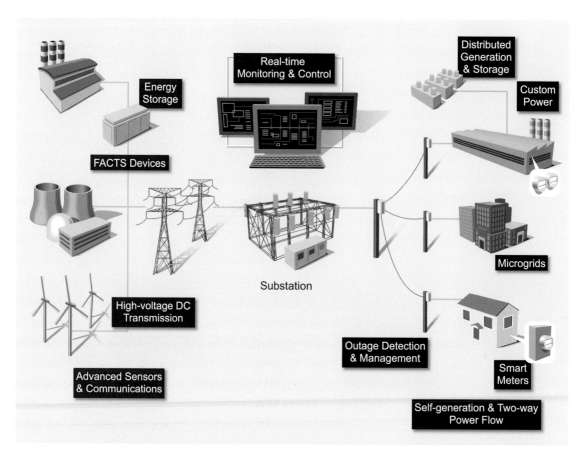

FIGURE 9.8 *Components of a modern T&D system.*
Source: Courtesy of NETL Modern Grid Team.

goals and metrics. A set of technologies that could be integrated as part of a modern grid is shown in Figure 9.8 and discussed in the following section.

KEY TECHNOLOGIES FOR A MODERN ELECTRIC T&D SYSTEM

Many of the technologies needed for a modern T&D system already exist, and some, to a limited extent, are already deployed in parts of the T&D systems. However, many technologies will need to be deployed in a systematic and integrated way to realize maximum benefits from a modernized T&D system. These technologies can be roughly divided into three categories: (1) advanced equipment and components; (2) measurements, communications, and controls; and

(3) improved decision-support tools. The major technologies within each category are discussed in the following sections, and the annex provides more detail.

Advanced Equipment and Components

Advanced equipment and components include technologies for improving and controlling power flows, enabling greater efficiency in long-distance transmission, storage of electrical energy (to be dispatched into the grid as needed), and grid operation. Advanced electronic equipment is also being used for smart metering and control in the distribution networks. The status of these technologies, likely future technology improvements, and potential for deployment into the T&D system are addressed in the five subsections below.

Power Electronics

A T&D system requires power-flow control and protection against overloads and instability. The electromechanical devices currently used for these purposes are slow and cannot react quickly enough to handle rapid transients, but modern solid-state power electronics can overcome this problem. Power electronics are not new, but their deployment has been limited to particular applications in which their higher cost is offset by their benefits to investors. Power electronics can be used in the transmission system (for both AC and HVDC applications),[17] and in the distribution system.

Power electronics on the AC transmission system are referred to as flexible alternating current transmission system (FACTS) devices.[18] FACTS devices can control both real and reactive power flows along transmission corridors, thereby maintaining the stability of transmission voltage. FACTS devices can also increase the power transfer capability of transmission lines and improve overall system reliability by reacting virtually instantaneously to disturbances. FACTS can enable wholesale markets, increase security, enable self-healing capacity, and optimize the use of system assets by controlling the flow of power, and they can help to

---

[17]The transmission system in the United States is almost entirely AC. Transmitting electricity via HVDC involves converting AC to DC, transmitting the electricity in DC, and converting it back to AC at the receiving end.

[18]Annex 9.A describes specific flexible alternating current transmission system (FACTS) devices and their applications in more detail.

integrate variable renewables by managing reactive power. FACTS devices are currently available and are already deployed in limited applications.

Power electronics also can be used for lower-voltage applications on distribution systems, where the equivalent to FACTS is known as custom power. Custom power devices can provide for significant improvement in power quality on the customer side by controlling voltage and frequency distortions. High power quality is needed for many modern applications, especially in industries with automated production, which could benefit from more economical local solutions to improved power quality. Power electronics also plays an important role in smart metering with two-way power flow (to encourage local power generation) and in real-time pricing (to shift loads away from expensive peak demand periods).

Custom power technologies that offer such solutions exist now, but their application is restricted to situations where their high cost is offset by significant benefits. R&D could help reduce costs and expand their use by 2020.

AC and DC Lines and Cables

A cost-effective way to obtain extra transmission capacity is to upgrade transmission lines and corresponding substations along existing corridors. Transmission capacity can be increased by "reconductoring" existing lines (using materials such as composite conductors that can carry higher current). These materials are presently available but not widely deployed; taking lines out of service for reconductoring is difficult, and new materials are expensive. In addition, all overhead lines can carry current higher than their nominal rating when weather conditions are favorable, and real-time rating that could be continuously adjusted would increase available capacity.[19]

HVDC becomes cost-effective at long distances, where the reduced capital costs of the lines and reduced energy losses can compensate for the cost of the converters.[20] For example, long-distance, high-power HVDC transmission could

---

[19]The nominal current rating of overhead lines is based on assumed worst seasonal conditions. Conductors have some resistance (except for superconductors) and heat is produced as current flows through them. If the line gets too hot, it expands and sags excessively. High air temperature with no wind is usually the design condition. Under less severe conditions, more current can be carried, but existing transmission controls cannot account for this.

[20]HVDC lines may be warranted for overhead lines longer than 800–1000 kilometers and underground or underwater lines longer than 60–80 kilometers. A 65-mile long undersea and underground HVDC cable began commercial operation in 2007, carrying 660 MW of power from New Jersey to Long Island.

aid in the deployment of large-scale wind generation and, potentially, solar generation by 2020;[21] these energy sources are regional, intermittent, and often far away
from major population centers. As the down periods for wind vary from region to
region, long-distance transmission would help to pool such resources for transmitting the power to load centers. HVDC also is less expensive than AC if lines need
to be underground—for example, when passing through pristine areas. Several
HVDC lines already exist in the country, and, if planning is started within the
next few years, several more large lines[22] could be completed before 2020.

Further R&D on advanced materials and nanotechnology could lead to
improved lightweight insulators, high-temperature low-sag conductors, and lightweight high-strength structures after 2020. In the longer term, breakthroughs in
superconducting materials are needed for superconducting cable technology to
become widespread. This is unlikely to occur until after 2030.

Storage

Cost-effective storage would be useful both on transmission and on distribution systems. Transmission systems require large-scale storage capacity with high
power ratings (on the order of hundreds of megawatts) and long discharge times
(hours to days). The variable power output of renewable resources is currently
managed by standby generation, but as large-scale and remote wind or solar generation facilities are built, such storage technologies would be very beneficial for
the transmission system that must deliver the power. Today, this type of storage
is largely limited to pumped hydro storage, where water is pumped uphill into a
reservoir and released to power turbines when needed. Another technology that
has been demonstrated and is currently available for commercial deployment is
compressed air energy storage (CAES).[23] A CAES plant stores energy by using
electricity (from off-peak hours) to compress air into an underground geologic
formation (or potentially in aboveground tanks). The energy is recovered when a
combustion turbine burns natural gas in this compressed air in lieu of operating

---

[21]These electricity sources are discussed in detail in Chapter 6.

[22]For example, lines might connect wind resources in Wyoming to California, or deliver wind
power in the Dakotas to Chicago. Such lines might account for a large fraction of the cost of that
electricity.

[23]CAES has been demonstrated at a pilot plant in Alabama as well as at locations in Germany.

its own compressor.[24] CAES is now a viable option for providing 100–300 MWe or more of electric power for up to 10 hours. Before 2020, CAES will be the only viable option, aside from pumped hydro, for storing hundreds to thousands of megawatts of energy. Both are dependent on specific features being available (caverns or hills where reservoirs can be built), which greatly limit their applicability.

For distribution systems, storage at lower power ratings (10 MW and below) and lower discharge times (hours to minutes, depending on the application) can be used to improve power quality and security. Distributed storage can help to regulate the system and improve system stability, including reducing the risk of system collapse by supporting islanding and restoration following a disruption. Some battery-storage technologies for these applications, such as lead-acid and sodium-sulfur batteries, have been demonstrated and are currently available for deployment (Bjelovuk, 2008). Batteries are modular and not site specific, meaning they can be located close to intermittent generation sites, near the load, or at T&D substations. However, current battery technologies are expensive and have high losses and reliability issues.

In the longer term, battery storage technology at larger capacities (in the 100 MW range) may help to accommodate variable renewable energy sources, but further R&D is needed before more widespread deployment is likely. Given the large potential in the electric vehicle market for lithium-ion, nickel metal hydride, and other types of batteries, much R&D is now in progress. Advanced batteries with lower cost, high energy density, and higher charge-discharge cycles could also be used for storage in the T&D systems. They may be available for deployment in T&D systems after 2020.

Other longer-term possibilities for energy storage in the grid include supercapacitors, superconducting energy storage, and flywheels. None of these technologies is currently suitable for grid use because of high costs and low energy-storage density. Flywheel storage units are being installed for first-of-a-kind experience with power capacity in the MW range that can smooth out short variations of wind power. However, the technology is a long way from economic deployment on a large scale that would affect daily peaks and day-to-day variations. If advances are made, particularly in materials, all these technologies may become

---

[24]Conventional gas turbines use about two-thirds of their output to operate their compressors; thus only a third of the turbine's output power is available to produce electricity. By moving the compression to off-peak hours when power costs are low, output of the turbine can be approximately tripled and sold at the much higher peak rate.

suitable for energy storage in distribution systems after 2020. Pumped hydroelectric power underground storage, which requires a deep underground water reservoir or aquifer and construction of a power plant deep underground as well, is not considered an effective solution in general because of high costs, but it may be suitable for some sites.

Distribution Transformers

From generation to the customer's meter, power typically flows through four transformer stages,[25] accumulating about 4 percent losses in total. The last transformer in the chain is the distribution transformer for residential/small commercial customers, and because there are so many in the distribution system, they account for a large portion of these losses. Improved materials used to form the transformer's core can reduce the losses. In the past, grain-oriented steel was universally used as the core material, and there has been sustained but slow progress in reducing its losses. A new material, amorphous steel, has become commercially available in significant quantities over the last 10 years. Transformers made with amorphous steel have about one-third the core loss of those made with grain-oriented steel. The market for amorphous steel transformers has been small, however, primarily because of their higher cost. This material may become more competitive economically as a result of new DOE standards regarding distribution-transformer efficiency for new equipment.[26]

Potential for Future Deployment

Many of the technologies needed to implement a modern T&D system, such as FACTS and custom power devices, are presently available for commercial deployment. While R&D is needed to reduce costs and improve performance, no breakthroughs are necessary to start using them in large quantities. In addition, some higher-voltage long-distance lines and substations could be deployed before 2020,

---

[25]Electrical transformers are used to increase or decrease AC voltage. For example, a transformer near the generating plant increases the electrical voltage ("steps it up") at the transmission line, and a transformer at the distribution substation decreases the voltage ("steps it down") from transmission voltages to voltages appropriate for distribution. Others are used within the distribution system to deliver the power at levels appropriate to end users.

[26]These standards are discussed in more detail in the subsection "Economic Benefits" within the section "Potential Benefits of a Modern T&D System."

and dynamic thermal rating of power lines could increase capacity along existing lines.

Some storage technologies will be ready for deployment before 2020; however, significant room for improvement remains. At larger scales that may be needed to support large quantities of intermittent renewable energy sources, pumped hydroelectric power and CAES will be the only viable options before 2020. Batteries may also be used for large-scale storage in the T&D systems but are unlikely to be available for deployment at the hundreds-of-MW scales until after 2020. On a smaller scale (around 10 MW), batteries are already being deployed to enable islanding and load-leveling in the distribution system. Newer technologies (including ultracapacitors and flywheels) may not be ready for wide-scale use before 2035.

Measurements, Communications, and Controls

A modern electric T&D system will need measurement, communications, and control technologies to gather real-time data on the state of the grid, communicate those data, and process them to enhance system controlability. These technologies, including associated software, are the basis for "intelligence in the grid." The following subsections discuss the status of several of these technologies, likely technology improvements, and the potential for their deployment in the U.S. T&D system.

Sensing and Measurements

Understanding and acting on the current state of the U.S. T&D system requires measuring the power characteristics at numerous points. The basic measurements needed are current (amperes) and voltage (volts) at every electrical connection and the status of all switches (on or off). These data provide information on the grid's electrical condition and connectivity.[27]

Measurements are made at each T&D substation and are used to drive its controls and protective devices (relays).[28] Supervisory control and data acquisi-

---

[27]Connectivity of the electrical network can be changed by selectively opening or closing its many circuit breakers.

[28]Protective devices can detect short circuits and isolate the faulty equipment by opening circuit breakers.

tion (SCADA) systems collect and transmit this information to control centers.[29] In most existing substations, the data can be sampled every few seconds, entered into a remote terminal unit (RTU), polled by the SCADA, and sent to the control center over relatively slow communications channels—usually microwave. In modern substations, some of which are already in place, the substation control and protection system is digital and the connectivity is through a local area network (LAN) within the substation. Data can be sampled many times per second, rather than once every few seconds. Most of the substation's controllers and protection systems, known collectively as intelligent electronic devices (IEDs), are based on microprocessors, as are recording systems such as fault recorders and sequence-of-events recorders.

Monitoring of the state of the transmission system is best if the high-voltage substations are equipped with measurement systems that sample at rates of 60–120 times per second[30] and incorporate global positioning system (GPS) signals.[31] Although the individual hardware costs of these measurement units are now very modest,[32] the cost of retrofitting them into the thousands of existing substations will be significant.

There are approximately four times more low-voltage distribution substations than there are high-voltage substations. Although the sampling speed does not need to be as large, high-bandwidth communication will be needed in order to use these data for system control.

Existing customer billing meters could be replaced with microprocessor-based meters which could provide the customer with new buying options such as time-of-day pricing, and could increase end user efficiency. These meters could also allow control signals from the power company to be brought directly into appliances and equipment on the customer side for load management.

---

[29]For further explanation of SCADA systems, see Annex 9.A.

[30]Automatic control action to stabilize the power system after a disturbance has to be taken in well under a second, thus requiring measurement sampling of around 60 times a second. The available phasor measurement units (PMUs) routinely provide measurement sampling at 30 or 60 Hz, and faster sampling rates are already appearing in the market.

[31]Global positioning system (GPS) signals and the associated absolute-time references allow accurate phase shifts in AC quantities to be measured between widely separated substations.

[32]PMUs were priced at around $50,000 when first introduced in the 1980s, but they cost less than $10,000 today; moreover, other substation equipment such as protective relays today can perform this function at almost no incremental cost.

Integrated Communications

Real-time measurements can be used to monitor and control the T&D system, but the measurement data must be transmitted to a location where they can be processed. To appropriately process the data, a fully integrated communications system with universal standards (protocols) must be developed, along with real-time data handling software that can collect and move the data to where they are needed.

If the measurement technologies described above are fully implemented, each control center will need to process approximately one million data points per second.[33] The existing communication channels between the control centers and the substations, many dating from the 1960s, cannot handle these data rates. They are currently being replaced with high-bandwidth optical fiber. However, even with increased bandwidth, the present system (in which all data from substation RTUs are collected at the control center SCADA) cannot handle the expected proliferation of real-time measurement data.

An alternative to this communications architecture is shown in Figure 9.9. Each substation has its own data-gathering system connected internally by a LAN. A gateway server connects these data to the rest of the system through a high-speed network of switching routers, which can move the needed data efficiently to monitoring and control applications. These applications require coordination across several substations, either regionally or over the entire interconnection. Such applications are often referred to as wide-area controls or special protection schemes. Today's local controls are contained within a substation and will remain part of the substation automation design.

Communication systems must be able to handle a wide range of speed and data flow requirements, and the switching network and distributed database will have to be designed. Although similar systems exist today (e.g., cellular telephone systems), the communications needs for the power grid are unique; specialized software will have to be designed and developed. Such a communications system should be ready for deployment by 2020, possibly continuing into the 2020–2030 time period.

---

[33]For a sense of scale, each of the approximately 100 control centers in the Eastern Interconnection oversees about 100 high-voltage substations on average.

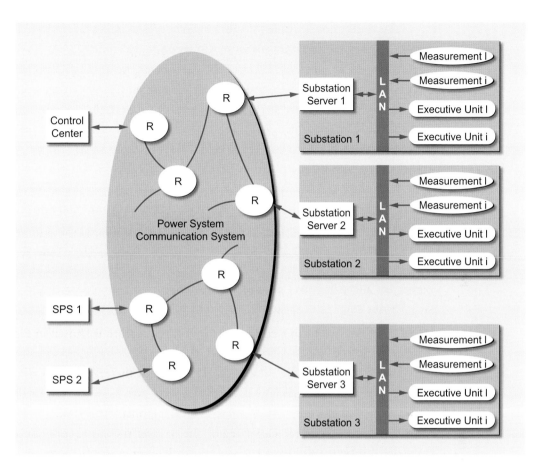

FIGURE 9.9 *An alternative measurement architecture for the transmission system. Each substation (shown on the right) takes measurements that are collected by its own data-gathering system. These measurements are communicated internally by a local area network (LAN). A substation server communicates these data to the rest of the system through a high-speed network of switching routers (shown as circles) that can move the data efficiently as needed to specific monitoring and control applications.*

Advanced Control Methods

Measurement and communication technologies create a picture of the state of the systems, which control technology can use for greater reliability and security (including self-healing following a disruption) and more efficient operation and optimization of assets. If the T&D system is equipped with new measurement sensors, a high-speed communication network, and power electronics, fast wide-area controllers can be designed and installed with software only. This will enable the evolution of better controls to make the grid increasingly reliable and efficient.

The thousands of mostly local controllers in existing T&D systems are slow; response times typically are measured in seconds. In contrast, FACTS devices (already in use as fast local controllers) can control voltages and power flows with response times measured in milliseconds. Moreover, fast wide-area controls, combining rapid communications with remotely controlled FACTS devices, are becoming feasible. Time-stamped measurements will make multiple inputs available to the controller, which can then send out multiple output signals to several FACTS controllers simultaneously.

With these technologies, many types of grid monitoring and control will become possible.[34] Digitized measurements that incorporate data at high sampling rates will allow faster and more frequent calculation of the state of the transmission system. This can provide better predictions of the T&D system's behavior under contingencies (natural, human error, or malicious), thus enabling automatic corrective and preventive actions. Cascading failures can be predicted, and defensive actions such as islanding can prevent the spread of the disturbance. Advanced distribution controls can accommodate two-way power flow from distributed generation by balancing the load on all the distribution feeders. In addition, demand-side responses can be efficiently coordinated if appropriate sensors and communications are in place. Such control technologies could begin to be deployed by 2020.

Potential for Future Deployment

Measurement, communications, and control technologies are already being deployed to a modest degree and could be fully deployed by 2030. About 15,000 transmission substations will require new sensors, measurement systems, and LANs. To add high-bandwidth communications hardware (mainly fiber-optic cables) across the transmission system of approximately 200,000 miles of network and 20,000 switches, investment in both hardware and software will be needed. The costs of developing the needed software to operate the hardware for control will be significant.

Technologies for distribution systems are different in character from transmission. Sensing, monitoring, and communications technologies will need to be

---

[34]Although some one-of-a-kind controllers and special protective schemes have been built to handle unique problems in parts of the T&D system, these are expensive installations because everything—from the sensors to the communication channels to the controllers—is special.

installed in the approximately 60,000 distribution substations and their associated feeders. These controls will be particularly important as smart metering is introduced into distribution networks. Additional investment will also be needed for coordination between transmission level-controls and distribution-level controls.

Improved Decision-Support Tools

The T&D system in the United States is managed by a large number of private and public entities that have long used computer-based decision-support tools both for commercial and for engineering decisions. These tools need to be further improved because of the massive amounts of data that are available in real time and the need to use these data in system control. This section examines improved decision-support technology (IDST), including those tools necessary for split-second decision making by system operators during emergencies as well as for long-term decision making on investments needed in the grid itself.

System Operations

A recurring theme in blackout investigations has been the need for better visualization capabilities and decision-support tools over a wide geographic area. In many circumstances, a human operator will require at least some seconds to make a decision, but automatic controls operate on the order of milliseconds. IDST enables grid operators and managers to make faster decisions by converting the complex power-system data into information that can be understood at a glance. Improved visualization interfaces and decision-support technologies will increase reliability, decrease outages due to natural causes and human error, and enhance asset management.

IDST covers three general systems-operations categories:

- *Grid visualization.* Real-time analysis of system stability will require online analytical tools that process the vast amount of data and automatically determine what actions should be taken to prevent an incipient disturbance from spreading. This objective requires completing the analysis within a fraction of a second and presenting it visually in a control room for fast responses to deteriorating conditions. The algorithms have not been developed to perform these functions, but they could be deployed by 2020 and would be continually improved in the 2020–2035 timeframe and beyond.

- *Decision support.* Decision-support technologies can identify existing, emerging, and predicted problems and provide analysis to support solutions. By analyzing the consequences of each contingency and its probability of occurrence, decision-support systems can quantify relative risk and severity. These relative risks can be integrated into a composite risk factor and presented to the operator to assist in decision making. Further work on decision-support algorithms will be needed to make them available for deployment before 2020, with continuing improvements in the 2020–2035 timeframe and beyond.
- *Systems operator training.* Advanced simulators currently under development will give operators a real-time, faster-than-real-time, or historic view of the power system and its parameters. These dynamic simulators, together with industry-wide certification programs, will significantly improve the skill sets and performance of system operators. Such simulators could be ready for deployment by 2020, as soon as the visualization and decision-support algorithms are in place. IDST, together with system-operator training, will then need to be continuously evaluated and improved.

As the systems become more complex, R&D on software and artificial intelligence will be needed to improve the operator's ability to control a wide-area transmission system as well as an ever more complicated distribution system. Improved software and artificial intelligence for IDST could begin to be deployed by 2020, and deployment is likely to continue into the 2020–2035 timeframe.

Operations Planning and Design

Decision tools are also needed for decisions that occur over longer timescales than do real-time operating decisions. Applications include next-day planning decisions for the power market, planning for adequate generation, and design of T&D substations as well as distribution feeders.

Operations-planning decisions set the schedules of how the T&D system will be operated over the next day. Decisions include forecasting the load, scheduling dispatchable generation and long-term contracts to meet the load, conducting auction markets, using power contracts to check on possible congestion on the transmission system, and modifying the power contracts if congestion is indicated. The decision tools needed for these tasks are mostly new or have been significantly modified in recent years.

Longer-term planning for new generation and transmission capacity must deal with considerable uncertainty, especially where the industry has been restructured and no one organization holds the ultimate responsibility for building adequate generation. Transmission is still regulated, but transmission planning is dependent on knowing where the new generating plants are going to be located. Computerized planning decision tools must be improved to handle increased uncertainty for the 20- to 30-year time horizon. It is anticipated that renewables will present unique challenges, and the addition of probabilistic methods not in use today may help system operators respond to the changing generation mix.

T&D substations are designed by computerized tools that need to be further coordinated with asset-management tools—inventory management for spare parts and maintenance of all components for example—used by utilities. The two tool sets should be seamlessly coordinated with one another and connected to the operations and operations-planning databases so that customer trouble calls can be coordinated with maintenance crews, spare part inventories, and system operations.

Potential for Future Deployment

Several major conditions must be met before IDST can be effectively implemented. First, modern measurement, communications, and control technologies must be implemented along with the power electronics technologies needed to enable automated controls. In addition, development is needed in applications that integrate advanced visualization technologies with geospatial tools to improve the speed of comprehension and decision making. Some of these technologies could begin to be implemented well before 2020.

Integrating Technologies to Create a Modern Electric T&D System

The key technologies discussed above are in various stages of development, with many already having been deployed in a limited way. However, the primary challenge will be the integrated deployment of these technologies to achieve the desired characteristics and performance of a modern grid. For example, the capabilities of power electronics would be maximized by coupling them with real-time measurement, communications, control, and decision-support tools. Smart meters with two-way communications tied to wireless controllers within the customer's premises will be needed on distribution systems to maximize the benefits of a modernized transmission system.

It is important to note that even though most modernization technologies are available now, further R&D is very important. All these technologies can be improved upon and would benefit from cost reduction. A few, such as large-scale storage, are simply impractical now. In addition, the nation is facing a critical shortage of power engineers, the very people who will be needed to implement modernization. University R&D funding is vital in persuading students to embark on careers in power engineering.

COSTS OF MODERNIZATION

Projecting the costs of modernizing the U.S. T&D systems is complex, given the expansive and interconnected nature of the system, the difficulty of estimating development costs (especially for software), and uncertainties over technology readiness. Complicating matters further, costs have been escalating sharply in recent years for large-scale T&D construction, as for other energy projects. Transmission investment is anticipated to continue to increase to meet load growth and replace aging equipment, but additional investment will be needed over the next few decades to modernize the T&D system.

A comprehensive discussion of the costs of modernization was published in 2004 by the Electric Power Research Institute (EPRI 2004).[35] The AEF Committee reviewed the assumptions made by EPRI in this report and largely agrees with its estimates, with two exceptions for transmission. First, EPRI projected that superconducting cables would be added to the system over the next 20 years, but the committee concluded that high costs and slow technological development would preclude commercial deployment before 2030. The committee has thus modified EPRI's cost estimates to reflect this judgment.[36] Second, the cost of developing and deploying software on the transmission system is routinely underestimated; it is likely that more investment will be required for this purpose.

---

[35]Estimates by the Electric Power Research Institute (EPRI), originally in 2002 dollars, were escalated to 2007 dollars for the committee's analysis. In addition, recent real escalation in materials and construction costs were accounted for by using the national average transmission and distribution indexes (33 percent for transmission, 40 percent for distribution). These changes are described in Annex 9.A.

[36]The investment in superconducting cables has been removed from the total investment needed for the transmission system. It has also been removed from the synergies calculation. These changes are described in detail in Annex 9.A.

The improvements needed in the T&D system could be completed over the next 20 years, with significant progress by 2020. Modifying the EPRI results as detailed in Annex 9.A suggests that a total investment of $225 billion will be required for transmission systems, and modern distribution systems are likely to require a total of $640 billion.[37] If the T&D system is not modernized but simply expanded to meet growing loads, transmission would require $175 billion and distribution $470 billion. Thus the incremental costs of modernization are $50 billion for transmission and $170 billion for distribution spread over the next 20 years.[38] Modernization would actually cost about twice this amount, but less expansion would be needed to meet projected loads, and the savings from these synergies would account for the difference. For example, existing lines could carry greater loads if improved control systems prevented overloading, so some new lines would not be needed.

A more recent analysis was performed by the Brattle Group, which built on the EPRI analysis (Brattle, 2008). Estimated costs in the two studies are very similar.[39] The Brattle study does not distinguish between investments to meet increased load demands and investments in modernizing the system, but it estimates that $233 billion will be needed for the transmission system (compared to EPRI's $225 billion).[40] Brattle estimates that $675 billion will be needed for distribution versus EPRI's $640 billion.

Neither report explicitly accounted for the construction of new transmission lines to bring power from remote wind or other renewable energy sources to load centers. These lines could be longer than those from conventional power sources and carry power at a lower capacity factor, thus increasing costs. According to a DOE report on achieving 20 percent of U.S. electricity from wind power, an estimated 12,000 miles could be constructed for $20 billion (DOE, 2008). Actual expenditures will be highly dependent on the routes chosen and the capacities of the lines, but additional costs on the order of tens of billions of dollars seem plausible. Large-scale power generation from photovoltaics or solar thermal technology is a longer-term possibility if cost reductions are achieved. Much of this power

---

[37]These estimates are in 2007 dollars.

[38]The cost of implementing a T&D system is less than the cost of meeting load growth plus the cost of adding intelligence to the systems because of synergies, which are discussed in Annex 9.A.

[39]A comparison between these two studies is made in more detail in Annex 9.A.

[40]Again, these estimates are in 2007 dollars.

could be generated in the Southwest region, which would require additional long-distance transmission. Construction of such lines will depend on the regulatory environment and government policy; the transmission technology is available, although further improvements would be beneficial.

T&D expenditures are unlikely to be linear over the next 20 years. Representatives of EPRI and the Edison Electric Institute (EEI), which funded the Brattle study, have suggested that the split would be approximately one-third during the first 10 years and the remaining two-thirds over the second 10 years.[41] The AEF Committee has assumed that 40 percent of the expenditures should be made before 2020, with the remaining 60 percent between 2020 and 2030. Thus investments averaging $9 billion per year would be needed in the transmission system from 2010 to 2020, with approximately $2 billion per year of this total dedicated to modernization.[42] From 2020 to 2030, an average of approximately $14 billion per year will be needed, including $3 billion per year for modernization.

As discussed in the section "Barriers to Deploying a Modern T&D System," utilities and transmission operators may find it more difficult to raise the relatively small amount needed for modernization than to raise the more substantial amount needed for expansion. If modernization were not included, however, utilities would have to continue using existing technologies for control, sensing, and monitoring equipment, and the nation would be deprived of the many benefits discussed here.

For distribution systems, an investment of $26 billion per year would be needed from 2010 to 2020 ($19 billion for expansion, $7 billion for the modernization increment). From 2020 to 2030, approximately $38 billion per year would be needed (including $10 billion for modernization).

POTENTIAL BENEFITS OF A MODERN T&D SYSTEM

A modern T&D system would offer significant benefits. Costs to consumers could be reduced through more efficient electricity markets; national security could be

---

[41]This estimate was confirmed by personal communications with experts at EPRI and EEI.

[42]Transmission investments were $7.8 billion in 2007, according to EEI. Very little of this amount was for modernization, so the total would have to be increased slightly from the committee's target annual expenditures of $9 billion to include the modernization portion of $2 billion.

enhanced because of greater reliability and reduced vulnerability to major disruptions; greater capacity to accommodate renewables would improve the environment; and public safety would be enhanced. Although these benefits are even harder to estimate than are the costs (as described in the previous section), estimates of the benefits of a modern T&D system significantly outweigh the costs. For example, EPRI estimates these benefits to U.S. society to be about $640–800 billion over the next 20 years for a cost of implementation of $165 billion; that is a cost-benefit ratio of 1:4 (EPRI, 2004). On a more local scale, a recent study by the University of San Diego considered the value of a modern distribution system to the San Diego area. The investigators found about $1.4 billion in system benefits, plus another $1.4 billion in societal benefits, producing an internal rate of return of at least 26 percent (San Diego, 2006).

While the benefits of modernizing the T&D system in the United States are potentially very large—possibly several times the investment—this study had neither the time nor the resources to examine the assumptions and modeling needed for a reliable estimate. The following sections, however, provide some specific examples of potential benefits from modernizing the system.

Economic Benefits

A blackout in a single area can cost approximately $1 billion, and major regional blackouts can cost $10 billion. EPRI estimates the annual cost of power disturbances to the U.S. economy to be between $80 billion and $100 billion (EPRI, 2004). Disturbances to power quality add to the total, and inefficiency and congestion in the current T&D systems also have significant economic costs. These costs could be reduced significantly by a modernized T&D system. Thus improving grid reliability and efficiency could result in substantial economic benefits.

A modern T&D system will benefit the U.S. economy in less direct ways as well, such as by allowing cost information to be made available to buyers and sellers of electricity in real time. These energy price signals will allow customers to more effectively participate in the electricity market, based on current supply-and-demand influences. Overall, markets will be more efficient when consumer decisions are based on realistic prices. There will also be a reduction in grid congestion and forced power outages. A modernized grid will enable a wide array of new options for load management, distributed generation, energy storage, and revenue opportunities for those who choose to participate, such as residential self-generators.[43]

Security Benefits

A modern grid can contribute to energy security by reducing the energy system's vulnerability to terrorist attacks and natural disasters, thereby reducing the risk of a devastating long-duration blackout. In addition, the ability to handle a high level of electricity generated from domestic renewable energy sources has national security benefits as well as the environmental benefits discussed in the next section.[44] In other words, the enhanced controllability that a modern T&D system could provide, and the broad penetration of distributed generation, will make the transmission system far more difficult to disrupt. Moreover, sophisticated analytical capabilities can detect and prevent or mitigate the consequences of an attack or disaster, and probabilistic analytical tools can identify inherent weaknesses in the grid so that they can be integrated into an overall national security plan.

However, as T&D systems become increasingly dependent on computer-driven communications and control networks, physical attacks will not be the only concern. Guidelines for cyber-security are already in place, but these may not be adequate for a fully deployed communications and control system. Cyber-security has to be an integral part of modernizing the grid.

A modern grid could improve the diversity of energy supplies by allowing larger proportions of renewable energy into the U.S. energy supply. Coal is the source of about half the nation's electricity, but although domestic coal reserves will be sufficient for many decades to come, concerns about carbon emissions may affect its future use. Natural gas is also a concern because projected rates of consumption may lead to importing increased amounts of liquefied natural gas, some from politically unstable areas of the world. Little oil is used for electric generation, but the modern grid will reduce oil imports by helping to make electric vehicles commercially viable.

Environmental Benefits

Modernizing the power delivery system is an essential step in reducing emissions of carbon dioxide and other pollutants such as SO_x, NO_x, and mercury:

---

[43]Self-generation refers to electricity generation that the end user owns and controls.
[44]See Chapter 6 for further discussion.

- A modern T&D system can allow for greater penetration of large-scale intermittent renewable electricity sources as well as distributed generation and self-generation, thereby reducing the amount of coal that must be burned.[45]
- Modern demand-response technologies (such as grid-friendly appliances that can be controlled by the utility to shift load to off-peak times) can be better accommodated, thereby reducing demand that must be met by inefficient generating equipment.
- Battery electric vehicles (BEVs) can be better accommodated, particularly after 2030.
- Efficiency can be improved in the T&D system as well as in end-uses, reducing the need for new generation and the siting of new transmission lines.

A modern T&D system can enable intermittent renewable electricity sources (particularly wind power) to contribute substantially to the U.S. energy supply. The electricity provided by wind power varies significantly over the course of a day and over the year because of natural variations in wind speed. As a general rule, a power-delivery system can handle the loss of 10–20 percent of the local generating capacity as long as adequate reserve capacity is available.[46] Grid operators normally require generating companies to have spinning reserve (generators that can increase their output very quickly) equivalent to the largest unit on the system; if that unit fails it can be replaced without disrupting delivery of power. Because intermittent sources cannot be depended on, the spinning reserve has to include a significant fraction of the renewable capacity in addition to the largest unit of conventional power.[47] Above 10–20 percent, a rapid loss of wind power could cause system instability unless the system was modernized.[48] Even at lower

---

[45]Self-generation is a special case of distributed generation. End users generate some portion of their own energy needs, utilizing, for example, rooftop solar panels. Under some conditions, any excess power may be sold to the utility.

[46]Grid modernization is not needed for integrating intermittent renewable-electricity sources in relatively small percentages of the overall electricity supply. This is discussed in greater detail in Chapter 6.

[47]Wind and solar power are the main intermittent renewable-energy sources. Other renewables, such as hydropower, geothermal, and biofuels, are not intermittent.

[48]The changes needed to accommodate renewables are discussed in more detail in the technology section of this chapter; they involve large-scale storage as well as high-voltage long-distance transmission.

levels, modernization would help integrate wind power and reduce the need for spinning reserve.

In addition to intermittency, the location of many renewable resources (remote or distributed) poses a challenge that a modern grid could address better than the current T&D system does. Many high-quality renewable resources, such as wind in the Dakotas and solar resources in the deserts of the Southwest, are located far from population centers. More transmission capacity will be required to bring electricity from these locations to areas of high demand, potentially using technologies such as HVDC transmission. Other low-emission and renewable resources are likely to be used as distributed generation (e.g., natural-gas-fired micro-turbines, small wind turbines, and solar panels on residential and commercial rooftops). The modern grid will enable better integration of these resources by incorporating two-way power flow and smart metering on the distribution system.

Many modern demand-response technologies can be regulated in response to grid conditions. With the implementation of time-of-day pricing, such technologies could allow for more cost-effective and efficient electric power generation (i.e., by running primarily at off-peak times, when the price of electricity is lower and generating capacity of greater efficiency is available).

A modern T&D system can also assist in the integration of BEV (including plug-in hybrids), thus reducing the consumption of petroleum fuels for transportation.[49] BEVs could result in an overall decrease in greenhouse gas emissions even though some of the electricity is generated at coal-fired power plants. Rapid growth of BEVs might significantly increase the demand on T&D systems, but this is unlikely before 2020, when the use of advanced meters could enable controlled battery charging. With the addition of such technologies, the impact on the grid could also be small even by 2035.

A modern grid can operate more efficiently, reducing the need for construction of new generators and transmission lines. Approximately 10 percent of the total power produced in the United States is lost in the process of delivering it to the end user. For example, reactive power flow over a transmission line not only increases losses in the transmission line but also significantly reduces the power-carrying capacity of the line; the use of power electronics, however, can reduce such flow of reactive power. In addition, power electronics can reduce losses by shifting power flow to the most advantageous transmission paths and by the use

---

[49]Plug-in hybrid vehicles are discussed in further detail in Chapter 4.

of more efficient distribution transformers.[50] The committee estimates that T&D losses could potentially be reduced by as much as 10–20 percent, resulting in an efficiency improvement in the overall electric system of about 1–2 percent, which in turn would produce significant economic benefits.

Public Safety Benefits

The American Public Power Association reports that about 1000 fatalities and 7000 flash burns occur annually in the electric utility business (Trotter, 2005). Improved monitoring and decision-support systems would quickly identify problems and hazards. For example, the ability to identify equipment that is on the verge of failure is certain to save lives and reduce severe injuries. Also, the modern T&D system would need less maintenance, which means less exposure to accidents and increased safety of maintenance workers. In addition, by reducing the risk of long-term outages following terrorist attacks or natural disasters, modernization could help prevent public health and safety catastrophes.

BARRIERS TO DEPLOYING A MODERN T&D SYSTEM

It should be clear from the previous section that a modernized electric grid is very much in the nation's best interest. The benefits would be substantial and quite likely to far outweigh the costs. Nevertheless, modernization is unlikely to happen unless it is also in the interests of those who must implement it.

Several barriers have the potential to impede this implementation. First, the technologies that utilities would employ to modernize the grid entail additional costs and uncertainties—particularly regarding how well they will work relative to older technologies. Second, some utilities may be reluctant to invest the additional funds required for modernization even when it would appear to make sense to do so. Third, there is a lack of regulatory and political support that could provide incentives for modernization. Finally, there is difficulty in communicating the need for modernization to the public and to regulatory and political decision makers.

---

[50]In January 2010, a DOE standard will take effect requiring higher efficiency in all new distribution transformers. The DOE estimates that between 2010 and 2038 the energy saved by this measure will be equivalent to the energy used by 27 million households in the United States in a single year. Given the expected life of distribution transformers, 5 percent are expected to be replaced each year under this new standard (DOE, 2007).

Technical Barriers

Most of the technologies needed for modernization of the T&D system are available now, but some technical hurdles, such as energy storage, remain. In addition, some technologies are still expensive; R&D is necessary to reduce costs and improve performance. Yet the rate of technology research, development, and deployment in the power industry is low compared to that of other industries.

Utilities, at least in part because they are regulated, are relatively risk averse and may be reluctant to deploy new T&D technologies—particularly new transmission technologies—until they are fully proven. Also, modernization technologies must be deployed in unison to achieve their full benefits, posing challenges in integrating technologies. For example, universal communications standards as well as a common architecture that promotes interoperability are needed. However, the security issues that are involved in an open system must be met with industry-approved and -adapted standards and protocols.

Investment Barriers

Modernization will cost more than simply building more transmission lines and replacing aging equipment. Even though the additional investment would eventually pay off, financial markets and regulatory constraints drive utilities to minimize investments.

In addition, some of the benefits of modern grid technologies are societal (higher quality, more reliable power) and not typically internalized in a company's decision making. The companies, however, must bear the full cost of modernizing the parts of the grid that serve their customers. This barrier is more significant for the transmission system, which is inherently interconnected: many entities own and regulate different parts of it. Cooperation will be needed among utilities and regulatory agencies.

Regulatory and Legislative Barriers

As noted above, utilities are cautious about adopting new technologies that may involve some risk. This is especially true when familiar technologies have lower first costs and utilities are given no incentive to invest more than the minimum required to maintain operations. If modernization is to occur and produce all the advantages it offers, legal and regulatory changes are likely to be necessary.

Legislators and regulators have not taken a strong leadership role regarding grid modernization, nor have they adopted a clear and consistent vision for the modern grid. There has been significant focus in recent years on individual technologies and on energy-related issues such as environmental impact, but less attention has been paid to developing a vision that integrates technologies, solves the various grid-related issues, and provides the desired benefits to stakeholders and society. The Energy Policy Act of 2005 and the Energy Independence & Security Act of 2007 have been positive steps in the right direction, but much more is needed to directly address the regulatory and policy factors that create significant impediments to the modernization of the U.S. T&D system. For example, a wholesale pricing structure that recognizes the value of reliability and signals when transmission system upgrades are necessary would help provide investment predictability.

In addition, policies regarding the grid are often inconsistent because they are set by multiple groups—individual states (state energy policies and public utility commissions [PUCs]); the Federal Energy Regulatory Commission (FERC); and environmental agencies. Inconsistent policies among states and between state and federal regulators, for example, prevent effective collaboration across transmission regions.

Also, time-of-day rates for consumers that reflect actual wholesale market conditions are not yet widely implemented, thereby preventing the level of demand-side involvement needed in the modern grid. Net metering policies that provide customers with retail credit for energy generated by them are also not widely deployed, which reduces the incentive for end users to install rooftop photovoltaics or other generating technologies. Finally, regulatory policies often do not reward customers for investments that provide substantial societal benefits, such as credits for local storage that has been made dispatchable.[51]

A reduction in R&D expenditures by utilities, an unintended result of restructuring, has impacted the development and deployment of newer T&D technologies. A more predictable regulatory environment that accounts for societal costs and benefits in the rate structure and supports R&D will be needed.

Cultural and Communication Barriers

The fundamental value of the T&D system in general and the societal and economic benefits of a modernized grid and the costs associated with antiquated sys-

---

[51]Dispatchable energy storage is a set of technologies for storing electricity to be deployed quickly (dispatched) into the grid when other power sources become unavailable.

tems in particular have not been adequately communicated to decision makers and the general public. Some electric utility executives assert that their customers value lower rates more than the benefits of a modernized grid, which would increase costs in the short term (NETL, 2007b).

In order to overcome this barrier, significant efforts need to be made to communicate the benefits of a modern grid to all stakeholders. Improved communication with the public is also necessary regarding the costs and benefits associated with the current transmission system in particular, which is experiencing ever increasing congestion and needs expansion. It is difficult to site new transmission lines. Many proposals for new lines generate considerable opposition, usually based on aesthetic, property value, or health and safety concerns. For example, American Electric Power, a large Midwest utility, recently experienced a 12-year approval process for a new 90-mile 765-kV transmission line.[52]

DEPLOYING A MODERN T&D SYSTEM

Many of the technologies needed to modernize the grid can be deployed before 2020, but most of the technical challenges will involve seamlessly integrating these technologies. Not only must multiple technologies work in concert across a huge and sprawling system, but the system is owned and operated by numerous (often regional) stakeholders with diverse perspectives, incentives, and constraints.

Given these factors, a broad vision and an accompanying road map are required to achieve consensus on common goals and to guide the integrated deployment of modern technologies that meet the performance requirements of the modern grid, as described previously in this chapter.

The complexity of the transmission system suggests that the development of clear metrics to measure societal benefits will be essential to measuring progress. The types of metrics that may be considered include reductions in electricity demand forecasting error (from 6 percent to, say, less than 0.5 percent); reductions in maintenance cost (by as much as 4 times); reductions in average recovery from major outages (from hours/days to minutes/seconds); reductions in average annual customer outages (from 100 minutes to, say, 3 seconds); and increases in

---

[52]These reasons included a redesignation by Congress of 19 miles of the New River in Virginia as wild and scenic, problematic interfaces between the states and federal agencies, and public opposition.

the use of power electronics (FACTS, custom power technologies), which benefits wider areas than just where they are installed. Thus a major challenge for the PUCs, which play an essential role in regulating the rates and services of the utilities, is to translate policy-level performance criteria into the metrics they need to analyze the overall economics and determine the merits of modernizing the T&D system (Centolella, 2008). For example, PUCs could look for methods to establish accountability for transmission availability, to measure and internalize the value of lost load and power quality, and to measure and appropriately reward utilities for contributions to efficiency improvement and market transformation.

With such a vision in place, modern technologies could be seamlessly deployed across regions. For example, they would be incorporated whenever new facilities were built, while control centers could be gradually modernized. Communications and control software, as well as tools for improved decision support, could then begin to be implemented.

In contrast, the modernization of distribution systems can occur on a regional level, and programs are emerging in the United States as well as around the world. Pilot projects involving smart meters have begun in many areas. For example, American Electric Power (AEP) is designing an advanced meter infrastructure network involving two-way communications with system-control devices and remote connect/disconnect, time-of-use, and demand-management capabilities. AEP expects to have all 5 million of its customers on this system by 2015 (Bjelovuk, 2008). Other countries that have already implemented partial distribution-system modernization programs report very positive results. For example, Italy's ENEL Telegestore Project is the largest metering program in the world, with over 27 million meters networked. Smart meters can come with a wide range of capabilities, and it will be important to determine what is needed to achieve specific goals and how they will be integrated into a utility's system.

The committee judges that a T&D system can be modernized within a 20- to 30-year timeframe, assuming that the resources and a strategy for the modernization are in place. As discussed previously in this chapter, modernizing and expanding the T&D system will require a comprehensive national vision and investment; however, the investment needed is not much greater than the amount that industry has already proposed to be invested in T&D systems. The key components of the modern grid (FACTS devices, custom power, HVDC and HVAC technologies, and storage) have largely been developed, as noted earlier, and measurement, communications, and control technologies to manage these components will be deployable on a large scale, along with the associated decision-support tools, before

2030. While R&D will be important in reducing costs and improving equipment performance, the main challenges involve integrating the diverse technologies. Development of a nationwide strategy to modernize the U.S. T&D system will be an important first step.

FINDINGS AND CONCLUSIONS

The U.S. electric T&D infrastructure remains dependent upon technologies developed and deployed in the 1950s and 1960s. Recently, many factors, including changes in the regulatory structure of the power industry, have lowered the reliability of this critical national infrastructure.

While it is encouraging that the industry has greatly increased its T&D investment in the past several years, more is still needed to implement a modern grid capable of meeting future challenges—such as enabling power markets, intermittent renewable-electricity sources, and modern efficiency technologies—while maintaining reliability and security in the systems. The following findings relate to the issues that need to be addressed to modernize today's electric T&D system so as to best serve our national needs over the coming decades:

Performance: **The T&D system in the United States is not adequate to manage the reliability, peak loads, and diverse sources of power that will be needed to meet U.S. electrical needs over the next 20 years.** However, many technologies capable of meeting these challenges are currently available or will be available before 2020. Significant progress in modernizing the systems could be achieved by 2020, and T&D system could be fully modernized by 2030.

Technology: Many advanced T&D technologies, including the following, are ready for deployment:

- *Advanced equipment.* **Many power electronics devices and transmission line technologies are currently commercially available and can be deployed before 2020.** These technologies are not widely deployed at present.
- *Measurements, communications, and control.* **Most measurement, communications, and control technologies are currently available and can begin to be deployed before 2020; however, software development is still needed.** Further work is needed to establish a standard communications protocol. Such a protocol could be deployable before 2020.

- *Improved decision-support tools.* Improved decision-support technologies could begin to be deployed before 2020; however, they will require the co-deployment of modern measurements, communications, and controls, as well as power electronics, to be effective. Further work is needed to develop and implement algorithms for rapid decision making and advanced search and optimization. This software is likely to be deployable before 2020.
- *Infrastructure.* Shortages of trained personnel and needed equipment could form a barrier to modernization of the T&D system. In particular:
  —A growing global demand for T&D technologies (as nations such as China build up their infrastructures) and a decline in U.S. equipment designers and manufacturers may lead to short-term bottlenecks in acquiring needed equipment.
  —A significant shortage in the skilled T&D workforce over the next 5 to 10 years is expected unless efforts are instituted quickly to address this issue. The number of university programs in power engineering, as well as R&D support, has decreased markedly.

Deployment:

- *Transmission.* The modernization of the transmission system will benefit greatly from a comprehensive national vision based on consensus among the many stakeholders. The transmission system is national in scale, and the major benefits of a modern system come from the operation of many technologies in concert across the entire system rather than from technologies deployed in isolation. State, regional, and national planning is needed on how the nation will deliver 20 percent of its energy and beyond from renewables, especially wind and solar. If such a vision is established and it addresses the many barriers to modernization, the transmission system could be modernized by 2030.
- *Distribution.* Smart meters and related technologies can improve the efficiency and economics of distribution. Modernization of the distribution system can occur regionally, allowing for rapid parallel deployment while encouraging experimentation to develop best practices. This modernization is already occurring in limited areas; however, it would benefit from a nationwide consensus on best practices such as standardization of communication methods (to better enable smart meters) and

grid-friendly appliances. The distribution systems could be modernized by 2030 if such a consensus is reached nationwide.

Costs: The estimated cost to modernize the T&D system is modest relative to the investments that will be required simply to meet load growth and replace or upgrade aging equipment.

- *Transmission.* At a minimum, an investment of $9 billion per year would be needed in the transmission system from 2010 to 2020, and $14 billion per year from 2020 to 2030. Of these investments, $2 billion per year from 2010 to 2030 and $3 billion per year from 2020 to 2030 would be the incremental costs of modernizing the transmission system. In comparison, utilities and organizations that operate transmission systems spent $7.8 billion in 2007.

- *Distribution.* An investment of $26 billion per year between 2010 and 2020 and $38 billion per year from 2020 to 2030 will be needed for the distribution system. Of these sums, $7 billion per year from 2010 to 2020 and $10 billion per year from 2020 to 2030 would be devoted to modernization.

Barriers: The committee has identified the major barriers to T&D modernization as follows:

- *Technical.* Current high costs of advanced technologies, as well as challenges of systematically integrating existing technologies, constitute a barrier to modernizing the T&D system. This situation is further compounded by the risk-averse nature of the electric utility industry.

- *Investment.* The exclusion of societal benefits (such as avoiding costs to the public from widespread blackouts) in the return on investment for the transmission system is a barrier to industry investment in modern transmission technologies.

- *Regulatory and legislative.* The lack of a comprehensive national vision for the transmission system could form a barrier to transmission modernization. In particular:

  —There is limited multiregional planning and coordination of improvements to the transmission system. Overarching consensus-based standards for grid modernization are necessary but do not currently exist. An open-protocol communications architecture and mechanisms for developing, implementing, and integrating advanced technologies should be part of the standards.

—There are no clear guidelines for measuring progress toward T&D modernization, particularly regarding societal benefits. Such guidelines can help each state's PUC to analyze the overall economics and determine the merits of modernizing its T&D system.

- *Cultural and communications.* Active public opposition stemming from environmental or cost concerns could form a barrier to construction of new transmission lines.

Integrating renewables: Renewable-electricity sources present additional challenges for the T&D system:

- **To integrate renewable sources such as wind and solar on a large scale, the transmission system will need to accommodate their variability.** This objective can be met with backup generation (such as gas-fired power plants) or by large-scale storage technologies, such as compressed air energy storage (CAES). Backup generation or CAES could be deployed before 2020.

- **Many renewables are likely to be deployed as distributed generation (such as rooftop PV panels), which will require two-way power flow capability.**

- **Transmitting power from high-quality renewable resources to population centers creates economic challenges.** These challenges include securing the rights of way for the needed corridors and making a business case for the transmission lines.

R&D: **Many of the technologies needed to modernize the grid are available now, but additional R&D is needed to reduce costs to encourage more rapid deployment. In addition, the current level of R&D investment is inadequate for developing new technologies that may be needed to meet future challenges (such as enabling a broad systems approach to managing the network).** The level of technology research, development, and deployment in the U.S. power industry is quite modest compared to other industries. In particular, the current level of R&D funding for the nation's T&D system is at an all-time low. University power-engineering programs have been badly hurt by low R&D funding, and the lack of graduates qualified to manage the future of the grid is becoming a serious issue.

REFERENCES

Bjelovuk, G. 2008. Presentation to the AEF Committee, February.

Brattle. 2008. Transforming America's Power Industry: The Investment Challenge 2010-2030. Washington, D.C.: The Edison Foundation.

Centolella, P. 2008. Ohio PUC. Presentation to the T&D Subgroup of the AEF Committee, January.

DOE (U.S. Department of Energy). 2007. 10 CFR 431. Energy Conservation Program for Commercial Equipment: Distribution Transformers Energy Conservation Standards, Final Rule. Federal Register 72(197), October 12. Available at www1.eere.energy.gov/buildings/appliance-standards/commercial/pdfs/distribution-transformers_fr_101207.pdf. Accessed July 2009.

DOE. 2008. 20% Wind Energy by 2030: Increasing Wind Energy's Contribution to U.S. Electricity Supply. May. Available at www1.eere.energy.gov/windandhydro/pdfs/41869.pdf. Accessed May 8, 2009.

EEI (Edison Electric Institute). 2005. EEI Survey of Transmission Investment: Historical and Planned Capital Expenditures (1999-2008). Washington, D.C. May.

EIA (Energy Information Administration). 2008. Annual Energy Outlook 2008. DOE/EIA-0383(2008). Washington, D.C.: U.S. Department of Energy, Energy Information Administration.

EPRI (Electric Power Research Institute). 2003. Electricity Technology Roadmap—Power Delivery and Markets. Palo Alto, Calif.

EPRI. 2004. Power Delivery System of the Future: A Preliminary Study of Costs and Benefits. Palo Alto, Calif.

NAE (National Academy of Engineering). 2003. A Century of Innovation: Twenty Engineering Achievements That Transformed Our Lives. Washington, D.C.: The National Academies Press.

NERC (North American Electric Reliability Corporation). 1994. 1994 System Disturbances: Review of Selected Electric System Disturbances in North America. Available at www.nerc.com/files/disturb94.pdf. Accessed July 2009.

NERC. 1995. 1995 System Disturbances: Review of Selected Electric System Disturbances in North America. Available at www.nerc.com/files/disturb95.pdf. Accessed July 2009.

NERC. 1996. 1996 System Disturbances: Review of Selected Electric System Disturbances in North America. Available at www.nerc.com/files/disturb96.pdf. Accessed July 2009.

NERC. 1997. 1997 System Disturbances: Review of Selected Electric System Disturbances in North America. Available at www.nerc.com/files/disturb97.pdf. Accessed July 2009.

NERC. 1998. 1998 System Disturbances: Review of Selected Electric System Disturbances in North America. Available at www.nerc.com/files/disturb98.pdf. Accessed July 2009.

NERC. 1999. 1999 System Disturbances: Review of Selected Electric System Disturbances in North America. Available at www.nerc.com/files/disturb99.pdf. Accessed July 2009.

NERC. 2000. 2000 System Disturbances: Review of Selected Electric System Disturbances in North America. Available at www.nerc.com/files/disturb2000.pdf. Accessed July 2009.

NERC. 2001. 2001 System Disturbances: Review of Selected Electric System Disturbances in North America. Available at www.nerc.com/files/disturb2001.pdf. Accessed July 2009.

NERC. 2002. 2002 System Disturbances: Review of Selected Electric System Disturbances in North America. Available at www.nerc.com/files/disturb2002.pdf. Accessed July 2009.

NERC. 2003. 2003 System Disturbances: Review of Selected Electric System Disturbances in North America. Available at www.nerc.com/files/disturb2003.pdf. Accessed July 2009.

NERC. 2004. 2004 System Disturbances: Review of Selected Electric System Disturbances in North America. Available at www.nerc.com/files/disturb2004.pdf. Accessed July 2009.

NERC. 2005. 2005 System Disturbances: Review of Selected Electric System Disturbances in North America. Available at www.nerc.com/files/disturb2005.pdf. Accessed July 2009.

NERC. 2006. 2006 System Disturbances: Review of Selected Electric System Disturbances in North America. Available at www.nerc.com/files/disturb2006.pdf. Accessed July 2009.

NERC. 2007. 2007 Long Term Reliability Assessment: 2007-2016. Princeton, N.J. Available at www.pserc.org/cgi-pserc/getbig/publicatio/specialepr/workforcec/etra2007.pdf. Accessed July 2009.

NETL (National Energy Technology Laboratory). 2007a. A Systems View of the Modern Grid—Enables Markets. Available at www.netl.doe.gov/moderngrid/docs/ASystemsViewoftheModernGrid_Final_v2_0.pdf. Accessed May 11, 2009.

NETL. 2007b. The NETL Modern Grid Initiative Powering Our 21st Century Economy—Barriers to Achieving the Modern Grid. Available at www.netl.doe.gov/moderngrid/docs/Enables%20Markets_Final_v2_0.pdf. Accessed July 2009.

NETL. 2007c. Modern Grid Initiative Powering Our 21st Century Economy—Modern Grid Benefits. August. Available at www.netl.doe.gov/moderngrid/docs/Modern%20Grid%20Benefits_Final_v1_0.pdf. Accessed July 2009.

NETL. 2007d. A Vision for the Modern Grid. Available at www.netl.doe.gov/moderngrid/docs/A%20Vision%20for%20the %20Modern%20Grid_Final_v1_0.pdf. Accessed July 2009.

Nevius, D. 2008. Presentation to the T&D Subgroup of the AEF Committee, February.

Owens, D. 2008. Personal communication with the T&D Subgroup of the AEF Committee; data from EEI surveys.

San Diego. 2006. San Diego Smart Grid Study: Final Report. Available at www.sandiego. edu/epic/publications/documents/061017_SDSGStudyES_FINAL.pdf. Accessed July 2009.

Trotter, J. 2005. Safety programs that work. Presentation at the American Public Power Association National Conference, Anaheim, Calif., June 18–22.

U.S.-Canada Power System Outage Task Force. 2004. Final Report on the August 14, 2003, Blackout in the United States and Canada: Causes and Recommendations. Available at www.reports.energy.gov/BlackoutFinal-Web.pdf. Accessed July 2009.

ANNEX 9.A: SUPPORTING INFORMATION

This annex provides selected additional information to support the material in the main text of Chapter 9. The first section provides information on reliability measures. Next is a more detailed description of the characteristics of a modern grid than could be discussed in the section "A Modern Electric T&D System" of the chapter, followed by a more detailed description of some of the technologies discussed in the section "Key Technologies for a Modern Electric T&D System." Finally, the cost analysis in the section "Costs of Modernization" in the main text is elaborated upon.

Reliability Measures in the Distribution System

The reliability of the distribution system is often measured using three indexes: the Customer Average Interruption Duration Index (CAIDI); the System Average Interruption Frequency Index (SAIFI); and the System Average Interruption Duration Index (SAIDI). CAIDI tracks the average duration (typically expressed in minutes) of customer interruptions over a given time period. SAIFI tracks the average number of customer interruptions in power service in a given period of time. SAIDI tracks the average number of customer interruptions in power service in a given time period. However, unlike CAIDI, the SAIDI average is calculated across the total number of customers served, rather than the number of customer interruptions. Results of applying these indexes are shown in Figures 9.A.1, 9.A.2, and 9.A.3 for the state of Ohio, which is roughly representative of the nation as a whole.

Characteristics of a Modern Electric Grid

The modern grid must meet the ever expanding needs of society and at the same time be reliable, secure, economic, efficient, environmentally friendly, and safe. In order to realize all these elements of modernity, our nation's T&D system should achieve the following goals:

- *Emergency response.* A modern grid provides advanced analysis for predicting problems before they occur and assessing problems as they develop. This capability allows actions that respond more effectively and minimize disruptions.

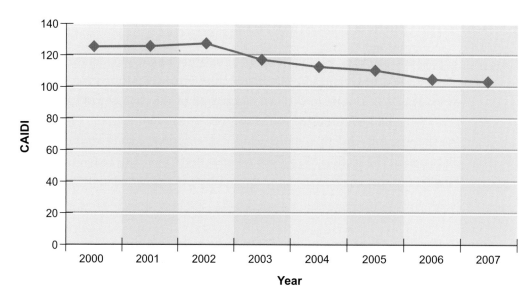

FIGURE 9.A.1 *Customer Average Interruption Duration Index (CAIDI) for the state of Ohio, 2000–2007.*
Source: Ohio Public Utility Commission.

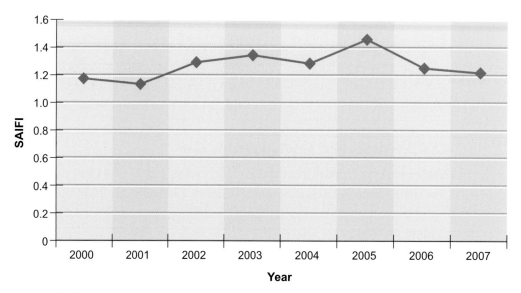

FIGURE 9.A.2 *System Average Interruption Frequency Index (SAIFI) for the state of Ohio, 2000–2007. As shown, Ohio's average SAIFI has been holding approximately steady over the last 7 years.*
Source: Ohio Public Utility Commission.

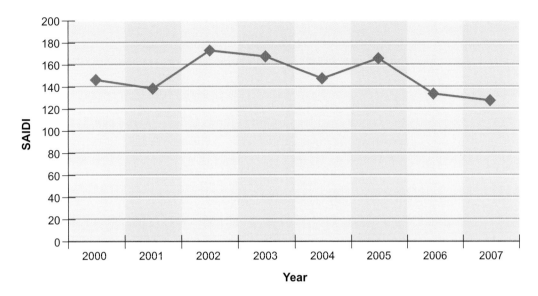

FIGURE 9.A.3 *System Average Interruption Duration Index (SAIDI) for the state of Ohio, 2000–2007. Expressed in minutes, the average SAIDI in the state of Ohio has been holding approximately steady over the last 7 years.*
Source: Ohio Public Utility Commission.

- *Restoration.* It can take days or weeks to return today's grid to full operation after an emergency. As better information, control, and communication tools become available to assist the operators and field personnel of a modern grid, it can be restored much faster and at lower cost.
- *Routine operations.* With the help of advanced visualization and control tools, fast simulations, and decision-support systems, the operators of a modern grid can better understand its real-time state and trajectory, provide recommendations for secure operations, and allow appropriate controls to be initiated. These capabilities could help achieve significant reduction of the system peak-to-average ratio, thereby saving resources.
- *Optimization.* The modern grid provides advanced tools for comprehending conditions, evaluating options, and exerting a wide range of control actions to optimize grid performance, whether from reliability, environmental, efficiency, or economic perspectives.
- *System planning.* Grid planners must analyze projected growth in supply and demand to guide their decisions about where to build, what to

build, and when to build. The data-mining and data-modeling capabilities of a modern grid will provide much more accurate information for answering those questions while potentially realizing significant savings.

In order to meet these goals, the modern T&D system will need to display the seven characteristics listed in the chapter, which were adapted from a sequence of 2007 reports by the National Energy Technology Laboratory (NETL, 2007a-e). To acquire these characteristics (each of which is discussed in turn below), it will not be enough simply to add isolated technologies to the existing system. Technologies will need to be integrated with one another and also have a common basis for communication across regions. Thus creating a transmission system that displays these characteristics will require a multiregional effort based on consensus among all of the key stakeholders and reflecting a common approach to deployment across the various transmission regions. (Given the regional nature of the distribution system, such a common vision for distribution is less essential.)

Such a vision for the future of the transmission system will need to be developed and proposed at a high enough level to become the basis for support from state and federal regulators, owners, and operators of the T&D system. This would encourage stakeholders to begin the planning process and eventual investments for the deployment of the technologies needed to modernize the T&D system. The following is an in-depth discussion of the seven characteristics.

Accommodating All Generation and Storage Options

The transmission system must be designed to accommodate large baseload generation, such as nuclear and coal, as well as sources that do not typically operate in baseload mode, such as renewables. In addition, the distribution system must accommodate smaller distributed-energy sources. Large-scale baseload generation resources may require backup generation and, possibly, also power electronics to ensure that power flows are accommodated. Also, both the transmission and the distribution system must accommodate the intermittency of wind and solar generation.

Because electricity for consumption is produced on demand, the T&D system must be able to allow for the sudden loss of a generation source or an increase in demand. For transmission, this requirement translates into the ability to withstand the loss of the largest single generator on the system. The variability of small amounts of renewables on the current system can be accommodated reasonably well, but as intermittent renewables grow to a significant portion of the system's total generating capacity—say, 10 to 20 percent—new measures will be required to maintain reliability (EWIS, 2007).

The modern transmission system can meet the challenges posed by remotely located renewable-energy sources in large part through the use of technologies such as high-voltage direct current (HVDC) and power electronics. Dispatchable energy storage or backup generation can help to smooth intermittent generation, but the cost of the backup generation, storage, and power electronics and the actual cost of the transmission line and substations will need to be incorporated into the overall economics (cost of power) of, for instance, a proposed wind farm.

For distributed renewable power systems, the situation is somewhat different. For small amounts of power, net metering schemes and two-way power flow will adequately support them. If distributed renewable power becomes significant, however, storage to buffer it will be required. Such storage carries a secondary benefit of improving power reliability. As the local distribution becomes smarter, the easier it will be to accommodate renewable power.

Enabling Power Markets

The transmission system is being pressed into a new mode, in which wholesale power is bought and sold across wide areas. Although some modifications have been made, the systems are woefully short of the flexibility and intelligence required to accommodate wholesale power markets (EPRI, 2004; NETL, 2007d). Better knowledge of the transmission grid, including its available capacity and potential congestion locations in real time, can make the generation market more efficient. (While many of the most pressing needs in this area are related to the transmission system, changes will also be needed on the distribution side. For example, it is difficult at present for consumers to respond to increases in price to seek lower-cost products.)

Major improvements must be made to the transmission system to achieve well-designed and operating markets, especially as industrial, commercial, and even residential consumers will generate and sell power. These contributors will be enabled by emerging generation and storage technologies; the modern grid will allow for two-way power flow on the distribution system and thereby provide self-generation opportunities for the end user to also participate in power markets. Thus transmission capacity needs to be increased, and communication and control between regions must be expanded to accommodate the vast amount of information flow required in real time.

Some of these needed improvements will require little in the way of new or advanced technologies, but they will depend more on policy decisions and implementation. For example, improvements in regulations, the training of participants in the market, and significant capital investments are all important.

Self-Healing

Power outages can cause significant financial losses for U.S. industrial and commercial customers, as much as $80–100 billion annually, and can be very inconvenient or even dangerous for people. As discussed in the main text of this chapter, the 2003 blackout in the North American Eastern Interconnection occurred because a small problem in one part of the system resulted in cascading failures throughout the system. A self-healing transmission system could minimize such occurrences.

Self-healing actions are defined as automatic responses by the system such that system collapse will not occur and that, at worst, "graceful degradation"—which involves minimal interruption of service—will result. For example, faulty equipment or lines can be isolated when necessary to prevent problems from spreading. A self-healing system should be capable of being restored to normal operation with little or no human intervention. This means that both the transmission and the distribution system will have the ability to sense the state of the system as well as communicate this information to other parts of the system and take appropriate action. A wide variety of new measures will need to be implemented to create a self-healing T&D system. Many of these measures will be technological, while others will involve the development of software and standards. For example, research and development (R&D) is still needed on many of the algorithms involved (J. Eto, personal communication, 2008). In addition, integrating the new technologies will be a major challenge.

For transmission, the needed measures include: effective and advanced monitoring; methods for very quickly determining the cause and location of a fault or instability; probability-based contingency analysis; rapid system alignment for the next contingency; effective use of flexible alternating current transmission system (FACTS) devices and HVDC to stabilize system voltages and power flows; remotely dispatchable storage near generators and load centers; effective use of customer-generated power and storage; intelligent load-shedding; effective islanding; fast restoration means; strict reliability standards; and predictive maintenance of key components (NETL, 2007b). Many of these approaches are described in the sections that follow.

On the distribution side, measures to enable self-healing include distribution automation; alternate feeders with power-electronics-based transfer switching; micro-grids and meshed distribution systems; high impedance fault location; automatic switching off of nonessential loads; and effective use of local, generally customer-owned power and storage.

Motivating and Involving the Customer

Customers are not just consumers of electricity; they may also participate in generation and storage options as well as interactively respond to price signals. One way to optimize the use of electricity resources, in fact, is to motivate the customer to make wise end-use decisions (PNNL, 2007). Implementing new technologies (such as smart two-way meters and wireless communications with a residence's major appliances) empowers consumers to make sound choices about their electricity use, thereby contributing greatly to a robust, efficient, and reliable distribution system (CECA, 2003; NETL, 2007c).

For example, providing electricity-pricing information to customers has been shown to reduce peak demand and assist in levelizing power demand. In addition, better information of this kind can enable the distribution-system operator to utilize the system more efficiently. The challenge primarily addresses the distribution system; however, it must ultimately include transmission, as decisions on the consumer side will, when taken in bulk, affect power markets and energy trading.

The aggressive introduction of technologies such as intelligent metering and real-time pricing could create incentives to shift energy use to off-peak times, thereby reducing demand for peak-load power generation and decreasing stress on the T&D system overall. For example, 20 percent of California's electricity demand is used to move water, which can be done predominantly at night. Similarly, technologies could automate industrial and residential electricity-use decisions so that energy-intensive equipment and appliances could be run at night (or on weekends) rather than during peak-load hours. Utilities may be able to reduce demand in this way, at virtually any time of day and in real time, by communicating with end users and even directly with their appliances.

Detailed information on energy use and costs empowers individuals to take more proactive actions in their best interests. Programs to enhance consumers' understanding of their pricing options will be critical in order to fully utilize the potential of the peak-shaving capability of demand response and grid-friendly appliances.

Resisting Physical and Cyber Attacks

Terrorist threats to the transmission system, whether physical or cyber, are serious; a widespread attack against the electric-system infrastructure cannot be ruled out. The transmission system is the primary focus: an attack on the distribution system would have only local impact, while an attack on the transmission system could affect millions. Because the aging system's infrastructure was never designed to handle well-organized acts of terrorism, it is critical that increased security be a requirement for all of the transmission system's elements.

Resilience must be built in to each element, and the overall system must be designed to deter, detect, respond to, and recover from human-induced (as well as natural) disruptions. Moreover, in order to reduce the threat of attack, the modern transmission system must conceal design vulnerabilities; disperse, eliminate, or reduce single-point failures; and protect key assets from both physical and cyber assaults. The modern transmission system must also reduce the consequences of a successful attack by devoting resources to recovery.

Many of the technologies described previously for advanced components, measurements, communications and controls, and improved decision-support technology (IDST) will help to guard the T&D system against physical and cyber attacks. In order to make the most significant difference, they will need to be deployed in an integrated manner, with an eye toward maximizing the system's reliability and resiliency.

For example, the T&D system should be able to implement self-healing (as described above) and "islanding" (the autonomous operation of selected grid elements). These capabilities would allow the systems to respond to attack by rerouting to unaffected segments, isolating the affected portion, and thus preventing the disturbance from spreading. In addition, predictive models and decision-support tools could help operators respond to impending disruptions in real time and preempt further disruption. Providing greater automation, wide-area monitoring, and remote control of electrical distribution systems would enable all of these measures.

In order to further increase security, it is also important to acquire and position spares for key elements, such as high-voltage transformers and breakers, and to ensure that added equipment and control systems do not create additional opportunities for attack.[1]

---

[1]These issues are discussed in greater depth in NRC (2002).

Providing Power Quality for 21st-Century Needs

Forty percent of the power used in this country today is regulated through microchips of various types that run a wide range of equipment. Over the coming decade, this proportion could grow to 60 percent. In order to accommodate these microchips, the utility supply voltage must be reasonably free from harmonics, and any voltage variations should be within acceptable limits. The utility supply voltage deviates from the ideal because of events such as faults; the switching of lines, loads, or system equipment; overloads and light loads; and loads that inject harmonics into the utility system. Providing high-quality power is primarily an issue for the distribution system, as it affects the end user of the electricity and not its long-distance transmission.

It is estimated that problems with power quality cost tens of billions of dollars annually. Accordingly, many industrial and commercial users install equipment—such as uninterruptible power supplies, alternate utility feeders with high-speed transfer switches, standby generators, or a variety of power electronics devices, depending on cost and benefit—to attain the needed power quality. But with proper monitoring of the network condition and anticipation of changes, power-quality problems can be avoided at the system level through the use of existing technologies. Flexible AC transmission systems with superconducting condensers can reduce sags, which are the biggest customer power-quality problem. Fault current limiters can reduce the voltage depressions; synchronous switching can eliminate transient over-voltages.

Mitigating these problems in a fundamental way will require setting and enforcing proper standards applying to utilities' power quality and to users' loads.

Optimizing Assets and Operating Efficiently

In order to make optimal use of T&D systems, losses must be reduced and lines utilized as efficiently as possible. This is not currently the case, particularly on the transmission side. Losses in the T&D system account for about 10 percent of the electricity generated in the United States, or 390 billion kWh (McDonald, 2008). Reducing these losses by just 10 percent would be equivalent to adding seven new 800 MW power plants operating 80 percent of the time.

Average loads are much lower than peak loads, but the system must be sized to accommodate peak loads along with adequate safety margins to allow for failure contingences. Over the course of a year, the transmission system carries only about 50 percent of its full load capacity, and that fraction is dropping.

This trend will necessarily drive the cost of electricity upward because the full cost of the system must be borne by the average transmission. Further, the growth in intermittent renewable sources of energy may shift baseload generation capacity to standby power, resulting in more capacity that is not fully utilized. Reducing transmission losses is important not just for the cost of the losses but also for increasing available transmission capacity. Several options could be considered:

- *Reducing flow of reactive power over the lines.* In principle, a deficit or surplus of reactive power should be corrected at or near where it occurs—namely, generators, transmission lines, and loads or load areas. Reactive power flow over a transmission line not only increases losses in the transmission line but also significantly reduces the line's power-carrying capacity.

- *Power flow through parallel paths.* Currents flow through all parallel paths and are distributed according to their impedances, which, if not carefully selected, may cause losses or reduce transmission capacity. In general, losses can be decreased by appropriate adjustment of impedances—for example, through series capacitor compensation of some lines or phase-shifting transformers.

- *Evaluation of transformer losses.* Most utilities evaluate load and no-load losses as part of their evaluation of procurement price. Nevertheless, for cash flow or other reasons there is a temptation to procure transformers on a first-cost basis. Appropriate regulations could ensure that they are purchased with appropriate loss evaluation.

In a modern electric T&D system, asset optimization does not mean that each asset will reach its maximum operational limit. Rather, it means that each asset will be coordinated with all other assets to maximize the overall function. For example, load-sharing would routinely adjust the loads of transformers or lighten the loads of transmission-line sections, thereby allowing for more efficient operation of the transmission system. Optimized maintenance will be possible when, for example, equipment monitors send a "wear" signal as part of a predictive maintenance program or a direct malfunction signal to a condition-based maintenance program.

Such ends may be accomplished because modern T&D systems will include many sensors and enhanced communications capability needed to monitor equipment conditions in real time. This information may be gathered as a direct reading of the condition of a component or piece of equipment, for example, by means of a vibration monitor, temperature sensor, hydrogen monitor on a transformer, or a derived estimation using a wear algorithm. Automated analysis, such as comparing the wear to the threshold value, would enable the signaling of an exceeded threshold to the asset manager, who would then perform maintenance. Today, operators know the condition of equipment only when they perform scheduled maintenance or when a failure occurs.

In the operation of a modern grid, optimization can extend to the identification of untapped capacity, thus avoiding the start-up of more costly generation resources. Dynamic real-time data reveal when and where such unused generating capacity is available. The use of excess capacity also applies to transformers, transmission lines, and distribution lines. For example, deploying a costly distributed energy resource could be avoided if the operator knew that the distribution system was capable of carrying a greater load from the substation.

As the sensors of a modern T&D system provide more data, asset planning is also enhanced. Decision makers can decide more economically where, what, and how to invest in future grid improvements. Whether from optimizing assets or operating efficiently, the real-time information from the modern grid sensors, coupled with communicating it widely and processing it effectively, will significantly enhance the system.

Detailed Discussion of Selected Technologies

Flexible Alternating Current Transmission System

The Flexible Alternating Current Transmission System (FACTS) is a collection of mostly power-electronics-based devices that are applied, depending on the need, to control one or more AC transmission parameters—such as current, voltage, active power, and reactive power—in order to enhance power-transfer capability and stability. FACTS devices will be needed, in several ways, to meet the challenges of modernized T&D systems. They will improve power quality and increase efficiency by enabling high-speed control of power systems, power-flow control over lines, control of voltages, and reactive-power management. They will also be of value in the prevention of system collapse and restoration. FACTS technology helps meet many of the challenges outlined previously: enabling the connection of remote and asynchronous sources of power such as wind, solar, fuel cells, and microturbines; supporting wholesale power markets through power-flow control; stabilizing power swings; making the system more secure and self-healing; and optimizing the use of available assets.

There are three basic applications of FACTS devices, each with high-speed control. The first can be characterized as adding voltage in series with a line, while controlling the line's reactive and active current. The second application is injection of current in shunt, which enables control of the line voltage. The third application is a combination of voltage injection in series and current injection in shunt both for active/reactive power and voltage control. All FACTS devices can help stabilize the power system and enhance the usable capacity of lines.

Within these three basic types of FACTS devices there are many specific device concepts, and several FACTS devices are commercially available. The ones most used are the following:

- *Static volt-amperes reactive compensators (SVCs),* which together with variable shunt capacitors or inductors help to control shunt current and reactive power, are used primarily for controlling the line voltage and stabilizing the power system.
- *Thyristor-controlled series capacitors (TCSCs),* which control the magnitude of current flow through the line, are used primarily for controlling the current and stabilizing the power system.
- *Static shunt compensators (STATCOMs),* which are voltage-sourced converters connected in shunt with a line for controlled injection of lagging or leading reactive current (and hence for controlling reactive power), are used primarily for controlling the voltage and stabilizing the power system.
- *Variable frequency transformers (VFTs)* are used primarily to control active and reactive power flow through a line as well as to adjust frequency drift.

Hundreds of SVCs, and a few STATCOMs, TCSCs, and VFTs, are currently deployed in the T&D system. More FACTS devices will be available with further R&D and can be deployed by 2020.

Custom Power

Custom power is very much like FACTS, but it is designed for lower voltages and for use in the distribution system. Custom-power devices, inserted between the utility and the customer, can achieve significant improvement in power quality by controlling voltage dips and harmonics and allowing for high-speed switching to alternate feeders.

These capabilities address the need for power quality in 21st-century applications. They would provide relief to many present users, such as banks, that have to employ expensive uninterruptible power supplies, along with standby generation, because any power interruption and voltage dip would be unacceptable. In other industries, annual losses because of power-quality issues amount to billions of dollars. Therefore many electricity consumers—in particular, companies with automated production—would appreciate low-cost solutions that provide substantial improvement in the number and duration of voltage dips and power outages.

Several custom-power devices are commercially available to control voltage or current. However, they are still too expensive for widespread use. The present market is about 100 devices (greater than 1 MW rating) per year, representing about $50–100M. Further R&D could help to decrease these costs.

High-Voltage Direct Current

DC lines have several advantages over AC lines that make them preferable under certain circumstances. While power on an AC line automatically follows the path of least resistance, DC current is controllable. Therefore, a DC line carrying power from a distant generating plant is considered to have the same reliability as a local plant, a significant advantage when an ISO is determining required reserve margins. DC lines can be less expensive per mile than AC lines are, especially for underground transmission. DC lines require two cables, while AC requires three. A DC underground or submarine line can carry 2–3 times the power of a comparably sized AC line. Finally, heat dissipation in an underground DC cable is less of a problem because the lower voltage allows a solid insulator rather than one containing fluid, which raises concerns about possible leaks and damage to groundwater.

Because the U.S. transmission system today is almost entirely AC, transmitting electricity via HVDC involves converting AC to DC, transmitting the DC electricity, and then converting it back to AC at the other end. Most HVDC projects to date have been based on current source converter technology, in which the DC current flows in the same direction and power reversal involves reversal of voltage. These converters, assembled with thyristors,[2] have been operated at

---

[2] A thyristor is a semiconductor power device that turns on with a gate pulse but has no gate turnoff.

a power rating of 6000 MW and 800 kV per two-conductor DC line. Recently, voltage source converter technology has become available in which the DC voltage has the same polarity and the power reversal involves reversal of current. This technology, which offers the advantages of low harmonic levels, a reactive power supply, and easier multiterminal HVDC (in which more than two converters are connected to one line), is available at ratings of up to 1000 MW.

Storage

Storage of electrical energy would offer many benefits to the T&D system, especially if it included significant input from intermittent sources such as wind and solar. Storage would provide improved system stability and efficiency by enabling load-leveling, system regulation, instantaneous reserve power, and the dispatch of reactive power to the system.

Pumped hydroelectric power, currently the only proven means of large-scale energy storage, is unlikely to be expanded greatly because few sites are both economically and environmentally acceptable. Other near-term candidates are compressed-air energy storage (CAES) and, for lower power levels, battery storage. Longer-term candidates include ultracapacitor storage, flywheel storage, and superconducting energy storage.

CAES technology has already been demonstrated and will be available for deployment in the near future. A CAES plant stores energy by using electricity (typically from off-peak hours) to compress air into an underground geologic formation (or, in some cases, in aboveground tanks). The energy is released by sending the compressed air to a combustion turbine, where it is mixed with natural gas and burned, increasing the efficiency of the gas turbine by as much as a factor of three.[3]

The compressed air can be stored in several types of underground sites, including porous rock formations, depleted natural gas or oil fields, and caverns in salt or rock formations. Considerable energy can be stored in underground geologic formations, and such facilities are much less expensive to build than are pumped hydroelectric plants. The compressed air can also be stored in

---

[3]In a conventional gas turbine plant, the turbine runs its own compressor simultaneously with driving the generator, so that only a third of the turbine's total power is available to produce electricity.

aboveground or near-surface pressured-air pipelines, which can be cost-effective for about 2 to 4 hours of energy.

EPRI studies have found that approximately three-fourths of the United States has geology that is potentially suitable for locating reliable underground CAES systems (EPRI, 2008). The Alabama Electric Cooperative built (with EPRI support) the first U.S.-based CAES plant, with a capacity of 110 MW for 26 hours. Because the plant was the first of its kind, the cost was high ($800/kW). With new CAES plants projected to cost in the range of $500–600/kW, CAES will be a viable option for providing the backup power that compensates for the electrical-output variability, for example, of a large wind farm. A 300 MWe CAES storage facility can utilize wind power to compress air, and then, during low wind periods, the compressed air can provide the combustion air for a natural-gas-fired combustion turbine providing up to 10 hours of backup capability for the wind farm.

Another possible storage technology for use in the grid is batteries, which rely on electrochemical processes to store electricity. There is a wide variety of battery types with potential for large- to small-scale dispatchable storage. Examples include lithium ion, sodium sulfur, zinc bromide, nickel metal hydride, and vanadium. In general, present battery technologies are expensive ($400/kW for 2 hours), incur high losses as the batteries are charged and discharged, and have reliability issues. In addition, battery storage requires AC/DC converters, which at present add $100–150/kW to the cost and about 4 percent to the in-out losses. However, more R&D and the mass production of standard power-electronics building blocks should bring converters' costs and losses down to less than half by 2020 and to one-quarter by 2035. Also, in converter-based FACTS applications, batteries can be added at little extra converter cost.

There are significant advantages to battery storage. Batteries are modular and non-site-specific, meaning they can be located close to intermittent-generation sites, near the load, or at T&D substations. Battery storage technology can provide needed reliability and flexibility to the T&D system if it can be economically developed in the 100 MW range. Some battery storage technologies, such as sodium sulfur batteries, have been demonstrated and should be available for deployment before 2020. For example, American Electric Power plans to increase reliability by deploying 25 MW of sodium sulfur batteries in its distribution system by 2010 (Bjelovuk, 2008). Meanwhile, extensive R&D is in progress on lithium ion, nickel metal hydride, and other types of batteries. These technologies have promise for lower cost and higher energy density. It is likely that these batteries would be available for deployment in the T&D systems after 2020.

In addition to batteries and CAES, there are several other possibilities for energy storage. For example, supercapacitors have been used as energy storage devices in power-quality and similar short-time applications, including HVDC and FACTS. They have very long life as well as very high efficiency compared to batteries. A second example is superconducting energy storage (SES), whereby energy is stored in a magnetic field created by circulating DC current through a coil made of superconducting material. SES, which has high in-out efficiency and cycle life, has been demonstrated for stabilizing power systems and used for power-quality applications, but its application for storage will require more advances in materials science. Energy can also be stored in flywheels, which are particularly suitable for power-quality applications and have a very long cycle life. Given their high costs and low energy-storage density, none of these three technologies is currently suitable for storage in the grid. However, if advances are made, particularly in materials, they may become suitable for use in distribution systems during the 2020–2035 and post–2035 time periods. Because no one type of storage fits all applications, R&D is needed for all of these technologies.

At the distribution and customer levels, the loads being protected or leveled are generally much smaller in size (a few kilowatts to a few megawatts). Thus devices such as ultracapacitors, flywheels, batteries, and uninterruptible power supplies can be used. The choice will normally be determined by the load characteristics. Figure 9.A.4 shows the various types of storage and their applications.

Transformers

Electrical transformers are devices used to raise or lower AC voltage. For example, a transformer near the generating plant increases the voltage (steps it up) at the transmission line, and a transformer at the distribution substation decreases the voltage (steps it down) from transmission levels to those appropriate for the distribution system. This voltage is subsequently reduced as the power travels to the consumer. All told, power from the point of generation to the customer's meter may flow through four transformers stages, causing total energy losses of about 4 percent in the process. Though utilities procuring transformers generally take estimates of such losses into account, there is always a trade-off between capital costs and operating costs which can push the buyer toward lower first cost. Thus

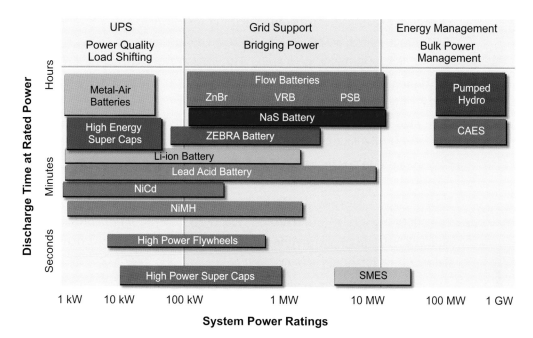

FIGURE 9.A.4 *Energy storage options.*
Note: CAES = compressed-air energy storage; Caps = capacitor; Li-ion = lithium ion; NaS = sodium sulfur; NiCd = nickel-cadmium battery; NiMH = nickel metal hydride battery; PSB = polysulfide bromide battery; SMES = superconducting magnetic energy storage; UPS = uninterruptible power supplies; VRB = vanadium redox battery; ZnBr = zinc bromide.
Source: Adapted from a presentation by Dan Rastler, Electric Power Research Institute, to the Panel on Electricity from Renewable Resources, March 11, 2008.

setting standards for transformer efficiency can be important in lowering the T&D losses.

The last transformer in the chain is the distribution transformer for residential/small commercial customers, which incurs about 1–2 percent losses. These devices experience about 0.2–0.5 percent constant core losses (in the magnetic material) and load losses that vary according to the load. Core losses are important because they occur all the time, whether the transformer is fully or lightly loaded; the installed capacity of distribution transformers may be two times the total load, causing core losses to add up to a significant and continuous amount.

Grain-oriented steel has generally been used as the core material, though there has been sustained but slow progress both toward improving it and developing alternatives. Transformers with amorphous steel, which have been commercially available in limited quantities for better than 10 years now, have about one-third the core loss of transformers with grain-oriented steel. This material is made by running molten metal on a fast-moving belt, thereby solidifying it rapidly without producing grains in tape form. The market for amorphous steel transformers has been quite low, however—fewer than 10,000 units per year—mainly because of their higher cost.

The U.S. Department of Energy (DOE) has established standards for distribution-transformer efficiency that will become effective in 2010 (DOE, 2007). The DOE estimates that the cost of the standard will be $463 million per year in increased equipment and installation costs, while the annualized benefits will be $602 million.[4] This standard could help to make amorphous steel transformers, as well as advanced grain-oriented steel transformers, more competitive. Given the typical service lives of distribution transformers, it is expected that 5 percent of them will be replaced each year.

Sensing and Measurements

Understanding and acting on the current state of the T&D system require measuring their power characteristics at many points. The basic measurements that need to be made are the current (amperes) and voltage (volts) at every electrical connection and the status of all switches (on/off). The first two measures indicate the electrical condition of the electric T&D system—although the derived value of power flow (watts, VARs) is often preferred for monitoring. Whether the switches are on or off provides information on the connectivity of the T&D systems, such as which components are connected and which ones are switched out.

These measurements, made at each substation, are used to drive controls and protective relays. In the early days, all the measurements and controls were hardwired within the substation, and a few—very few—of the measurements from high-voltage transmission substations were hardwired all the way back to a central control center. From the 1960s on, the control center was based on the digital computer's supervisory control and data acquisition (SCADA) system, and the data from substations could be transmitted over slow communication channels, usually microwave, to the control center. Within the substation, the measured data could be sampled every few seconds and put into a remote terminal unit (RTU) that could be polled by the SCADA system over the microwave channel

---

[4]A 7 percent discount rate is used in this calculation. Alternatively, using a 3 percent discount rate, the cost of the standard is $460 million per year and the benefits are $904 million per year.

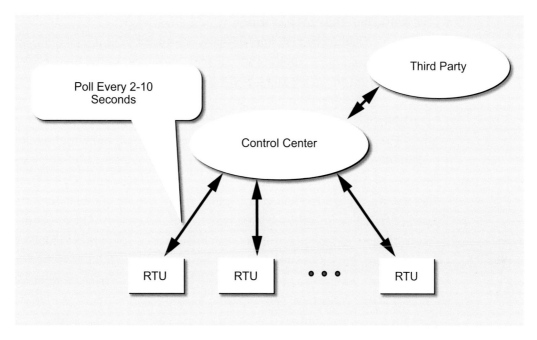

FIGURE 9.A.5 *The SCADA at the control center collects real-time data from each substation remote terminal unit (RTU) every few seconds.*

(Figure 9.A.5). This configuration remains the architecture of most control centers in place today.

More recently, most modern high-voltage substation control and protection systems are digital and the connectivity is through a local area network (LAN). Most of the recent and all future controllers and protection systems in the substation are based on digital processing. In fact, all the recording systems—for example, fault recorders and sequence of events recorders—are also based on digital processors.

Given that the currents and voltages measured are all AC, the phase differences between these values reveal the stability of the power flow in the transmission system. Phase differences were not a problem to measure within one substation when the measurements were hardwired and continuous. However, the values sent to the control center were limited to current and voltage magnitudes, as there was no way to measure phase differences between values at widely separated substations. This situation has recently changed because of the availability of GPS signals, which can provide an absolute time reference to all substations on the continent. Thus both magnitudes and phase angles of AC currents and voltages can be measured and stored today, although the number of measurements taken that incorporate phase (phasor measurements) in the first place is still very low. However, the availability of digitized phasor measuring at high sampling rates raises the possibility of many new and fast control applications not previously available.

The modern transmission system should have all of its high-voltage sub stations equipped with measurement systems that will be sampling critical data at rates of 30 to 120 times per second (and even faster for localized applications) with an absolute GPS time reference, allowing a more complete picture to be created of the current real-time state stability of the system. Although the hardware costs of these measurement units themselves are modest, they have to be retrofitted into the thousands of existing substations at significant cost. In this regard, developing countries such as China have an advantage over more established industrialized countries. They are able to leapfrog directly to the latest technologies for substation automation as they expand their electric grids.

On the distribution side, there are about four times as many lower-voltage substations as there are transmission substations. Distribution systems can use measurement instrumentation with slower sampling rates than those needed for transmission systems, but the flood of data requires high-bandwidth communication to use these data for control. Also, synchronizing measurements by using GPS at the low-voltage substations is not yet considered cost-effective. With regard to end users, there is a move toward replacing the existing kilowatt-hour meters for billing with intelligent (i.e., microprocessor-based) meters that can provide the customer with new buying options, such as time-of-day pricing. These meters can also bring control signals from the power company directly into appliances and other equipment on the customer side.

The ubiquity of more and faster measurements throughout the T&D system raises the issue of how to handle this proliferation of measurement data. Certainly they can be stored at the substations where they are collected and then used for various local engineering analyses as needed. However, a higher value of these real-time measurements is in helping to monitor and control the overall T&D system more efficiently and reliably. Such applications require the development of real-time data-handling software that can collect and move these data where they are needed.

Integrated Communications

The Eastern Interconnection has approximately 10,000 high-voltage (above 100 kV) substations, overseen by about 100 control centers. If fully instrumented, each substation could have about 100 measurement points (currents, voltages, powers, switch statuses), each of which may be sampled about 100 times per second. This arrangement would require each control center to process about a million data points per second. In addition, the center should be aware of what is going on in the neighboring parts of the T&D system and perhaps in the whole interconnection.

The Eastern Interconnection also has about 10 second-level control centers, known as reliability coordinators, that supervise larger areas of the T&D systems; each of these facilities has to process data at rates that are an order of magnitude higher than those of the substations. But these data rates cannot be handled by the communication system used today between control centers and substations.

A basic problem is that the existing communication channels between high-voltage substations and the control center, many dating from the 1960s, are slow. They are being replaced with high-bandwidth optical fiber. But even with the high bandwidth, the present architecture—wherein all data from substation RTUs are collected at the control center's SCADA—cannot handle the expected proliferation of real-time measurement data. Moreover, it does not make sense to centralize this large amount of data. Automatic controllers need not be physically located in one place either but can be sited conveniently according to their input sources and output destinations.

The actual data needs for particular applications to monitor and control the transmission network will vary widely. For example, there will always be control centers where human operators are monitoring a region of the system. The number of measurement data points needed at such a control center will be very large, but the sampling of the data can be as slow as once a second, as the human eye cannot follow much faster changes. However, the processing of these data for checking limits, warning, predicting, and visualization at the control center will be very large. And while an automatic control such as a special protection scheme for islanding a portion of the electric T&D system to keep a disturbance from cascading will require only a few measurements, they involve very high sampling rates and speeds.

On the distribution side, the communication needs are localized to neighborhoods, but the sheer number of substations, feeders, and customers requires low-cost alternatives such as radio-frequency and power-line carrier systems. Smart meters, time-varying rates, the handling of customer generation, demand-side management, and other such applications require ubiquitous communications.

Thus the communication system for both the transmission and the distribution system will need to be able to handle a wide range of speed and quantity requirements. Although such systems exist today—for example, cellular telephone networks—the communication needs of the power grid are unique; its software will thus have to be custom designed and developed. The cost of this communication infrastructure, which can begin to be deployed by 2020, is partly for the physical fiber-optic cables and switching computers, but mostly for the software.

Costs of Modernizing an Electric T&D System

The AEF Committee's cost estimates are based on a study published by the Electric Power Research Institute (EPRI) in 2004 (EPRI, 2004). EPRI's projected costs are summarized in Tables 9.A.1 and 9.A.2.[5] The committee modified EPRI's estimates to reflect its conclusion that superconducting cables, which account for $30 billion of the total in Table 9.A.1, are unlikely to be deployed during the next 20 years. If indeed they are not available, the costs of alternative technologies are likely to be higher than that amount and/or the benefits of modernizing the grid could be lower.

The committee also considered the investment that would be required to meet load growth and replace aging equipment. The annual level of investment in transmission over the 20 years prior to 1985 averaged around $5 billion per year, but from 1985 to 1999 only about $3 billion per year was invested. That $30 billion shortfall meant that the transmission system failed to keep pace with load growth. EPRI assumes that load growth will continue in the future, as it has in recent decades, and that an investment of $5 billion per year continues to be needed to meet it. In addition, $1.5 billion per year for 20 years will be required to make up for the 1985–1999 investment shortfall. Thus EPRI estimates that $6.5 billion (in 2002 dollars) will annually be needed for transmission systems over the next 20 years simply to meet load growth and to correct deficiencies in the current system, in addition to implementing advanced technologies to modernize the transmission system.

---

[5]EPRI's cost estimates are in 2002 dollars.

TABLE 9.A.1 EPRI Cost Estimates for Modernizing the Transmission System

Technology Category	Cost (billion 2002$)
Communications and sensors	4
Hardware improvements to substations (includes transformers and other substation equipment)	5
Substation automation	10
Other equipment (power electronics, storage, HV lines and equipment, superconducting lines)	55
Emergency operation and restoration tools and equipment	12
IDST[a] software	3
Dynamic thermal circuit rating	1
Predictive maintenance	20
Total	110

[a] IDST = Improved decision-support technology.
Source: EPRI, 2004.

TABLE 9.A.2 EPRI's Cost Estimates for Modernizing Distribution Substations and Feeders

	Component Cost per Substation Feeder (2002$)	Individual Cost per Substation Feeder (billion 2002$)	Number to Be Upgraded	Total Cost (billion 2002$)
Upgrading distribution substations		600,000	40,000	24
Communications	75,000			
Hardware improvements	350,000			
Sensors and monitoring	75,000			
Advanced controls and diagnostics	100,000			
Upgrading distribution feeder circuits		540,000	320,000	173
Communications	60,000			
Hardware improvements	170,000			
Sensors and monitoring	100,000			
Advanced controls and diagnostics	210,000			
Integrating consumer systems with the grid				62
Total				259

Source: EPRI, 2004.

As shown in Table 9.A.1, EPRI projects the cost to modernize the transmission system to be $110 billion over 20 years, or approximately $5.5 billion per year. Similarly, EPRI estimated the expenditures for the distribution system over 20 years to be $340 billion to meet load growth, $6 billion to correct deficiencies, and $259 billion to modernize the distribution system. Table 9.A.2 summarizes the costs to modernize distribution.

Summing the T&D system expenditures needed to meet load growth and to correct deficiencies with the expenditure needed for the modern T&D system is likely to overestimate the total investment needed. When new lines are built (or rebuilt) to meet load growth, the additional investment to install modern technologies is less significant. In addition, technologies can meet multiple purposes. For example, dynamic thermal circuit rating can help to meet load growth by increasing the capacity of existing lines, but this is also an important part of a modern transmission system. Correcting for such overlaps (synergies), EPRI's estimate of the total investment needed in the T&D system is shown in Table 9.A.3.

It should be noted that in order to achieve the full benefits of synergies on the transmission side, equipment throughout the system would need to be deployed in an integrated way. This is unlikely to occur until after 2020. EPRI estimated the synergies for T&D to be $72 billion and $132 billion, respectively, over the 20-year time horizon of their study. The AEF Committee (as previously stated) dropped $30 billion from the $110 billion to modernize the transmission system (by eliminating superconducting cables), which also required dropping $30 billion from the transmission synergies. The net result was that the committee estimated $80 billion to modernize the transmission system, with synergies of $42 billion when incorporating the expenditures to meet load growth and to correct deficiencies. The elimination of superconducting cables was assumed to negate an equal benefit (synergies) in meeting load growth and correcting deficiencies. These details are shown in Table 9.A.3.

In the committee's analysis, EPRI's cost estimates were escalated to 2007 dollars. The committee accounted for recent real escalation in materials and construction costs by using the national average T&D indexes.[6] In 2007 dollars, the investment needed in the T&D systems over the next 20 years will be about $225 billion for transmission and $640 billion for distribution. These estimates

---

[6]The national average transmission index increased by about 33 percent between 2002 and 2007. The national average distribution cost index has increased by about 40 percent during that same period (Brattle Group, 2007).

TABLE 9.A.3 Costs to Implement Modern T&D Systems

	EPRI[a]		Brattle Group[b]		AEF Committee Adjusted	
	Transmission (billion 2002$)	Distribution (billion 2002$)	Transmission (billion 2007$)	Distribution (billion 2007$)	Transmission (billion 2007$)	Distribution (billion 2007$)
Investment to meet load growth	100	330				
Investment to correct deficiencies	30	6	233	675	175[c]	470[c]
Modern T&D systems	80	259	N/A[d]	N/A[d]	105[e]	365[e]
Total	210	595	233[d]	675[d]	280	835
Synergies	42	132	N/A	N/A	55	195
Total minus synergies	168	463	N/A	N/A	225	640

[a]EPRI, 2004.

[b]Brattle Group, 2008.

[c]EPRI's estimates, originally in 2002 dollars, were escalated to 2007 dollars for the committee's analysis. Recent real escalations in materials and construction costs were accounted for by using the national average T&D indexes (33 percent for transmission, 40 percent for distribution).

[d]Brattle Group numbers include investments needed for the business-as-usual case but do not identify costs of deploying the modern T&D systems.

[e]The $30 billion (in 2002 dollars) that EPRI estimated for investment in superconducting cables has been removed from the total investment needed for the transmission system. This quantity has also been removed from the synergies calculation.

include investments needed to meet load growth, to replace aging equipment, and, additionally, to implement modernization. Implementation of the modern T&D system alone makes up a small portion of this total, as shown in Table 9.A.3: $50 billion for transmission and $170 billion for distribution.[7]

The committee assumed that 40 percent of the transmission improvements involved in implementing the modern grid, meeting load growth, and correcting deficiencies would be made before 2020, while the remaining 60 percent would need to be implemented between 2020 and 2030. Thus an investment of $9 billion per year would be needed in the transmission system from 2010 to 2020

[7]The $50 billion for transportation and $170 billion for distribution are the total costs for a modern T&D system, less the listed "synergies."

($2 billion per year for modernizing the grid). From 2020 to 2030, approximately $13.5 billion per year would be needed ($3 billion for the modernization alone).

On the distribution side, as with the transmission system, the committee has assumed that 40 percent of the improvements would be made by 2020 and the remaining 60 percent from 2020 to 2030. Thus an investment of $26 billion per year would be needed for the distribution system from 2010 to 2020 ($7 billion per year for modernization). From 2020 to 2030, approximately $38 billion per year would be needed ($10 billion for modernization), and an investment of $32 billion per year would be needed for the distribution system over the 20 years beyond 2030. Such an investment would be more than returned in the form of benefits from the improved system.

References for Annex 9.A

Bjelovuk, G. 2008. Presentation to the AEF Committee, Washington, D.C., February 21.

Brattle Group. 2007. Rising Utility Construction Costs: Sources and Impacts. Prepared by M.W. Chupka and G. Basheda for the Edison Foundation. September. Available at www.edisonfoundation.net/Rising_Utility_Construction_Costs.pdf. Accessed July 2009.

Brattle Group. 2008. Transforming America's Power Industry: The Investment Challenge 2010-2030. Prepared by M.W. Chupka, R. Earle, P. Fox-Penner, and R. Hledik for the Edison Foundation. Available at www.edisonfoundation.net/Transforming_Americas_Power_Industry.pdf. Accessed July 2009.

CECA (Consumer Energy Council of America). 2003. Positioning the Consumer for the Future: A Roadmap for an Optimal Electric Power System. Washington, D.C. Available at www.cecarf.org/publications/RestExecSummary.pdf. Accessed July 2009.

DOE (U.S. Department of Energy). 2007. 10 CFR 431. Energy Conservation Program for Commercial Equipment: Distribution Transformers Energy Conservation Standards, Final Rule. Federal Register 72(197), October 12. Available at www1.eere.energy.gov/buildings/appliance-standards/commercial/pdfs/distribution-transformers_fr_101207.pdf. Accessed July 2009.

EPRI (Electric Power Research Institute). 2004. Power Delivery System of the Future: A Preliminary Study of Costs and Benefits. Palo Alto, Calif.

EPRI. 2008. Compressed Air Energy Storage Scoping Study for California. Prepared for the California Energy Commission. CEC-500-2008-069. Available at www.energy.ca.gov/2008publications/CEC-500-2008-069/CEC-500-2008-069.pdf. Accessed July 2009.

EWIS (European Wind Integration Study). 2007. Towards a Successful Integration of Wind Power into European Electricity Grids. European Transmission System Operators. Brussels, Belgium.

McDonald, John (General Electric Co.). 2008. Discussion of electric T&D technologies and costs. Presentation to the T&D Subgroup of the AEF Committee, April 24.

NETL (National Energy Technology Laboratory). 2007a. A Systems View of the Modern Grid. Prepared for the U.S. Department of Energy. Available at www.netl.doe.gov/moderngrid/resources.html. Accessed July 2009.

NETL. 2007b. A Systems View of the Modern Grid. Appendix A1: Self-Heals. Prepared for U.S. DOE. Available at www.netl.doe.gov/moderngrid/resources.html. Accessed July 2009.

NETL. 2007c. A Systems View of the Modern Grid. Appendix A2: Motivates and Includes the Consumer. Prepared for U.S. DOE. Available at www.netl.doe.gov/moderngrid/resources.html. Accessed July 2009.

NETL. 2007d. A Systems View of the Modern Grid. Appendix A6: Enables Markets. Prepared for U.S. DOE. Available at www.netl.doe.gov/moderngrid/resources.html. Accessed July 2009.

NETL. 2007e. A Vision for the Modern Grid. Available at www.netl.doe.gov/moderngrid/docs/A%20Vision%20for%20the%20Modern%20Grid_Final_v1_0.pdf.resources.html. Accessed July 2009.

NRC (National Research Council). 2002. Making the Nation Safer: The Role of Science and Technology in Countering Terrorism. Washington, D.C.: The National Academies Press.

PNNL (Pacific Northwest National Laboratory). 2007. Testbed Demonstration Projects. Part II: Grid Friendly Appliance Project. October. Available at www.gridwise.pnl.gov/docs/gfa_project_final_report_pnnl17079.pdf. Accessed July 2009.

APPENDIXES

A Committee and Staff Biographies

COMMITTEE

HAROLD T. SHAPIRO (*Chair*), a member of the Institute of Medicine, is president emeritus of Princeton University and a professor of economics and public affairs at the Woodrow Wilson School. He served as president of the University of Michigan from 1980 to 1988. Dr. Shapiro's expertise is in the intersection of public policy, science policy, and bioethics. Widely recognized for his astute judgment in policy situations, he has chaired the National Bioethics Advisory Committee under President Bill Clinton and served as vice chair of the President's Council of Advisors on Science and Technology under President George H.W. Bush. Other distinctions include his chairing of the Association of American Universities and service on the board of directors of the National Bureau of Economic Research, Inc., and the board of trustees of the Universities Research Association, Inc. He has also served on numerous National Research Council committees, including the Committee on the Organizational Structure of the National Institutes of Health. In 2006, the American Association for the Advancement of Science awarded Dr. Shapiro the William Carey Medal for Lifetime Achievement in Science Policy. In 2008, he was awarded the Clark Kerr Medal for lifetime achievement in higher education. He received a bachelor's degree from McGill University in 1956 and a Ph.D. in economics from Princeton in 1964.

MARK S. WRIGHTON (*Vice Chair*) is chancellor and professor of chemistry at Washington University in St. Louis. Prior to moving there in 1995, he had been a faculty member at the Massachusetts Institute of Technology since 1972. Dr. Wrighton was head of the Department of Chemistry at MIT from 1987 until

1990, when he was appointed provost. He was elected a fellow of the American Academy of Arts and Sciences in 1988 and of the American Association for the Advancement of Science in 1986. In 2001, he was elected to membership in the American Philosophical Society. From 2000 to 2006, Dr. Wrighton was a presidential appointee to the National Science Board (NSB), which serves as a science policy advisor to the president and Congress and is the primary advisory board of the National Science Foundation. While serving on the NSB, he chaired the Audit and Oversight Committee. Dr. Wrighton earned a B.S. from Florida State University in 1969 and a Ph.D. in chemistry from the California Institute of Technology in 1972.

JOHN F. AHEARNE, a member of the National Academy of Engineering, is Executive Director Emeritus of Sigma Xi, The Scientific Research Society; an adjunct scholar at Resources for the Future; and an adjunct professor of civil and environmental engineering at Duke University. His professional interests are reactor safety, energy issues, resource allocation, and public policy management. He has served as commissioner and chair of the U.S. Nuclear Regulatory Commission, system analyst for the White House Energy Office, deputy assistant secretary of energy, and principal deputy assistant secretary of defense. Dr. Ahearne currently is vice chair of the Department of Energy's Nuclear Energy Advisory Committee. He is a fellow of the American Physical Society, the Society for Risk Analysis, the American Association for the Advancement of Science, and the American Academy of Arts and Sciences; he is a member the American Nuclear Society; and he has been active in several National Research Council committees that examined issues in risk assessment. Dr. Ahearne received a Ph.D. in physics from Princeton University.

ALLEN J. BARD, a member of the National Academy of Sciences, is a professor of chemistry and biochemistry and holds the Norman Hackerman/Welch Regents Chair in Chemistry at the University of Texas at Austin. He has published widely and is the winner of numerous honors and awards, including the Priestley Medal and the Welch and Wolf Prizes. Dr. Bard was president of the International Union of Pure and Applied Chemists (IUPAC) and editor in chief of the *Journal of the American Chemical Society* from 1982 to 2001. He has served on the National Research Council's Energy Engineering Board (EEB), been chair of the Board on Chemical Sciences and Technology, chaired the EEB Committee on Potential Applications of Concentrated Solar Photons, and served as president of the U.S.

National Committee for IUPAC. His research interests include electro-organic chemistry, photoelectrochemistry, electrogenerated chemiluminescence, electroanalytical chemistry, and fuel cells. His policy interests include issues related to electrochemical and renewable-energy sources. Dr. Bard received a Ph.D. in chemistry from Harvard University.

JAN BEYEA, chief scientist of Consulting in the Public Interest, consults on nuclear physics and other energy/environmental topics for numerous local, national, and international organizations, including the National Audubon Society. He has served as chief scientist and vice president of the National Audubon Society and has held positions at Holy Cross College, Columbia University, and Princeton University's Center for Energy and Environmental Studies. Dr. Beyea has been a member of numerous advisory committees and panels, including the National Research Council's Board on Energy and Environmental Systems, Energy Engineering Board, Committee on Alternative Energy R&D Strategies, and Committee to Review DOE's Fine Particulates Research Plan. He has also served on the Secretary of Energy Advisory Board's Task Force on Economic Modeling, been a member of the policy committee of the Recycling Advisory Council, and advised various studies of the Office of Technology Assessment. Dr. Beyea has expertise in energy technologies and associated environmental and health concerns, and he has written numerous articles on energy and the environment. He received a B.A. from Amherst College and a Ph.D. in physics from Columbia University.

WILLIAM F. BRINKMAN is vice president of physical sciences research for Lucent Technologies, formerly AT&T Bell Laboratories; he held the same position at AT&T Bell Laboratories. He was vice president of the Sandia National Laboratories in 1984–1987, director of the Chemical Physics Research Laboratory in 1981–1984, head of the Infrared Physics and Electronics Research Department of Bell Laboratories in 1972–1974, and a resident fellow at Oxford University in 1965–1966. Dr. Brinkman received his B.S. (1960), M.S. (1962), and Ph.D. (1965) in physics from the University of Missouri at Columbia. He received an honorary D.H.L. from the same institution in 1987. He is a member of the National Academy of Sciences.

DOUGLAS M. CHAPIN, a member of the National Academy of Engineering (NAE), until recently was principal officer and director of MPR Associates, Inc., in Alexandria, Virginia. He has extensive experience in electrical, chemical, and nuclear engineering, with particular application to nuclear and conventional power plants. He has worked in areas such as instrumentation and control systems, nuclear fuels, fluid mechanics, heat transfer, pumps, advanced analysis methods, test-facility design, and electrical systems and components. Dr. Chapin has participated in projects such as the Japan/Germany/United States research program on loss-of-coolant accidents; served as project leader for the design, construction, and testing of the loss of fluid test facility; been a member of the Electric Power Research Institute's (EPRI's) Utility Review Committee on Advanced Reactor Designs; and worked with the Utility/EPRI Advanced Light Water Reactor Program that defines utility requirements for future nuclear power plants. He was chair of the National Research Council's Committee on Application of Digital Instrumentation and Control Technology to Nuclear Power Plant Operations and Safety. He is currently a member of the NRC's Committee on Review of DOE's Nuclear Energy R&D Program, chair of the NRC's Board on Energy and Environmental Systems, and a member of the NAE's Committee on Membership. He formerly served as a member of the NAE's Electric Power/Energy Systems Engineering Peer Committee. He is also a fellow of the American Nuclear Society. Dr. Chapin has a B.S. in electrical engineering from Duke University, an M.S. in applied science from George Washington University, and a Ph.D. in nuclear studies in chemical engineering from Princeton University.

STEVEN CHU,[1] a recipient of the Nobel Prize for Physics (1997), was appointed by President Obama as Secretary of Energy and sworn into office on January 21, 2009. Dr. Chu has devoted his recent scientific career to the search for new solutions to our energy challenges and to stopping global climate change—a mission he continues with even greater urgency as Secretary of Energy. He is charged with helping to implement President Obama's ambitious agenda to invest in alternative and renewable energy, end U.S. addiction to foreign oil, address the global climate crisis, and create millions of new jobs. Prior to his appointment, Dr. Chu was director of the U.S. Department of Energy's Lawrence Berkeley National Laboratory and a professor of physics and professor of molecular and cell biology at the University of California, Berkeley. He has successfully applied the techniques he developed in atomic physics to molecular biology and, motivated by his deep interest in climate change, has in recent years led the Lawrence Berkeley National Lab in pursuit of new alternative and renewable energies. Previously, he held posi-

---

[1]Dr. Chu resigned from the committee on January 21, 2009.

tions at Stanford University and AT&T Bell Laboratories. Dr. Chu's research in atomic physics, quantum electronics, polymers, and biophysics includes tests of fundamental theories in physics, the development of methods to laser-cool and trap atoms, atom interferometry, and the manipulation and study of polymers and biological systems at the single-molecule level. While at Stanford, he helped start Bio-X, a multidisciplinary initiative that brings together the physical and biological sciences with engineering and medicine. Dr. Chu is a member of the National Academy of Sciences, the American Philosophical Society, the Chinese Academy of Sciences, Academica Sinica, the Korean Academy of Sciences and Technology, and numerous other professional and civic organizations. He holds an A.B. in mathematics and a B.S. degree in physics from the University of Rochester, a Ph.D. in physics from the University of California, Berkeley, and honorary degrees from 10 universities.

CHRISTINE A. EHLIG-ECONOMIDES, a member of the National Academy of Engineering, is a professor in the Harold Vance Department of Petroleum Engineering at Texas A&M University and holder of the Albert B. Stevens Chair in Petroleum Engineering. Before returning to academia, she worked for Schlumberger for 20 years. Dr. Ehlig-Economides is a distinguished member of the Society of Petroleum Engineers (SPE) and has held a variety of leadership positions in the society. In 1982 she was named the Alaska SPE Engineer of the Year and received the SPE Distinguished Achievement Award for Petroleum Engineering Faculty. She received the SPE Formation Evaluation Award in 1995 and the society's Lester C. Uren Award in 1997, and was named distinguished lecturer in 1997. Dr. Ehlig-Economides is also a member of Sigma Xi, The Scientific Research Society. She received a bachelor's degree in math-science from Rice University, a master's degree in chemical engineering from the University of Kansas, and a Ph.D. in petroleum engineering from Stanford University.

ROBERT W. FRI is a visiting scholar and senior fellow emeritus at Resources for the Future (RFF), where he served as president from 1986 to 1995. From 1996 to 2001, he was director of the National Museum of Natural History at the Smithsonian Institution. Before joining RFF, Mr. Fri served in both the public and the private sectors, specializing in energy and environmental issues. In 1971 he became the first deputy administrator of the U.S. Environmental Protection Agency. In 1975, President Ford appointed him deputy administrator of the U.S. Energy Research and Development Administration. He served as acting administrator of both agencies for extended periods. From 1978 to 1986, Mr. Fri headed his own company, Energy Transition Corporation. He began his career with McKinsey & Company, where he was elected a principal. A senior advisor to private, public, and nonprofit organizations, Mr. Fri is currently a member of the National Petroleum Council and of the Advisory Council of the Electric Power Research Institute. He is also vice chair of the National Research Council's (NRC's) Board on Energy and Environmental Systems. He has chaired several NRC committees, most recently the Committee on Review of DOE's Nuclear Energy R&D Program. Mr. Fri is a member of Phi Beta Kappa and Sigma Xi, The Scientific Research Society. He received a B.A. in physics from Rice University and an M.B.A. (with distinction) from Harvard University.

CHARLES H. GOODMAN has had a long career in electric utility research and development at Southern Company, primarily in establishing and improving coal-to-energy processes and in addressing the public policy issues associated with coal utilization. His contributions span heat transfer, emission controls, environmental science, and advanced generation technologies. Prior to retirement in 2007 he was the senior vice president for generation policy, with responsibilities that included serving as chair of the board for the FutureGen Industrial Alliance. Earlier, he was senior vice president for research and environmental policy—Southern Company's chief environmental officer. In that capacity he directed environmental research and development, environmental policy, and compliance-strategy efforts for Southern Company as it initiated cleaner and more efficient ways to meet the energy needs of its customers. Dr. Goodman served for many years on the Electric Power Research Institute's Research Advisory Committee and was chair of its Environment Sector Council. He is a member of the National Research Council's Board on Energy and Environmental Systems, the Energy and Environment Directorate Review Committee at Pacific Northwest National Laboratory, and the R&D Advisory Council for the Babcock and Wilcox Company. He has chaired the Environmental Staff Committee of the Business Roundtable, and he was a member of the U.S. Environmental Protection Agency's Clean Air Act Advisory Committee. His responsibilities included oversight of the Power Systems Development Facility—the United States' premier clean-coal-technology research center—in cooperation with the U.S. Department of Energy (DOE). In addition, he led the development and execution of four DOE Clean Coal Technology projects that provided new emission-control options, which have now been applied to the industry's conventional power plants. He is a life fellow in the American Society of

Mechanical Engineers. Dr. Goodman received an undergraduate degree from the University of Texas at Arlington and a master's degree and Ph.D. in mechanical engineering from Tulane University.

JOHN B. HEYWOOD, a member of the National Academy of Engineering, is Sun Jae Professor of Mechanical Engineering and director of the Sloan Automotive Laboratory at the Massachusetts Institute of Technology (MIT). Dr. Heywood's research has focused on engine combustion, pollutant formation, the operating emissions characteristics and fuel requirements of automotive and aircraft engines, and on reducing transportation's petroleum consumption and greenhouse gas emissions. He has served on a number of National Research Council committees, including the Committee on Review of the Research Program of the Partnership for a New Generation of Vehicles, and has consulted for many companies in the automotive and petroleum industries and for governmental organizations. Among the many awards he has received for his research contributions, Dr. Heywood was honored by the American Society of Mechanical Engineers, the British Institution of Mechanical Engineers, and the Society of Automotive Engineers. He has a Ph.D. in mechanical engineering from MIT, an Sc.D. from Cambridge University, and honorary doctorates from Chalmers University of Technology (Sweden) and City University (UK). He is a fellow of the American Academy of Arts and Sciences.

LESTER B. LAVE, a member of the Institute of Medicine, is the Harry B. and James H. Higgins Professor of Economics, a university professor, director of the Green Design Initiative, and codirector of the Electricity Industry Center at Carnegie Mellon University. Dr. Lave's teaching and research interests include applied economics, political economy, quantitative risk assessment, safety standards, modeling the effects of global climate change, public policy concerning greenhouse gas emissions, and issues surrounding the electric transmission and distribution system. A recipient of the Distinguished Achievement Award of the Society for Risk Analysis, he is a member of the National Research Council's Committee on Prospective Benefits of DOE's Energy Efficiency and Fossil Energy R&D Program Phase 2; he is also chair of the NRC's Panel on Benefits of Sequestration R&D. Dr. Lave has a B.S. in economics from Reed College and a Ph.D. in economics from Harvard University.

JAMES J. MARKOWSKY, a member of the National Academy of Engineering, is retired executive vice president of American Electric Power (AEP) Service Corporation, where he led the power-generation group and was responsible for providing overall administrative, operational, and technical direction to the AEP System's fossil and hydropower generating facilities, including fuel procurement and transportation, coal mining, planning, licensing, environmental engineering, design, construction, maintenance, and integrated operation of the fossil and hydro generation fleet. Dr. Markowsky served as chair of the National Research Council's Committee to Review DOE's Vision 21 R&D Program, Phase 1, and he was chair of the Committee on R&D Opportunities for Advanced Fossil-Fueled Energy Complexes. He was also a member of the NRC's Board on Energy and Environmental Systems and of its Energy Engineering Board. Dr. Markowsky received a B.S. in mechanical engineering from the Pratt Institute, master's degrees from Cornell University and the Massachusetts Institute of Technology, and a Ph.D. in mechanical engineering from Cornell University.

RICHARD A. MESERVE, a member of the National Academy of Engineering, is president of the Carnegie Institution for Science. He previously was chair of the U.S. Nuclear Regulatory Commission (USNRC; the federal agency with responsibility for ensuring public health and safety in the operation of nuclear power plants and usage of nuclear materials). He served as chair during the terms of Presidents Bill Clinton and George W. Bush and led the USNRC in responding to the terrorism threat after the 9/11 attacks. Before joining the USNRC, Dr. Meserve was a partner in the Washington, D.C., law firm of Covington & Burling, and he now serves as Senior of Counsel to the firm. Early in his career, he served as legal counsel to the president's science advisor and was law clerk to Justice Harry A. Blackmun of the U.S. Supreme Court and to Judge Benjamin Kaplan of the Massachusetts Supreme Judicial Court. Dr. Meserve has served on numerous legal and scientific committees over the years, including many established by the National Academy of Sciences and the National Academy of Engineering; currently he serves as chair of the Academies' Nuclear and Radiation Studies Board. He is chair of the International Nuclear Safety Group, which is chartered by the International Atomic Energy Agency, and serves as a member of the Harvard Board of Overseers. Among other affiliations, he is a member of the American Philosophical Society and a fellow of the American Academy of Arts and Sciences, the American Association for the Advancement of Science (AAAS), the American Physical Society, and Phi Beta Kappa. Dr. Meserve serves on the boards of directors of the

PG&E Corporation, Luminant Holding Co. LLC, the Universities Research Association, Inc., and the Council of the American Academy of Arts and Sciences. He has a bachelor's degree from Tufts University, a law degree from Harvard University, and a Ph.D. in applied physics from Stanford University.

WARREN F. MILLER, JR., a member of the National Academy of Engineering, is associate director of the Nuclear Security Science and Policy Institute at Texas A&M University. He has expertise in nuclear reactor analysis and theory, reactor design, radioactive waste management, transmutation of materials, and management of R&D programs. From 1974 to 2001 he held a number of positions at Los Alamos National Laboratory, including group leader for reactor and transport theory, deputy associate director for nuclear programs, associate laboratory director for energy programs, and deputy laboratory director for science and technology. Dr. Miller has held positions at the University of New Mexico, the University of Michigan, Howard University, the University of California, Berkeley, and Northwestern University. He is a fellow of the American Nuclear Society and a State of New Mexico Eminent Scholar (1989); he was honored as 2004 distinguished engineer by the National Society of Black Engineers. He has served on a variety of advisory groups and committees, including as vice chair of the National Research Council's Division of Earth and Life Sciences and as a member of the NRC's Committee on Long-Term Environmental Quality Research and Development. Dr. Miller was a member of the NRC's Nuclear and Radiation Studies Board and the NRC Committee on Review of DOE's Nuclear Energy R&D Program. He served on the U.S. Department of Energy's Nuclear Energy Research Advisory Committee from 1997 to 2006. He has a B.S. in engineering sciences from the United States Military Academy at West Point and M.S. and Ph.D. degrees in engineering sciences from Northwestern University.

FRANKLIN M. ("LYNN") ORR, JR., a member of the National Academy of Engineering, became director of the Precourt Institute for Energy at Stanford University upon its establishment in 2009. He served as director of Stanford's Global Climate and Energy Project from 2002 to 2008, was the Chester Naramore Dean of the university's School of Earth Sciences from 1994 to 2002, and has been a member of the faculty since 1985. Dr. Orr's research activities involve the flow of complex fluid mixtures in the porous rocks of Earth's crust; the design of gas-injection processes for enhanced oil recovery; and CO_2 storage in subsurface formations. He is a member of the board of directors of the Monterey Bay Aquarium

Research Institute and was a board member of the David and Lucile Packard Foundation from 1999 to 2008; he now chairs the foundation's Science Advisory Committee. Dr. Orr received a B.S. in chemical engineering from Stanford University and a Ph.D. in chemical engineering from the University of Minnesota.

LAWRENCE T. PAPAY, a member of the National Academy of Engineering, is currently a consultant with a variety of clients in electric power and other energy areas. His expertise and knowledge span a wide variety of electric system technologies, including production, transmission and distribution, utility management and systems, and end-use. He has served as senior vice president for the integrated solutions sector of Science Applications International Corporation and as senior vice president and general manager of Bechtel Technology and Consulting. Dr. Papay also held several positions at Southern California Edison, including senior vice president, vice president, general superintendent, and director of research and development, with responsibilities for bulk power generation, system planning, nuclear power, environmental operations, and development of the organization and plans for the company's R&D efforts. Among his other professional affiliations, past and present, are the Electric Power Research Institute's Research Advisory Committee; the Atomic Industrial Forum; the U.S. Department of Energy's Energy Research Advisory Board, Lab Operations Board, and Environmental Management Advisory Board; the Department of Homeland Security's Science and Technology Advisory Board; numerous National Academies' boards and committees, including the National Academy of Engineering's Board of Councillors; and the Renewable Energy Institute. Dr. Papay received a B.S. in physics from Fordham University and S.M. and Sc.D. degrees in nuclear engineering from the Massachusetts Institute of Technology.

ARISTIDES A.N. PATRINOS is president of Synthetic Genomics, Inc. (SGI), a privately held company founded in 2005 that is devoted to applying genomic-driven commercial solutions to global energy and environmental challenges. Prior to joining SGI, he was instrumental in advancing the scientific and policy framework underpinning key governmental energy and environmental initiatives while serving as director of the Office of Biological and Environmental Research in the Department of Energy's (DOE) Office of Science. Dr. Patrinos oversaw the department's research activities in human and microbial genome research, structural biology, nuclear medicine, and climate change. Previously Dr. Patrinos worked at several DOE National Laboratories and the University of Rochester. The recipient of numerous awards and honorary degrees, including three presidential-rank awards for meritorious and distinguished service and two Secretary of Energy gold medals, Dr. Patrinos is a fellow of the American Association for the Advancement of Science and the American Meteorological Society, and is a member of the American Society of Mechanical Engineers, the American Geophysical Union, and the Greek Technical Society. Dr. Patrinos received a diploma in mechanical and electrical engineering from the National Technical University of Athens and a Ph.D. in mechanical and astronautical sciences from Northwestern University.

MICHAEL P. RAMAGE, a member of the National Academy of Engineering, is a retired executive vice president of ExxonMobil Research and Engineering Company. Previously he was executive vice president, chief technology officer, and director of Mobil Oil Corporation. Dr. Ramage held a number of positions at Mobil, including research associate, manager of process research and development, general manager of exploration and producing research and technical services, vice president of engineering, and president of Mobil Technology Company. He has broad experience in many aspects of the petroleum and chemical industries. Dr. Ramage has served on a number of university visiting committees, was a director of the American Institute of Chemical Engineers, and now is a member of Secretary of Energy Chu's Hydrogen Technical Advisory Council. He is a member of several professional organizations and serves on the Energy Advisory Board of Purdue University. Dr. Ramage was a member of the National Academies' Government-University-Industry Research Roundtable. He chaired the National Research Council (NRC) committees responsible for the reports *The Hydrogen Economy: Opportunities, Costs, Barriers, and R&D Needs* and *Resource Requirements for a Hydrogen Economy*. He is currently chairing the NRC Panel on Alternative Liquid Transportation Fuels. Dr. Ramage has B.S., M.S., Ph.D., and H.D.R. degrees in chemical engineering from Purdue University.

MAXINE L. SAVITZ, vice president of the National Academy of Engineering, is a director of the Washington Advisory Group. A former deputy assistant secretary for conservation at the U.S. Department of Energy (DOE), she received the department's Outstanding Service Medal in 1981. Prior to her DOE service, she was a program manager for research applied to national needs at the National Science Foundation. Following her government service, Dr. Savitz held executive positions in the private sector—including president of the Lighting Research Institute, assistant to the vice president for engineering at the Garrett Corporation, and general manager of AlliedSignal Ceramic Components. She recently retired from the position of general manager for technology partnerships at Honeywell. Dr. Savitz serves on advisory bodies for Sandia National Laboratories and Pacific Northwest National Laboratory. She serves on the board of directors of the Draper Laboratory and the American Council for an Energy Efficient Economy. She was recently appointed to the President's Council of Advisors for Science and Technology. Dr. Savitz received a B.A. in chemistry from Bryn Mawr College and a Ph.D. in organic chemistry from the Massachusetts Institute of Technology.

ROBERT H. SOCOLOW is a professor of mechanical and aerospace engineering at Princeton University, where he has been a faculty member since 1971. He was previously an assistant professor of physics at Yale University. Dr. Socolow currently codirects Princeton's Carbon Mitigation Initiative, a multidisciplinary investigation of fossil fuels in a future carbon-constrained world. From 1979 to 1997, he directed Princeton's Center for Energy and Environmental Studies and contributed significantly to progress in energy efficiency technologies, policy, and applications. Dr. Socolow has served on many National Research Council boards and committees, including the Committee on R&D Opportunities for Advanced Fossil-Fueled Energy Complexes, the Committee on Review of DOE's Vision 21 R&D Program, and the Board on Energy and Environmental Systems. He is a fellow of the American Physical Society and the American Association for the Advancement of Science. Dr. Socolow has B.A., M.A., and Ph.D. degrees in physics from Harvard University.

JAMES L. SWEENEY, Stanford University, is director of Stanford University's Precourt Energy Efficiency Center, professor of management science and engineering, senior fellow of the Stanford Institute for Economic Policy Research, and senior fellow of the Hoover Institution. His professional activities focus on economic policy and analysis, particularly regarding energy, natural resources, and the environment. Dr. Sweeney served as chair of the Stanford Department of Engineering-Economic Systems, chair of the Department of Engineering-Economic Systems and Operations Research, director of the Energy Modeling Forum, chair of the Institute for Energy Studies, and director of the Center for Economic Policy Research. He was a founding member of the International Association for Energy Economics, served as director of the Office of Energy Systems Modeling and Forecasting of the U.S. Federal Energy Administration, has been a member of numerous committees of the National Research Council, and is a lifetime National Associate of the National Academies. Dr. Sweeney is a senior fellow of the U.S. Association for Energy Economics and a council member and senior fellow of the California Council on Science and Technology; he is also a member of the External Advisory Council of the National Renewable Energy Laboratory and a member of Governor Arnold Schwarzenegger's Council of Economic Advisors. He holds a B.S. in electrical engineering from the Massachusetts Institute of Technology and a Ph.D. in engineering-economic systems from Stanford University.

G. DAVID TILMAN, a member of the National Academy of Sciences, is Regents' Professor and McKnight Presidential Chair in Ecology at the University of Minnesota. His research explores how to meet human needs for energy, food, and ecosystem services sustainably. He is a member of the American Academy of Arts and Sciences, a J.S. Guggenheim Fellow, and a recipient of the Ecological Society of America's Cooper Award, the ESA's MacArthur Award, the Botanical Society of America's Centennial Award, and the Princeton Environmental Prize. He has written two books, edited three others, and published more than 200 scientific papers, including more than 30 in *Science*, *Nature*, and the *Proceedings of the National Academy of Sciences*. For the past 18 years, the Institute for Scientific Information has ranked him as the world's most-cited environmental scientist. In 2008, the emperor of Japan presented him with the International Prize for Biology.

C. MICHAEL WALTON, a member of the National Academy of Engineering, is a professor of civil engineering and holds the Ernest H. Cockrell Centennial Chair in Engineering at the University of Texas at Austin. In addition, he holds a joint academic appointment in the Lyndon B. Johnson School of Public Affairs. He is a past chair and member of the Transportation Research Board (TRB) Executive Committee. As the National Research Council chair of the TRB Division he serves as an ex-officio member of the Governing Board of the NRC. He is a past chair of the board of the American Road and Transportation Builders Association, past member of the Board of Governors of the Transportation and Development Institute of the American Society of Civil Engineers, and a founding member and past chair of the board of the Intelligent Transportation Society (ITS) of America. Dr. Walton has published widely and received numerous honors and awards for his research in the areas of ITS, freight transport, and transportation engineering, planning, policy, and economics. Dr. Walton has a B.S. from the Virginia Military Institute and M.S. and Ph.D. degrees from North Carolina State University, all in civil engineering.

STAFF

KEVIN D. CROWLEY (*Study Director*) is senior board director of the Nuclear and Radiation Studies Board, which advises the National Academies on the design and conduct of studies on radiation health effects, radioactive-waste management and environmental cleanup, and nuclear security and terrorism. The board also provides scientific support to the Radiation Effects Research Foundation in Hiroshima, Japan, a joint U.S.-Japanese scientific organization that investigates the health effects arising from exposures to ionizing radiation among World War II atomic-bombing survivors. Dr. Crowley's professional interests and activities focus on the safety, security, and technical efficacy of nuclear and radiation-based technologies. He has directed or codirected some 20 National Research Council (NRC) studies, including *Safety and Security of Commercial Spent Nuclear Fuel Storage* (2005); *Going the Distance: The Safe Transport of Spent Nuclear Fuel and High-Level Radioactive Waste in the United States* (2006); and *Medical Isotope Production without Highly Enriched Uranium* (2009). Before joining the NRC staff, Dr. Crowley held teaching/research positions at Miami University of Ohio, the University of Oklahoma, and the U.S. Geological Survey. He received his Ph.D. in geology from Princeton University.

PETER D. BLAIR is executive director of the Division on Engineering and Physical Sciences of the National Academies and is responsible for overall management of the America's Energy Future portfolio of studies. At the time of his appointment in January 2001 he was executive director of Sigma Xi, The Scientific Research Society. From 1983 to 1996, he served in several capacities at the Congressional Office of Technology Assessment, concluding as assistant director of the agency and director of the Division of Industry, Commerce and International Security. Dr. Blair has served on the faculties of the University of Pennsylvania (1976–1996) and the University of North Carolina at Chapel Hill (1997–2001). He was cofounder in 1978 and principal of Technecon Research, Inc., an engineering-economic consulting and power generation projects firm in Philadelphia, Pennsylvania, acquired by the Reading Energy Corporation in 1985. Dr. Blair holds a B.S. in engineering from Swarthmore College (1973), an M.S.E. in systems engineering (1974) and M.S. (1975) and Ph.D. (1976) degrees in energy management and policy from the University of Pennsylvania. He is the author or coauthor of three books and more than 100 technical articles in the areas of energy and environmental policy, electric power systems, operations research, regional science, and economic systems.

SARAH C. CASE joined the National Research Council in December 2007 and is currently a program officer in the Nuclear and Radiation Studies Board. In that capacity she has worked primarily with the study committee on America's Energy Future, facilitating the committee's work on nuclear energy and the electric transmission and distribution systems. Before arriving at the NRC, Dr. Case conducted research in condensed-matter physics, studying the collective behavior of ordinary materials such as fluids and granular material at the point of transition between states. Her research focused primarily on the physics of fluid topological transitions (such as droplet coalescence and drop snap-off). She has also conducted research in experimental high-energy particle physics, primarily in "beyond the Standard Model" particle searches and in neutrino physics. She was an NRC Christine Mirzayan Science and Technology Policy Fellow in the fall of 2007. Dr. Case received an A.B. in physics from Columbia University and M.S. and Ph.D. degrees in physics from the University of Chicago.

ALAN T. CRANE is a senior program officer at the National Research Council. He has directed projects that analyzed fuel-cell vehicle development, electric power systems, alternatives to the Indian Point nuclear power station, and fuel-economy standards for cars and light trucks. He has also contributed to other projects on energy R&D and on countering terrorism against energy systems and urban infrastructure. Prior to his current position, Mr. Crane was an independent consultant on energy, environmental, and technology issues for government and private-sector clients. He was also a senior associate at the Congressional Office of Technology Assessment, where he directed projects on energy policy and international technology transfer. During sabbaticals from OTA he served as director of energy and environmental studies at the European Institute of Technology, visiting researcher at the Oak Ridge National Laboratory, and visiting professor at Dartmouth College. His earlier work included engineering and managerial positions in the nuclear power industry. Mr. Crane has a B.S. from Haverford College and an M.S.M.E. from New York University.

GREG EYRING received a B.S. in chemistry from Stanford University in 1976 and a Ph.D. in chemistry from the University of California, Berkeley, in 1981. After doing 3 years of postdoctoral research at Stanford University, he joined the congressional Office of Technology Assessment (OTA), where he directed several studies related to advanced materials and environmental aspects of the use of materials. After the demise of OTA in 1995, Dr. Eyring worked as an independent consultant before joining the National Research Council in 2006. His work at the NRC has included studies on chemical weapons, explosives, and military- and intelligence-related technologies.

K. JOHN HOLMES has served as a study director at the National Research Council for the past 10 years. In this position he has been responsible for directing committee studies on contentious environmental and energy issues, particularly those related to motor vehicles, energy, air quality, and the quantitative analysis of policy impacts. Dr. Holmes is currently a senior staff officer at the Board on Energy and Environmental Systems, where he is responsible for the NRC Committee on Fuel Economy Technologies for Light Duty Vehicles. Dr. Holmes received his B.S. from Indiana University, an M.S.E. from the University of Washington, and a Ph.D. from the Johns Hopkins University. His doctoral dissertation focused on integrated assessment modeling of climate change and other environmental system impacts.

THOMAS R. MENZIES is a senior program officer in the Transportation Research Board's (TRB) policy studies unit. In this capacity, he manages studies on transportation-related programs and policies called for by the U.S. Congress and sponsored by the U.S. Department of Transportation, U.S. Department of Homeland Security, National Aeronautics and Space Administration, and other federal agencies. Since joining TRB in 1987 he has staffed more than two dozen projects examining the economic, safety, security, environmental, and energy performance of the aviation, rail, maritime, transit, trucking, and automotive sectors. Reports from relevant studies of energy performance include *Tires and Passenger Vehicle Fuel Economy*, *Toward a Sustainable Future: Addressing the Long-term Effects of Motor Vehicle Transportation on Climate and Ecology*, and an ongoing assessment of policy options for reducing energy use and greenhouse gas emissions from transportation. He has published numerous articles in technical journals and has made presentations on study results, and he serves on the editorial board of *TRNews*. He earned a bachelor's degree in economics from Colby College and an M.A. in public policy and public finance from the University of Maryland.

EVONNE P.Y. TANG is a senior program officer at the National Research Council. She has served as study director for multiple projects, on subjects ranging from science policy to research and development, since she joined the National Academies in 2002. Dr. Tang's areas of expertise include ecology, genomics, and biofuels. Among her recently completed projects are the studies *Liquid Transportation Fuels from Coal and Biomass* (2009), *Achievements of the National Plant Genome Initiative and New Horizons in Plant Biology* (2008), *Protecting Building Occupants and Operations from Biological and Chemical Airborne Threats* (2007), and *Status of Pollinators in North America* (2007). Dr. Tang received a B.Sc. from the University of Ottawa, an M.Sc. from McGill University, and a Ph.D. from Laval University, Canada. Her doctoral dissertation focused on the ecophysiology of cyanobacteria and the use of cyanobacteria in tertiary wastewater-treatment systems. After completion of her doctorate, she received postdoctoral fellowships from the Smithsonian Institution, the National Research Council Canada, and the Quebec Ministry of Education.

MADELINE G. WOODRUFF, a senior program officer at the National Research Council's Board on Energy and Environmental Systems, is responsible for the AEF Panel on Energy Efficiency Technologies. Prior to joining the NRC she spent 8 years as a senior analyst and project manager at the International Energy Agency in Paris, France, focusing on evaluation of energy technology R&D policy and programs, both domestic and international, and assessment of the potential for energy technology to contribute to reducing greenhouse gas emissions. Earlier, Ms. Woodruff was a senior analyst at the Pacific Northwest National Laboratory, where she managed or contributed to projects on nuclear energy regulatory policy, storage of plutonium recovered from retired nuclear weapons, regulation of mixed radioactive and chemical wastes, industrial energy efficiency, and energy technology R&D. Ms. Woodruff received an M.S. in nuclear engineering and an M.S. in technology and policy from the Massachusetts Institute of Technology, where she was a National Academy of Sciences Graduate Fellow.

JAMES J. ZUCCHETTO is director of the Board on Energy and Environmental Systems, National Research Council. Since joining the NRC in 1985, Dr. Zucchetto has been involved in a variety of multidisciplinary studies related to energy technologies, engineering, the environment, research and development programs, and public policy. In his work at the NRC, he has contributed to numerous studies and reports with an important influence on federal programs and policies, including on technologies for improving the fuel economy of light-duty and heavy-duty vehicles and for producing liquid fuels from a variety of fossil and nonfossil resources; hydrogen production; fuel-cell vehicles; and electricity generation, transmission and distribution, as well as related policy analyses and issues. Prior to joining the NRC, he was on the faculty of Arts and Sciences, Department of Regional Science, University of Pennsylvania; a guest researcher at the Institute of Marine Ecology and Zoologiska Institutionen, University of Stockholm; an associate in engineering, Department of Environmental Engineering Sciences, University of Florida; and a member of the technical staff, Bell Telephone Laboratories. He serves on the editorial advisory board of the International Journal of Ecological Modelling and Systems Ecology and is a former member of the editorial advisory board of Ecological Economics. In addition to work and research on energy technologies and associated environmental, economic, and policy implications since the early 1970s, he has also worked in the area of systems ecology and ecological modeling. He has published approximately 50 articles in refereed journals, books, and conference proceedings, two monographs, and one book. He has a Ph.D. in environmental engineering sciences from the University of Florida, an M.S.M.E. from New York University, and a B.S.M.E. from the Polytechnic Institute of Brooklyn (Polytechnic University).

Editorial Consultant

STEVEN J. MARCUS, an independent editor specializing in science, technology, and health policy, edited the America's Energy Future report. Prior to establishing his own practice in 2001, he was editor in chief of MIT's *Technology Review,* editor in chief of the National Academies' *Issues in Science and Technology*, executive editor of *High Technology*, science/medicine editor of the *Minneapolis Star Tribune,* and technology reporter for the *New York Times.* Prior to becoming a journalist, Dr. Marcus worked as a systems engineer for the MITRE Corporation and as an environmental engineering consultant. Under a Fulbright Lecturer grant, he taught courses on environmental issues at the University of Paris. He holds a bachelor's degree in electrical engineering from the City College of New York and a Ph.D. in environmental sciences and engineering from Harvard University.

B

Meeting Participants

The following individuals provided information for this study through their participation in subgroup meetings of the America's Energy Future Committee and in the Summit on America's Energy Future (see Appendix C).

ALTERNATIVE TRANSPORTATION FUELS SUBGROUP MEETINGS

Rich Bain, National Renewable Energy Laboratory (NREL)
Bruce Dale, Michigan State University
Otto Doering, Purdue University
Jonathan Foley, University of Wisconsin, Madison
Amory Lovins, Rocky Mountain Institute
Maggie Mann, NREL
James Newcomb, Rocky Mountain Institute
Robert Perlack, Oak Ridge National Laboratory
Sam Tabak, Exxon Mobil
Samuel Tam, Headwaters
Theodore Wegner, U.S. Department of Agriculture Forest Service
Robert Williams, Princeton University

ELECTRICITY TRANSMISSION AND DISTRIBUTION SUBGROUP MEETINGS

David Andrejcak, Federal Energy Regulatory Commission (FERC)
George Bjelovuk, American Electric Power
Paul Centolella, Ohio Public Utility Commissioner
Joe Eto, Lawrence Berkeley National Laboratory (LBNL)
Gerald FitzPatrick, National Institute of Standards and Technology
Craig Glazer, PJM Interconnection
Chris Gomperts, Siemens
Patricia Hoffman, U.S. Department of Energy (DOE)
Lawrence Jones, Areva Transmission and Distribution
Stephen Lee, Electric Power Research Institute (EPRI)
Ron Litzinger, Southern California Edison
Richard Lordan, EPRI
John McDonald, General Electric (GE)
Ken Nemeth, Southern States Energy Board
Dave Nevius, North American Electrical Reliability Council
Dave Owens, Edison Electric Institute
Steve Pullins, Horizon Energy Group
Edmund O. Schweitzer III, Schweitzer Engineering Laboratories, Inc.
Le Tang, ABB, Inc.

ENERGY EFFICIENCY SUBGROUP MEETINGS

Jonathan Creyts, McKinsey and Company
John Heywood, Massachusetts Institute of Technology (MIT)
Kathleen Hogan, U.S. Environmental Protection Agency
Revis W. James, EPRI
Douglas Kaempf, DOE
Mark Levine, LBNL
Fred Moore, The Dow Chemical Company
Steve Nadel, American Council for an Energy Efficient Economy
Jaana Remes, McKinsey Global Institute
David Rodgers, DOE
Lee Schipper, World Resources Institute Center for Sustainable Transport
Steven Smith, Pacific Northwest National Laboratory (PNNL)

Appendix B   661

FOSSIL ENERGY SUBGROUP MEETINGS

Carl Bauer, National Energy Technology Laboratory
Jim Dooley, PNNL
Julio Friedmann, Lawrence Livermore National Laboratory
James Katzer, ExxonMobil Research and Engineering Company (retired)
Granger Morgan, Carnegie Mellon University
John Novak, EPRI
Scott Tinker, University of Texas, Austin

NUCLEAR ENERGY SUBGROUP MEETINGS

Jim Asselstine, Lehman Brothers
Ralph Bennett, Idaho National Laboratory (INL)
Tom Cochran, National Resources Defense Council
Philip Finck, INL
Jim Harding, Consultant
Adrian Heymer, Nuclear Energy Institute (NEI)
Valentin Ivanov, State Duma Energy Committee, Russia
Revis James, EPRI
Elizabeth King, NEI
Paul Lisowski, DOE
Arjun Makhijani, Institute for Energy and Environmental Research
Michael Mariotte, Nuclear Information and Resource Service
Ernest J. Moniz, MIT
Richard Myers, NEI
John Parsons, MIT
Per Peterson, University of California, Berkeley
Dennis Spurgeon, DOE
Gordon Thompson, Institute for Resource and Security Studies

RENEWABLE ENERGY SUBGROUP MEETINGS

Dan Arvizu, NREL
Alan Beamon, Energy Information Administration (EIA)

Jacques Beaudry-Losique, DOE
Peter Bierden, GE
J. Michael Canty, DOE
Steve Chalk, DOE
Craig Cornelius, DOE
Mike Grable, ERCOT
Imre Gyuk, DOE
Pat Hoffman, DOE
Christopher King, U.S. House of Representatives Science and Technology
    Committee Staff
Martha Krebs, California Energy Commission
Ben Kroposki, NREL
Steve Lindenberg, DOE
Ann Miles, FERC
JoAnn Milliken, DOE
Christopher Namovicz, EIA
Pedro Pizarro, Southern California Edison
Dan Rastler, EPRI
Adam Rosenberg, U.S. House of Representatives Science and Technology
    Committee Staff
J. Charles Smith, The Utility Wind Integration Group
Steven Smith, PNNL
Jeff Tester, MIT
Ryan Wiser, LBNL

SUMMIT ON AMERICA'S ENERGY FUTURE

Jeff Bingaman, U.S. Senate
Samuel W. Bodman, DOE
Jon Creyts, McKinsey and Company
Ged Davis, World Economic Forum
Jose Goldemberg, International Institute for Applied Systems Analysis
John P. Holdren, Harvard University
Reuben Jeffery III, U.S. Department of State
Amory Lovins, Rocky Mountain Institute
Robert Marlay, DOE

Ernest J. Moniz, MIT
Rod Nelson, National Petroleum Council
Raymond L. Orbach, DOE
Paul R. Portney, University of Arizona
Dan W. Reicher, Google.org
James R. Schlesinger, MITRE Corporation and Lehman Brothers
Steven R. Specker, EPRI
Charles M. Vest, National Academy of Engineering

America's Energy Future Project

In 2007, the National Academies initiated the America's Energy Future (AEF) project (Figure C.1) to facilitate a productive national policy debate about the nation's energy future. The Phase I study, headed by the Committee on America's Energy Future and supported by the three separately constituted panels whose members are listed in this appendix, will serve as the foundation for a Phase II portfolio of subsequent studies at the Academies and elsewhere, to be focused on strategic, tactical, and policy issues, such as energy research and development priorities, strategic energy technology development, policy analysis, and many related subjects.

PANEL ON ENERGY EFFICIENCY TECHNOLOGIES

LESTER B. LAVE, Carnegie Mellon University, *Chair*
MAXINE L. SAVITZ, Honeywell, Inc. (retired), *Vice Chair*
R. STEPHEN BERRY, University of Chicago
MARILYN A. BROWN, Georgia Institute of Technology
LINDA R. COHEN, University of California, Irvine
MAGNUS G. CRAFORD, LumiLeds Lighting
PAUL A. DeCOTIS, Long Island Power Authority
JAMES DeGRAFFENREIDT, JR., WGL Holdings, Inc.
HOWARD GELLER, Southwest Energy Efficiency Project
DAVID B. GOLDSTEIN, Natural Resources Defense Council
ALEXANDER MacLACHLAN, E.I. du Pont de Nemours & Company (retired)

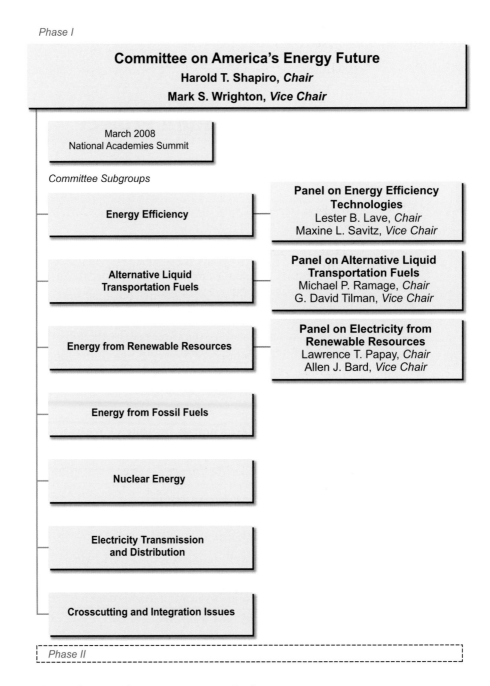

FIGURE C.1 *America's Energy Future Project.*

WILLIAM F. POWERS, Ford Motor Company (retired)
ARTHUR H. ROSENFELD, California Energy Commission
DANIEL SPERLING, University of California, Davis

PANEL ON ALTERNATIVE LIQUID TRANSPORTATION FUELS

MICHAEL P. RAMAGE, ExxonMobil Research and Engineering Company (retired), *Chair*
G. DAVID TILMAN, University of Minnesota, St. Paul, *Vice Chair*
DAVID GRAY, Noblis, Inc.
ROBERT D. HALL, Amoco Corporation (retired)
EDWARD A. HILER, Texas A&M University (retired)
W.S. WINSTON HO, Ohio State University
DOUGLAS R. KARLEN, U.S. Department of Agriculture, Agricultural Research Service
JAMES R. KATZER, ExxonMobil Research and Engineering Company (retired)
MICHAEL R. LADISCH, Purdue University and Mascoma Corporation
JOHN A. MIRANOWSKI, Iowa State University
MICHAEL OPPENHEIMER, Princeton University
RONALD F. PROBSTEIN, Massachusetts Institute of Technology
HAROLD H. SCHOBERT, Pennsylvania State University
CHRISTOPHER R. SOMERVILLE, Energy Biosciences Institute
GREGORY STEPHANOPOULOS, Massachusetts Institute of Technology
JAMES L. SWEENEY, Stanford University

PANEL ON ELECTRICITY FROM RENEWABLE RESOURCES

LAWRENCE T. PAPAY, Science Applications International Corporation (retired), *Chair*
ALLEN J. BARD, University of Texas, Austin, *Vice Chair*
RAKESH AGRAWAL, Purdue University
WILLIAM L. CHAMEIDES, Duke University
JANE H. DAVIDSON, University of Minnesota, Minneapolis
J. MICHAEL DAVIS, Pacific Northwest National Laboratory
KELLY R. FLETCHER, General Electric

CHARLES F. GAY, Applied Materials, Inc.

CHARLES H. GOODMAN, Southern Company (retired)

SOSSINA M. HAILE, California Institute of Technology

NATHAN S. LEWIS, California Institute of Technology

KAREN L. PALMER, Resources for the Future, Inc.

JEFFREY M. PETERSON, New York State Energy Research and Development Authority

KARL R. RABAGO, Austin Energy

CARL J. WEINBERG, Pacific Gas and Electric Company (retired)

KURT E. YEAGER, Galvin Electricity Initiative

Principal Units and Conversion Factors

This report uses a variety of units to describe the supply and consumption of energy. Although these units are in common usage throughout the energy industry, they are generally not well understood by nonexperts. This appendix describes the principal supply and consumption units and provides some useful conversion factors. The Department of Energy–Energy Information Administration's website (see www.eia.doe.gov/basics/conversion_basics.html) provides additional information about energy units and conversion factors, including easy-to-use energy conversion calculators.

ELECTRICITY

- **Electrical generating capacity** is expressed in units of *kilowatts* (kW), *megawatts* (MW = 10^3 kW), and *gigawatts* (GW = 10^6 kW). It is defined as the maximum electrical output that can be supplied by a generating facility operating at ambient conditions. Coal power plants typically have generation capacities of about 500 MW; nuclear plants about 1000 MW (1 GW); intermittent sources (e.g., natural gas peaking plants and wind plants) about one to a few megawatts; and residential roof-top installations of solar photovoltaics about a few kilowatts.
- **Electricity** supply and consumption are expressed in units of *kilowatt-hours* (kWh), *megawatt-hours* (MWh = 10^3 kWh), *gigawatt-hours* (GWh = 10^6 kWh), and *terawatt-hours* (TWh = 10^9 kWh). One kilowatt-hour is equal to the energy of 1000 watts (the typical amount of elec-

tricity that is consumed by a handheld hair dryer) supplied or consumed over a period of 1 hour. Annual total delivered electricity in the United States is about 4,000 TWh, and the average annual electricity consumption per U.S. household is about 11,000 kWh.

FOSSIL FUELS AND OTHER LIQUID FUELS

- **Coal** supply and consumption are usually expressed in units of *metric tons* (tonnes); 1 metric ton is equal to about 2205 pounds. A typical coal-fired power plant consumes about 2 million tonnes of coal per year, and annual coal consumption in the United States is about 1 billion tonnes per year. Coal prices are expressed in units of *dollars per gigajoule* ($/GJ). A tonne of coal contains about 23.5 GJ of energy.

- **Petroleum and gasoline** supply and consumption are expressed in units of *barrels* (bbl); a barrel contains 42 U.S. gallons or 159 liters. Units of barrels of gasoline equivalent (bbl gasoline eq) represent the energy content of **other liquid fuels** (e.g., ethanol) in terms of the energy content of a barrel of motor gasoline. The United States consumes about 9 million barrels of motor gasoline per day and over 7 billion barrels of liquid fuels (crude oil, finished products, and other liquid fuels) per year.

- **Natural gas** supply and consumption are expressed in units of *trillion cubic feet* (Tcf). The United States consumes about 23 Tcf of natural gas each year.

- **Biomass** supply for liquid fuels production is expressed in units of dry tonnes; 1 dry tonne is equal to about 2205 pounds. The dry ton equivalent is 2000 pounds.

ENERGY CONVERSION FACTORS

- **Total energy** supply and consumption are expressed in *British thermal units* (Btu) and *quads* (quadrillion Btu = 10^{15} Btu). A Btu is defined as the amount of energy (in the form of heat) needed to raise the temperature of 1 pound of water by 1 degree Fahrenheit. The energy content of electricity and natural gas, liquid, and coal fuels can be quantified in terms of Btu using the following approximate factors:

1 kilowatt-hour electricity = 3,412 Btu
1 barrel crude oil = 5,800,000 Btu
1 barrel gasoline = 5,200,000 Btu
1 barrel fuel ethanol = 3,500,000 Btu
1 cubic foot of natural gas = 1,028 Btu
1 tonne coal = 22,230,000 Btu

The United States consumes about 100 quads (100×10^{15} Btu) of energy per year (see Figure 1.2 in Chapter 1).

GREENHOUSE GASES

- **Carbon dioxide (CO_2)** emissions from energy production and use are expressed in tonnes. The term *tonnes of CO_2 equivalent* (CO_2 eq) indicates the global warming potential of other greenhouse gases (e.g., methane) in terms of CO_2 quantities. The United States emits about 7 billion tonnes of CO_2 eq per year, about 6 billion of which is CO_2 arising primarily from energy production and use (see Figure 1.3 in Chapter 1). Average CO_2 emissions in the United States are about 20 tonnes per person.

Select Acronyms and Abbreviations

3D	three-dimensional
AAAS	American Association for the Advancement of Science
ABWR	advanced boiling water reactor
AC	alternating current
ACA	Australian Coal Association
ACEEE	American Council for an Energy-Efficient Economy
AECL	Atomic Energy Canada Ltd.
AEF	America's Energy Future
AEO	Annual Energy Outlook
AEP	American Electric Power Corporation
AESO	Alberta Electric System Operator
AFCI	Advanced Fuel Cycle Initiative
AFDC	allowance for funds used during construction
AFV	alternative fuel vehicle
AHAM	Association of Home Appliance Manufacturers
AHTR	advanced high temperature reactor
ANWR-1002	Arctic National Wildlife Refuge 1002
AISI	American Iron and Steel Institute
ALA	American Lung Association
Am	americium
APS	American Physical Society
APWR	advanced pressurized water reactor
ARRA	American Recovery and Reinvestment Act

ASES	American Solar Energy Society
ASHRAE	American Society of Heating, Refrigerating, and Air-Conditioning Engineers
ASME	American Society of Mechanical Engineers
ASP	alkaline surfactant polymer
AWEA	American Wind Energy Association
bbl	barrel
bbl/d	barrel(s) per day
BCG	Boston Consulting Group
BEV	battery-electric vehicle
BLM	U.S. Bureau of Land Management
BOF	basic oxygen furnace
BOP	balance of plant
BTP	biomass-to-power
BTL	biomass-to-liquid
Btu	British thermal unit
BWR	boiling water reactor
C	Celsius
CAES	compressed air energy storage
CAFE	corporate average fuel economy
CAIDI	Customer Average Interruption Duration Index
CAIR	Clean Air Interstate Rule
CANDU PHWR	Canada deuterium uranium pressurized heavy water reactor
Caps	capacitor
CAR	ceramic autothermal recovery
CAREM	advanced small nuclear power plant
CBTL	coal-and-biomass-to-liquid
CBTP	coal-and-biomass-to-power
CBFT	coal-and-biomass-to-liquid fuel, Fischer-Tropsch
CBMTG	coal-and-biomass-to-liquid fuel, methanol-to-gasoline
CCE	cost of conserved energy
CCS	carbon capture and storage
CDF	core damage frequency
CEC	California Energy Commission
CECA	Consumer Energy Council of America

CEF	Clean Energy Future, Scenarios for a (study)
CERCLA	Comprehensive Environmental Response, Compensation, and Liability Act
CEO	chief executive officer
Cf	californium
CFC	chlorofluorocarbons
CFL	compact fluorescent light
CFR	Code of Federal Regulations
CFT	coal-to-liquid fuel, Fischer-Tropsch
CH_4	methane
CHP	combined heat and power
CI	compression-ignition
Cm	curium
CMTG	coal-to-liquid fuel, methanol-to-gasoline
CNG	compressed natural gas
CO	carbon monoxide
CO_2	carbon dioxide
CO_2 eq	carbon dioxide equivalent
COL	construction and operating license
CRP	Conservation Reserve Program
Cs	cesium
CSP	concentrating solar power
CTL	coal-to-liquid fuel
CVT	continuously variable transmission
CWIP	construction work in progress
CZMA	Coastal Zone Management Act
DBT	Design Basis Threat
DC	direct current
DI&C	digital instrumentation and control
DME	dimethyl ether
DNA	deoxyribonucleic acid
DOE	U.S. Department of Energy
DSM	demand-side management
EAF	electric-arc furnace
ECBM	enhanced coal bed methane

EEB	Energy Engineering Board
EEI	Edison Electric Institute
EERE	Office of Energy Efficiency and Renewable Energy
EGR	enhanced gas recovery
EGS	enhanced geothermal system
EHV	extra-high-voltage
EIA	Energy Information Administration
EIO	economic input/output
EIS	environmental impact statement
EISA	Energy Independence and Security Act
EOR	enhanced oil recovery
EPA	U.S. Environmental Protection Agency
EPAct05	Energy Policy Act of 2005
EPR	European pressurized water reactor
EPRI	Electric Power Research Institute
EPRI-TAG	Electric Power Research Institute—Technical Assessment Guide
ERCOT	Electric Reliability Council of Texas
ESBWR	economic simplified boiling-water reactor
EU	European Union
EWEA	European Wind Energy Association
EWIS	European Wind Integration Study
F	Fahrenheit
FACTS	Flexible Alternating Current Transmission System
FCV	fuel-cell vehicle
FEMP	Federal Energy Management Program
FERC	Federal Energy Regulatory Commission
FFB	Federal Financing Bank
FT	Fischer-Tropsch
FTBR	fast thorium breeder reactor
FY	fiscal year
g	gram
GaAs	gallium arsenide
GAO	United States Government Accountability Office
GCR	gas-cooled reactor
GDP	gross domestic product

GE	General Electric
GHG	greenhouse gas
GIF	Generation IV International Forum
GJ	gigajoule
GNEP	Global Nuclear Energy Partnership
GPS	global positioning system
GREET	Greenhouse gases, Regulated Emissions, and Energy use in Transportation (model)
GT	gas turbine
Gt	gigatonne
GTL	gas-to-liquid
GW	gigawatt
GW_e	gigawatt-electric
GWh	gigawatt-hour
h	hour
H_2	hydrogen
H_2O	water
H_2S	hydrogen sulfide
HEV	hybrid-electric vehicle
HFCV	hydrogen fuel-cell vehicle
Hg	mercury
HHV	higher heating value
HLW	high-level waste
HV	high voltage
HVDC	high-voltage direct current
Hz	hertz
IAEA	International Atomic Energy Agency
ICE	internal-combustion engine
IDST	improved decision-support technology
IEA	International Energy Agency
IED	intelligent electronic device
IGCC	integrated gasification combined cycle
IMF	inert matrix fuel
INL	Idaho National Laboratory
INVAP	Investigaciones Aplicadas Sociedad del Estado

IOU	investor-owned utility
IPCC	Intergovernmental Panel on Climate Change
IPP	independent power producer
IRGC	International Risk Governance Council
IRR	internal rates of return
IPST	Institute of Paper Science and Technology, Georgia Institute of Technology
ISFSI	independent spent fuel storage installation
ISL	in situ leach
ISO	independent system operator
ITER	International Thermonuclear Experimental Reactor
ITM	ion transport membrane
ITS	Intelligent Transportation Society
IUPAC	International Union of Pure and Applied Chemists
J	joule
JCSP	Joint Coordinated System Plan
KAERI	Korean Atomic Energy Research Institute
kg	kilogram
km	kilometer
kV	kilovolt
kW	kilowatt
kWh	kilowatt-hour
LAN	local area network
LBNL	Lawrence Berkeley National Laboratory
LCA	life-cycle analysis
LCCR	levelized capital charge rate
LCOE	levelized cost of electricity
LDV	light-duty vehicle
LED	light-emitting diode
LES	Louisiana Energy Services Limited Partnership
LFR	lead-cooled fast reactor
Li-ion	lithium ion
LLC	limited liability corporation
LLNL	Lawrence Livermore National Laboratory

LLW	low-level waste
LNG	liquefied natural gas
low-E	low-emissivity
LPG	liquefied petroleum gas
LWR	light-water reactor
m^3	cubic meter
MECS	Manufacturing Energy Consumption Survey
MISO	Midwest Independent Transmission System Operator
MIT	Massachusetts Institute of Technology
MMS	United States Minerals Management Service
MOGD	methanol-to-olefins, gasoline, and diesel
MOF	metal organic framework
MOX	mixed-oxide
MSR	molten salt reactor
MTG	methanol-to-gasoline
MTU	metric tons of uranium
MW	megawatt
MWd	megawatt-day
MWh	megawatt-hour
MW_e	megawatt-electric
MW_t	megawatt-thermal
NAS	National Academy of Sciences
NaS	sodium-sulfur
NAE	National Academy of Engineering
NBI	New Buildings Institute
NDE	nondestructive examination
NEAC	U.S. Department of Energy Nuclear Energy Advisory Committee
NEI	Nuclear Energy Institute
NEMS	National Energy Modeling System
NERAC	U.S. Department of Energy Nuclear Energy Research Advisory Committee
NERC	North American Electric Reliability Corporation
NETL	National Energy Technology Laboratory
NGCC	natural gas combined cycle
NGNP	Next Generation Nuclear Plant

NGV	natural gas vehicle
NiCd	nickel-cadmium
NiMH	nickel-metal hydride
NIOSH	National Institute of Occupational Safety and Health
NOAA	National Oceanic and Atmospheric Administration
NO_x	nitrogen oxides
Np	neptunium
NPC	National Petroleum Council
NPRA	National Petroleum Reserve-Alaska
NPV	net present value
NRC	National Research Council
NRDC	National Resources Defense Council
NREL	National Renewable Energy Laboratory
NSB	National Science Board
NSPS	National Source Performance Standards
NSR	National Source Review
O_2	oxygen
OCRM	Office of Ocean and Coastal Resource Management
OCS	outer continental shelf
OECD	Organisation for Economic Co-operation and Development
OECD–NEA	Organisation for Economic Co-operation and Development–Nuclear Energy Agency
O&M	operation and maintenance
OMB	U.S. Office of Management and Budget
OPEC	Organization of the Petroleum Exporting Countries
ORNL	Oak Ridge National Laboratory
OSHA	Occupational Safety and Health Administration
OTA	Office of Technology Assessment
OTM	oxygen transport membrane
PBMR	pebble-bed modular reactor
PC	pulverized coal
PEI	Princeton Environmental Institute
PGC	Potential Gas Committee
PHEV	plug-in hybrid vehicle
PMU	phasor measurement unit

PNNL	Pacific Northwest National Laboratory
PRA	probabilistic risk assessment
PSEG	Public Service Enterprise Group
PTC	Federal Renewable Electricity Production Tax Credit
Pu	plutonium
PUC	public utility commission
PUREX	plutonium and uranium extraction
PURPA	Public Utilities Regulatory Policies Act
PV	photovoltaic
PWR	pressurized water reactor
quads	quadrillion Btu
RAND	Research and Development Corporation
RCRA	Resource Conservation and Recovery Act
R&D	research and development
RD&D	research, development, and demonstration
RFF	Resources for the Future
RFS	Renewable Fuel Standard
RNA	ribonucleic acid
RPS	renewables portfolio standard
RTO	regional transmission operator
RTU	remote terminal unit
SAGD	steam assisted gravity drainage
SAIDI	System Average Interruption Duration Index
SAIFI	System Average Interruption Frequency Index
SCADA	supervisory control and data acquisition
SCWR	supercritical water reactor
SEIA	Solar Energy Industries Association
SERI	Solar Energy Research Institute
SES	superconducting energy storage
SFR	sodium-cooled fast reactor
SGI	Synthetic Genomics, Inc.
Shell	Shell Frontier Oil and Gas, Inc.
SHGC	solar heat-gain coefficient
SI	spark-ignition

SiC	silicon carbide
SMART	system integrated modular advanced reactor
SMCRA	Surface Mine Control and Reclamation Act
SMES	superconducting magnetic energy storage
SNF	spent nuclear fuel
SNG	synthetic natural gas
SO_2	sulfur dioxide
SO_3	sulfur trioxide
SO_x	sulfur oxides
S&P	Standard and Poor's
SPE	Society of Petroleum Engineers
SPP	Southwest Power Pool
SPS	special protection schemes
Sr	strontium
SSTAR	small, sealed, transportable autonomous reactor
STATCOM	static shunt compensator
SVC	static volt-amperes reactive compensator
SWU	separative work units
synfuel	synthetic transportation fuel
t	tonne
Tc	technetium
TCE	total cash expended
Tcf	trillion cubic feet
TCSC	thyristor-controlled series capacitor
T&D	transmission and distribution
Th	thorium
ThO_2	thorium oxide
TLR	transmission loading relief
TOU	time-of-use
TPC	total plant cost
TPI	total plant investment
TCR	total capital requirement
TRB	Transportation Research Board
TRU	transuranic elements
TSCA	Toxic Substances Control Act
TVA	Tennessee Valley Authority

TWh	terawatt-hour
U	uranium
U_3O_8	triuranium octoxide
UCS	Union of Concerned Scientists
UF_6	uranium hexafluoride
UPS	uninterruptible power supplies
UREX	uranium extraction
USABC	U.S. Advanced Battery Consortium
USEC	U.S. Enrichment Corporation, Inc.
USEPR	U.S. evoluiontary power reactor
USGS	U.S. Geological Survey
USNRC	U.S. Nuclear Regulatory Commission
USPC	ultrasupercritical pulverized coal
U-value	heat-transfer coefficient
UWIG	Utility Wind Integration Group
V	carbon dioxide vented
VAR	volt-amperes reactive
VFT	variable frequency transformer
VHTR	very-high-temperature reactor
VRB	vanadium redox battery
VRI	vehicle-to-refueling-station index
W	watt
WRI	World Resources Institute
WGA	Western Governors' Association
Wh	watt-hour
WinDS	Wind Development System
ZEBRA battery	sodium nickel chloride battery
ZnBr	zinc-bromine

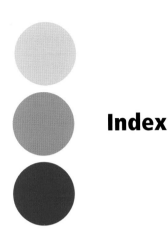

Index

C

sulfur oxides, xi, 189, 249, 288, 301, 313, 333, 369, 408

water supply/pollution, 101, 301, 313, 408, 409-410, 539-541

weather/climate, 301

wildlife and ecosystems, 301-302

Ethanol. *See also* Cellulosic ethanol; Corn ethanol

energy produced compared to gasoline, 223

greenhouse gas emissions, 72

supply, 37, 241-242

European Union, 11, 20, 144, 157, 170, 506. *See also individual countries*

Exports of energy, trends, 20. *See also individual sources*

ExxonMobil, 226

F

Federal Renewable Electricity Production Tax Credit, 95, 99, 101, 273, 274, 298, 299, 307, 308, 317, 447 n.7, 468, 489

Feedstocks. *See* Biomass; Coal

Finland, 446, 453, 504, 505

Fischer-Tropsch process, 67, 72, 93, 220, 226, 228, 229, 230-231, 235, 236, 237, 238, 239, 240, 242, 252, 261

Fish and Wildlife Service, 403 n.35

Florida, 151, 193, 393

FMC, 163

Ford administration, 26, 458 n.34

Ford Motor Co., 163

Fossil fuels. *See also* Coal; Natural gas; Oil; Petroleum

access issue, 350-355

carbon capture and storage, 2, 300-301, 406-407

CO_2 emissions, 11, 16, 25, 300-301, 333, 358-359

consumption, 14, 332

cost comparisons, 369-379

dependence on, 14-15, 25, 333

economic importance, 14, 331-332

electric power generation, 3, 16, 104-107, 358-396, 418-419

environmental and safety issues, 11, 16, 25-26, 108-109, 300, 301, 333, 403-410

findings, 2, 356-358, 394-396, 401-402, 410

geologic storage of CO_2, 396-402, 406-407

prices, 2, 14, 28, 365-366

resources and reserves, 13, 334-358, 415-418

supply and demand, 25, 357-358

for transportation, 108

water use, 409-410

world resources, 334-335

France

energy consumption, 140

geothermal projects, 286

greenhouse gas emissions, 533

nuclear fuel production and recycling, 519, 522, 533

nuclear power, 446, 453, 457 n.30, 504, 505, 508, 510 n.10, 519

Freight transportation

air, 86, 170

commercial HEVs, 156, 159-160

consumption of fuel, 156, 157, 171

intermodal transfers, 157, 171

potential energy efficiency improvements, 45, 171, 174

rail, 86, 170-171

truck transport, 85-86, 156, 170, 171

waterborne, 171

Fuel-cell technologies, 4, 5, 6, 30, 44, 45, 67, 73, 74, 75, 84, 85, 86, 94, 158, 160, 161, 164, 165, 166, 173, 174, 179, 195, 258, 263-267, 431, 621

Fuels. *See* Alternative transportation fuels

G

Gasification technologies

biomass, 226, 227, 228, 229, 232, 361-362 n.15, 386, 422

carbon capture and storage, 360, 387, 397, 406, 421, 425, 431

cellulosic ethanol production, 221-222

coal, 227, 228-229, 230-231, 360, 368, 386, 397, 406

co-fed coal and biomass, 3-4, 91, 227, 228, 229, 230, 231, 232, 235, 237-238, 242, 245-246, 422

high-pressure systems, 231, 425

H

Z